Molecular and Cellular Biology of Platelet Formation

Harald Schulze • Joseph Italiano
Editors

Molecular and Cellular Biology of Platelet Formation

Implications in Health and Disease

Springer

Editors
Harald Schulze
Institute of Experimental Biomedicine I
University Hospital Würzburg
Würzburg
Germany

Joseph Italiano
Vascular Biology Programm
Harvard Medical School
Boston
Massachusetts
USA

ISBN 978-3-319-39560-9 ISBN 978-3-319-39562-3 (eBook)
DOI 10.1007/978-3-319-39562-3

Library of Congress Control Number: 2016960044

Printed on acid-free paper

This Springer imprint is published by Springer Nature
The registered company is Springer International Publishing AG
The registered company address is Gewerbestrasse 11, 6330 Cham, Switzerland

Preface

All cells circulating in the peripheral blood of mammals are derived from a small pool of hematopoietic cells (HSCs) residing in the bone marrow. These HSCs have the potential to develop into all different blood cell lineages, differentiating into different cell types such as hemoglobin-filled red blood cells for gas transport, leukocytes with various functions in innate and adaptive immune responses, or the subcellular fragments – designated blood platelets – with their main function for hemostasis and to prevent blood loss. This differentiation occurs in the presence (or absence) of hematopoietic growth factors like erythropoietin or G-CSF. A similar specific growth factor for platelet formation (as well as for the differentiation from their immediate precursor cells, the megakaryocytes) was already designated as a "thrombopoietin." For almost 30 years, scientists were searching for this elusive factor. And then finally in 1994, there was a major breakthrough when three seminal discoveries launched the research of this neglected blood lineage: Several groups succeeded finally in identifying and cloning thrombopoietin and thus allowed its recombinant production for biomedical research. This major finding allowed us to begin culturing megakaryocytes and study how they make platelets. One year later, the transcription factor NF-E2 was identified as a critical factor indispensable for the final step of platelet biogenesis from megakaryocytes. In 1999, time-lapse microscopy allowed us to catch megakaryocytes in the act of making platelets and to begin to understand the fundamental role of platelet biogenesis

Now, about two decades later, research on platelet biogenesis has flourished and led to many new insights, from basic cell biology to genetic variants that underline clinical diseases. We are very happy that we could recruit the following internationally renowned experts to contribute to this book on Platelet Formation in Health and Disease.

In the first part *Megakaryocytes and Thrombopoiesis*, J.E. Italiano (Chap. 1) will provide an overview on our current understanding of platelet formation with a focus on how the cytoskeleton, which functions as the bones and muscle of the megakaryocyte, drives platelet formation. M. Tijssen, T. Moreau, and C. Ghevaert (Chap. 2) will describe the transcription factors involved in this process. They will address their interactions and the networks formed and discuss how this knowledge can be translated to engineer platelet production in vitro. The impact of mutations and altered expression of transcription factors in disease and leukemia is explained by A. Cantor (Chap. 3). Novel insights from the bone marrow matrix, where platelet biogenesis occurs in vivo are given by H. Schulze, D. Semeniak, and A. Balduini (Chap. 4). This chapter comprises our current knowledge on bioreactors for platelet production in vitro. Finally, evidence that platelets can form progeny is given by M.T. Rondino and H. Schwertz (Chap. 5). They will discuss how extramedullary niches might impact platelet morphogenesis.

The second part addresses *Platelet Biology: Signals and Functions*. The platelet contents are outlined by R. Flaumenhaft and S. Koseoglu (Chap. 6). They will summarize the knowledge on alpha and dense granules and the platelet membrane systems. The signaling pathways in platelets and megakaryocytes are summarized by A. Mazharian and Y. Senis (Chap. 7). The authors give a rundown of the signaling events in megakaryocytes and compare them to platelets and focus on inhibitory signaling cascades. The novel model of "gas and brake pedals" for megakaryocyte and platelet function by the Rap GTPases is depicted by W. Bergmeier and E.C. O'Shaughnessy (Chap. 8). Here, similarities and differences between megakaryocytes and platelets are also carved out. The platelet life span and the underlying clock proteins in megakaryocytes are discussed by A. Au, M. Lebois, I. Pleines, and E. Josefsson (Chap. 9). The summary of mouse models lacking pro- and anti-apoptotic proteins will give insights on how an anucleate cell has a "timer." Platelet functions beyond hemostasis and thrombosis are presented by R. Kapur and J. Semple (Chap. 10). They will elaborate on the how platelets play a role in pathogen recognition and how the communication between platelets and target cells occurs. The importance and limitations of mouse models for platelet production and function are explained by M. Bender and B. Nieswandt (Chap. 11). These authors describe well-established mouse models as well as novel reporter mice that will help to decipher open questions in research on platelet biogenesis.

The third part addresses *Platelets in Health and Disease*. C. Dame, V. Lorenz, and M. Sola-Visner (Chap. 12) will present aspects of fetal and neonatal megakaryopoiesis and platelet biology. The authors will highlight that even though in newborns and preterm babies platelets might look the same as in adults, they are derived from a distinct megakaryocytic origin and thus harbor many different aspects. The role of platelets in ischemic stroke will be characterized by D. Cherpokova and B. Nieswandt (Chap. 13). This chapter will address thrombotic aspects of platelet biogenesis and function and summarize current experimental models in mice to study this aspect. The increasing inherited and acquired causes that contribute to altered platelet number and/or function are outlined in two chapters: First, K. Freson (Chap. 14) will give insights into how genetic variants and mutations affect megakaryopoiesis and platelet formation by inherited thrombocytopenias. Second, T. Bakchoul and A. Greinacher (Chap. 15) will discuss the current knowledge on acquired thrombocytopenias. They report on causes of platelet reduction to hemodilation or surgery and the role of infections and will focus on immune thrombocytopenia and heparin-induced thrombocytopenia. G. Schönrich and M. Raftery (Chap. 16) will focus on how certain viral infections modulate megakaryocytes and platelet biogenesis. They will report on viral receptors on platelets and megakaryocytes with a focus on HIV and hantaviruses. In the last chapter, T. Bakchoul and H. Schulze (Chap. 17) will provide an overview of how inherited and acquired platelet disorders are currently diagnosed and where the future challenges are to be overcome. They will also discuss the options and limitations of mutations and variants identified by next-generation sequencing-based approaches.

The last part covers *Future Perspectives for Platelet Biogenesis*. C.J. Braun and C. Klein (Chap. 18) will outline cellular therapies for Wiskott-Aldrich syndrome. These authors will report on supportive therapies, stem cell transplantation, and gene transfer approaches in clinical trials. The use of induced pluripotent stem cells to generate blood is outlined by P. Karagiannis, H. Endi, and K. Eto (Chap. 19). They will describe the current understanding of this technology and focus on the generation of erythrocytes and platelets. The last chapter by L. Latorre and U. Modlich (Chap. 20) leaves us with strategies for the gene modification of megakaryocytes and platelets. They will give an overview of established and novel vector systems to manipulate specifically the megakaryocytic lineage to generate modified platelets in the living organism.

Being fully aware that there are further important and exciting topics in "platelet formation with implications in health and disease" and that we had to restrict to the total amount, we consider the selection of topics and chapters packaged in this book as a comprehensive overview of the current understanding of platelet biogenesis. With this, we would like to thank all authors and contributors.

Boston, MA, USA — Joseph E. Italiano
Würzburg, Germany — Harald Schulze

Contents

Part I

Megakaryocytes and Thrombopoiesis

Megakaryopoiesis and Platelet Biogenesis

1

Joseph E. Italiano Jr.

Abstract

Blood platelets are 2–3 um anucleate fragments that are formed from the megakaryocyte cytoplasm and have a distinctive discoid shape. To generate and release platelets, megakaryocytes undergo endomitosis to become polyploid and follow a maturation program that results in the transformation of the bulk of their cytoplasm into multiple long processes called *proplatelets*. To generate 1000 to 2000 platelets, a megakaryocyte may extend multiple proplatelets, each of which begins as a thick pseudopodia that over time elongates and branches repetitively. Platelets form predominantly at the tips of proplatelets. As platelets mature, their content of organelles and granules is delivered to them in a flow of individual cargo moving from the cell body of the megakaryocyte to the assembling platelets at the proplatelet ends. Platelet generation can be indiscriminately divided into two stages. The first stage takes days to complete and requires megakaryocyte-specific cytokines, such as thrombopoietin. Substantial nuclear proliferation to 16 to 32 × N and expansion of the megakaryocyte cytoplasm occur as the platelet is packed with platelet-specific granules, cytoskeletal proteins, and abundant membrane to complete the platelet assembly phase. The second stage is relatively fast and can be completed in hours. During this phase, megakaryocytes generate platelets by reorganizing their cytoplasm first into proplatelets, then preplatelets, which undergo fission to generate platelets. Each day, 100 billion platelets must be generated from megakaryocytes to sustain the normal platelet count of 2 to 3×10^8/ml.

J.E. Italiano Jr.
Hematology Division, Brigham and Women's Hospital,
Karp Research Bldg. 5th Floor, Room 214, Boston, MA 02115, USA
e-mail: jitaliano@partners.org

H. Schulze, J. Italiano (eds.), *Molecular and Cellular Biology of Platelet Formation*,
DOI 10.1007/978-3-319-39562-3_1

1.1 Introduction

Platelets are small anucleate fragments that are developed from the cytoplasm of megakaryocytes and have a distinctive disc shape. To assemble and release platelets, megakaryocytes become polyploid by endomitosis and follow a maturation program that results in the conversion of the majority of their cytoplasm into multiple long processes called *proplatelets*. To produce its quota of 1,000–2,000 platelets, a megakaryocyte may protrude multiple proplatelets, each of which begins as a blunt protrusion that over time thins and branches repeatedly. Platelets form selectively at the ends of proplatelets. As platelets develop, their content of organelles is delivered to them in a stream of individual particles moving from the megakaryocyte cell body to the assembling platelet buds at the proplatelet ends. Platelet formation can be arbitrarily divided into two phases. The first phase takes days to complete and requires megakaryocyte-specific growth factors, such as thrombopoietin. Massive nuclear proliferation to 16–32 × N and enlargement of the megakaryocyte cytoplasm occur as the platelet is filled with cytoskeletal proteins, platelet-specific granules, and abundant membrane to complete the platelet assembly process. The second phase is relatively rapid and can be completed in hours. During this stage, megakaryocytes produce blood platelets by reorganizing their cytoplasm into proplatelets, then preplatelets, which go through fragmentation to give rise to individual platelets. One hundred billion platelets need to be generated daily from megakaryocytes to uphold the average platelet count of $2–3 \times 10^8$/ml. This chapter is separated into two sections that review the mechanisms by which megakaryocytes undergo cytoplasmic maturation, and the structural reorganizations that power platelet production.

1.2 Megakaryocyte Development

Megakaryocytes are highly differentiated bone marrow cells that function to generate and release platelets into the blood. Studying the process by which megakaryocytes form and release platelets has intrigued scientists for over a century. Megakaryocytes are derived from hematopoietic stem cells and undergo a process called endomitosis, which involves several DNA replications in the absence of cell divisions. During endomitosis, megakaryocytes start a rapid expansion of their cytoplasm distinguished by the formation of a highly invaginated demarcation membrane network and the buildup of proteins and granules fundamental for platelet biology. During the final phase of maturation, the cytoplasm of the megakaryocyte goes through a massive remodeling into beaded cytoplasmic processes called proplatelets. The proplatelets function as the assembly lines of platelet production and ultimately give rise to individual platelets.

1.2.1 Commitment to the Megakaryocyte Lineage

Megakaryocytes arise from hematopoietic stem cells, which are in charge of the continuous supply of almost all circulating blood cells. Hematopoietic stem cells are classified by surface markers, their ability to reconstitute in host animals, as well as colony assays that indicate their developmental potential. Hematopoietic stem cells are infrequent and make up less than 0.1 % of cells in the bone marrow. Hematopoietic stem cell development into megakaryocytes includes a series of steps in which the developmental competence of the progenitor cells becomes increasingly more limited. Mouse hematopoietic stem cells are characteristically recognized by the surface markers Lin-Sca-1+c-kit^{high} [1, 2]. A model of hematopoiesis has come into sight from studies investigating the influence of hematopoietic growth factors and cytokines on marrow cells that are cultured on semisolid media. Two major lineages arise from hematopoietic stem cells, a common lymphoid progenitor that can expand into lymphocytes, and a myeloid progenitor that can develop into myeloid, macrophage, eosinophil, erythroid, and megakaryocyte lineages. A common erythroid-megakaryocytic progenitor develops from the myeloid lineage [3]. However, more recent experiments also indicate that hematopoietic stem cells may directly expand into erythroid-megakaryocyte progenitors [4]. All hematopoietic progenitors express CD41 and CD34 on their surface, and the commitment to the megakaryocyte lineage is typified by integrin CD61 expression and increased CD41 levels. From the committed myeloid progenitor cell (CFU-GEMM), there is solid support for a bipotential progenitor situated between the pluripotential stem cell and the committed precursor that can form biclonal colonies composed of erythroid and megakaryocytic cells. The transcriptional factors and signaling pathways that direct the erythroid and megakaryocyte lineages to separate from the bipotential progenitor are not fully understood. Diploid precursors that are committed to the megakaryocyte lineage have characteristically been divided into two colonies based on their function. The megakaryocyte burst-forming cell is a primitive progenitor that has a very high proliferation ability and gives rise to large megakaryocyte colonies. Under specific in vitro culture conditions, the megakaryocyte burst-forming cell can form 40–500 megakaryocytes within a single week. The colony-forming cell is a more cytoplasmically mature megakaryocyte progenitor that gives rise to a colony containing from 3 to 50 mature megakaryocytes that are variable in their potential to proliferate. Megakaryocyte progenitors can be readily observed in marrow by acetylcholinesterase and immunoperoxidase labeling. Although both human megakaryocyte colony-forming and burst-forming cells express the CD34 antigen, only colony-forming cells express the HLA-DR antigen.

Various mechanisms based on histochemical staining, biochemical markers, as well as morphological features have been employed to classify different stages of megakaryocyte development. Three types of morphologies and structures can be identified in bone marrow. The promegakaryoblast is the first recognizable megakaryocyte precursor. The megakaryoblast, or Stage I MK, has a distinct morphology and is more cytoplasmically mature [5]. The megakaryoblast has a nucleus with a kidney shape with two sets of chromosomes (4N). The megakaryoblast is 10–50 μm in diameter and presents deeply basophilic in Romanovsky-stained marrow sections given the large numbers of ribosomes, even though the cytoplasm at this stage does not contain alpha and dense granules. The megakaryoblast presents with a high nuclear-to-cytoplasmic ratio and, in rodents, is positive for acetylcholinesterase activity. The promegakaryocyte, or Stage II megakaryocyte, is 20–80 μm in diameter with a cytoplasm that is less basophilic than the megakaryoblast and now contains forming alpha and dense granules.

1.2.2 Polyploidization

Megakaryocytes, in contrast to most other cells, undergo endomitosis and become polyploid through repetitive cycles of DNA replication in the absence of cell division [6–10]. At the end of the proliferation stage, mononuclear megakaryocyte precursor cells leave the diploid state to differentiate and undergo endomitosis, producing a cell that now contains multiples of a normal diploid chromosome content (i.e., 4N, 16N, 32N, 64N). While the number of endomitotic cycles can vary from 2 to 6, most of the megakaryocytes undergo three endomitotic cycles to obtain a DNA content of 16N. Nevertheless, a few megakaryocytes can attain a DNA content as high as 256N. Megakaryocyte endomitosis is caused by a functional gene amplification whose most probable function is an increase in protein synthesis parallel with an increase in cell size [11]. The molecular mechanisms that power endomitosis are not fully understood. It was first proposed that endomitosis may be a consequence of a lack of mitosis after every round of DNA replication. However, more current studies of primary cultured megakaryocytes suggest that endomitosis does not result from a total absence of mitosis, but more likely a prematurely terminated mitosis [11–13]. Megakaryocyte progenitors begin the cycle and initiate a short G1 phase, an average 6–7 h S phase for DNA synthesis, and then a short G-2 phase followed by endomitosis. Megakaryocytes initiate the mitotic cycle and advance from prophase to anaphase A but do not enter anaphase B, telophase, or cytokinesis. Throughout megakaryocyte polyploidization, the nuclear envelope breaks down and an unusual spherical mitotic spindle forms. Each spindle attaches chromosomes that align to a position equidistant from the spindle poles (metaphase). Sister chromatids separate and start to move toward their respective poles (anaphase A). Nevertheless, the spindle poles do not migrate apart and do not undergo the segregation normally observed during anaphase B. Individual chromatids do not move to the poles and then a nuclear envelope reforms around the complete set of sister chromatids, assembling a single inflated but lobed nucleus with several chromosome copies. The cell then omits telophase and cytokinesis to enter G1. This failure to completely separate sets of daughter chromosomes may stop the assembly of a nuclear envelope around each individual set of chromosomes. In most cell types, feedback controls as well as checkpoints ensure that DNA replication and cell division are coordinated. Megakaryocytes seem to be the exception to this rule as they have developed a mechanism to deregulate this process. Experiments from several laboratories have focused on establishing the signals that direct endomitosis in megakaryocytes [14]. It has been suggested that endomitosis may be a result of a decrease in mitosis-promoting factor (MPF) activity, a complex of proteins composed of Cdc2 and cyclin B [15, 16]. MPF has a kinase activity that is required for entry of cells into mitosis. In most cells, recently formed cyclin B attaches to Cdc2 and generates active MPF, while cyclin breakdown at the conclusion of mitosis inactivates MPF. Strains of yeast with conditional mutations that inhibit either cyclin B or cdc2 cause them to go through an extra round of DNA replication with no mitosis [17, 18]. Furthermore, experiments using a human erythroleukemia cell line have shown that these cells contain inactive cdc2 during endomitosis, and experiments with phorbol ester-induced Meg T cells have shown that cyclin B is absent in this cell line during endomitosis [19–21]. Nevertheless, it has been difficult to establish the function of MPF activity in stimulating endomitosis because these cell lines have a limited ability to undergo polyploidization. In addition, studies using normal megakaryocytes in vitro have established normal levels of cyclin B and cdc2 with functional mitotic kinase activity in megakaryocytes undergoing mitosis, demonstrating that endomitosis can be regulated by regulatory pathways other than MPF. Cyclins appear to play an essential role in regulating endomitosis, although a triple knockout of cyclins D1, D2, and D3 does not seem to affect the progression of megakaryocyte development [22]. Nevertheless, mice lacking cyclin E do exhibit a pronounced defect in the development of megakaryocytes [23]. It has been shown that the molecular mechanisms involved in polyploidization are characterized by the aberrant localization or lack of at least two essential regulators of mitosis, survivin as well as the chromosomal passenger proteins Aurora-B/AIM-1 [24]. While knockout of the APC/C cofactor Cdc20 causes mitotic arrest and severe low platelet count, lack of the kinases Aurora-B, Cdk1, or Cdk2 does not affect endomitosis of megakaryocytes or platelet levels. Deletion of Cdk1 causes a change to endocycles without mitosis, whereas polyploidization in the absence of cdk1 and cdk2 occurs in the presence of aberrant replication events. Particularly, deletion of these kinases rescues defects in Cdc20 null megakaryocytes. These observations imply that endomitosis can be functionally replaced by other polyploidization mechanisms in vivo [25].

1.2.3 Cytoplasmic Maturation

During polyploidization, the megakaryocyte begins a maturation phase in which the cytoplasm quickly expands with organelles, proteins, and membrane systems that will eventually be subdivided and packaged into platelets. During maturation, the megakaryocyte increases intensely, and the cytoplasm attains its unique structural characteristics, including the formation of a demarcation membrane system (DMS), production of granules, and the generation of a dense tubular system. Throughout this phase of development, the cytoplasm acquires a large amount of rough endoplasmic reticulum and ribosomes, where protein synthesis predominates. One of the most outstanding features of a mature megakaryocyte is its extensive DMS, an elaborate system of membrane channels made up of flattened cisternae and tubules. The organization of the cytoplasm of the megakaryocyte into membrane-defined platelet territories was initially suggested by Kautz and DeMarsh [26], and a high-resolution account of this membrane system by Yamada came soon after [27]. The DMS is observed in early promegakaryocytes but becomes most noticeable in mature megakaryocytes where it pervades the entire cytoplasm of the megakaryocyte, except for a thin rim of cytoplasm in the cortex, from which it is excluded. It has been suggested that the DMS originates from the plasma membrane in the form of tubular invaginations [28]. The DMS is in direct contact with the external milieu and can be stained with extracellular tracers, such as lanthanum salts, tannic acid, and ruthenium red [28–30]. The precise role of this extensive, invaginated membrane system has been a topic of debate for many years. Originally, the DMS was proposed to participate in platelet production by creating preformed "platelet territories" within the cytoplasm of the megakaryocyte. Nevertheless, newer experiments strongly indicate that the DMS serves mainly as a membrane reservoir for proplatelet production and elaboration. Experiments by Eckly and colleagues have started to provide some insights into the mechanics by which the DMS originates and matures [31]. To generate the DMS, the megakaryocyte plasma membrane enfolds at defined locations and a perinuclear pre-DMS is formed. Next, the pre-DMS is extended into its mature form by membrane added from Golgi-derived vesicles and endoplasmic reticulum-mediated transfer of lipids. These findings are consistent with experiments on platelet glycosyltransferases, which are delivered early in the assembling DMS and ultimately translocate to the megakaryocyte and platelet surface membrane [32]. Just a few proteins have been established to contribute to DMS generation based on changing of its ultrastructure in specific knockout murine models. Membrane-deforming proteins that employ F-BAR domains to bend membranes, or GTP as an energy source to bud vesicles from membranes, appear to be required for normal megakaryocyte development and platelet production. Gross disruptions in DMS structure are found in megakaryocytes isolated from either filamin A knockout, dynamin 2 knockout, or Cdc42-interacting protein 4 (CIP4) knockout mice. CIP4 is an F-BAR protein that causes membrane tubulation and preferentially binds to membranes via its BAR domain and interacts with the Wiskott-Aldrich syndrome protein [33]. CIP4-/- mice have a moderately reduced platelet count with a 25 % decrease. Although megakaryocyte numbers and ploidy are fairly normal in CIP4-/- knockout mice, the megakaryocytes cultured from these mice are less effective in generating proplatelets in vitro. Dynamins are an extremely well-conserved large mechanochemical GTPases involved in vesicular transport and endocytosis, and mutations in dynamin 2 have been linked with low platelet count in humans. Dynamin-2-dependent endocytosis is essential for the development of megakaryocytes in mice [34]. The DMS has, in addition, been suggested to mature into the platelet open canalicular system, which operates as a channel for the release of platelet alpha- and dense-granule contents. However, megakaryocytes from bovine, which have a highly developed DMS, generate platelets that do not form an OCS, suggesting the OCS is not necessarily a leftover membrane system of the DMS [30].

1.3 Platelet Formation

The precise mechanisms by which platelets are generated have been studied for well over a century. In 1906, James Homer Wright initiated a detailed study of how megakaryocytes give birth to platelets [35]. Numerous ideas have been proposed to explain how megakaryocytes generate platelets. The DMS, explained in detail by Yamada in 1957, was originally hypothesized to demarcate preformed "platelet territories" within the megakaryocyte cytoplasm [27]. Microscopists documented that mature megakaryocytes became packed with membranes and platelet-specific organelles and proposed that these membranes developed a system that defined fields for developing nascent platelets [36]. Release of individual platelets was postulated to take place by an enormous fragmentation of the megakaryocyte cytoplasm along DMS fracture lines located between these fields. The DMS model suggests that platelets assemble via an extensive internal membrane remodeling process [37]. Tubular membranes, which may initiate from invagination of the megakaryocyte plasma membrane, are expected to interconnect and branch, forming a continuous system throughout. The fusion of neighboring tubules has been proposed as a means to produce a flat membrane that eventually envelopes the cytoplasm of an assembling platelet. Models trying to use the DMS to justify how the megakaryocyte cytoplasm becomes partitioned into platelet volumes and surrounded by its own membrane have lost support because of multiple problems. For example, platelet territories inside the megakaryocyte cytoplasm do not contain microtubule rings, one of hallmark features of platelets in the resting state. If platelets are assembled within the cytoplasm of the megakaryocyte via the DMS, then platelet fields should show signs of structural features of resting platelets, which is not the case [38]. Furthermore, there are no experiments on living megakaryocytes directly showing that platelet fields explosively shatter into bona fide platelets. On the contrary, experiments that focused on the DMS of megakaryocytes before and after retraction of proplatelets triggered by microtubule disassembling drugs suggest this highly specialized membrane network may function mainly as a membrane reservoir that evaginates to deliver plasma membrane for the elaborate extension of proplatelets [39]. Radley and Haller have proposed that the name DMS may be a misnomer and have recommended "invagination membrane system" as a more appropriate name to depict this membrane system.

1.3.1 The Proplatelet Theory

The bulk of evidence supports the proplatelet hypothesis of platelet biogenesis. The term "proplatelet" is typically used to describe long cytoplasmic processes extending from megakaryocytes [40]. These processes are distinguished by numerous platelet-sized swellings linked together by thin cytoplasmic bridges and are believed to correspond to intermediate structures as megakaryocytes transition to platelets. In many ways, proplatelets can be thought of as the assembly lines of platelet production. The actual idea of platelets forming from these pseudopodia-like extensions took place when Wright recognized that platelets originate from megakaryocytes and described "the detachment of plate-like fragments or segments from pseudopods" extending from megakaryocytes [35]. Thiery and Bessis [41] and later Behnke [42] described the structure of these cytoplasmic processes emanating from megakaryocytes during platelet production in more detail. The classic "proplatelet theory" was established by Becker and De Bruyn, who suggested that megakaryocytes generate long pseudopod-like extensions that later break apart to produce single platelets [40]. In this initial model, the DMS was still hypothesized to subdivide the megakaryocyte cytoplasm into platelet territories. Radley and Haller later described the "flow model" which proposed that platelets originated entirely from the interconnected platelet-sized beads and proposed that the DMS did not function mainly to demarcate platelet fields but as a reservoir of surface membrane to be evaginated throughout proplatelet production [39]. Assembling platelets were presumed to become enclosed by megakaryocyte plasma membrane only as proplatelets were generated.

The majority of scientific data now supports a modified proplatelet model of platelet production. Proplatelets have been visualized: (1) in a wide variety of mammals, including guinea pigs, cows, mice, rats, dogs, and humans [43–46]; (2) both in vivo and in vitro, and maturation of proplatelets produces platelets that manifest structural and functional similarities to blood platelets [43, 47]; (3) emanating from megakaryocytes in the bone marrow through seams in the endothelial lining where they have been proposed to be released into the blood circulation and undergo additional fragmentation into individual platelets [48–50]; and (4) to be missing in mice in which two individual hematopoietic transcription factors have been knocked out. These mice do not form proplatelets in culture and have a very low platelet count [51–53]. Altogether, these observations support an essential role for proplatelet formation in platelet biogenesis.

1.3.2 Morphogenesis of Proplatelets

The discovery of thrombopoietin, the principal physiological controller of platelet generation, and the development of megakaryocyte cultures that faithfully reconstitute platelet formation in vitro has made available systems to visualize and study megakaryocytes in the act of producing proplatelets. Direct visualization of living megakaryocytes reveals both spatial and temporal changes that lead to the assembly of proplatelets (Fig. 1.1) [54]. A massive transformation of the megakaryocyte

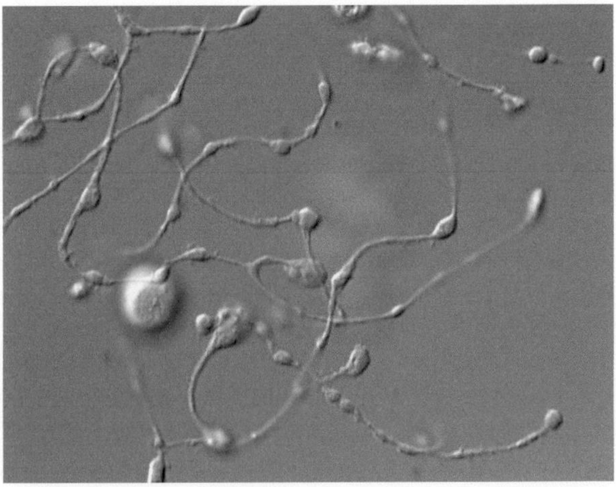

Fig. 1.1 Structural features of proplatelets. Light microscope image of proplatelets formed by a megakaryocyte in culture. The proplatelets can reach millimeters in length and contain platelet-sized swellings that decorate their length, ultimately giving them a beads-on-a-string appearance

cytoplasm concentrates almost all of the intracellular components into proplatelet processes and their platelet-sized swellings, which during the final stages appear as beads linked by very thin cytoplasmic strings. This striking morphogenesis unfolds over 5–10 h and initiates with the erosion of one pole (Fig. 1.1) of the cytoplasm of the megakaryocyte. Thick extensions initially develop and then extend into thin tubes of a regular diameter of 2–4 μm. These slender tubules, in turn, go through a dynamic bending and branching event and form periodic densities along their length. Ultimately, the megakaryocyte is converted into a "naked" nuclei besieged by an extensive network of proplatelet extensions. Megakaryocyte maturation finishes when a rapid retraction divorces the proplatelet extensions from the cell body liberating the fragments into culture (Fig. 1.1). The ensuing rupture of the cytoplasmic bridges between platelet-sized swellings is thought to liberate individual platelets into the blood circulation.

1.3.3 The Cytoskeletal Engine of Platelet Biogenesis

The cytoskeleton of the resting platelet, which functions as the bones and muscle of the cell, plays an essential role in preserving the disc shape of the mature, resting platelet and is the key driver of the shape change that transpires during the activation of platelets. This same set of cytoskeletal molecules supplies the force to power morphological changes associated with the maturation of megakaryocytes and production of platelets [55]. Three cytoskeletal polymer systems exist in megakaryocytes: actin, tubulin, and spectrin. These proteins reversibly polymerize into filaments. A body of experimental data supports a model of platelet generation in which microtubules, actin filaments, and the spectrin membrane skeleton play key roles.

1.3.3.1 Microtubules Power Proplatelet Extension

Microtubules were first extensively studied in resting platelets. The marginal microtubule coil is one of the most hallmark features of the resting platelet (Fig. 1.2b) [56]. αβ tubulin dimers polymerize into hollow microtubules under physiological

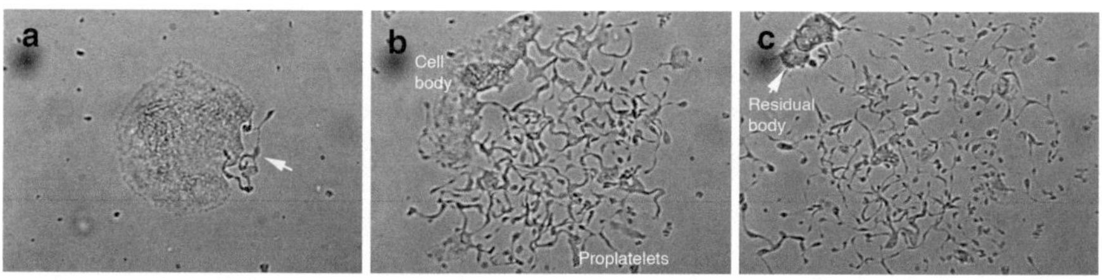

Fig. 1.2 Production of proplatelets by a megakaryocyte. These images show a time-lapse sequence of a mouse megakaryocyte demonstrating the events that lead to generation of proplatelets in culture. (**a**) Platelet generation begins when the megakaryocyte cytoplasm begins to erode at one pole. (**b**) The majority of the cytoplasm has been transformed into multiple extensions that continue to elongate and develop platelet-sized swellings that decorate their length. These extensions are very dynamic and exhibit both bending and branching. (**c**) After the majority of the megakaryocyte cytoplasm has been transformed into proplatelets, the process ends in a massive retraction that causes the proplatelets to separate from the residual cell body containing a naked nuclei

conditions, and tubulin is evenly separated between polymer and dimer fractions in resting platelets. In most cells, the αβ tubulin dimers are in a dynamic equilibrium with polymer such that reversible cycles of assembly disassembly of microtubules are observed. Microtubules are hollow polymers that are 24 nm in diameter and provide the force for many types of cell movements, such as the segregation of chromosomes during mitosis and the vesicular translocation of organelles and cargo throughout the cell. The ring of microtubules of the mature, resting platelet, initially described in the late 1960s by James White, has been depicted as a single microtubule approximately 100 μm long and is coiled 8–12 times in the cortex of the platelet [56]. The main role of the microtubule coil is to maintain the disc shape of the resting platelet. Depolymerization of platelet microtubules with pharmaceuticals such as nocodazole, vincristine, or colchicine causes platelets to become spherical and lose their disc shape [56]. Chilling platelets to 4 °C also causes depolymerization of the microtubule coil and loss of the disc shape. Moreover, molecular experiments show that mice lacking the major hematopoietic β-tubulin isoform (β1 tubulin) contain platelets that lack the hallmark disc shape and have defective microtubule coils. Knockout of the β1-tubulin gene in mice results in low platelet count with mice having platelet counts below 50 % of normal. β1-tubulin-deficient platelets are spherical in shape, and this appears to be caused by defective microtubule coils with fewer microtubule coilings. Although normal platelets possess a microtubule coil that is made up of 8–12 coils, β1-tubulin knockout platelets contain only 2–3 coils [57]. Human β1-tubulin functional substitution (AG>CC) inducing both structural and functional platelet changes has been described [58]. The Q43P β1-tubulin variation was observed in 10.6 % of the normal population and in 24.2 % of 33 unrelated patients with undefined congenital macrothrombocytopenia. Ultrastructural evaluation by electron microscopy revealed enlarged spherical platelets with an abnormal marginal band and structure. Platelets with the Q43P β1-tubulin variant showed a mild platelet phenotype, with reduced adenosine triphosphate (ATP) secretion, thrombin-receptor-activating peptide-induced aggregation, and impaired attachment to collagen under flow conditions [59]. One possibility is that it could confer an evolutionary advantage and protective cardiovascular role given that a more than twofold prevalence of the β1-tubulin variant has been observed in healthy individuals not undergoing ischemic events. The microtubules that that are within the coil are decorated with proteins that regulate the stability of the polymer [60]. The microtubule motor proteins kinesin and cytoplasmic dynein are present in platelets, but their precise functions in resting and activated platelets are not fully understood.

Proplatelet production is dependent on microtubule-based forces as exposure of megakaryocytes to drugs that disassemble microtubules, such as vincristine or nocodazole, prevents proplatelet production. Microtubules are the main cytoskeletal structural components of the engine that powers proplatelet extension. Direct visualization of the microtubule cytoskeletons of proplatelet-producing megakaryocytes provides insights as to how microtubules power platelet generation (Fig. 1.3) [55]. The microtubule cytoskeleton in megakaryocytes undergoes a spectacular reorganization during proplatelet production. In immature, round megakaryocytes that have

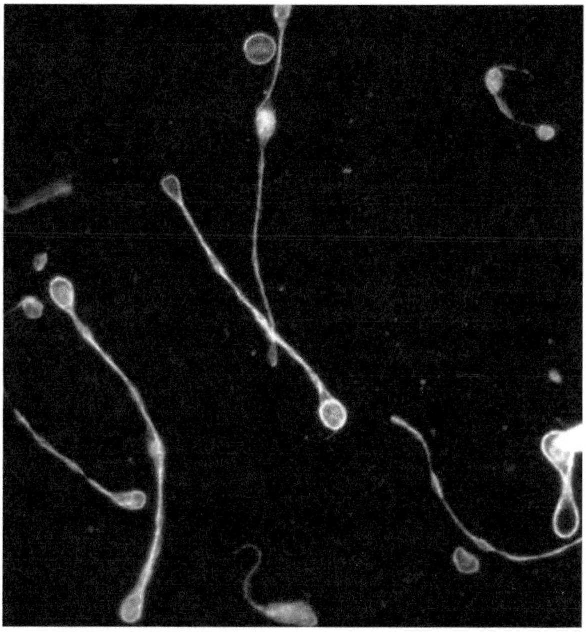

Fig. 1.3 Microtubules provide the force to elongate proplatelets. Immunofluorescence staining of mouse megakaryocytes grown in culture and labeled with anti-β1-tubulin antibodies shows the localization of microtubules along proplatelets. The characteristic features of proplatelets, including the swelling, shaft, tip, and branch points, are highlighted

not started proplatelet production, microtubules extend from the cell center to the cortex. As thick extensions form during the beginning stages of proplatelet production, membrane-associated and cortical microtubules merge into thick bundles positioned just under the plasma membrane of these structures. And once pseudopodia start to extend (at an average rate of 1 μm/min), microtubules form thick linear arrays that run along the whole length of the proplatelet processes (Fig. 1.3). The bundles of microtubules are thickest in the region of the proplatelet closest to the cell body of the megakaryocyte but thin down to bundles of approximately seven microtubules near proplatelet ends. The distal end of each proplatelet always has a platelet-sized bead that contains a microtubule bundle that loops just under the plasma membrane and reenters the shaft to form a teardrop-shaped structure. Because microtubule coils similar to those seen in resting blood platelets are observed only at the tips of proplatelets and not within the platelet-sized swellings found along the length of proplatelets, mature platelets are assembled principally at the ends of proplatelets.

Direct observation of the dynamics of microtubules in live megakaryocytes using green fluorescent protein (GFP) methodology has offered insights into how microtubules provide the force to extend proplatelets [55]. End-binding protein three (EB3), a protein that is associated only with growing plus ends of microtubules, fused to GFP was retrovirally expressed in mouse megakaryocytes and used as a probe to track the plus end dynamics of microtubules. Immature megakaryocytes without proplatelets use a centrosome microtubule nucleation reaction, which appears as a projecting starburst pattern when observed with EB3-GFP. Microtubules polymerize only from the centrosomes and extend outward into the cell edge where they turn and run along the edge of the cell. But, just before proplatelet production starts, centrosome assembly ceases and microtubules start to concentrate into the cortex. Real-time fluorescence microscopy of proplatelet-producing megakaryocytes expressing EB3-GFP shows that as proplatelets extend, microtubule assembly occurs constantly along the entire length of the proplatelet, including the swellings, shaft, and tip. The microtubule assembly rates (average of 10.2 μm/min) are approximately tenfold more rapid than the proplatelet elongation rate, suggesting assembly and proplatelet elongation are not tightly linked. The EB 3-GFP experiments also revealed that microtubules assemble in both directions in proplatelets, e.g., both toward the tips and cell body, showing that the microtubules that make up the bundles have a mixed polarity.

While microtubules are constantly assembling in proplatelets, assembly does not provide the force for proplatelet extension. Proplatelets continue to extend even when microtubule assembly is blocked by drugs that inhibit net microtubule polymerization, suggesting an alternate mechanism for proplatelet extension [55]. In line with this thinking, proplatelets have an intrinsic microtubule sliding mechanism. Cytoplasmic dynein, a minus end microtubule molecular motor protein, concentrates along the microtubules of the proplatelet and appears to directly provide force for microtubule sliding, since inhibition of cytoplasmic dynein through disruption of the dynactin complex prevents proplatelet production. The sliding of microtubules can also be reactivated in proplatelets that have been permeabilized with detergent. When ATP, which can provide the chemical energy to drive the enzymatic activity of molecular motors, is added, it activates proplatelet extension in permeabilized proplatelets that contain both cytoplasmic dynein and dynactin, its regulatory complex. A recent analysis has suggested six different types of behaviors characterize the elaboration of proplatelets: elongation, branching, pausing, fusions, fragmentations, and retractions. Although the average extension rate for proplatelets over time is 1 μm/min, elongation typically occurs in bursts and pauses. Surge rates greatly exceed the average rates and under shear flow, rates of >30 μm/min have been documented. These rates correlate well with the sliding rates of microtubules within the proplatelet microtubule bundles. Fluorescence recovery after photobleaching studies has shown that microtubule sliding powers proplatelet extension and is dependent on cytoplasmic dynein [61]. Thus, dynein-mediated microtubule sliding appears to be the major event in powering proplatelet elongation.

1.3.3.2 Transport of Organelles and Granules Along Proplatelets into Assembling Platelets

In addition to functioning as the major engine that powers proplatelet extension, the microtubules within proplatelets have a secondary function – the transport of organelles, granules, and membrane into proplatelets and nascent platelets at proplatelet tips. Single, discrete organelles are transported from the body of the megakaryocyte into the proplatelets where they move back and forth until they are captured at proplatelet ends [62]. Electron microscopy and immunofluorescence studies demonstrate that organelles are closely associated with microtubules, whereas actin poisons do not reduce organelle movement. Thus, organelle motility seems to require microtubule-based mechanisms. Bidirectional organelle movement is transmitted in part by the mixed polarity of microtubules within the proplatelet, as kinesin-coated beads move bidirectionally over the microtubule bundles of skinned proplatelets. Of the two major microtubule-based motors, cytoplasmic dynein and kinesin, only kinesin is localized in an arrangement comparable to organelles and is likely responsible for moving these cargo along microtubules [62]. It appears that a twofold mechanism of organelle motility occurs during proplatelet production. First, organelles and granules translocate along microtubules, and second, the microtubules can also slide bidirectionally in relation to other microtubules to indirectly move organelles along proplatelets in a piggyback fashion.

1.3.3.3 Actin-Driven Bending and Branching of Proplatelets

Actin, at a very high concentration of 0.5 mM, is the most abundant of all the platelet proteins with approximately two million molecules per platelet [63]. Similar to tubulin, actin is in a dynamic polymer-monomer equilibrium. Approximately, 40 % of the actin subunits assemble to form the 2,000–5,000 actin filaments observed in the resting cell [64]. The rest of the cytoplasmic actin in the platelet is maintained in sequestration as a 1 to 1 complex with β4-thymosin [65] and is converted to filaments during platelet activation to power cell spreading. A body of data indicates that the filaments of the resting platelet are cross-linked at various points into a rigid cytoplasmic network, as platelets contain very high amounts of proteins that cross-link actin including filamin and α-actinin [66, 67]. Both filamin and α-actinin exist as homodimers in solution. Filamin subunits are elongated molecules composed mainly of 24 repeats, each ~100 amino acids in length that are folded into IgG-like beta barrels [68, 69]. There are three filamin genes on chromosomes 3, 7, and X. Filamin A (X) [70] and filamin B are expressed in platelets with Filamin A present at greater than tenfold excess to filamin B. Filamin is known to be a prototypical scaffolding protein that draws binding partners and arranges them next to the plasma membrane [71]. Some of the partners that bind filamin members include: the small GTPase, RalA, Rac, Rho, and Cdc42 with RalA binding in a GTP-dependent manner [72]; the exchange factors Trio and Toll; and kinases such as PAK1. Critical to the structural organization of the resting platelet is a linkage between filamin and the cytoplasmic tail of the GPIbα subunit of the GPIb-IX-V complex. The second rod domain (repeats 17–20) of filamin has an attachment site for the cytoplasmic tail of GPIbα 33, and experiments have demonstrated that the majority of platelet filamin (≥90 %) is in complex with GPIbα [73]. This connection has three major consequences. First, because a large fraction of filamin is bound to actin, it aligns the GPIb-IX-V complexes into rows on the surface of the platelet over the underlying filaments. Second, it positions filamin's self-association domain and associated partner proteins at the membrane at the same time as presenting filamin's actin binding sites into the cytoplasm. Third, because the filamin connections between GPIb-IX-V complex and actin filaments go through the holes of the spectrin lattice, it restrains the movement of the spectrin strands in this network and restrains the lattice in compression. The FLN-GPIbα linkage is critical for the production of discoid platelets from megakaryocytes as platelets lacking this connection are fragile and large and produced in very low numbers. However, the role of the filamin-GPIbα linkage in platelet generation per se is not fully understood. Given that a small amount of Bernard-Soulier platelets assemble and release from megakaryocytes, it can be argued that this linkage is a late event in the production process and is not per se essential for platelet release.

Each megakaryocyte has been predicted to produce thousands of platelets [74–77]. Time-lapse imaging of platelet production from megakaryocytes cultured in vitro has shown that proplatelet ends are increased in an extremely dynamic manner that repeatedly bends and branches the proplatelet shaft [54]. End amplification starts when a proplatelet shaft is bent into a sharp kink, which then folds back on itself, developing a loop in the microtubule bundle. The nascent loop then elongates, forming a new proplatelet shaft branching from the side of the original proplatelet. Loops lead the proplatelet end and establish the location where nascent platelets will develop and where platelet-specific contents are transported. In clear difference to the microtubule-based motor that extends proplatelets, actin-based power is used to bend the proplatelet in end amplification. Megakaryocytes exposed to the actin inhibitors, latrunculin or cytochalasin, can only extend lengthy, unbranched proplatelets that contain very few swellings along their length. Notwithstanding extensive analysis of actin filament dynamics during platelet activation, how actin functions in this process and the molecular signals that direct bending have yet to be defined. Ultrastructural analysis via phalloidin staining and rapid freeze electron microscopy of megakaryocytes undergoing proplatelet production have shown that actin filaments are dispersed throughout the proplatelet and are very abundant within swellings and at proplatelet branch points. One possibility is that proplatelet bending and branching is powered by the actin nonmuscle motor myosin II. A genetic mutation in the nonmuscle myosin heavy chain-A gene MYH9 in humans results in a group of diseases comprising May-Hegglin anomaly and the Sebastian, Fechtner, and Epstein syndromes, characterized by low platelet count with giant platelets. Experiments also suggest that protein kinase Cα (PKCα) associates with concentrations of actin filaments in megakaryocytes that are exhibiting proplatelet production and inhibition of PKCα or integrin signaling pathways stop actin filament aggregation and proplatelet production in megakaryocytes [78]. Since proplatelets extend but do not branch in the presence of the actin disassembly drug cytochalasin B, it is unexpected that the deletion of specific actin associated proteins from the megakaryocyte lineage leads to macrothrombocytopenia. It seems likely that the removal of actin-modulating proteins alters and or increases filamentous actin and the cytoskeletal structure appears to have a dominant inhibitory effect on proplatelet generation and release. The deletion of the actin cross-linking protein Filamin A in conditional mice whose megakaryocyte and platelet lineage lack Filamin A results in giant platelets in which platelet counts are reduced by 80–90 % [79]. Conditional mice in which the actin filament turnover protein cofilin-1 has been knocked out in the megakaryocyte lineage also have macrothrombocytopenia with a platelet count reduced 60–80 % of normal. In contrast, mice in which the actin filament turnover protein ADF has been deleted have normal platelet counts and structure [80]. However, when ADF-/- mice are bred with cofilin-1 knockout mice, platelet generation is drastically reduced and the structure and shape of platelets are highly variable. Mice that contain megakaryocytes in which profilin 1, a small protein that stimulates polymerization of actin filaments, has been deleted have giant platelets with decreased platelet counts [81]. Profilin-null platelets have an enlarged microtubule coil with microtubules that are hyper-acetylated, and in a number of ways, the profilin 1 knockout phenotype is very comparable to the phenotype of platelets in the Wiskott-Aldrich syndrome or in WASp knockout mice. Defective proplatelet generation has also been described in mice in which the small regulatory GTPases Rho, Cdc42, and Rac have been deleted in the megakaryocyte lineage [82, 83].

1.3.3.4 The Spectrin Membrane Skeleton

While the functions of actin filaments and microtubules in proplatelet generation have been significantly studied, our understanding of the role of the membrane skeleton has only recently been appreciated. The plasma membrane and OCS membrane of the resting platelet are supported by an intricate cytoskeletal system. The platelet is the only other cell in addition to the red blood cell whose membrane skeleton has been visualized at high resolution. Like the red blood cell, the platelet membrane skeleton is also a self-assembly of extended spectrin strands that interconnect through their attachment to actin filaments creating triangular pores. Platelets contain approximately 2,000 spectrin molecules [64, 84, 85] that assemble into a network that coats the cytoplasmic surface of both the OCS and plasma membrane. While much less is known about how the spectrin-actin network assembles and is connected to the plasma membrane in the platelet relative to the red blood cell, specific differences between the two membrane skeletons have been established. First, the spectrin strands that make up the platelet membrane skeleton interconnect using the ends of long actin filaments instead of short actin oligomers [64]. These ends arrive at the membrane initiating from filaments in the cytoplasm. Hence, the spectrin lattice is organized into a continuous network by its association with actin filaments. Second, tropomodulins are not found at sufficiently high levels, if at all, to have a role in the capping of the pointed ends of the platelet actin filaments; instead, cell biological experiments have shown that a substantial number (~2,000) of these ends are free in the resting platelet. Third, although little tropomodulin protein is expressed, adducin is amply expressed and appears to cap many of the barbed ends of the filaments composing the resting actin cytoskeleton [86]. Adducin is a major component of the membrane skeleton forming a triad complex with actin and spectrin. Capping of barbed filament ends by adducin also functions to target them to the spectrin-based membrane skeleton, as the affinity of spectrin for adducin-actin complexes is much greater than for either actin or adducin alone [87–90]. High-resolution ultrastructural examination via electron microscopy shows that proplatelets contain a spectrin-based membrane skeleton comparable in structure to that of platelets [91]. Spectrin tetramer assembly is critical for the generation of the DMS and proplatelet extension, as expression of a spectrin tetramer-disrupting peptide in megakaryocytes blocks the progression of both processes. In addition, incorporation of this spectrin-disrupting peptide construct into a permeabilized model system very quickly destabilizes proplatelet structure, causing massive swelling as well as blebbing. Spectrin tetramers also appear to stabilize the barbell-shaped structures that give rise to individual platelets. Overall, these findings suggest an essential role for spectrin in distinct stages of megakaryocyte development by its participation in the generation of the DMS and in the preservation of proplatelet structure.

1.3.4 Release of Individual Platelets

In vivo, proplatelets project into bone marrow sinusoids, where they are released and enter the blood. Junt et al. as well as Massberg and colleagues have used intravital multiphoton microscopy to observe proplatelet generation in the opened cranial marrow cavity of live mice [92, 93]. Fluorescently labeled megakaryocytes could be observed to extend proplatelets and liberate fragments of megakaryocytes into the marrow sinusoids of live mice. Remarkably, these megakaryocyte fragments generally exceed the dimensions of platelets, implying that platelet morphogenesis persists in the blood circulation. Consistent with these findings, we have discovered a new intermediate stage in platelet production and release, which we called the preplatelet [94]. Preplatelets, which appear as "giant platelets," are classified as discoid cells (3–10 μm) that still maintain the ability to convert into barbell-shaped proplatelets and undergo fission into platelets. Inhibitors of microtubule assembly prevent the conversion of preplatelets to barbells. Thus, the conversion of preplatelets to barbell proplatelets is driven by microtubule-powered forces. It is appealing to speculate that the preplatelet fission reaction is a major regulator of platelet size and that some giant platelet disorders (macrothrombocytopenias) may represent a failure to convert preplatelets into barbell proplatelets. Force constraints resulting from cortical microtubule band diameter and thickness appear to regulate barbell proplatelet generation, and platelet size is possibly limited by microtubule bundling, elastic bending, and actin-myosin-spectrin cortex forces [95]. It was recently shown that individual human platelets have the innate ability to duplicate and form new cell bodies that undergo fission into platelets [96]. The morphological resemblance between platelets that form new cell bodies and preplatelets is remarkable. Whether the newly released platelets exhibit a preplatelet phenotype, which may permit them to generate barbell shapes and divide again, is unclear.

References

1. Li CL, Johnson GR (1995) Murine hematopoietic stem and progenitor cells: I. Enrichment and biologic characterization. Blood 85(6):1472–1479
2. Weissman IL, Anderson DJ, Gage F (2001) Stem and progenitor cells: origins, phenotypes, lineage commitments, and transdifferentiations. Annu Rev Cell Dev Biol 17:387–403
3. Akashi K et al (2000) A clonogenic common myeloid progenitor that gives rise to all myeloid lineages. Nature 404(6774):193–197

4. Adolfsson J et al (2005) Identification of Flt3+ lympho-myeloid stem cells lacking erythro-megakaryocytic potential a revised road map for adult blood lineage commitment. Cell 121(2):295–306
5. Long MW, Williams N, Ebbe S (1982) Immature megakaryocytes in the mouse: physical characteristics, cell cycle status, and in vitro responsiveness to thrombopoietic stimulatory factor. Blood 59(3):569–575
6. Ebbe S (1976) Biology of megakaryocytes. Prog Hemostasis Thromb 3:211–229
7. Ebbe S, Stohlman F (1965) Megakaryocytopoiesis in the rat. Blood 26:20–34
8. Odell T, Jackson CJ, Reiter R (1968) Generation cycle of rat megakaryocytes. Exp Cell Res 53:321
9. Odell TT Jr, Jackson CW, Friday TJ (1970) Megakaryocytopoiesis in rats with special reference to polyploidy. Blood 35(6):775–782
10. Therman E, Sarto G, Stubblefiels P (1983) Endomitosis: a reappraisal. Hum Genet 63:13–18
11. Raslova H et al (2003) Megakaryocyte polyploidization is associated with a functional gene amplification. Blood 101:541–544
12. Nagata Y, Muro Y, Todokoro K (1997) Thrombopoietin-induced polyploidization of bone marrow megakaryocytes is due to a unique regulatory mechanism in late mitosis. J Cell Biol 139:449–457
13. Vitrat N et al (1998) Endomitosis of human megakaryocytes are due to abortive mitosis. Blood 91:3711–3723
14. Ravid K et al (2002) Roads to polyploidy: the megakaryocyte example. J Cell Physiol 190(1):7–20
15. Gu M et al (1999) Analysis of the roles of 14-3-3 in the platelet glycoprotein Ib-IX-mediated activation of integrin aIIbb3 using a reconstituted mammalian cell expression model. J Cell Biol 147:1085–1096
16. Wang Z et al (1995) Cyclin D3 is essential for megakaryocytopoiesis. Blood 86:3783–3788
17. Broek D, Bartlett R, Crawford K (1991) Involvement of p34cdc2 in establishing the dependency of S phase on mitosis. Nature 349:388–393
18. Hayles J, Fisher D, Woodlard A (1994) Temporal order of S phase and mitosis in fission yeast is determined by the state of the p34cdc22-mitotic B cyclin complex. Cell 78:813–822
19. Brieher W et al (2006) Rapid actin monomer-insensitive depolymerization of Listeria actin comet tails by cofilin, coronin, and Aip1. J Cell Biol 175:315–324
20. Datta NS et al (1996) Novel alterations in CDK1/cyclin B1 kinase complex formation occur during the acquisition of a polyploid DNA content. Mol Biol Cell 7:209–223
21. Zhang Y, Wang Z, Ravid K (1996) The cell cycle in polyploid megakaryocytes is associated with reduced activity of cyclin B1-dependent cdc2 kinase. J Biol Chem 271:4266–4272
22. Kozar K (2004) Mouse development and cell proliferation in the absence of D-cyclins. Cell 118:477–491
23. Geng Y (2003) Cyclin E ablation in the mouse. Cell 114:431–443
24. Zhang Y et al (2004) Aberrant quantity and localization of Aurora-B/AIM-1 and survivin during megakaryocyte polyploidization and the consequences of Aurora-B/AIM-1-deregulated expression. Blood 103(10):3717–3726
25. Trakala M et al (2015) Functional reprogramming of polyploidization in megakaryocytes. Dev Cell 32(2):155–167
26. Kautz J, De Marsh QB (1955) Electron microscopy of sectioned blood and bone marrow elements. Rev Hematol 10(2):314–323; discussion, 324–44
27. Yamada F (1957) The fine structure of the megakaryocyte in the mouse spleen. Acta Anat 29:267–290
28. Behnke O (1968) An electron microscope study of megakaryocytes of rat bone marrow. I. The development of the demarcation membrane system and the platelet surface coat. J Ultrastuct Res 24:412–433
29. Nakao K, Angrist AA (1968) Membrane surface specialization of blood platelet and megakaryocyte. Nature 217(5132):960–961

30. Zucker-Franklin D, Benson KA, Myers KM (1985) Absence of a surface-connected canalicular system in bovine platelets. Blood 65(1):241–244
31. Eckly A et al (2014) Biogenesis of the demarcation membrane system (DMS) in megakaryocytes. Blood 123(6):921–930
32. Wandall HH et al (2012) The origin and function of platelet glycosyltransferases. Blood. 120(3):626–635
33. Chen Y et al (2013) Loss of the F-BAR protein CIP4 reduces platelet production by impairing membrane-cytoskeleton remodeling. Blood 122(10):1695–1706
34. Bender M et al (2015) Dynamin 2-dependent endocytosis is required for normal megakaryocyte development in mice. Blood 125(6):1014–1024
35. Wright J (1906) The origin and nature of blood platelets. Boston Med Surg J 154:643–645
36. Shaklai M, Loskutoff D, Tavassoli M (1978) Membrane characteristics of cultured endothelial cells: identification of gap junction. Isr J Med Sci 14(3):306–313
37. Kosaki G (2005) In vivo platelet production from mature megakaryocytes: does platelet release occur via proplatelets? Int J Hematol 81(3):208–219
38. Radley J, Hatshorm M (1987) Megakaryocyte fragments and the microtubule coil. Blood Cells 12:603–608
39. Radley JM, Haller CJ (1982) The demarcation membrane system of the megakaryocyte: a misnomer? Blood 60:213–219
40. Becker RP, DeBruyn PP (1976) The transmural passage of blood cells into myeloid sinusoids and the entry of platelets into the sinusoidal circulation; a scanning electron microscopic investigation. Am J Anat 145:1046–1052
41. Thiery JP, Bessis M (1956) Mechanism of platelet genesis; in vitro study by cinemicrophotography. Rev Hematol 11(2):162–174
42. Behnke O (1969) An electron microscope study of the rat megakaryocyte. II. Some aspects of platelet release and microtubules. J Ultrastruct Res 26:111–129
43. Choi E, Nichol JL, Hokom MM, Hornkohl AC, Hunt P (1995) Platelets generated in vitro from proplatelet-displaying human megakaryocytes are functional. Blood 85:402–413
44. Handagama P et al (1987) In vitro platelet release by rat megakaryocytes: effect of metabolic inhibitors and cytoskeletal disrupting agents. Am J Vet Res 48:1142–1146
45. Leven R (1987) Megakaryocyte motility and platelet formation. Scanning Microsc 1:1701–1709
46. Tablin F, Castro M, Leven R (1990) Blood platelet formation in vitro. The role of the cytoskeleton in megakaryocyte fragmentation. J Cell Sci 97:59–70
47. Cramer E et al (1997) Ultrastructure of platelet formation by human megakaryocytes cultured with the Mpl ligand. Blood 89:2336–2346
48. Lichtman M et al (1978) Parasinusoidal location of megakaryocytes in marrow: a determinant of platelet release. Am J Hematol 4:303–312
49. Scurfield G, Radley JM (1981) Aspects of platelet formation and release. Am J Hematol 10:285–296
50. Tavassoli M, Aoki M (1989) Localization of megakaryocytes in the bone marrow. Blood Cells 15(1):3–14
51. Lecine P, Villeval J-L, Paresh V, Swencki B, Xu Y, Shivdasani R (1998) Mice lacking transcription factor NF-E2 provide in vivo validation of the proplatelet model of thrombocytopoiesis and show a platelet production defect that is intrinsic to megakaryocytes. Blood 92:1608–1616
52. Shivdasani RA et al (1997) A lineage-selective knockout establishes the critical role of transcription factor GATA-1 in megakaryocyte growth and platelet development. EMBO J 16:3965–3973
53. Shivdasani RA et al (1995) Transcription factor NF-E2 is required for platelet formation independent of the actions of thrombopoietin/MGDF in megakaryocyte development. Cell 81:695–704
54. Italiano JE Jr et al (1999) Blood platelets are assembled principally at the ends of proplatelet processes produced by differentiated megakaryocytes. J Cell Biol 147(6):1299–1312
55. Patel S et al (2005) Differential roles of microtubule assembly and sliding in proplatelet formation by megakaryocytes. Blood 106:4076–4085

56. White J, Krivit W (1967) An ultrastructural basis for the shape changes induced in platelets by chilling. Blood 30:625–635
57. Schwer H et al (2001) A lineage-restricted and divergent b tubulin isoform is essential for the biogenesis, structure and function of mammalian blood platelets. Curr Biol 11:579–586
58. Freson K et al (2005) The b1-tubulin Q43P functional polymorphism reduces the risk of cardiovascular disease in men by modulating platelet function and structure. Blood 106:2356–2362
59. Browne K et al (2000) Filamin (280-kDa actin-binding protein) is a caspase substrate and is also cleaved directly by the cytotoxic T lymphocyte protease granzyme B during apoptosis. J Biol Chem 275:39262–39266
60. Kenney D, Linck R (1985) The cytoskeleton of unstimulated blood platelets: structure and composition of the isolated marginal microtubular band. J Cell Sci 78:1–22
61. Bender M et al (2015) Microtubule sliding drives proplatelet elongation and is dependent on cytoplasmic dynein. Blood. 125(5):860–868
62. Richardson J et al (2005) Mechanisms of organelle transport and capture along proplatelets during platelet production. Blood 106:4066–4075
63. Nachmias VT, Yoshida K-i (1988) The cytoskeleton of the blood platelet: a dynamic structure. Adv Cell Biol 2:181–211
64. Hartwig J, DeSisto M (1991) The cytoskeleton of the resting human blood platelet: structure of the membrane skeleton and its attachment to actin filaments. J Cell Biol 112:407–425
65. Safer D, Nachmias V (1994) Beta thymosins as actin binding peptides. Bioessays 16:473–479
66. Rosenberg S, Stracher A (1982) Effect of actin-binding protein on the sedimentation properties of actin. J Cell Biol 94:51–55
67. Rosenberg S, Stracher A, Burridge K (1981) Isolation and characterization of a calcium-sensitive α-actinin-like protein from human platelet cytoskeletons. J Biol Chem 256:12986–12991
68. Fucini P et al (1997) The repeating segments of the F-actin cross-linking gelation factor (ABP-120) have an immunoglobulin-like fold. Nat Struct Biol 4(3):223–230
69. Gorlin J et al (1990) Human endothelial actin-binding protein (ABP-280, non-muscle filamin): a molecular leaf spring. J Cell Biol 111:1089–1105
70. Gorlin J et al (1993) Actin-binding protein (ABP-280) filamin gene (FLN) maps telomeric to the colar vision locus (R/GCP) and centromeric to G6PD in Xq28. Genomics 17:496–498
71. Stossel T et al (2001) Filamins as integrators of cell mechanics and signalling. Nat Rev 2:138–145
72. Ohta Y et al (1999) The small GTPase RalA targets filamin to induce filopodia. Proc Natl Acad Sci U S A 96:2122–2128
73. Kovacsovics T et al (1995) Phosphoinositide 3-kinase inhibition spares actin assembly in activating platelets, but reverses platelet aggregation. J Biol Chem 270:11358–11366
74. Harker L (1978) Platelet survival time: its measurement and use. Prog Hemostatis Thromb 4:321–347
75. Harker LA, Finch CA (1969) Thrombokinetics in man. J Clin Investig 48:963–974
76. Kaufman RM, Airo R, Pollack S, Crosby WH (1965) Circulating megakaryocytes and platelet release in the lung. Blood 26:720–728
77. Trowbridge EA, Martin JF, Slater DN, Kishk YT, Warren CW, Harley PJ, Woodcock B (1984) The origin of platelet count and volume. Clin Phys Physiol Meas 5:145–156
78. Rojnuckarin P, Kaushansky K (2001) Actin reorganization and proplatelet formation in murine megakaryocytes: the role of protein kinase c alpha. Blood 97:154–161
79. Begonja AJ et al (2011) FlnA-null megakaryocytes prematurely release large and fragile platelets that circulate poorly. Blood 118:2285–2295
80. Bender M et al (2010) ADF/n-cofilin-dependent actin turnover determines platelet formation and sizing. Blood 116(10):1767–1775
81. Bender M et al (2014) Megakaryocyte-specific Profilin1-deficiency alters microtubule stability and causes a Wiskott-Aldrich syndrome-like platelet defect. Nat Commun 5:4746
82. Pleines I et al (2013) Defective tubulin organization and proplatelet formation in murine megakaryocytes lacking Rac1 and Cdc42. Blood 122(18):3178–3187

83. Pleines I et al (2012) Megakaryocyte-specific RhoA deficiency causes macrothrombocytopenia and defective platelet activation in hemostasis and thrombosis. Blood 119(4):1054–1063
84. Fox J, Aggerbeck L, Berndt M (1988) Structure of the glycoprotein Ib-IX complex from platelet membranes. J Biol Chem 263:4882–4890
85. Fox J et al (1987) Spectrin is associated with membrane-bound actin filaments in platelets and is hydrolyzed by the Ca^{2+}-dependent protease during platelet activation. Blood 69:537–545
86. Barkalow K et al (2003) a-Adducin dissociates from F-actin filaments and spectrin during platelet activation. J Cell Biol 161:557–570
87. Kaiser H, O'Keefe E, Bennett V (1989) Adducin: Ca++-dependent association with sites of cell-cell contact. J Cell Biol 109:557–569
88. Kuhlman P et al (1996) A new function for adducin. Calcium/calmodulin-regulated capping of the barbed ends of actin filaments. J Biol Chem 271:7986–7991
89. Matsuoka Y, Li X, Bennett V (1998) Adducin is an in vivo substrate for protein kinase C: phosphorylation in the MARCKS-related domain inhibits activity in promoting spectrin-actin complexes and occurs in many cells, including dendritic spines of neurons. J Cell Biol 142:485–497
90. Matsuoka Y, Li X, Bennett V (2000) Adducin: structure, function, and regulation. Cell Mot Life Sci 57:884–895
91. Patel-Hett S et al (2011) The spectrin-based membrane skeleton stabilizes mouse megakaryocyte membrane systems and is essential for proplatelet and platelet formation. Blood 118(6):1641–1652
92. Junt T et al (2007) Dynamic visualization of thrombopoiesis within bone marrow. Science 317(5845):1767–1770
93. Zhang L et al (2012) A novel role of sphingosine 1-phosphate receptor S1pr1 in mouse thrombopoiesis. J Exp Med 209(12):2165–2181
94. Thon JN et al (2010) Cytoskeletal mechanics of proplatelet maturation and platelet release. J Cell Biol 191(4):861–874
95. Thon JN et al (2012) Microtubule and cortical forces determine platelet size during vascular platelet production. Nat Commun 3:852
96. Schwertz H et al (2010) Anucleate platelets generate progeny. Blood 115(18):3801–3809

Transcriptional Regulation of Platelet Formation: Harnessing the Complexity for Efficient Platelet Production In Vitro

2

Marloes R. Tijssen, Thomas Moreau, and Cedric Ghevaert

Abstract

It is now common knowledge that specific repertoires of transcription factors (TFs) determine a cell's protein content and thereby its phenotype. The expression of a given TF is not necessarily cell specific, and many TFs play a pivotal role in several different cell types. For example, TAL1, FLI1, RUNX1, ERG and GATA2 are important regulators of stem cells, but also play a vital role in megakaryopoiesis. Although the megakaryocyte (MK) and its closest relative, the red blood cell, share key TFs like GATA1 and NFE2, the bifurcation between the two lineages has been associated with pairs of TFs that act as a toggle switch (such as FLI1 and KLF1). This chapter will summarise the current knowledge of key transcriptional regulators of MK differentiation and how some of these TFs, despite being expressed in several cell types, can impose MK cell identity. Since the discovery of TPO in 1994, our knowledge of MK biology and differentiation has increased exponentially, but we still lack a deep understanding of what triggers the transition from MK growth and maturation to proplatelet formation. We describe how some well-known TFs control the expression of proteins that play a pivotal role in the dramatic cytoplasmic and cytoskeletal events that accompany proplatelet formation. Finally, we show how TFs can be harnessed in a powerful way to produce MKs and, potentially, platelets in vitro for future clinical applications.

The original version of this chapter was revised. The copyright holder has been updated to The Author(s). An erratum to this chapter can be found at https://doi.org/10.1007/978-3-319-39562-3_21

M.R. Tijssen • T. Moreau • C. Ghevaert (✉)
Department of Haematology, NHS Blood and Transplant Centre, University of Cambridge, Cambridge, UK
e-mail: mrt37@cam.ac.uk; tm384@medschl.cam.ac.uk; cg348@cam.ac.uk

H. Schulze, J. Italiano (eds.), *Molecular and Cellular Biology of Platelet Formation*, DOI 10.1007/978-3-319-39562-3_2

2.1 Introduction

Although the role of platelets clearly extends well beyond their primary role in homeostasis, the main symptom of thrombocytopenia (low platelet count) or platelet dysfunction is bleeding. Human peripheral blood contains 150–450 $\times$ 10^9 platelets/L, and since platelets have a life span of 10 days, there is a substantial demand for continuous platelet production in the bone marrow (estimated to be around 10^{11} platelets per day). This process is tightly regulated, and the body has means to increase it when more platelets are needed, for example, in the case of increased platelet consumption following a haemorrhage or infection. All blood cells originate from multipotent haematopoietic stem cells (HSCs) that can self-renew and differentiate into specialist cells. Platelets are derived from large polynucleated cells called the megakaryocytes (MKs) that are generated from HSCs. Although there is still a debate whether MKs descend directly from the HSCs, a specific population of stem cells, or go through the stage of a common myeloid progenitor first (see Sect. 2.2), we do have extensive knowledge on the transcription factors (TFs) involved in enforcing an MK cell fate. Once an MK has matured, it will ultimately release thousands of platelets into the bloodstream. Despite researchers rather efficiently recapitulating the process of MK formation and maturation from stem cells in the laboratory, we can at present only harvest five to ten functional platelets per MK in vitro. It is possible that these in vitro-derived MKs are not given the right environment to produce higher numbers of platelets or that the in vitro-produced platelets are somehow activated and therefore non-functional and lost from the final harvest. Alternatively, we may be missing crucial cues in vitro for the final step of MK maturation or platelet formation. Therefore, the knowledge of the transcriptional regulation of MK differentiation is extremely relevant in order to understand how platelets are produced and to ultimately harness that knowledge and translate it into the efficient in vitro production of platelets for clinical applications.

2.2 Haematopoietic Stem Cell Differentiation: About Heterogeneity, Stochastic and Directed Fate Decisions

The accepted central concept of "a stem cell" is that of a cell that possesses the ability to self-renew and is able to differentiate into lineage-restricted cells that become increasingly specialised until they reach the terminally differentiated fully mature cells. This concept was first demonstrated in the haematopoietic system in the 1950s through mouse models of bone marrow transplantation. Low numbers of stem cells could repopulate the bone marrow of a mouse with a compromised haematopoietic system and regenerate all haematopoietic cell types. The multiple observations resulting from this large body of work have produced strong support for a hierarchical model in which the long-term self-renewal potential of the HSCs is restricted upon differentiation into lineage progenitors that will eventually produce the terminally differentiated mature blood and immune cells [1].

However, the advent of cell sorting technologies and single-cell assays has allowed us to realise that even a well-defined HSC population is heterogeneous, with each cell possessing specific differentiation patterns and self-renewal properties. Fascinatingly, these properties appear to be largely intrinsically stable within the original HSC and its progeny (reviewed in [2]), although external stimuli (usually described under the blanket term of "niche") can influence HSC behaviour (reviewed in [3]). This heterogeneity of HSC has been brought to light specifically in terms of MK differentiation by the demonstration by two independent groups of a previously undescribed subpopulation of HSCs which is "primed" to preferentially differentiate to the MK lineage [4, 5]. This suggests HSCs do not necessarily need to go through the multipotent progenitor stage.

Despite a massive body of research into stem cell traits, the field is still divided with regard to the mechanism triggering the cell to stop self-renewing and to commit to a certain cell fate. There is obvious evidence for both a "directed" decision (i.e. under the influence of both external stimuli and intrinsic cell characteristics) and a stochastic process. However, a stochastic cell fate decision might only appear to be so as a consequence of our current inability to look into the minute molecular details of each individual cell in order to identify the probabilistic outcome of cell differentiation. The orchestration of the outcome of differentiation is ultimately driven at the cellular level by the precise arrangement of TFs in a given progenitor which will trigger the expression of lineage-specific gene regulatory networks.

This paradigm is illustrated by the bipotent MK-erythroid progenitors (MEPs) that can develop either into an erythroblast or an MK for which they need to switch on a completely different gene repertoire. Erythrocytes reduce their cell size and condense their nucleus, which is finally expelled. MKs on the other hand become bigger, increase their DNA content, and form specific "granules". Although the demonstration that pairs of TF "switches" may regulate the MEPs cell fate decision, the process by which the cell decides which TF to favour (and thereby the choice of pathway) is still unclear.

2.3 Regulation of Gene Expression: Multilayer Complexity

Each cell of the body contains the same genetic information, whilst its specific function and identity are defined by the restricted protein set that it contains and thereby its pattern of gene expression. Regardless of whether the cell fate decision is stochastic or directed, it will be executed by changes in the gene expression profile of the cell. The regulation of each protein concentration in a specific cell is subject to a very complex multilayered regulation at the transcriptional, translational and protein structure levels. This includes epigenetic regulation which relates to the three-dimensional organisation of the chromatin and its accessibility to DNA-binding proteins that regulate gene transcription [6], non-coding RNA such as miRNA (which regulate degradation of mRNA or repress translation) [7] and long non-coding RNAs whose role in haematopoiesis (including erythroid/Mk maturation) is gradually being uncovered [8]. For example, although the level of TFs involved in late MK differentiation and platelet formation (such as GATA1, FOG1, FLI1, TAL1, RUNX1 and NFE2) does not increase with polyploidisation, the transcripts of their target genes are upregulated [9] indicating that several nonmutually exclusive mechanisms (translational and/or post-translational regulation) might be at play. In addition, a recent publication showed that an extra dimension is added to this transcriptional regulation by the lineage-specific expression of different isoforms from the same gene, conferring specific biological activity to the protein within a given cell type in the haematopoietic tree [10]. The mechanisms underlying lineage-specific alternative promoter use or splicing are still to be fully understood. Intriguingly, recent studies have shown that whilst devoid of nuclei, platelets are still endowed with gene expression control mechanisms including mRNA splicing in response to external cues and effective RNA interference machinery associated with miRNA expression [11, 12]. This chapter will, however, concentrate specifically on the role of TFs in MK differentiation and, ultimately, platelet production.

Expression of actively transcribed genes is generally initiated by binding of ubiquitously expressed general TFs to the TATA sequence in the promoter of these genes. This leads to the recruitment of the transcription machinery consisting of other regulatory cofactors and RNA polymerases [13]. Specificity is introduced by the tight regulation of the expression of a repertoire of cell-type-specific TFs. These TFs recognise certain sequences in the DNA called motifs. These motifs are usually not long and appear many times in the DNA. Whether a TF will bind to its motif depends on the accessibility of the DNA and collaborative association with cofactors. To accommodate the large eukaryotic genome in the nucleus, the DNA is packed into chromatin. It is organised in nucleosomes by histone proteins and repeating units of nucleosomes make up the chromatin. The cell uses modifications of the histone proteins to regulate the tightness of the packaging and thereby DNA access to TFs and the transcriptional machinery. Interestingly, these modifications are inherited by daughter cells and are thus one of the ways to regulate cell-type specificity [14].

Complexes of different TFs can bind a single DNA element inducing transcriptional activation or repression depending on the constituents of the complex [15] (Fig. 2.1). The crucial role of TFs in lineage specification from HSC to MK and subsequent MK maturation is well documented [16], but the exact role that these TFs may play in enabling MKs to form platelets is not as clear. Some evidence can however be gleaned from existing data and will be discussed for individual key TFs in Sect. 2.5.

It was long thought that once a cell had gone down the path of differentiation, this decision was irreversible. However, as early cell nuclei transfer experiments suggested [17] and the birth of Dolly the sheep in 1996 spectacularly demonstrated, the transfer of a somatic nucleus into an egg can reverse programme the cell back into development to make it pluripotent. More recently, the discovery that this reversal of a cell's differentiation status (so-called reprogramming) can also be accomplished by overexpression of four TFs (thereby creating an "induced" pluripotent stem cell) earned Shinya Yamanaka a Nobel Prize [18]. It has become clear that overexpressing key TFs can not only reprogramme a somatic cell into a pluripotent cell, but that a somatic cell can be pushed into an alternative cell identity without necessarily going through the pluripotent stage (lateral programming) and that a similar approach can

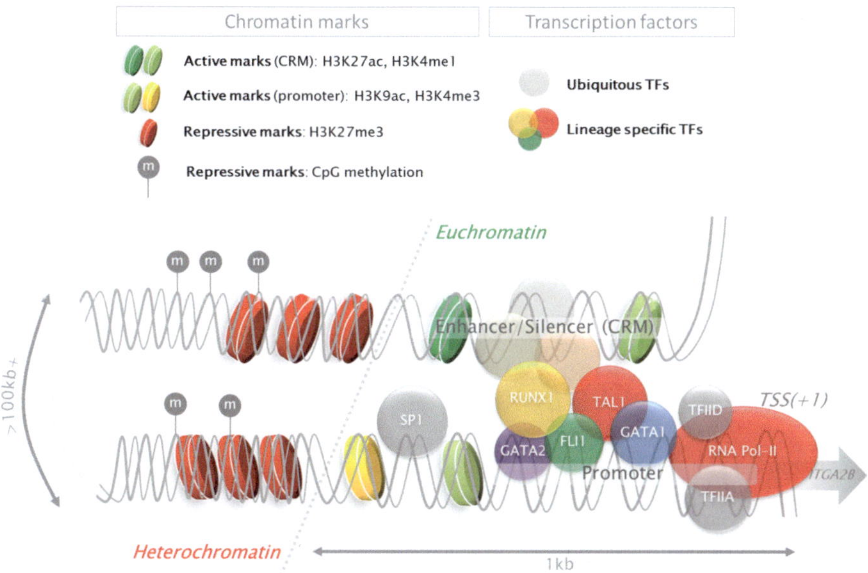

Fig. 2.1 Transcriptional regulation of gene expression. The specificity for spatio-temporal regulation of gene expression is acquired by combinatorial binding of multiple transcription factors (TFs). TFs and co-activators bind at the promoter of a gene proximal to the transcription start site (TSS). The subsequent recruitment of the RNA polymerase-II and ubiquitous transcriptional machinery factors will initiate transcription. Sets of TFs binding on a distal DNA element are called cis-regulatory modules (CRMs). CRMs can be located several hundreds of kilobases (kb) or even more away from the promoter and still function as enhancers/repressors of gene expression through long-range DNA loop interaction. Gene expression is further regulated by the chromatin structure (opened euchromatin and closed heterochromatin) guided by epigenetic modifications involving histone modifications (e.g. acetylation and methylation of lysine residues on H3) and DNA methylation on cytosine which ultimately controls accessibility of TFs to their DNA-binding sequence. These mechanisms are illustrated here for the key megakaryocyte surface receptor gene integrin alpha-2b (*ITGA2B*) based on ChIP-Seq data [206, 244]

force a pluripotent stem cell towards a specific stem cell fate (forward programming). The latter has great potential for the generation of cells from patients (enabling disease modelling) or the production of cells for clinical use and will also be discussed in the context of megakaryopoiesis and in vitro platelet production.

2.4 Making Platelets: The Megakaryocyte, Its Niche and Its DNA Content

2.4.1 MK Specification and Differentiation

The main driver of MK differentiation is the binding of thrombopoietin (TPO) to its receptor MPL [19]. The subsequent dimerisation of the receptor induces the autophosphorylation of the Janus kinase 2 (JAK2). JAK2 phosphorylates a number of downstream substrates leading to the activation of multiple signalling pathways including mitogen-activated protein kinases (MAPK), phosphoinositol-3 kinase (PI3K) and signal transducers and activators of transcription (STATs). The ultimate effect of the activation of these signalling pathways is induction and repression of gene expression and MK differentiation.

Besides TPO, other cytokines, chemokines and extracellular matrix proteins influence megakaryopoiesis [20, 21]. Noteworthy is the residual level of platelets (about 15 %) in *Tpo* or *Mpl* knockout mice probably driven by the interaction between the MK progenitor and the BM endothelial niche [22].

2.4.2 Triggers for Proplatelet Formation and Platelet Release: The Little Bit We Do Know

As the process of platelet formation is still debated, so is the trigger for this process. Although the role of TPO in MK maturation is undisputed, TPO seems dispensable for platelet formation and maybe even inhibits this process [23, 24]. Removal of growth factors commonly triggers cell death. Therefore, the dispensability of TPO may be related to the fact that proplatelet formation has been described as a form of compartmentalised caspase-dependent cell death. Overexpression of antiapoptotic molecules (BCL2 and BCL2L1 [BclxL]) or reduced expression of proapoptotic members of the BCL2 family (BCL2L11 [Bim]) reduces platelet formation by MKs [25–27] as does pharmacological inhibition of caspases [28, 29]. A recent study confirmed the central role played by BCL2L1 in platelet release from mature MKs, although in this particular study, mice with combined deletion of both proapoptotic BAK1 and BAX proteins showed no alteration of platelet formation [30]. Also, a lack of caspase-9 or BCL2, in contrast to overexpression, does not seem to affect steady-state platelet formation [31, 32]. Together, these studies suggest that proplatelet formation does not require the activation of the intrinsic (mitochondrial) apoptosis pathway.

The location of the MK and interaction with its environment seem to be crucial for platelet formation. It is known that the relocation of immature MKs from the osteoblastic niche to the endothelial niche driven by chemotactic agents is essential for maturation and platelet production [21]. Contact with bone marrow endothelial cells (BMECs) induces further maturation and platelet formation [33]. The chemokine CXCL12 (SDF1α) and growth factor FGF4 promote both MK migration and interaction with the BMECs, thereby promoting platelet production [22]. Interestingly, this interaction is enhanced by inflammatory cytokines, such as interleukin-1 beta (IL-1B) [33, 34]. Furthermore, stimulation of the pro-inflammatory VEGFR1 pathway leads to an upregulation of the CXCL12 receptor CXCR4 and increased in vivo platelet formation [35]. These findings suggest that increased interaction between MKs and stimulated endothelium is responsible for the thrombocytosis often observed in inflammation [36]. It is also clear that when MK migration is impaired, platelet formation is impaired too. This is true for Wiskott-Aldrich syndrome (WAS) where actin polymerisation is disturbed [37] and for thrombocytopenia induced by compounds like the kinase inhibitor dasatinib that decrease MK migration [38].

In addition to this direct MK-to-endothelial cell contact, extracellular matrix proteins influence proplatelet formation. Type I collagen inhibits proplatelet formation through integrin αIIβI [39]. Fibrinogen on the other hand promotes proplatelet formation through interaction with glycoprotein IIB/IIIA (GPIIB/IIIA, integrin αIIbβ3). Mutations causing constitutive GPIIB/IIIA activation interfere with proplatelet formation and lead to the production of very large platelets (macrothrombocytopenia) [40, 41]. There is also evidence that von Willebrand factor (vWF) binding to the GPIB/V/IX complex regulates proplatelet formation, which may be the reason for the macrothrombocytopenia observed in patients with Bernard-Soulier syndrome who suffer from a defect in the GPIB/V/IX complex [42]. Platelet production seems therefore under control of several adhesive interactions between the MK, the BM vasculature and the extracellular matrix. However, these do not trigger the protrusion of the proplatelet extensions into the sinusoids. For this, one would expect the presence of a compelling factor in the serum, which could be sphingosine 1-phosphate [43, 44].

2.4.3 MK DNA Content: Does Ploidy Really Matter?

Maturation of MKs is accompanied by successive rounds of DNA replication without cytokinesis (endomitosis), resulting in large polyploid cells with a lobulated nucleus. It has been hypothesised that polyploidisation is required to meet the MK's vast need of protein synthesis and cell growth, as well as production of the demarcation membrane system which provides extra cell membrane supply for platelet formation [45]. This is supported by a study by Raslova et al. showing that all alleles of a series of MK-specific genes in cells ranging from 4N to 32N are functional and not epigenetically silenced [9]. However, a subsequent study, by the same authors, showed that multiple genes involved in platelet formation and DNA proliferation are regulated differently depending on ploidy levels [46]. Thus, specific gene regulatory processes are at work at different levels of ploidy to ensure the MK develops into a platelet-producing cell. Indeed, murine MKs with high ploidy have downregulated genes involved in DNA replication and upregulated genes involved in cytoskeletal dynamics, cell migration, G-protein signalling and platelet function [47].

Surprisingly, proplatelet formation is not strictly within the remit of the higher ploidy MKs. Both in vivo and in culture 2N or 4N MKs form proplatelets [48–50], and platelet numbers are similar in mouse backgrounds with different levels of BM MK polyploidy [51]. Therefore, although some cytoskeletal (TUBB1, MYH9, RAC1, RAP1B), transmembrane (GPIIIA, GP1BA and B) and signalling proteins (LAT and SRC family kinases) known to be involved in proplatelet formation are slightly more abundant in high-ploidy MKs, this appears not to affect platelet shedding. However, instead of having an effect on its ability to produce platelets, the ploidy level of an individual MK might influence protein content. Platelets originating from high-ploidy MKs might be more easily activated than platelets generated from MKs with a lower ploidy [52].

In keeping with the idea that platelet shedding is not necessarily coupled with ploidy, genes regulating platelet production, such as nuclear factor erythroid-derived 2 (NFE2, discussed below), are not involved in the regulation of polyploidisation [53], and inversely, genes that increase MK ploidy, such as cyclin D3 (CCND3), do not modify the platelet count [54].

2.5 Transcription Factors Controlling the Megakaryocyte Cell Fate

2.5.1 The GATAs

2.5.1.1 GATA1 and GATA2

The six members of the GATA (globin transcription factor) family of TFs recognise [5′-(A/T)GATA(A/G)-3′] motifs for which they are named [55]. The defining feature of this family is the presence of two highly conserved Cys4 zinc fingers. The C-terminal finger binds DNA, whilst the N-terminal finger mediates the binding of cofactors [56] and may be involved in stabilising DNA binding [57]. The GATA N-terminal domain regulates target genes depending on the context of the cell type through its binding to co-activators. Interestingly, genes that have low sensitivity to GATA activation are more prominently affected by mutations affecting binding to co-activators [58].

GATA1–3 are known as the haematopoietic GATA factors although they have also been found to be expressed in other tissues. GATA3 is expressed in HSCs and lymphoid cells, but not in MKs. Both GATA1 and GATA2 are expressed in the MK lineage, erythroid progenitors, eosinophils and mast cells. Importantly, GATA factors are also endowed with pioneer activity – i.e. they can access their DNA target sites in compact chromatin and presumably bind to the genome before the binding of other factors, a phenomenon notably at play in the process of lineage anticipation [59].

GATA2 is essential for the appearance of early haematopoietic progenitor cells [60]. It is part of the "stem cell heptad" (together with TAL1, LYL1, LMO2, RUNX1, FLI1 and ERG) defined by overlapping occupancy in chromatin immunoprecipitation (ChIP) experiments combined with deep whole genome sequencing (ChIP-Seq) in the mouse haematopoietic progenitor cell-7 (HPC-7) cell line [61]. GATA2 is necessary for the early commitment to the red cell lineage. Subsequent induction of GATA1 expression in the red cell lineage leads to a rapid decrease of expression of GATA2 (the "GATA switch") which is necessary to achieve terminal red cell maturation [62]. In the red cell lineage, GATA1-binding sites appear "de novo" in the maturing erythroblasts, whilst in MKs, 40 % of GATA1-binding sites correspond to sites already occupied by GATA2 in HSCs. These sites are enriched for TAL1 and strongly associated with gene induction [63]. Unlike the red cell lineage, GATA2 expression is maintained alongside GATA1 in the developing MKs with evidence that GATA2 can compensate and reinforce MK identity in the absence of GATA1 [64, 65]. The

GATA1- and GATA2-bound sites in mature MKs do not overlap, however, with GATA1 sites found mostly in platelet-specific genes (and enriched for ETS factor motifs) and GATA2 sites corresponding to genes expressed in stem cells that are repressed in MKs [63].

GATA1 is expressed primarily in mature cell types, including red cells and MKs. GATA1 deficiency leads to a complete block of erythroid maturation, but megakaryopoiesis is not completely absent, presumably due to some compensation by overexpression of GATA2. However, GATA1 deficiency leads to a decrease in ploidy and abnormalities of certain features that are characteristic of MK terminal maturation (such as cytoplasmic granules) [66, 67]. Overexpression of some downstream targets of GATA1 can restore, at least in part, some of the MK phenotype: cyclin D1 (CCND1) overexpression increases ploidy, but not cytoplasmic maturation [68], whereas overexpression of signal transducer and activator of transcription 1 (STAT1) or IFN regulatory factor-1 (IFR1, its downstream effector) not only increases DNA content but also expression of platelet-specific genes.

A key role for the regulation of MK-specific genes expression by GATA1 is illustrated by human pathologies: for example, a point mutation in a GATA-binding site in the glycoprotein1B β subunit (*GP1BB; CD42b*) promoter proximal region causes a form of Bernard-Soulier syndrome [69].

The *GATA1* gene locus is on the X chromosome (Xp11.23). X-linked mutations of *GATA1* present with a wide range of phenotypes. Patients have bleeding tendencies ranging in severity, associated in some instances, with red cell abnormalities. These patients show not only a decrease in the number of platelets, but their platelets also lack key features such as alpha granules [70]. The variety of phenotypes observed with *GATA1* mutation relates to whether the mutation affects binding of GATA1 to the DNA itself or to cofactors such as FOG1 [71] and TAL1 [72].

2.5.1.2 FOG1 and Other Partners of GATA1

Friend of GATA1 (FOG1, encoded by the gene *ZFPM1* on chromosome 16q24.2) does not bind to DNA, but binds to GATA1 and GATA2. It was originally identified as a GATA1 co-partner in yeast two-hybrid studies and shown to be a co-activator of transcription for haematopoietic genes such as NFE2 and to play a role in both erythroid and MK development [56]. This was confirmed in *Fog1* knockout mice that die at mid-gestation with severe anaemia and absence of platelet production [73]. The loss of FOG1 leads to a decrease of GATA1 binding to its DNA targets and, crucially, prevents the association of the GATA1/FOG1 complex with the nucleosome remodelling and deacetylase (NURD) complex. The formation of the "GATA1/FOG1/NURD" complex leads to the efficient regulation of gene transcription necessary to enforce the lineage commitment and cell maturation [74, 75]. Loss of *Fog1* in a conditional Mx-Cre mouse model leads to a much more extensive phenotype than the loss of *Gata1* using the same Mx-Cre conditional approach. This is due to a deficient repression of *Gata2* by GATA1 and a complete loss of erythroid/MK commitment in progenitors and a skewing towards myeloid differentiation [76, 77]. Interestingly FOG1 can also inhibit

GATA1-driven transcription at certain promoter sites, which is mediated by binding of FOG1 to the corepressor C-terminal binding protein 2 (CTBP2) [78]. Overexpressing either wild-type GATA1 or a mutant GATA1V205G (which does not bind FOG1) in an erythroid-MK cell line deficient in endogenous GATA1 (G1ME), Chlon et al. showed that GATA1V205G bound to genomic regions distinct from those bound by wild-type GATA1. In particular, in the absence of FOG1, wild-type GATA1 was less likely to bind to promoter regions, more likely to bind into regions of "closed chromatin" and less likely to bind to regions where concurrent ETS factor (see Sect. 2.5.4)-binding motifs are found. Crucially, overexpression of GATA1V205G failed to induce transcription of MK genes, but instead led to overexpression of genes belonging to the myeloid lineage (particularly mast cell genes). Therefore, binding of GATA1 to its cofactor FOG1 maintains erythroid/MK lineage fidelity [79].

In addition to FOG1, GATA1 combines with other TFs such as leukaemia/lymphoma-related factor (LRF, encoded by *ZBTB7A*), TAL1 and CBFA2T3 (core-binding factor, runt domain, alpha subunit 2; translocated to 3, also known as ETO2). Interestingly, the association of TAL1 with GATA1 is only seen at genomic loci where GATA1 acts as an activator not as a repressor [75]. CBFA2T3 acts as a corepressor of GATA1-bound genes. It is only expressed in immature MKs where it presumably represses inappropriate early expression of a subset of genes expressed in terminally differentiated MKs [80], and therefore *Cbfa2t3* knockdown promotes MK maturation. ChIP-Seq studies have been carried out in primary murine cells in order to look specifically at the erythroid/MK lineage bifurcation and have shown some key features of GATA1 association with other TFs. GATA1-binding sites in MKs are mostly distinct to those seen in erythroblasts. The genomic region around these GATA1-binding sites is enriched for ETS and RUNX motifs in the former, whilst they are enriched for KLF1-binding motifs in the latter. Whilst clearly the combination of GATA1 and FLI1 binding identifies genes that are actively transcribed in MKs, the combination of GATA1 and ETS factor elements results in gene silencing in the erythroid lineage [63].

2.5.1.3 Down Syndrome and Megakaryopoiesis

Foetuses with trisomy 21 (Down syndrome) have haematological abnormalities characterised by an expansion of MK-erythroid progenitors in the foetal liver during the second trimester [81]. This expansion is dramatically increased in 5 % of children with Down syndrome and develops into a transient clonal myeloproliferative disorder (TMD) characterised by an expansion of immature MKs which often undergoes spontaneous remission. This is the result of an acquired mutation on the X chromosome resulting in the production of a shorter isoform of GATA1, so-called GATA1s [82]. GATA1s lack the N-terminal transactivation domain and is normally co-expressed with full-length GATA1 [83]. Mice that express *Gata1s* only show MK proliferation with maturation arrest in the prenatal period [84], which is thought to be driven by the removal of the inhibitory activity of the GATA1 N-terminal domain on E2F [85]. Down syndrome patients can develop a specific type of childhood megakaryocytic leukaemia (DS AMKL) that has a much better prognosis than other forms of AMKL. It is thought that it is the combination of *GATA1s* with over-expression of other genes (such as *ERG*, see Sect. 2.5.4.2) through the trisomy 21 that leads to DS AMKL.

2.5.2 TAL1

T-cell acute lymphocytic leukaemia 1 (TAL1, also called as SCL) is a basic-helix-loop-helix (bHLH) TF (encoded on chromosome 1p32). TAL1 functions as an obligate heterodimer with the ubiquitously expressed bHLH TF E2A. It binds to its consensus sequence, the so-called E-box (CANNTG). In red cells it functions in a multiprotein complex, which also includes LMO2 (LIM domain only 2) and LIM domain-binding 1 (LDB1). Depending on other factors recruited to this complex (such as EP300, GFI1B, CBFA2T3 and KDM1A), it can function as an activator or repressor of transcription (summarised in [86]). As mentioned above, TAL1 binding in conjunction with GATA1 distinguishes active from repressed transcription sites. Furthermore, evolutionary conserved association of E-box and GATA motifs separated by 9–12 base pairs has been documented in regulatory regions of several erythroid/MK genes including GATA1 itself [67]. A study of TAL1-binding sites in primary mouse erythroid and MK cells has shown very distinctive patterns of binding, which is conferred by the state of the chromatin (active/inactive) and interactions with other TFs, particularly GATA1. The phenomenon of lineage anticipation was also clearly demonstrated in this study, with a significant number of MK genes bound by TAL1 in mature MKs, already bound by TAL1 in the HPC-7 haematopoietic progenitor cell line [87].

TAL1 is essential for the emergence of all haematopoietic lineages, which has been demonstrated both in vivo (*Tal1* knockout mice die between embryonic day 8.5–10.5 due to a complete lack of blood formation) [88] and in vitro [89]. Although TAL1 is necessary for the emergence of HSCs, the use of Mx-Cre and Tie2-Cre conditional models has shown that TAL1 is dispensable for the maintenance of long-term repopulating stem cells and their multipotency, but is absolutely required for differentiation of the erythro-MK lineage [90, 91]. Knockdown of TAL1 translates into a lack of proliferation, polyploidisation and cytoplasmic maturation of the MKs and a reduced platelet number [92]. One of the targets of TAL1 is the cell-cycle regulator cyclin-dependent kinase inhibitor 1A (CDKN1A, p21) which is overexpressed upon knockdown of TAL1 expression [92]. Crucially, knockdown of CDKN1A in TAL1-mutant MKs restores the endomitotic cell-cycle progression, but only partially restores the cytoplasmic maturation necessary for the production of fully functional platelets. Thus, other targets of TAL1 are also responsible for the defects in these late stages of MK maturation. One of these targets may be myocyte enhancer factor 2C (MEF2C), as mice lacking MEF2C in the haematopoietic lineage have reduced numbers of platelets with larger size and abnormal shape and granularity [93].

Other TAL1-interacting proteins include epigenetic modifier enzymes such as LSD1 (histone 3 lysine 4 demethylase), HDAC1 and HDAC2 (histone deacetylases) and the corepressor molecule CoREST [94]. CoREST (REST corepressor 1, encoded by the *RCOR1* gene) was recently identified as part of a genome-wide association study as a locus linked to platelet count in humans [95], and its role in haematopoietic lineage specification has been recently confirmed in a zebrafish model [96].

2.5.3 RUNX1

2.5.3.1 RUNX1 and Haematopoiesis

RUNX1 (runt-related transcription factor 1, aka AML1) is a member of the RUNT family of TFs and together with its cofactor core-binding factor, beta subunit (CBFB) represents the most common mutational target in human acute leukaemia. RUNX1 plays an essential role in definitive HSC emergence from the aorta-gonad-mesonephros region during embryogenesis [97]. Therefore complete absence of RUNX1 in mice is embryonic lethal [98]. However, using a murine Mx-Cre conditional system which allows the deletion of *Runx1* from adult haematopoiesis, a clear role for RUNX1 was shown for the development of lymphocytes and MKs, whilst myelopoiesis and HSC function were preserved [99], similar to what is observed for TAL1. *Runx1* deletion resulted in the accumulation of small immature MK progenitors within the bone marrow and a marked decrease in polyploidisation and cytoplasmic development of MKs, similarly to what is observed for GATA1 deficiency.

2.5.3.2 Familial Platelet Disorder/AML

The complex role of RUNX1 in MK differentiation is illustrated by the autosomal dominant human syndrome familial platelet disorder with propensity to develop acute myeloid leukaemia (FDP/AML) in which germline heterozygous *RUNX1* mutations lead not only to thrombocytopenia but also to impaired platelet function, in addition to a high risk for the development of myelodysplasia and leukaemia [100]. In MKs cultured from FPD/AML patients' stem cells, expression of non-muscle myosin is perturbed with persistent expression of MYH10 (non-muscle myosin IIb) and decreased expression of MYL9 and MYH9 (non-muscle myosin IIa) [101, 102]. Actin and myosin control branching of the proplatelet elongations and MYH9 alongside its regulator RHOA (RAS homolog family member A) restrain platelet formation. Consequently, mutations that reduce myosin IIa activity such as seen in the MYH9-related May-Hegglin syndrome lead to inappropriate platelet shedding causing macrothrombocytopenia [103, 104]. MYH10 specifically localises to the contractile ring separating the cell during mitosis, and its silencing by RUNX1 contributes to the transition from mitosis to endomitosis as the MKs mature [105]. Thus, RUNX1 regulates key constituents of the MK and platelet cytoskeleton and thereby affects features of late megakaryopoiesis such as polyploidisation and platelet formation.

2.5.3.3 RUNX1 in Partnership

The role of RUNX1 in megakaryopoiesis is a perfect illustration of how TFs are part of a network of regulatory proteins with differential effects depending on the cellular stage and crucially other TF partners. One publication analysed the dynamics of gene expression regulation during megakaryopoiesis using three cell line models: the erythroleukaemia K562 cell line, K562 cells induced towards the MK lineage using 12-O-tetradecanoylphorbol-13-acetate (TPA) and the megakaryoblastic CMK cell line, which reportedly represents a more mature MK type than the K562 cell line. Although the results have to be interpreted with caution due to the use of cell lines, the authors report that constitutive RUNX1-binding sites (i.e. present in K562 cells prior to induction) were enriched for GATA1 motifs, whilst in the TPA-induced K562, de novo RUNX1-binding sites regulating the expression of MK-specific genes were enriched for the AP1 complex binding motif. In the CMK line, looking at genes more specifically expressed in the later stages of MK differentiation (such as RAB27B), the authors showed enrichment for the ETS binding motif. They therefore proposed a model whereby RUNX1 controls early MK differentiation genes in collaboration with GATA1, whilst for genes which are more typically expressed in later stages, RUNX1 regulation is effected in collaboration with FOS/JUN family members and ETS TFs [106]. Another example of how RUNX1 can regulate the expression of the same gene, but with different partners, is exemplified by the analysis of the promoter for myeloproliferative leukaemia (*MPL*, virus oncogene, the TPO receptor) where RUNX1 interacts with the SIN3A corepressor complex in haematopoietic stem and progenitor cells, whilst it forms a complex with the transcription activator EP300 (E1A-binding protein p300) on the same promoter in MKs [107]. The co-occupancy of RUNX1 and EP300 on the promoter region of multiple key MK genes was later confirmed in a genome-wide ChIP-Seq study using primary murine MKs [108].

2.5.4 ETS Factors

2.5.4.1 FLI1

The E26 transformation-specific (ETS) family of TFs contains more than 20 helix-loop-helix domain TFs. Friend leukaemia virus integration 1 (FLI1, encoded on chromosome 11q24) is involved in several types of malignancies such as Ewing's sarcoma in humans [109] and erythroleukaemia in mice [110]. Evidence for its contribution to haematopoiesis comes from knockout mice. These mice die at day 11.5 of embryogenesis of cerebral bleeds caused by a lack of vascular integrity. In those animals, megakaryopoiesis was clearly affected with an increase in MK forming colonies, accompanied by the production of small MKs that lack expression of maturity markers [111]. Forced expression of FLI1 induces MK differentiation in K562 cells [112], and the overexpressed FLI1 binds to MK-specific gene promoters such as the promoter of *ITGA2B* (GPIIB, see Sect. 2.4.2) where it acts in synergy with GATA1 and FOG1 [113]. Inducible deletion of FLI1 in the haematopoietic compartment showed a marked increase in bipotent MEPs, which, in colony assays, were biased towards the erythroid compartment [114]. This observation is consistent with the concept that FLI1 and EKLF act as a toggle switch in MEPs each favouring differentiation towards the MK or erythroid lineage, respectively [115].

A lack of FLI1, however, does not seem to affect early commitment to the MK lineage as the number of ITGA2B-positive early MKs was only marginally decreased in knockdown studies [111]. It has in fact been shown that FLI1 acts in tandem with another ETS family TF GABPA (GA-binding protein TF alpha subunit). GAPBA is composed of a DNA-binding α-subunit and a β-subunit responsible for nuclear localisation. As MK maturation progresses, the ratio of FLI1/GABPA increases. In keeping with this observation, GABPA regulates expression of early MK genes (including *ITGA2B* and *MPL*), whilst FLI1 binds to both early and late (*GPIBA*, *GP9* [both part of the von Willebrand factor receptor complex GPIb-V-IX] and *PF4* [platelet factor 4]) MK genes [116].

The key role of FLI1 in the terminal differentiation of MKs is illustrated in the Paris-Trousseau syndrome, an inherited disorder associated with an 11q chromosome deletion with thrombocytopenia and an increased tendency to bleed. Hemizygous loss of FLI1 due to the deletion underlies the disease [117]. Patients show a maturation block with microMKs on BM smears and typical abnormal granule formation [117, 118].

2.5.4.2 ERG

ERG (v-ets avian erythroblastosis virus E26 oncogene homolog) is another member of the ETS protein family and is the most closely related to FLI1. In the multipotent murine HPC-7 line, ERG genome-wide occupancy paralleled that of FLI1 [61]. Studies of ERG mutant mice have shown that it is essential for the establishment of definitive haematopoiesis and stem cell maintenance as well as production of platelets [119]. ERG is a well-known oncogene: its expression level correlates with bad prognosis in myeloid leukaemias. In murine stem cells, ERG overexpression leads to both lymphoid leukaemia and myeloid leukaemia with erythromegakaryocytic characteristics [120, 121].

ERG is encoded by a gene on chromosome 21q22.3 and has been implicated in the development of DS AMKL. In a mouse model, overexpressed ERG was shown to synergise with GATA1s to potentiate the expansion of foetal MK-erythroid progenitors [122]. These mice went on to develop frank leukaemia by 3 months of age. In a separate study, *Erg* overexpression was found to immortalise haematopoietic stem cells expressing *Gata1s*, but not full-length *Gata1* [123]. Interestingly ChIP-Seq profiling in haematopoietic cells overexpressing *Erg* identified potential therapeutic targets such as the PIM1 kinase. PIM1 inhibition could potentially be used to treat leukaemias when the leukaemic blasts overexpress *ERG* [121]. *AKT* is overexpressed in DS AMKL, and it has been recently shown that AKT-induced MK apoptosis is inhibited by both ERG and GATA1s and that in fact GATA1s block MK differentiation induced by AKT and sustain long-term proliferation, which makes the AKT pathway another potential target for therapy in DS AMKL [124].

2.5.4.3 ETV6

ETS variant 6 (ETV6, also called TEL) is another ETS family TF which is frequently rearranged and fused to other partners in chromosomal translocations seen in leukaemia. Although *Tel* knockout mice die of vascular malformations by day 11 post-fertilisation, blood formation in the embryo is largely unaffected [125]. However, in an inducible *Tel* knockout mouse model, deletion of *Tel* in the stem cell compartment using Mx-Cre resulted in a dramatic decrease of the HSCs content in the bone marrow, with a gradual recovery from the cells where the floxed allele was not excised. This indicates an absolute requirement for TEL in the maintenance of adult haematopoietic stem cells. Noticeably, following Mx-Cre induction, mature blood cells were unaffected except for a dramatic drop in the platelet count. In keeping with this observation, excision of *Tel* in the MK-erythroid compartment using Cre expression driven by the *Gata1* promoter had no effect on red cell production, but again caused a drop in the platelet count >50 %. This was accompanied by an increase in MK forming colonies, suggesting that TEL plays a role in MK maturation rather than lineage commitment [126].

2.5.5 EVI1

Ecotropic virus integration site 1 (EVI1) is a member of the SET/PR domain family of TFs, and it contains a total of ten zinc finger motifs organised in two discrete domains, located at the N terminus and towards the C terminus, comprising seven (ZF1) and three (ZF2) repeats, respectively, which have distinct DNA-binding specificities [127].

The first evidence that EVI1 plays a role in megakaryopoiesis comes from patients with myeloid leukaemias. The *MECOM* locus on chromosome 3, encoding EVI1, is implicated in translocations in 4–6 % of all AML cases. These so-called 3q21q26 syndrome leukaemias present with particular dysmorphic MKs and an elevated platelet count. This is thought to be caused by inhibition of cyclin A-dependent kinase inhibitor 2 (CDK2) expression mediated by the abnormal expression of EVI1 [128]. EVI1 is expressed in haematopoietic progenitor cells and MKs. Cell line experiments have shown that EVI1 regulates expression of MK-specific genes. Ectopic expression in the human megakaryoblastic UT-7/GM cell line changes these cells into polynuclear large cells that express PF4 [129]. Knockdown of EVI1 in K562 cells reduces *ITGA2B* and *ITGB3* expression after TPA treatment [130].

The evidence is further strengthened by animal models. *Evi1* knockout mice die by day 10.5 of gestation due to widespread hypocellularity of the paraxial mesenchyme and haemorrhages. Conditional knockdown in *Evi1* Mx-Cre mice has shown that EVI1 is essential for both embryonic and adult stem cell maintenance and repopulating activity. Lineage commitment appears conserved, but after Cre activation the mice developed mild thrombocytopenia and showed delayed platelet recovery after 5FU administration [131]. The role of EVI in stem cell (and MK) biology may be mediated through the expression of GATA2 and RUNX1, which are both regulated by EVI1 [132, 133].

Further evidence that EVI1 transcriptional regulation plays a key role in megakaryopoiesis comes from whole exome sequencing showing that a mutation creating an EVI1-binding site in the promoter of the RNA-binding motif protein 8A (*RBM8A*) gene underlies the thrombocytopenia with absent radii (TAR) syndrome [134]. In 80 % of a cohort of TAR patients, a single nucleotide polymorphism (SNP) was discovered in the 5′UTR of *RBM8A*. The SNP increases binding of EVI1 and leads to a reduction of transcription of *RBM8A* and the encoded protein Y14. TAR patients have low numbers of MKs in the BM, which seem to have a maturation defect [135]. In addition there is some evidence that platelet function may be abnormal in these patients [136–138]. Therefore, one might speculate that EVI1 regulation of *RBM8A* is crucial for late MK differentiation.

2.5.6 NFE2 and Related Transcription Factors

2.5.6.1 NFE2

NFE2 was first identified as a transcriptional regulator in erythroblasts where it binds to the β-globin locus in a region that contains the core symmetric AP-1 motif TCAT/C to which dimers of JUN and FOS subfamily proteins are bound. NFE2 is an obligate dimer between a large 45KDa subunit specifically expressed in haematopoietic cells (p45 NFE2, encoded on chromosome 12) and a choice of smaller, widely expressed, 18 KDa subunits belonging to the MAF family (p18). *MAFK* is the prominent p18 expressed in erythroblasts, whilst *MAFG* and *MAFF* predominate in MKs, which may dictate the binding site preference on DNA sequences and/or p45 NFE2 activity [139]. Remarkably *p45 Nfe2* knockout mice display a mild dyserythropoiesis, but completely lack circulating platelets. This is despite an apparent increased number of MKs in the BM [53]. It was therefore proposed that NFE2 acts as a regulator of proplatelet formation by promoting the final stage of maturation of MKs to the point where they are capable of platelet shedding. NFE2-deficient MKs can be grown in vitro in response to TPO, but are unusually large, and have a disorganised demarcation membrane and granules which are small and sparse indicating a late maturation block [140]. Knocking out *MafG* phenocopies the findings in the *p45 Nfe2* knockout mice, confirming its specific role in megakaryopoiesis [141].

Overexpression of NFE2 has been reported in patients with all three subtypes of myeloproliferative neoplasms, independent of the presence or absence of the JAK2V617F mutation [142, 143]. When expressed ectopically in BM cells, NFE2 also influences earlier stages of MK differentiation and allegedly enhances in vivo platelet production. However, transplantation of NFE2 overexpressing cells only accelerated platelet production and did not lead to an increase in the maximum level or total number of platelets detected in recipient blood [144].

The apparent role of NFE2 in proplatelet formation might be explained through the function of some of its direct transcriptional targets such as tubulin β1 (TUBB1) [145], RAB27b [146], caspase 12 (CASP12) [147] and HSD3B1 (3-beta-hydroxysteroid dehydrogenase) [148]. *Tubb1* knockout mice have thrombocytopenia with spherical platelets [149], and a mutation in *TUBB1* has been identified in a patient with congenital macrothrombocytopenia [150]. Mice with deficient RAB signalling have macrothrombocytopenia with few granules and abnormal MK morphology, and RAB27B may coordinate granule transport during proplatelet formation [146]. *Nfe2* null MKs fail to bind to fibrinogen in response to platelet agonists indicative of a defect in the signalling leading to activation of GPIIB/IIIA, and, similarly, *Casp12* null platelets have a defect in GPIIB/IIIA which is a well-documented positive regulator of proplatelet formation [151]. Finally, HSD3B1 (an enzyme implicated in oestrogen metabolism) rescues proplatelet formation in *Nfe2* null MKs [148]. The authors conclude that MKs may secrete autocrine estradiol that regulates proplatelet formation. Additional studies have shown that oestrogens can induce MK differentiation [152, 153], but to our knowledge this has not been successfully applied to increase in vitro platelet formation.

2.5.6.2 BACH1 and BACH2

The NFE2 heterodimer binds to MARE (musculoaponeurotic fibrosarcoma oncogene recognition), a cis-regulatory element that can be bound by a variety of heterodimers consisting of MAF and p45-related proteins such as BACH1 (BTB and CNC homology 1) and BACH2, both of which are repressors of transcription. *BACH1* (encoded on chromosome 21q22) is expressed in both the erythroid and MK compartment. In murine models, a lack of BACH1 does not affect platelet counts, but when *Bach1* is overexpressed specifically in the erythroblasts and MKs using the *Gata1* promoter, the mice showed thrombocytopenia, very similar to the *Nfe2*-deficient mice (including a defect in proplatelet formation), but without an effect on erythroid differentiation [154]. In this particular study, the authors showed that target genes of NFE2 (such as thromboxane A2 synthetase) were bound by BACH1 and their levels of expression reduced. Overexpression studies are difficult to interpret, but the authors propose that relative levels between p45 NFE2 and BACH1 may act as a rheostat favouring either erythroid or MK development.

2.5.7 SRF and MKL1

Serum response factor (SRF) is a ubiquitously expressed TF part of the MCM1-Agamous-Deficiens-SRF (MADS) family of TFs. SRF binds to CArG sites, the so-called serum response elements, which control more than 150 targets, including genes of the cytoskeleton [155]. Extracellular stimuli can activate SRF through two main signalling pathways: the MAP-kinase pathway through members of the ternary complex factor (TCF) such as ELK1, SAP1 and SAP2 and the small GTPase pathways (Rho family members such as RHO, RAC and CDC42) via two main cofactors from the myocardin-related family, megakaryocytic leukaemia translocation 1 (MKL1, also called MAL) and MKL2 (MAL16). MKL1 was originally described as part of an in-frame translocation t(1,22)(p13;q13) which fuses MKL1 (encoded on chromosome 22q13) with the protein RBM15 (also called OTT) and is frequently found in acute megakaryoblastic leukaemias occurring de novo in infancy or in children older than 1 year, but only occasionally in Down syndrome patients [156–158]. MKL1 activity is regulated by its subcellular localisation. MKL1 binds to monomeric actin (G-actin) which traps it into the cytoplasm. Upon actin polymerisation and formation of F-actin, the pool of G-actin is depleted which promotes MKL1 nuclear localisation [159]. MKL1 expression increases during MK differentiation [160], and its nuclear localisation increases in response to TPO or upon stimulation of the RhoGTPase activation [161, 162].

Srf knockout mice fail to form mesoderm and therefore die early in development. However specific deletion of *Srf* in the MK compartment using a PF4-Cre system leads to macrothrombocytopenia with accumulation of abnormal low ploidy early MK progenitors. *Mkl1* knockout mice also display thrombocytopenia [160, 163] although with a phenotype that is less severe than the one observed in *Srf* knockout mice, probably relating to a compensating role of Mkl2 [164].

In both cases, SRF and MKL1 depletion resulted in alterations in cytoskeletal proteins resulting in abnormalities of MK and platelet ultrastructures and functions. Matrix metalloproteinase 9 (MMP9) and MYL9 are both directly regulated by MKL1, and through dysregulation of the expression of these two proteins, MKL1 deficiency leads to abnormalities of cell migration and proplatelet formation, respectively [161]. MKL1 has also an effect on MK polyploidisation, through dysregulation of RhoA. MK polyploidisation is dependent on RhoA inhibition [165]. RhoA activity is regulated by guanine exchange factors (GEFs) such as ECT2 and GEF-H1. Both of these GEFs are downregulated during MK polyploidisation: the first endomitotic division (from 2N to 4N) requires low levels of GEF-H1, whilst subsequent endodivisions require low levels of ECT2. MKL1 null MKs express persistently high levels of GEF-H1 preventing polyploidisation, reflected in the increase in 2N immature MKs found in the bone marrow of these mice [166].

Of note, RBM15, the other partner involved in the t(1,22) megakaryoblastic leukaemias, is an RNA-binding protein, probably involved in transcriptional regulation. RBM15 has been shown to play a role in megakaryopoiesis in its own right. *Rbm15* conditional knockout mice show an increase in the myeloid and MK compartment [167]. RBM15 also plays a role in stem cell quiescence by controlling the production of an alternative dominant-negative isoform of the TPO receptor [168].

2.5.8 NFAT

The nuclear factor of activated T-cell (NFAT) family of TFs plays a role in a multitude of developmental pathways including in the haematopoietic system. Calcineurin is a calcium-dependent phosphatase which, upon a sustained rise in intracellular calcium, dephosphorylates NFATs, which allows their entry into the nucleus. The two main NFATs expressed in MKs are NFATC1 and NFATC2 [169]. Calcium entry into MKs in response to collagen, for example, has been shown to induce calcineurin/NFATs [170]. It is thought that NFATs negatively regulate megakaryopoiesis in two ways. First is the downregulation of cell cycling and induction of apoptosis. Pharmacological inhibition of calcineurin with cyclosporin A leads to thrombocytosis in mice with accumulation of MKs showing an increase in cell cycling driven by enhanced transcription of *CDK4* and G1 cyclins [171]. Second, NFATs have been shown to increase expression of FAS ligand (FASLG) on the surface of MKs, increasing apoptosis in the MKs and bystander cells expressing FAS [170].

Interestingly, NFATs may also be implicated in the pro-inflammatory role of platelets. FASLG expression on the surface of platelets has been shown to modulate inflammatory target cells [172]. The regulation of the expression of FASLG on platelets is effected through NFATC2, but modulated by another TF, namely, early growth factor 1 (EGR1) [173]. EGR1 itself has been shown to play a role in megakaryopoiesis including the regulation of the expression of the platelet signal transduction protein $G_{\alpha q}$ [174].

The NFAT pathway appears to be involved in several MK pathologies. Two genes known to inhibit calcineurin (*DSCR1* and *DYRK1A*) are situated on Chr21 and are overexpressed in Down syndrome, probably contributing to the development of the Down syndrome-associated TMD and AMKL [175, 176]. This was confirmed in a mouse model where overexpression of a single copy of *Dscr1* reproduced the results obtained with cyclosporin A inhibition of calcineurin [171]. NFATs are also implicated in myeloproliferative diseases. FKBP5, another calcineurin inhibitor, was found to be upregulated in myelofibrosis [177]. Furthermore, in one patient with essential thrombocythaemia, a three-way translocation which created a breakpoint in NFATC2 was identified possibly contributing to the overproliferation of MKs [178].

2.5.9 MEIS1 and Its Regulator GFI1B

Myeloid ecotropic viral integration site 1 (MEIS1) belongs to the TALE class of homeodomain TFs, characterised by three amino acid loop extensions between the α-helices of its homeodomain. It interacts with other homeodomain proteins, in particular PBX1 (pre-B-cell leukaemia homeobox 1) forming a DNA-binding heterodimer and cooperatively associates with HOX homeodomain proteins (such as HOXA9). *Meis1* knockout mice are embryonically lethal due to impaired vasculogenesis and haematopoiesis alongside limb and eye defects [179].

MEIS1 plays a key role at the stem cell/progenitor stages, promoting appearance of definitive haematopoiesis and cellular self-renewal [180, 181]. However, MEIS1 expression is also maintained specifically in the MK lineage [182]. Overexpression studies have shown a MEIS1-driven induction of a MK transcriptional programme [183] and a skewing towards the MK lineage rather than erythroid development [181]. The embryonic lethality and haemorrhages observed in the MEIS1-deficient animals are attributed to a defect in the separation of lymphatic from the blood vessels secondary to a complete lack in the production of platelets [184].

Further evidence of the role of MEIS1 in the production of platelets was found in a GWAS study which identified the existence of a variant in a binding site for MEIS1 in the gene *DNM3* associated with the regulation of platelet count and volume [95]. This MEIS1 binding site falls within a region acting as an alternative promoter for *DNM3* solely used in MKs [185]. DNM3 has since been confirmed as a key regulator of proplatelet formation [186].

The transcriptional repressor growth factor independence (GFI) has a clearly documented role in maintaining stem cell homeostasis [187], and its homolog GFI1B has been shown to play a role in the emergence of both erythroid and MK lineages in the foetal liver [188]. GFI1B and its cofactors lysine (K)-specific demethylase 1A (LSD1, encoded by the *Kdm1A* gene) and rest corepressor 1 (RCOR1) have been identified in ChIP-Seq studies as key regulators of *Meis1* [189]. The evidence that GFI1B plays a role in MK maturation has been recently further exemplified with the identification of a mutation creating a dominant-negative form of GFI1B leading to an autosomal dominant form of gray platelet syndrome [190].

2.5.10 MYB

MYB (myeloblastosis viral oncogene homolog, encoded on chromosome 6q22) clearly plays a role in megakaryopoiesis, but essentially at the early progenitor stage. Complete knockdown in the mouse leads to embryonic lethality due to a lack of red cell production. However, in a murine model where the level of MYB activity has been reduced to 10% of normal rather than abolished, animals can reach adulthood and present with anaemia accompanied by a marked thrombocytosis [191]. Erythroid maturation is impaired due to failure to progress from the BFU-E to erythroblasts stage due to inhibition of the KIT receptor and persistence of "stem cell"-like TFs such as GATA2 and FLI1 [192]. Analysis of the MK compartment, however, shows an increase in the proliferation of early progenitors, but low levels of c-Myb are still permissive of full MK maturation therefore leading to thrombocytosis. Interestingly, in dual *Gata1low Myblow* mice, the reduction in MYB only partially rescues the thrombocytopenia due to GATA1 deficiency, showing clearly separate, nonoverlapping roles for both TFs [193]. Recently, it was shown that one mechanism through which MYB drives erythroid versus MK differentiation is by the transactivation of miR-486-3p expression resulting in the downregulation of MAF (NFE2 subunit) [194].

2.5.11 It Is All About Networking

Although it is clear that the TFs described above play a role in MK differentiation, the level of complexity as to how they regulate gene expression is way beyond their individual function (Fig. 2.2) . Several of the TFs discussed above have been shown to interact with each other to regulate megakaryopoiesis, possibly in a linear hierarchy. For example, GATA1, GATA2 and TAL1 have been shown to regulate NFE2 [195, 196]. TF biology has to be understood in the context of networks where each TFs will bind with a series of partners that will influence not only its DNA-binding characteristics but also its effect on transcription, i.e. activation or repression as is the case for the regulation of MPL transcription by RUNX1 (see Sect. 2.5.3.3) or the

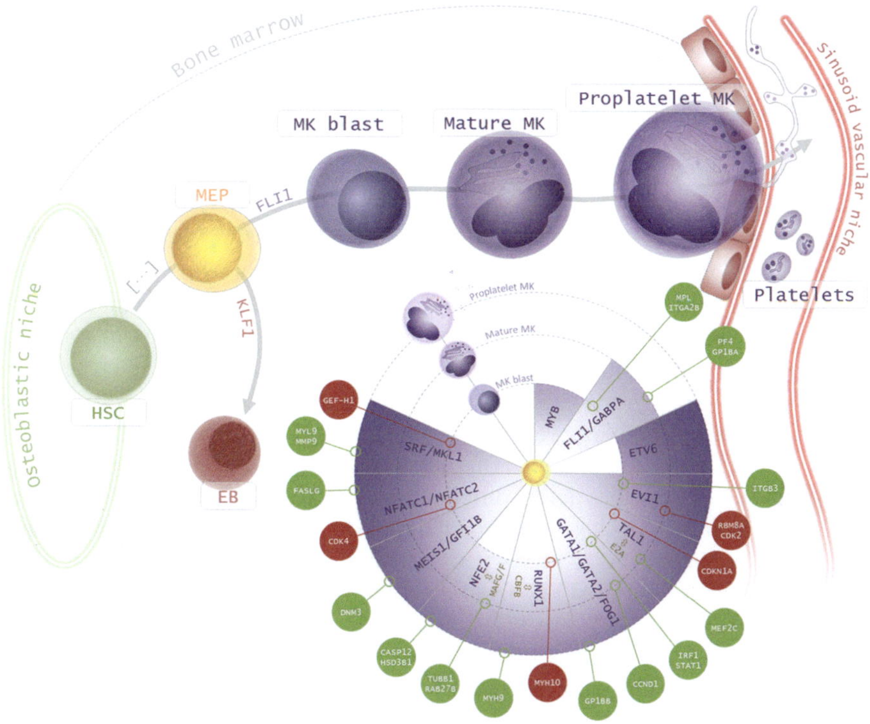

Fig. 2.2 Transcription factors (TFs) and downstream targets regulating megakaryocyte (MK) differentiation. MKs differentiate in the bone marrow from the common erythroid and MK progenitor (MEP), which is derived from the myeloid differentiation arm of the haematopoietic stem cell (HSC). The two TFs KLF1 and FLI1 act in the MEP as master switches to drive erythroid and MK commitment, respectively. Late MK differentiation involves endomitosis, cytoplasmic maturation and proplatelet formation which are accompanied by a progressive journey from the bone marrow to the vascular niche where platelets are released into the circulation upon contact with the endothelial layer of the sinusoids. The MK differentiation wheel depicts the TFs (*purple*, obligate cofactor in *brown*) shown to play a role in these processes and the identified downstream targets performing this role (peripheral circles; *green* and *red* indicate positive and negative regulation, respectively, by the associated TF). *Coloured* surfaces represent the differentiation stages the transcription factors have been implicated with. Further details can be found in the main text

selective binding of TAL1 to active GATA1 TF complexes (see Sect. 2.5.2) described above [107, 197]. In addition, TFs regulate not only expression of other partners with which they cooperate and physically associate but also their own expression levels. Even ubiquitously expressed TFs can regulate proteins specific for a given cell type probably through interactions with cell-type-specific transcription regulators, as a recent study for the ubiquitous TFs SP1 and SP3 indicates. SP1 and SP3 were shown to regulate proteins important for terminal MK maturation and platelet production and function [198]. Several approaches have been taken in order to get a glimpse of how these networks are organised, usually combining cellular data with bespoke statistical tools and in silico algorithms. One group took the approach of looking at genome-wide expression arrays in 38 distinct populations of human haematopoietic cells and identified 80 "modules" of highly co-expressed genes. A small number of these modules were used specifically within a certain cell type, but most were "reused" across multiple lineages reflecting functional requirement for certain pathways in various cell states. TFs were also found to belong to certain "modules", some of which were already known to play a regulatory role in the given cell state (e.g. a late erythroid module contains GATA1 and forkhead box O3 [FOXO3]) alongside novel TFs for the specific cell state (nuclear factor I/X [NFIX] and myelin transcription factor 1 [MYT1] in the case in point). In silico analysis of putative TF motifs found in the promoter regions of genes belonging to certain modules combined with expression levels of TFs in each cell state allowed the design of putative hierarchical networks responsible for a cell state and subsequent transition of one cell type to the other [199]. Another group used a more functional approach by directly measuring TF activity using lentiviral vectors where luciferase expression is driven by a TF-dependent responsive element. This approach enables us to build a dynamic picture of several TF activities during differentiation. This particular approach was used in K562 cells that were subsequently differentiated towards the red cell or MK lineage using hemin or TPA (aka PMA), respectively, recording TF activity for GATA1 and GATA2, FLI1, NFE2, MYB, TAL1 and SPI1 (spleen focus forming virus proviral integration oncogene 1, aka PU1). This data was used to build regulatory networks using bespoke software. The results, for example, confirmed the mutually activating interaction between TAL1 and GATA1 [200].

To add to this layer of complexity, just like other proteins, the level of expression of TFs is only one of the multiple facets by which cell fate is controlled. TF activity is also regulated by cell signalling. One of the prime examples of this is the phosphorylation of the STATs and MAP kinases in response to TPO, downstream of MPL and JAK2. The discovery by different groups in 2005 of a *JAK2V617F* mutation present in 50 % of patients with essential thrombocythaemia (ET) emphasised the importance of the JAK2 signalling pathway in MK growth and the production of platelets [201–203]. There is evidence that not only MK growth but also the last steps of maturation are altered in ET. One study showed that culture-derived MKs from patients with ET had an increased ability to form proplatelets and that the number of proplatelet forming MKs in culture correlated with the platelet count in the patient from whom the MKs were derived [204]. This has also been observed in a knock-in *Jak2V617F* mouse model of ET where higher phosphorylation of STATs and ERK1/2 was demonstrated in bone marrow-derived MKs leading to altered expression profiles [205].

2.6 Gene Regulation: A Gateway into New Biological Discoveries

The advent of new generation sequencing and techniques such as ChIP allows us to identify where TFs bind in the genome (Fig. 2.3). As TF binding tends to cluster on sites that are critical for gene regulation, looking at multiple TFs simultaneously greatly improves the ability to infer biological relevance from a binding event. In a

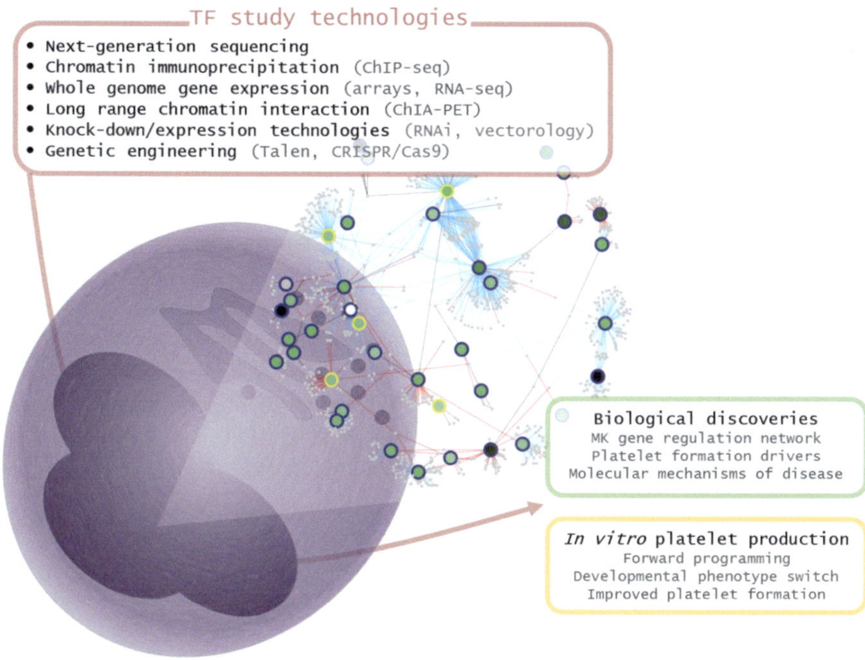

Fig. 2.3 The study of transcription factors (TFs) in megakaryocytes (MKs). The main technologies allowing the deciphering of gene regulatory networks and their control by TFs are indicated. Moving towards a better understanding of molecular mechanisms underlying the MK identity and platelet generation could be ultimately applied to improve in vitro platelet production for clinical development in transfusion medicine

published study, we looked at simultaneous binding of GATA1, GATA2, RUNX1, FLI1 and TAL1 in primary human MKs [206]. In keeping with the existence of TF networks that collaborate to regulate gene expression, simultaneous binding of all five in a given genomic region was particularly enriched, i.e. it occurred far more than would be expected by chance. The five TF-binding sites identified 151 "candidate" genes, some of which are already known to play a key role in megakaryopoiesis, whilst others are not. If one envisions a temporal hierarchy of TFs, it is likely that, amongst the list of genes for which expression is putatively controlled by these five TFs, there are proteins that are crucial in particular for the latter stages of MK maturation, including platelet formation. This is illustrated by the fact that, for seven out of nine genes selected from the 151 targets genes and not previously known to play a role in MKs, there was a clear thrombocyte phenotype upon morpholino knockdown in zebrafish. Just as the genes identified in the study described above are now the subject of ongoing research, regulators of the very late stages of thrombopoiesis, in particular of the proplatelet formation process will potentially be revealed by studies centred on additional genes that are controlled by TFs that have got a clear effect on proplatelet formation (such as NFE2).

The tumbling costs of sequencing also allows us to look into far more detail at the genetic events that might underlie inherited platelet disorders. In particular, whereas most sequencing studies of pedigrees have in the past focused on exomes, non-coding regions (containing putative regulatory motifs) are now sequenced on an unprecedented scale (www.genomicsengland.co.uk/the-100000-genomes-project). TAR and the GATA-binding-related Bernard-Soulier syndrome may be only two of a series of inherited disorders where the causative mutation lies not in a coding, but in a regulatory region of the genome. Understanding how these genetic variants can lead to disease will further inform our knowledge of how MK maturation is controlled at the transcriptional level.

The same reasoning applies to the understanding of how genetic variations affect gene transcription in healthy individuals. A recent GWAS study identified 68 loci associated with platelet count and volume [95]. Nine of these genes have TF activity, but are completely novel in the context of megakaryopoiesis. These novel factors may regulate genes that affect how platelets are formed.

2.7 Transcription Factors and the "Making" of Platelets for Clinical Use

2.7.1 The Clinical Need for Platelets Produced in the Laboratory

Thrombocytopenia in patients can be caused by bone marrow failure (inherited and acquired, such as post-cancer treatment) or peripheral bleeding (trauma or surgery), which potentially leads to life-threatening haemorrhages. Treatment relies on transfusion of ABO and rhesus D-matched platelet concentrates from voluntary donations [207]. The increase in high-dose cancer therapy, advanced surgical procedures and the ageing population has led to a rising demand for platelets with over 4.5 million platelet units transfused per year in Europe and the United States [208]. Platelet transfusion refractoriness in chronically transfused patients and multiparous women necessitates the provision of HLA class 1-matched platelet units, sourced from genotyped recallable donors. The dependence on donations combined with the short shelf life of platelet concentrates (5–7 days) represents a logistical, financial and biosafety challenge for health organisations worldwide.

2.7.2 Harvesting Platelets from In Vitro-Produced MKs

Universally the platelet harvest from in vitro-derived MKs has been in single figures [209], whilst MKs in vivo are estimated to produce thousands of platelets per cell [210, 211]. Maximising platelet production in vitro from these MKs will be a condition without which production of platelets on the scale necessary for a clinically relevant product (each platelet concentrate contains 3×10^{11} platelets) will remain a dream. The apparent shortfall in platelet formation might be due to our inability to re-create the niche signal (including shear) necessary for efficient proplatelet formation and platelet release, although several approaches using 3D bioreactors have been published to that effect [212–214]. It is also worth mentioning that the evidence that platelets released in vitro are functional has been gathered using descriptive rather than quantitative assays with usually no comparison to platelets freshly isolated from donors. One way to address this issue is to functionally analyse the platelet progeny from in vitro-produced MKs in a mouse transplantation model. Nonobese diabetic-severe combined immunodeficient-IL-2 receptor gamma knockout (NSG) mice completely lack an immune system, which makes them tolerant to human cells [215]. In vitro-grown MKs can be transplanted into irradiated NSG mice and have been shown to produce functional platelets that can be readily detected and functionally analysed by flow cytometry. Although some claim that platelet production happens in the first moments post-MK injection (seemingly in the lung vasculature), the particles detected are clearly heterogeneous, with a significant proportion of "platelets" not expressing the CD42 receptors [216] raising the question whether these particles represent cell debris. We have shown, however, that as from day 3 post-transplantation, a homogeneous population of human platelets appears in the murine peripheral blood, expressing both ITGA2B (CD41) and GP9 (CD42a) and showing the expected response to agonists [217]. Until we have mastered the production of platelets in vitro, the mouse transplantation model probably remains the most reliable way to interrogate the functionality of the platelet progeny from in vitro-cultured MKs.

2.7.3 Developmental Differences in Megakaryocytes and Platelets: Embryos Are Not Adults

The development of the haematopoietic system during embryogenesis, the foetal in utero period and after birth has been studied in detail. "Primitive" haematopoiesis first arises in the yolk sack and is followed by the emergence of "definitive" HSCs in the aorta-gonad-mesonephros region, the colonisation of the foetal liver by the haematopoietic system and finally, in the late stage of foetal development, the emergence of the bone marrow as the principal site of haematopoiesis [218]. The evidence that the transition from foetal to adult haematopoiesis is relevant to megakaryopoiesis is exemplified by clinical observations. TMD, characterised by a transient expansion of cells with megakaryoblastic features (see Sect. 2.5.1.3), is seen exclusively in the foetal/neonatal period in some children with Down syndrome. Patients with thrombocytopenia with absent radii (TAR) present in the neonatal period with skeletal abnormalities and low platelet count. The thrombocytopenia associated with TAR however often corrects in the first year of life, although for some patients, the correction can be only partial and occur much later on in life [219].

Understanding the differences in foetal and adult thrombopoiesis has also direct therapeutic implications. The lag time of platelet recovery following cord blood transplantation in comparison to transplantation with stem cells derived from adult donors is much longer. This is a prime example of the difference between neonatal and adult megakaryopoiesis and results in a much longer period of time during which patients are depending on platelet transfusions.

These differences are also found at the cellular level. It is well known that precursors isolated from the bone marrow have a higher proliferative potential when cultured in vitro for MK production [220, 221]. Also, neonatal MKs are smaller and have a lower ploidy than adult MKs (similar to what is seen for hPSC-derived MKs) [222]. Differences between foetal and adult MKs have also emerged at molecular level which may explain some of the cellular features and clinical findings described above. GATA1 was shown to be overexpressed in CB-MKs compared to adult-derived MKs, which also correlated with increased response to TPO through the mTOR pathway [85]. The increase level of GATA1 in foetal MKs might explain the hypersensitivity of neonatal MKs to GATA1 mutations that underlie TMD.

2.7.4 MKs and Platelets from Human Pluripotent Stem Cells

Human pluripotent stem cells (hPSCs) can be maintained in vitro whilst retaining the capacity to differentiate towards virtually any cell type upon adequate stimulation [223–225]. The in vitro production of platelets from genetically defined hPSC lines could revolutionise transfusion medicine by providing a controllable source of platelets [226]. Moreover, platelets are anucleate corpuscles that do not proliferate, thus allowing irradiation of the final transfusion product. This represents a marked safety advantage over other hPSC-derived cell therapies where there is an oncogenic risk associated with the chance of administering partially differentiated progenitor cells.

Two types of protocols are in existence to produce MKs from hPSCs. The first relies on external signals provided by cytokines or stromal cells to mimic embryonic development and direct sequential differentiation of hPSCs into MKs, a process designated as "directed differentiation" (DD) [227–231]. These protocols are impaired by a lack of efficiency, low purity of the MKs and often rely on serum and xenogenic feeder cells. Use of serum or xenogenic feeders can present major regulatory issues for the production of a clinical-grade product. The emergence of the knowledge that cell identities can be manipulated by enforcing expression of specific TFs [232] has led investigators to explore the potential of "forward programming" hPSCs towards the MK lineage. Forced expression of GATA1 or TAL1 alone in haematopoietic progenitors has been shown to bias differentiation towards MK and erythroid fates [233, 234]. Accordingly, exogenous expression of TAL1 in human embryonic stem cells has been reported to promote haematopoiesis and megakaryopoiesis [235, 236]. More recently, the combinatorial expression of TAL1 with GATA2 was found to induce a haemogenic endothelial phenotype biased towards erythro-MK differentiation from hPSCs [237]. Proceeding from a methodically curated list of candidate genes, we found that the minimal combination of GATA1, FLI1 and TAL1 results in the highly efficient production of MKs from an array of hPSC lines in chemically-defined serum-free conditions and with minimum cytokine input [238]. None of these three TFs by itself or a combination of two could efficiently induce forward programming and impose MK identity to hPSCs. The proof of the power of the synergistic effect of key transcriptional controllers on cell fate is exemplified by the fact that forward programmed MKs using these three TFs are >90 % pure, can be cryopreserved, maintained and amplified in vitro for up to 90 days exceeding a 100,000-fold cell yield. Gene expression arrays showed the acquisition of a genuine MK phenotype at the whole genome level and notably the appropriate expression of all key MK TFs and effectors discussed above.

Whilst very close phenotypically and at whole genome expression level to MKs derived from primary neonatal haematopoietic progenitors, MKs generated from hPSCs showed a distinct expression signature reminiscent of the "foetal MK profile" regardless of the method of production (directed differentiation or forward programming) [239]. Bluteau et al. showed that this translated into functional differences in particular a reduction in the rate of proplatelet formation in the hPSC-derived MKs. Whether this foetal MK signature has functional consequences on their platelet progeny is unknown. Recent studies in mice showed an extended life span of neonatal platelets and the existence of low ploidy platelet-forming MKs in the early embryo evocative of hPSC-derived MKs [240, 241]. There are also numerous studies on the function of platelets in newborns versus adult platelets (reviewed in [242]) which indicate that differences do exist, although the results are not consistent and the jury is still out as to the clinical significance of these differences. In the future, the manipulation of the TF expression pattern could become a powerful approach to control developmental phenotype switches in MKs produced from different stem cell sources.

2.7.5 Disease Modelling

Beyond the production of platelets for clinical use, the development of protocols to generate large amount of MKs from hPSCs has opened new avenues of research in disease modelling. hPSCs can be readily derived from patients' cells that carry specific mutations identified through genetic studies of pedigrees with inherited platelet disorders. The emergence of genome-editing technologies such as transcription activator-like endonucleases (TALENs) and Cas9/clustered regularly interspaced short palindromic repeats (CRISPR) also offers the opportunity to introduce specific mutations in the genome of an hPSC line from which the phenotype and MK output are known using a specific differentiation protocol. The latter approach has the advantage to eliminate potential bias of interpretation due to inter-donor cell line variability [243].

2.8 Concluding Remarks

Our knowledge of the key TFs that regulate megakaryopoiesis has grown exponentially over the last 20 years. This has in part been driven by the analyses of the phenotype of murine models where TFs have been knocked down or overexpressed, quite often initially designed to study other cells than MKs (e.g. haematopoietic stem cells). Another important driver has been the advent of sequencing technologies (first Sanger sequencing followed by new generation sequencing), which has allowed us to identify mutations in TFs in pedigrees of human patients with inherited platelet disorders. MKs represent only 0.01 % of the cells in the bone marrow, and it is only since TPO was discovered 20 years ago that we have been able to culture MKs in sufficient quantities to carry out the detailed cellular studies of how these TFs confer cell identity and which specific biological processes they control in the maturing MK. More recently, techniques such as ChIP-Seq have allowed us to get a glimpse of the true complexity of transcriptional regulation of the MK, which cannot be understood without putting each individual TF within large regulatory networks that interact with each other in order to confer cellular specificity on gene transcription. It is clear that understanding gene transcription in the MK will be key to unlock how specific cellular processes such as proplatelet formation are controlled which, in turn, will have great bearing on our ability to, at some point in the future, produce MKs and platelets in vitro for clinical use.

Acknowledgements The authors would like to thank Verity Kew for critically reading the text.

References

1. Doulatov S, Notta F, Laurenti E, Dick JE (2012) Hematopoiesis: a human perspective. Cell Stem Cell 10:120–136
2. Copley MR, Beer PA, Eaves CJ (2012) Hematopoietic stem cell heterogeneity takes center stage. Cell Stem Cell 10:690–697
3. Krause DS, Scadden DT, Preffer FI (2013) The hematopoietic stem cell niche – home for friend and foe? Cytometry B Clin Cytom 84:7–20
4. Sanjuan-Pla A et al (2013) Platelet-biased stem cells reside at the apex of the haematopoietic stem-cell hierarchy. Nature 502:232–236
5. Yamamoto R et al (2013) Clonal analysis unveils self-renewing lineage-restricted progenitors generated directly from hematopoietic stem cells. Cell 154:1112–1126
6. van den Heuvel A, Stadhouders R, Andrieu-Soler C, Grosveld F, Soler E (2015) Long-range gene regulation and novel therapeutic applications. Blood 125:1521–1525
7. Edelstein LC et al (2013) MicroRNAs in platelet production and activation. J Thromb Haemost JTH 11(Suppl 1):340–350
8. Paralkar VR et al (2014) Lineage and species-specific long noncoding RNAs during erythro-megakaryocytic development. Blood 123:1927–1937
9. Raslova H et al (2003) Megakaryocyte polyploidization is associated with a functional gene amplification. Blood 101:541–544
10. Chen L et al (2014) Transcriptional diversity during lineage commitment of human blood progenitors. Science 345:1251033
11. Landry P et al (2009) Existence of a microRNA pathway in anucleate platelets. Nat Struct Mol Biol 16:961–966
12. Denis MM et al (2005) Escaping the nuclear confines: signal-dependent pre-mRNA splicing in anucleate platelets. Cell 122:379–391
13. Lee TI, Young RA (2000) Transcription of eukaryotic protein-coding genes. Annu Rev Genet 34:77–137
14. Ng RK, Gurdon JB (2008) Epigenetic inheritance of cell differentiation status. Cell Cycle Georget Tex 7:1173–1177
15. Cooper G (2000) The cell: a molecular approach. Sinauer Associates, Sunderland
16. Doré LC, Crispino JD (2011) Transcription factor networks in erythroid cell and megakaryocyte development. Blood 118:231–239
17. Gurdon JB (1962) The developmental capacity of nuclei taken from intestinal epithelium cells of feeding tadpoles. J Embryol Exp Morphol 10:622–640
18. Takahashi K, Yamanaka S (2006) Induction of pluripotent stem cells from mouse embryonic and adult fibroblast cultures by defined factors. Cell 126:663–676
19. Geddis AE (2010) Megakaryopoiesis. Semin Hematol 47:212–219
20. Zheng C et al (2008) TPO-independent megakaryocytopoiesis. Crit Rev Oncol Hematol 65:212–222
21. Malara A, Balduini A (2012) Blood platelet production and morphology. Thromb Res 129:241–244

22. Avecilla ST et al (2004) Chemokine-mediated interaction of hematopoietic progenitors with the bone marrow vascular niche is required for thrombopoiesis. Nat Med 10:64–71
23. Ito T, Ishida Y, Kashiwagi R, Kuriya S (1996) Recombinant human c-Mpl ligand is not a direct stimulator of proplatelet formation in mature human megakaryocytes. Br J Haematol 94:387–390
24. Choi ES et al (1996) The role of megakaryocyte growth and development factor in terminal stages of thrombopoiesis. Br J Haematol 95:227–233
25. Ogilvy S et al (1999) Constitutive Bcl-2 expression throughout the hematopoietic compartment affects multiple lineages and enhances progenitor cell survival. Proc Natl Acad Sci U S A 96:14943–14948
26. Bouillet P et al (1999) Proapoptotic Bcl-2 relative Bim required for certain apoptotic responses, leukocyte homeostasis, and to preclude autoimmunity. Science 286:1735–1738
27. Kaluzhny Y et al (2002) BclxL overexpression in megakaryocytes leads to impaired platelet fragmentation. Blood 100:1670–1678
28. Clarke MCH, Savill J, Jones DB, Noble BS, Brown SB (2003) Compartmentalized megakaryocyte death generates functional platelets committed to caspase-independent death. J Cell Biol 160:577–587
29. De Botton S et al (2002) Platelet formation is the consequence of caspase activation within megakaryocytes. Blood 100:1310–1317
30. Josefsson EC et al (2011) Megakaryocytes possess a functional intrinsic apoptosis pathway that must be restrained to survive and produce platelets. J Exp Med 208:2017–2031
31. White MJ et al (2012) Caspase-9 mediates the apoptotic death of megakaryocytes and platelets, but is dispensable for their generation and function. Blood 119:4283–4290
32. Debrincat MA et al (2015) BCL-2 is dispensable for thrombopoiesis and platelet survival. Cell Death Dis 6, e1721
33. Hamada T et al (1998) Transendothelial migration of megakaryocytes in response to stromal cell-derived factor 1 (SDF-1) enhances platelet formation. J Exp Med 188:539–548
34. Avraham H, Cowley S, Chi SY, Jiang S, Groopman JE (1993) Characterization of adhesive interactions between human endothelial cells and megakaryocytes. J Clin Invest 91:2378–2384
35. Pitchford SC, Lodie T, Rankin SM (2012) VEGFR1 stimulates a CXCR4-dependent translocation of megakaryocytes to the vascular niche, enhancing platelet production in mice. Blood 120:2787–2795
36. Hoffman R (1989) Regulation of megakaryocytopoiesis. Blood 74:1196–1212
37. Sabri S et al (2006) Deficiency in the Wiskott-Aldrich protein induces premature proplatelet formation and platelet production in the bone marrow compartment. Blood 108:134–140
38. Mazharian A, Ghevaert C, Zhang L, Massberg S, Watson SP (2011) Dasatinib enhances megakaryocyte differentiation but inhibits platelet formation. Blood 117:5198–5206
39. Sabri S et al (2004) Differential regulation of actin stress fiber assembly and proplatelet formation by alpha2beta1 integrin and GPVI in human megakaryocytes. Blood 104:3117–3125
40. Ghevaert C et al (2008) A nonsynonymous SNP in the ITGB3 gene disrupts the conserved membrane-proximal cytoplasmic salt bridge in the alphaIIbbeta3 integrin and cosegregates dominantly with abnormal proplatelet formation and macrothrombocytopenia. Blood 111:3407–3414
41. Bury L, Malara A, Gresele P, Balduini A (2012) Outside-in signalling generated by a constitutively activated integrin $\alpha IIb\beta 3$ impairs proplatelet formation in human megakaryocytes. PLoS One 7, e34449
42. Balduini A et al (2009) Proplatelet formation in heterozygous Bernard-Soulier syndrome type Bolzano. J Thromb Haemost JTH 7:478–484
43. Golfier S et al (2010) Shaping of terminal megakaryocyte differentiation and proplatelet development by sphingosine-1-phosphate receptor S1P4. FASEB J Off Publ Fed Am Soc Exp Biol 24:4701–4710
44. Zhang L et al (2012) A novel role of sphingosine 1-phosphate receptor S1pr1 in mouse thrombopoiesis. J Exp Med 209:2165–2181
45. Eckly A et al (2014) Biogenesis of the demarcation membrane system (DMS) in megakaryocytes. Blood 123:921–930

46. Raslova H et al (2007) Interrelation between polyploidization and megakaryocyte differentiation: a gene profiling approach. Blood 109:3225–3234
47. Chen Z, Hu M, Shivdasani RA (2007) Expression analysis of primary mouse megakaryocyte differentiation and its application in identifying stage-specific molecular markers and a novel transcriptional target of NF-E2. Blood 109:1451–1459
48. Mattia G et al (2002) Different ploidy levels of megakaryocytes generated from peripheral or cord blood CD34+ cells are correlated with different levels of platelet release. Blood 99:888–897
49. Miyazaki R et al (2000) Comparative analyses of megakaryocytes derived from cord blood and bone marrow. Br J Haematol 108:602–609
50. Leysi-Derilou Y, Duchesne C, Garnier A, Pineault N (2012) Single-cell level analysis of megakaryocyte growth and development. Differ Res Biol Divers 83:200–209
51. Jackson CW, Steward SA, Chenaille PJ, Ashmun RA, McDonald TP (1990) An analysis of megakaryocytopoiesis in the C3H mouse: an animal model whose megakaryocytes have 32N as the modal DNA class. Blood 76:690–696
52. Brown AS, Martin JF (1994) The megakaryocyte platelet system and vascular disease. Eur J Clin Invest 24(Suppl 1):9–15
53. Shivdasani RA et al (1995) Transcription factor NF-E2 is required for platelet formation independent of the actions of thrombopoietin/MGDF in megakaryocyte development. Cell 81:695–704
54. Zimmet JM, Toselli P, Ravid K (1998) Cyclin D3 and megakaryocyte development: exploration of a transgenic phenotype. Stem Cells Dayt Ohio 16(Suppl 2):97–106
55. Merika M, Orkin SH (1993) DNA-binding specificity of GATA family transcription factors. Mol Cell Biol 13:3999–4010
56. Tsang AP et al (1997) FOG, a multitype zinc finger protein, acts as a cofactor for transcription factor GATA-1 in erythroid and megakaryocytic differentiation. Cell 90:109–119
57. Trainor CD et al (1996) A palindromic regulatory site within vertebrate GATA-1 promoters requires both zinc fingers of the GATA-1 DNA-binding domain for high-affinity interaction. Mol Cell Biol 16:2238–2247
58. Johnson KD, Kim S-I, Bresnick EH (2006) Differential sensitivities of transcription factor target genes underlie cell type-specific gene expression profiles. Proc Natl Acad Sci U S A 103:15939–15944
59. Zaret KS, Carroll JS (2011) Pioneer transcription factors: establishing competence for gene expression. Genes Dev 25:2227–2241
60. Tsai FY, Orkin SH (1997) Transcription factor GATA-2 is required for proliferation/survival of early hematopoietic cells and mast cell formation, but not for erythroid and myeloid terminal differentiation. Blood 89:3636–3643
61. Wilson NK et al (2010) Combinatorial transcriptional control in blood stem/progenitor cells: genome-wide analysis of ten major transcriptional regulators. Cell Stem Cell 7:532–544
62. Briegel K et al (1993) Ectopic expression of a conditional GATA-2/estrogen receptor chimera arrests erythroid differentiation in a hormone-dependent manner. Genes Dev 7:1097–1109
63. Pimkin M et al (2014) Divergent functions of hematopoietic transcription factors in lineage priming and differentiation during erythro-megakaryopoiesis. Genome Res 24:1932–1944
64. Chang AN et al (2002) GATA-factor dependence of the multitype zinc-finger protein FOG-1 for its essential role in megakaryopoiesis. Proc Natl Acad Sci U S A 99:9237–9242
65. Huang Z et al (2009) GATA-2 reinforces megakaryocyte development in the absence of GATA-1. Mol Cell Biol 29:5168–5180
66. Shivdasani RA, Fujiwara Y, McDevitt MA, Orkin SH (1997) A lineage-selective knockout establishes the critical role of transcription factor GATA-1 in megakaryocyte growth and platelet development. EMBO J 16:3965–3973
67. Vyas P et al (1999) Different sequence requirements for expression in erythroid and megakaryocytic cells within a regulatory element upstream of the GATA-1 gene. Dev Camb Engl 126:2799–2811
68. Muntean AG et al (2007) Cyclin D-Cdk4 is regulated by GATA-1 and required for megakaryocyte growth and polyploidization. Blood 109:5199–5207

69. Ludlow LB et al (1996) Identification of a mutation in a GATA binding site of the platelet glycoprotein Ibbeta promoter resulting in the Bernard-Soulier syndrome. J Biol Chem 271:22076–22080
70. Ciovacco WA, Raskind WH, Kacena MA (2008) Human phenotypes associated with GATA-1 mutations. Gene 427:1–6
71. Freson K et al (2002) Different substitutions at residue D218 of the X-linked transcription factor GATA1 lead to altered clinical severity of macrothrombocytopenia and anemia and are associated with variable skewed X inactivation. Hum Mol Genet 11:147–152
72. Campbell AE, Wilkinson-White L, Mackay JP, Matthews JM, Blobel GA (2013) Analysis of disease-causing GATA1 mutations in murine gene complementation systems. Blood 121:5218–5227
73. Tsang AP, Fujiwara Y, Hom DB, Orkin SH (1998) Failure of megakaryopoiesis and arrested erythropoiesis in mice lacking the GATA-1 transcriptional cofactor FOG. Genes Dev 12:1176–1188
74. Gao Z et al (2010) FOG-1-mediated recruitment of NuRD is required for cell lineage re-enforcement during haematopoiesis. EMBO J 29:457–468
75. Miccio A et al (2010) NuRD mediates activating and repressive functions of GATA-1 and FOG-1 during blood development. EMBO J 29:442–456
76. Chou ST et al (2009) Graded repression of PU.1/Sfpi1 gene transcription by GATA factors regulates hematopoietic cell fate. Blood 114:983–994
77. Mancini E et al (2012) FOG-1 and GATA-1 act sequentially to specify definitive megakaryocytic and erythroid progenitors: FOG-1 and GATA-1 in lineage commitment. EMBO J 31:351–365
78. Fox AH et al (1999) Transcriptional cofactors of the FOG family interact with GATA proteins by means of multiple zinc fingers. EMBO J 18:2812–2822
79. Chlon TM, Doré LC, Crispino JD (2012) Cofactor-mediated restriction of GATA-1 chromatin occupancy coordinates lineage-specific gene expression. Mol Cell 47:608–621
80. Hamlett I et al (2008) Characterization of megakaryocyte GATA1-interacting proteins: the corepressor ETO2 and GATA1 interact to regulate terminal megakaryocyte maturation. Blood 112:2738–2749
81. Chou ST et al (2008) Trisomy 21 enhances human fetal erythro-megakaryocytic development. Blood 112:4503–4506
82. Mundschau G et al (2003) Mutagenesis of GATA1 is an initiating event in Down syndrome leukemogenesis. Blood 101:4298–4300
83. Rainis L et al (2003) Mutations in exon 2 of GATA1 are early events in megakaryocytic malignancies associated with trisomy 21. Blood 102:981–986
84. Li Z et al (2005) Developmental stage-selective effect of somatically mutated leukemogenic transcription factor GATA1. Nat Genet 37:613–619
85. Klusmann J-H et al (2010) Developmental stage-specific interplay of GATA1 and IGF signaling in fetal megakaryopoiesis and leukemogenesis. Genes Dev 24:1659–1672
86. Kassouf MT et al (2010) Genome-wide identification of TAL1's functional targets: insights into its mechanisms of action in primary erythroid cells. Genome Res 20:1064–1083
87. Wu W et al (2014) Dynamic shifts in occupancy by TAL1 are guided by GATA factors and drive large-scale reprogramming of gene expression during hematopoiesis. Genome Res 24:1945–1962
88. Shivdasani RA, Mayer EL, Orkin SH (1995) Absence of blood formation in mice lacking the T-cell leukaemia oncoprotein tal-1/SCL. Nature 373:432–434
89. Porcher C et al (1996) The T cell leukemia oncoprotein SCL/tal-1 is essential for development of all hematopoietic lineages. Cell 86:47–57
90. Mikkola HKA et al (2003) Haematopoietic stem cells retain long-term repopulating activity and multipotency in the absence of stem-cell leukaemia SCL/tal-1 gene. Nature 421:547–551
91. Schlaeger TM, Mikkola HKA, Gekas C, Helgadottir HB, Orkin SH (2005) Tie2Cre-mediated gene ablation defines the stem-cell leukemia gene (SCL/tal1)-dependent window during hematopoietic stem-cell development. Blood 105:3871–3874
92. Chagraoui H et al (2011) SCL-mediated regulation of the cell-cycle regulator p21 is critical for murine megakaryopoiesis. Blood 118:723–735

93. Gekas C et al (2009) Mef2C is a lineage-restricted target of Scl/Tal1 and regulates megakaryopoiesis and B-cell homeostasis. Blood 113:3461–3471
94. Hu X et al (2009) LSD1-mediated epigenetic modification is required for TAL1 function and hematopoiesis. Proc Natl Acad Sci 106:10141–10146
95. Gieger C et al (2011) New gene functions in megakaryopoiesis and platelet formation. Nature 480:201–208
96. Bielczyk-Maczyńska E et al (2014) A loss of function screen of identified genome-wide association study Loci reveals new genes controlling hematopoiesis. PLoS Genet 10, e1004450
97. Liakhovitskaia A et al (2014) Runx1 is required for progression of CD41+ embryonic precursors into HSCs but not prior to this. Dev Camb Engl 141:3319–3323
98. Cai Z et al (2000) Haploinsufficiency of AML1 affects the temporal and spatial generation of hematopoietic stem cells in the mouse embryo. Immunity 13:423–431
99. Ichikawa M et al (2004) AML-1 is required for megakaryocytic maturation and lymphocytic differentiation, but not for maintenance of hematopoietic stem cells in adult hematopoiesis. Nat Med 10:299–304
100. Song WJ et al (1999) Haploinsufficiency of CBFA2 causes familial thrombocytopenia with propensity to develop acute myelogenous leukaemia. Nat Genet 23:166–175
101. Sun L, Gorospe JR, Hoffman EP, Rao AK (2007) Decreased platelet expression of myosin regulatory light chain polypeptide (MYL9) and other genes with platelet dysfunction and CBFA2/RUNX1 mutation: insights from platelet expression profiling. J Thromb Haemost JTH 5:146–154
102. Bluteau D et al (2012) Dysmegakaryopoiesis of FPD/AML pedigrees with constitutional RUNX1 mutations is linked to myosin II deregulated expression. Blood 120:2708–2718
103. Chang Y et al (2007) Proplatelet formation is regulated by the Rho/ROCK pathway. Blood 109:4229–4236
104. Chen Z et al (2007) The May-Hegglin anomaly gene MYH9 is a negative regulator of platelet biogenesis modulated by the Rho-ROCK pathway. Blood 110:171–179
105. Lordier L et al (2012) RUNX1-induced silencing of non-muscle myosin heavy chain IIB contributes to megakaryocyte polyploidization. Nat Commun 3:717
106. Pencovich N, Jaschek R, Tanay A, Groner Y (2011) Dynamic combinatorial interactions of RUNX1 and cooperating partners regulates megakaryocytic differentiation in cell line models. Blood 117:e1–e14
107. Satoh Y et al (2008) AML1/RUNX1 works as a negative regulator of c-Mpl in hematopoietic stem cells. J Biol Chem 283:30045–30056
108. Pencovich N et al (2013) Cell-autonomous function of Runx1 transcriptionally regulates mouse megakaryocytic maturation. PLoS One 8, e64248
109. Delattre O et al (1992) Gene fusion with an ETS DNA-binding domain caused by chromosome translocation in human tumours. Nature 359:162–165
110. Ben-David Y, Giddens EB, Letwin K, Bernstein A (1991) Erythroleukemia induction by Friend murine leukemia virus: insertional activation of a new member of the ets gene family, Fli-1, closely linked to c-ets-1. Genes Dev 5:908–918
111. Hart A et al (2000) Fli-1 is required for murine vascular and megakaryocytic development and is hemizygously deleted in patients with thrombocytopenia. Immunity 13: 167–177
112. Athanasiou M et al (1996) Increased expression of the ETS-related transcription factor FLI-1/ERGB correlates with and can induce the megakaryocytic phenotype. Cell Growth Differ Mol Biol J Am Assoc Cancer Res 7:1525–1534
113. Wang X et al (2002) Control of megakaryocyte-specific gene expression by GATA-1 and FOG-1: role of Ets transcription factors. EMBO J 21:5225–5234
114. Starck J et al (2010) Inducible Fli-1 gene deletion in adult mice modifies several myeloid lineage commitment decisions and accelerates proliferation arrest and terminal erythrocytic differentiation. Blood 116:4795–4805
115. Starck J et al (2003) Functional cross-antagonism between transcription factors FLI-1 and EKLF. Mol Cell Biol 23:1390–1402
116. Pang L et al (2006) Maturation stage-specific regulation of megakaryopoiesis by pointed-domain Ets proteins. Blood 108:2198–2206

117. Raslova H et al (2004) FLI1 monoallelic expression combined with its hemizygous loss underlies Paris-Trousseau/Jacobsen thrombopenia. J Clin Invest 114:77–84
118. Favier R et al (2003) Paris-Trousseau syndrome: clinical, hematological, molecular data of ten new cases. Thromb Haemost 90:893–897
119. Loughran SJ et al (2008) The transcription factor Erg is essential for definitive hematopoiesis and the function of adult hematopoietic stem cells. Nat Immunol 9:810–819
120. Carmichael CL et al (2012) Hematopoietic overexpression of the transcription factor Erg induces lymphoid and erythro-megakaryocytic leukemia. Proc Natl Acad Sci U S A 109:15437–15442
121. Goldberg L et al (2013) Genome-scale expression and transcription factor binding profiles reveal therapeutic targets in transgenic ERG myeloid leukemia. Blood 122:2694–2703
122. Birger Y et al (2013) Perturbation of fetal hematopoiesis in a mouse model of Down syndrome's transient myeloproliferative disorder. Blood 122:988–998
123. Stankiewicz MJ, Crispino JD (2009) ETS2 and ERG promote megakaryopoiesis and synergize with alterations in GATA-1 to immortalize hematopoietic progenitor cells. Blood 113:3337–3347
124. Stankiewicz MJ, Crispino JD (2013) AKT collaborates with ERG and Gata1s to dysregulate megakaryopoiesis and promote AMKL. Leukemia 27:1339–1347
125. Wang LC et al (1997) Yolk sac angiogenic defect and intra-embryonic apoptosis in mice lacking the Ets-related factor TEL. EMBO J 16:4374–4383
126. Hock H (2004) Tel/Etv6 is an essential and selective regulator of adult hematopoietic stem cell survival. Genes Dev 18:2336–2341
127. Delwel R, Funabiki T, Kreider BL, Morishita K, Ihle JN (1993) Four of the seven zinc fingers of the Evi-1 myeloid-transforming gene are required for sequence-specific binding to GA(C/T)AAGA(T/C)AAGATAA. Mol Cell Biol 13:4291–4300
128. Kilbey A et al (2005) The Evi1 proto-oncoprotein blocks endomitosis in megakaryocytes by inhibiting sustained cyclin-dependent kinase 2 catalytic activity. Br J Haematol 130:902–911
129. Shimizu S et al (2002) EVI1 is expressed in megakaryocyte cell lineage and enforced expression of EVI1 in UT-7/GM cells induces megakaryocyte differentiation. Biochem Biophys Res Commun 292:609–616
130. Maicas M et al (2012) Functional characterization of the promoter region of the human EVI1 gene in acute myeloid leukemia: RUNX1 and ELK1 directly regulate its transcription. Oncogene. doi:10.1038/onc.2012.222
131. Goyama S et al (2008) Evi-1 is a critical regulator for hematopoietic stem cells and transformed leukemic cells. Cell Stem Cell 3:207–220
132. Yuasa H et al (2005) Oncogenic transcription factor Evi1 regulates hematopoietic stem cell proliferation through GATA-2 expression. EMBO J 24:1976–1987
133. Senyuk V et al (2007) Repression of RUNX1 activity by EVI1: a new role of EVI1 in leukemogenesis. Cancer Res 67:5658–5666
134. Albers CA et al (2012) Compound inheritance of a low-frequency regulatory SNP and a rare null mutation in exon-junction complex subunit RBM8A causes TAR syndrome. Nat Genet 44(435–439):S1–S2
135. Letestu R et al (2000) Existence of a differentiation blockage at the stage of a megakaryocyte precursor in the thrombocytopenia and absent radii (TAR) syndrome. Blood 95:1633–1641
136. Sultan Y, Scrobohaci ML, Rendu F, Caen JP (1972) Abnormal platelet function, population, and survival-time in a boy with congenital absent radii and thrombocytopenia. Lancet 2:653
137. Day HJ, Holmsen H (1972) Platelet adenine nucleotide 'storage pool deficiency' in thrombocytopenic absent radii syndrome. JAMA J Am Med Assoc 221:1053–1054
138. Zahavi J, Gale R, Kakkar VV (1981) Storage pool disease of platelets in an infant with thrombocytopenic absent radii (TAR) syndrome simulating Fanconi's anaemia. Haemostasis 10:121–133
139. Lecine P, Blank V, Shivdasani R (1998) Characterization of the hematopoietic transcription factor NF-E2 in primary murine megakaryocytes. J Biol Chem 273:7572–7578
140. Lecine P et al (1998) Mice lacking transcription factor NF-E2 provide in vivo validation of the proplatelet model of thrombocytopoiesis and show a platelet production defect that is intrinsic to megakaryocytes. Blood 92:1608–1616

141. Motohashi H et al (2011) Molecular determinants for small Maf protein control of platelet production. Mol Cell Biol 31:151–162
142. Goerttler PS et al (2005) Gene expression profiling in polycythaemia vera: overexpression of transcription factor NF-E2. Br J Haematol 129:138–150
143. Wang W, Schwemmers S, Hexner EO, Pahl HL (2010) AML1 is overexpressed in patients with myeloproliferative neoplasms and mediates JAK2V617F-independent overexpression of NF-E2. Blood 116:254–266
144. Fock E, Yan F, Pan S, Chong BH (2008) NF-E2-mediated enhancement of megakaryocytic differentiation and platelet production in vitro and in vivo. Exp Hematol 36:78–92
145. Lecine P, Italiano JE Jr, Kim SW, Villeval JL, Shivdasani RA (2000) Hematopoietic-specific beta 1 tubulin participates in a pathway of platelet biogenesis dependent on the transcription factor NF-E2. Blood 96:1366–1373
146. Tiwari S et al (2003) A role for Rab27b in NF-E2-dependent pathways of platelet formation. Blood 102:3970–3979
147. Kerrigan SW, Gaur M, Murphy RP, Shattil SJ, Leavitt AD (2004) Caspase-12: a developmental link between G-protein-coupled receptors and integrin alphaIIbbeta3 activation. Blood 104:1327–1334
148. Nagata Y et al (2003) Proplatelet formation of megakaryocytes is triggered by autocrine-synthesized estradiol. Genes Dev 17:2864–2869
149. Schwer HD et al (2001) A lineage-restricted and divergent beta-tubulin isoform is essential for the biogenesis, structure and function of blood platelets. Curr Biol CB 11:579–586
150. Kunishima S, Kobayashi R, Itoh TJ, Hamaguchi M, Saito H (2009) Mutation of the beta1-tubulin gene associated with congenital macrothrombocytopenia affecting microtubule assembly. Blood 113:458–461
151. Shiraga M et al (1999) Primary megakaryocytes reveal a role for transcription factor NF-E2 in integrin alpha IIb beta 3 signaling. J Cell Biol 147:1419–1430
152. Bord S et al (2004) Estrogen stimulates differentiation of megakaryocytes and modulates their expression of estrogen receptors alpha and beta. J Cell Biochem 92:249–257
153. Fox SW, Chambers TJ (2006) The effect of oestrogen on megakaryocyte differentiation and platelet counts in vivo. Int J Cardiol 109:359–366
154. Toki T et al (2005) Transgenic expression of BACH1 transcription factor results in megakaryocytic impairment. Blood 105:3100–3108
155. Arsenian S, Weinhold B, Oelgeschläger M, Rüther U, Nordheim A (1998) Serum response factor is essential for mesoderm formation during mouse embryogenesis. EMBO J 17:6289–6299
156. Lion T et al (1992) The translocation t(1;22)(p13;q13) is a nonrandom marker specifically associated with acute megakaryocytic leukemia in young children. Blood 79:3325–3330
157. Carroll A et al (1991) The t(1;22) (p13;q13) is nonrandom and restricted to infants with acute megakaryoblastic leukemia: a Pediatric Oncology Group Study. Blood 78:748–752
158. Bernstein J et al (2000) Nineteen cases of the t(1;22)(p13;q13) acute megakaryoblastic leukaemia of infants/children and a review of 39 cases: report from a t(1;22) study group. Leukemia 14:216–218
159. Miralles F, Posern G, Zaromytidou A-I, Treisman R (2003) Actin dynamics control SRF activity by regulation of its coactivator MAL. Cell 113:329–342
160. Cheng E-C et al (2009) Role for MKL1 in megakaryocytic maturation. Blood 113:2826–2834
161. Gilles L et al (2009) MAL/SRF complex is involved in platelet formation and megakaryocyte migration by regulating MYL9 (MLC2) and MMP9. Blood 114:4221–4232
162. Smith EC et al (2013) Induction of megakaryocyte differentiation drives nuclear accumulation and transcriptional function of MKL1 via actin polymerization and RhoA activation. Blood 121:1094–1101
163. Ragu C et al (2010) The serum response factor (SRF)/megakaryocytic acute leukemia (MAL) network participates in megakaryocyte development. Leukemia 24:1227–1230
164. Smith EC et al (2012) MKL1 and MKL2 play redundant and crucial roles in megakaryocyte maturation and platelet formation. Blood 120:2317–2329

165. Lordier L et al (2008) Megakaryocyte endomitosis is a failure of late cytokinesis related to defects in the contractile ring and Rho/Rock signaling. Blood 112:3164–3174
166. Gao Y et al (2012) Role of RhoA-specific guanine exchange factors in regulation of endomitosis in megakaryocytes. Dev Cell 22:573–584
167. Raffel GD et al (2007) Ott1(Rbm15) has pleiotropic roles in hematopoietic development. Proc Natl Acad Sci U S A 104:6001–6006
168. Xiao N et al (2015) Ott1 (Rbm15) regulates thrombopoietin response in hematopoietic stem cells through alternative splicing of c-Mpl. Blood 125:941–948
169. Kiani A et al (2007) Expression analysis of nuclear factor of activated T cells (NFAT) during myeloid differentiation of CD34+ cells: regulation of Fas ligand gene expression in megakaryocytes. Exp Hematol 35:757–770
170. Arabanian LS, Kujawski S, Habermann I, Ehninger G, Kiani A (2012) Regulation of fas/fas ligand-mediated apoptosis by nuclear factor of activated T cells in megakaryocytes. Br J Haematol 156:523–534
171. Zaslavsky A et al (2013) The calcineurin-NFAT pathway negatively regulates megakaryopoiesis. Blood 121:3205–3215
172. Henn V et al (1998) CD40 ligand on activated platelets triggers an inflammatory reaction of endothelial cells. Nature 391:591–594
173. Crist SA, Elzey BD, Ahmann MT, Ratliff TL (2013) Early growth response-1 (EGR-1) and nuclear factor of activated T cells (NFAT) cooperate to mediate CD40L expression in megakaryocytes and platelets. J Biol Chem 288:33985–33996
174. Jalagadugula G, Dhanasekaran DN, Kim S, Kunapuli SP, Rao AK (2006) Early growth response transcription factor EGR-1 regulates Galphaq gene in megakaryocytic cells. J Thromb Haemost JTH 4:2678–2686
175. Kyttälä S, Habermann I, Minami T, Ehninger G, Kiani A (2009) Regulation of Down syndrome critical region 1 expression by nuclear factor of activated T cells in megakaryocytes. Br J Haematol 144:395–408
176. Malinge S et al (2012) Increased dosage of the chromosome 21 ortholog Dyrk1a promotes megakaryoblastic leukemia in a murine model of Down syndrome. J Clin Invest 122:948–962
177. Giraudier S et al (2002) Overexpression of FKBP51 in idiopathic myelofibrosis regulates the growth factor independence of megakaryocyte progenitors. Blood 100:2932–2940
178. Vieira L et al (2012) Three-way translocation (X;20;16)(p11;q13;q23) in essential thrombocythemia implicates *NFATC2* in dysregulation of *CSF2* expression and megakaryocyte proliferation. Genes Chromosomes Cancer 51:1093–1108
179. Hisa T et al (2004) Hematopoietic, angiogenic and eye defects in Meis1 mutant animals. EMBO J 23:450–459
180. Azcoitia V, Aracil M, Martínez-A C, Torres M (2005) The homeodomain protein Meis1 is essential for definitive hematopoiesis and vascular patterning in the mouse embryo. Dev Biol 280:307–320
181. Cai M et al (2012) Dual actions of Meis1 inhibit erythroid progenitor development and sustain general hematopoietic cell proliferation. Blood 120:335–346
182. Watkins NA et al (2009) A HaemAtlas: characterizing gene expression in differentiated human blood cells. Blood 113:e1–e9
183. Zeddies S et al (2014) MEIS1 regulates early erythroid and megakaryocytic cell fate. Haematologica 99:1555–1564
184. Carramolino L et al (2010) Platelets play an essential role in separating the blood and lymphatic vasculatures during embryonic angiogenesis. Circ Res 106:1197–1201
185. Nürnberg ST et al (2012) A GWAS sequence variant for platelet volume marks an alternative DNM3 promoter in megakaryocytes near a MEIS1 binding site. Blood 120: 4859–4868
186. Wang W, Gilligan DM, Sun S, Wu X, Reems J-A (2011) Distinct functional effects for dynamin 3 during megakaryocytopoiesis. Stem Cells Dev 20:2139–2151
187. Hock H et al (2004) Gfi-1 restricts proliferation and preserves functional integrity of haematopoietic stem cells. Nature 431:1002–1007

188. Saleque S, Cameron S, Orkin SH (2002) The zinc-finger proto-oncogene Gfi-1b is essential for development of the erythroid and megakaryocytic lineages. Genes Dev 16:301–306
189. Chowdhury AH et al (2013) Differential transcriptional regulation of meis1 by Gfi1b and its co-factors LSD1 and CoREST. PLoS One 8, e53666
190. Monteferrario D et al (2014) A dominant-negative GFI1B mutation in the gray platelet syndrome. N Engl J Med 370:245–253
191. Emambokus N et al (2003) Progression through key stages of haemopoiesis is dependent on distinct threshold levels of c-Myb. EMBO J 22:4478–4488
192. Vegiopoulos A, García P, Emambokus N, Frampton J (2006) Coordination of erythropoiesis by the transcription factor c-Myb. Blood 107:4703–4710
193. García P, Berlanga O, Vegiopoulos A, Vyas P, Frampton J (2011) c-Myb and GATA-1 alternate dominant roles during megakaryocyte differentiation. J Thromb Haemost JTH 9:1572–1581
194. Bianchi E et al (2015) MYB controls erythroid versus megakaryocyte lineage fate decision through the miR-486-3p-mediated downregulation of MAF. Cell Death Differ. doi:10.1038/cdd.2015.30
195. Takayama M et al (2010) Genetic analysis of hierarchical regulation for Gata1 and NF-E2 p45 gene expression in megakaryopoiesis. Mol Cell Biol 30:2668–2680
196. McCormack MP et al (2006) A critical role for the transcription factor Scl in platelet production during stress thrombopoiesis. Blood 108:2248–2256
197. Tripic T et al (2009) SCL and associated proteins distinguish active from repressive GATA transcription factor complexes. Blood 113:2191–2201
198. Meinders M et al (2015) Sp1/Sp3 transcription factors regulate hallmarks of megakaryocyte maturation and platelet formation and function. Blood 125:1957–1967
199. Novershtern N et al (2011) Densely interconnected transcriptional circuits control cell states in human hematopoiesis. Cell 144:296–309
200. Duncan MT et al (2014) Dynamic transcription factor activity profiles reveal key regulatory interactions during megakaryocytic and erythroid differentiation. Biotechnol Bioeng 111:2082–2094
201. Levine RL et al (2005) Activating mutation in the tyrosine kinase JAK2 in polycythemia vera, essential thrombocythemia, and myeloid metaplasia with myelofibrosis. Cancer Cell 7:387–397
202. James C et al (2005) A unique clonal JAK2 mutation leading to constitutive signalling causes polycythaemia vera. Nature 434:1144–1148
203. Baxter EJ et al (2005) Acquired mutation of the tyrosine kinase JAK2 in human myeloproliferative disorders. Lancet Lond Engl 365:1054–1061
204. Balduini A et al (2011) In vitro megakaryocyte differentiation and proplatelet formation in Ph-negative classical myeloproliferative neoplasms: distinct patterns in the different clinical phenotypes. PLoS One 6, e21015
205. Hobbs CM et al (2013) JAK2V617F leads to intrinsic changes in platelet formation and reactivity in a knock-in mouse model of essential thrombocythemia. Blood 122:3787–3797
206. Tijssen MR et al (2011) Genome-wide analysis of simultaneous GATA1/2, RUNX1, FLI1, and SCL binding in megakaryocytes identifies hematopoietic regulators. Dev Cell 20:597–609
207. Estcourt LJ et al (2012) Platelet transfusions in haematology patients: are we using them appropriately? Vox Sang 103:284–293
208. Stroncek DF, Rebulla P (2007) Platelet transfusions. Lancet 370:427–438
209. Takayama N et al (2008) Generation of functional platelets from human embryonic stem cells in vitro via ES-sacs, VEGF-promoted structures that concentrate hematopoietic progenitors. Blood 111:5298–5306
210. Harker LA, Finch CA (1969) Thrombokinetics in man. J Clin Invest 48:963–974
211. Kaufman RM, Airo R, Pollack S, Crosby WH (1965) Circulating megakaryocytes and platelet release in the lung. Blood 26:720–731

212. Nakagawa Y et al (2013) Two differential flows in a bioreactor promoted platelet generation from human pluripotent stem cell-derived megakaryocytes. Exp Hematol 41:742–748
213. Thon JN et al (2014) Platelet bioreactor-on-a-chip. Blood 124:1857–1867
214. Di Buduo CA et al (2015) Programmable 3D silk bone marrow niche for platelet generation ex vivo and modeling of megakaryopoiesis pathologies. Blood 125:2254–2264
215. Shultz LD et al (2005) Human lymphoid and myeloid cell development in NOD/LtSz-scid IL2R gamma null mice engrafted with mobilized human hemopoietic stem cells. J Immunol Baltim Md 1950(174):6477–6489
216. Fuentes R et al (2010) Infusion of mature megakaryocytes into mice yields functional platelets. J Clin Invest 120:3917–3922
217. Tijssen MR et al (2008) Transplantation of human peripheral blood CD34-positive cells in combination with ex vivo generated megakaryocytes results in fast platelet formation in NOD/SCID mice. Leukemia 22:203–208
218. Orkin SH, Zon LI (2008) Hematopoiesis: an evolving paradigm for stem cell biology. Cell 132:631–644
219. Hedberg VA, Lipton JM (1988) Thrombocytopenia with absent radii. A review of 100 cases. Am J Pediatr Hematol Oncol 10:51–64
220. Olson TA, Levine RF, Mazur EM, Wright DG, Salvado AJ (1992) Megakaryocytes and megakaryocyte progenitors in human cord blood. Am J Pediatr Hematol Oncol 14:241–247
221. Liu Z-J et al (2011) Developmental differences in megakaryocytopoiesis are associated with up-regulated TPO signaling through mTOR and elevated GATA-1 levels in neonatal megakaryocytes. Blood 117:4106–4117
222. Ma DC, Sun YH, Chang KZ, Zuo W (1996) Developmental change of megakaryocyte maturation and DNA ploidy in human fetus. Eur J Haematol 57:121–127
223. Thomson JA et al (1998) Embryonic stem cell lines derived from human blastocysts. Science 282:1145–1147
224. Takahashi K et al (2007) Induction of pluripotent stem cells from adult human fibroblasts by defined factors. Cell 131:861–872
225. Murry CE, Keller G (2008) Differentiation of embryonic stem cells to clinically relevant populations: lessons from embryonic development. Cell 132:661–680
226. Takayama N, Eto K (2012) Pluripotent stem cells reveal the developmental biology of human megakaryocytes and provide a source of platelets for clinical application. Cell Mol Life Sci CMLS 69:3419–3428
227. Klimchenko O et al (2009) A common bipotent progenitor generates the erythroid and megakaryocyte lineages in embryonic stem cell-derived primitive hematopoiesis. Blood 114:1506–1517
228. Takayama N et al (2010) Transient activation of c-MYC expression is critical for efficient platelet generation from human induced pluripotent stem cells. J Exp Med 207:2817–2830
229. Pick M, Azzola L, Osborne E, Stanley EG, Elefanty AG (2013) Generation of megakaryocytic progenitors from human embryonic stem cells in a feeder- and serum-free medium. PLoS One 8, e55530
230. Feng Q et al (2014) Scalable generation of universal platelets from human induced pluripotent stem cells. Stem Cell Rep 3:817–831
231. Gaur M et al (2006) Megakaryocytes derived from human embryonic stem cells: a genetically tractable system to study megakaryocytopoiesis and integrin function. J Thromb Haemost JTH 4:436–442
232. Cherry ABC, Daley GQ (2012) Reprogramming cellular identity for regenerative medicine. Cell 148:1110–1122
233. Iwasaki H et al (2003) GATA-1 converts lymphoid and myelomonocytic progenitors into the megakaryocyte/erythrocyte lineages. Immunity 19:451–462
234. Valtieri M et al (1998) Enforced TAL-1 expression stimulates primitive, erythroid and megakaryocytic progenitors but blocks the granulopoietic differentiation program. Cancer Res 58:562–569

235. Real PJ et al (2012) SCL/TAL1 regulates hematopoietic specification from human embryonic stem cells. Mol Ther J Am Soc Gene Ther 20:1443–1453
236. Toscano MG et al (2015) SCL/TAL1-mediated transcriptional network enhances megakaryocytic specification of human embryonic stem cells. Mol Ther J Am Soc Gene Ther 23:158–170
237. Elcheva I et al (2014) Direct induction of haematoendothelial programs in human pluripotent stem cells by transcriptional regulators. Nat Commun 5:4372
238. Moreau T, Evans AL, Vasquez L, Tijssen MR, Yan Y, Trotter MW, Howard D, Colzani M, Arumugam M, Wu WH, Dalby A, Lampela R, Bouet G, Hobbs CM, Pask DC, Payne H, Ponomaryov T, Brill A, Soranzo N, Ouwehand WH, Pedersen RA, Ghevaert C (2016) Large-scale production of megakaryocytes from human pluripotent stem cells by chemically defined forward programming. Nat Commun 7:11208.
239. Bluteau O et al (2013) Developmental changes in human megakaryopoiesis. J Thromb Haemost 11:1730–1741
240. Liu Z-J et al (2014) Expansion of the neonatal platelet mass is achieved via an extension of platelet lifespan. Blood 123:3381–3389
241. Potts KS et al (2014) A lineage of diploid platelet-forming cells precedes polyploid megakaryocyte formation in the mouse embryo. Blood. doi:10.1182/blood-2014-02-559468
242. Ferrer-Marin F, Stanworth S, Josephson C, Sola-Visner M (2013) Distinct differences in platelet production and function between neonates and adults: implications for platelet transfusion practice. Transfusion (Paris) 53:2814–2821; quiz 2813
243. Rouhani F et al (2014) Genetic background drives transcriptional variation in human induced pluripotent stem cells. PLoS Genet 10, e1004432
244. Dumon S et al. (2012) Itga2b regulation at the onset of definitive hematopoiesis and commitment to differentiation. PLoS One 7(8):e43300. doi: 10.1371/journal.pone.0043300

Megakaryocytic Transcription Factors in Disease and Leukemia

3

Alan B. Cantor

Abstract

An understanding of the transcriptional regulation of megakaryopoiesis has lagged behind that of other hematopoietic lineages due to the rarity of these cells and the relatively recent development of systems to culture large numbers of megakaryocytes. However, significant progress has been made over the past few decades resulting in the identification of many key transcription factors involved in megakaryocyte specification and maturation. A number of important principles have emerged including physical and functional interactions among a core set of transcription factors including GATA, ETS, and RUNX family members, cross antagonistic network interactions with key erythroid-specific factors in cell fate determination of bipotential erythroid-megakaryocytic progenitor cells, and a surprising overlap with hematopoietic stem cell transcriptional regulators. A high proportion of genes encoding megakaryocytic transcription factors are mutated in human thrombopoiesis disorders, and a number of these are associated with leukemia predisposition. This chapter reviews the current knowledge about transcription factors involved in megakaryopoiesis, how they interact, and how their activities are influenced by cell signaling events. It also highlights the important role that dysregulation of these factors play in certain human platelet biogenesis disorders and leukemogenesis.

A.B. Cantor
Division of Pediatric Hematology-Oncology,
Boston Children's Hospital/Dana-Farber Cancer Institute,
Harvard Medical School, 300 Longwood Ave. Karp 7,
Boston, MA 02115, USA

Harvard Stem Cell Institute, Cambridge, MA 02138, USA
e-mail: alan.cantor@childrens.harvard.edu

H. Schulze, J. Italiano (eds.), *Molecular and Cellular Biology of Platelet Formation*,
DOI 10.1007/978-3-319-39562-3_3

3.1 Introduction

An understanding of the transcriptional control of megakaryopoiesis has lagged behind that of other blood lineages. This is due to the extreme rarity of megakaryocytes within the bone marrow (representing only ~0.4 % of nucleated cells) and the relatively recent purification and cloning of the major megakaryocyte cytokine, thrombopoietin (TPO), which occurred in 1994. Earlier studies focused on megakaryocyte-specific gene promoter analysis using in vitro assays and transformed megakaryocytic cell lines. Despite the limitations of these approaches, it was recognized early on that closely spaced binding sites for GATA and ETS family transcription factors played important roles in megakaryocyte-specific gene expression. Application of gene-targeting analysis in mice later confirmed the requirement for specific GATA and ETS family transcription factors for megakaryopoiesis in vivo. Additional transcription factors and cofactors, including Friend of GATA-1 (FOG1; zfpm1), RUNX-1/CBFβ, SCL/TAL-1, GFI-1b, NF-E2 p45, MEIS1, and HOXA9, were subsequently identified and shown to play essential roles in megakaryopoiesis. Through biochemical and genetic studies, it became clear that some of these factors physically and functionally interact to form megakaryocyte-specific enhanceosome complexes and gene regulatory networks. In parallel to these basic studies, human genetic analyses revealed that many of the same transcription factors are mutated in familial platelet disorders (Table 3.1). Interestingly, mutations in a number of these factors also predispose to leukemia development. Close connections with hematopoietic stem cell (HSC) regulators were also noted. Current challenges to the field include better understanding how bone marrow niche cues and cell signaling pathways modulate megakaryocytic transcription factor complex assembly and activity. It will also be of interest to further understand why mutations in many megakaryocytic transcription factors predispose to leukemia and why there is such a large overlap between megakaryocyte and HSC transcriptional control.

This chapter will review the different known key megakaryocytic transcription factors and highlight where mutations lead to human disease. Interactions between sets of these factors and their roles in gene regulatory networks will be discussed. The influence of cell signaling pathways on megakaryocytic transcriptional control will be reviewed with particular emphasis on cues from bone marrow microenvironments. The role of megakaryocytic transcription factors in the cell fate decision of erythroid-megakaryocytic progenitor cells will also be highlighted. Lastly, future directions and current challenges to the field will be discussed.

3.2 Specific Transcription Factors

3.2.1 GATA Family Transcription Factors

GATA family transcription factors are a phylogenetically conserved class of DNA binding proteins characterized by the presence of two (and sometimes one) C_2H_2-type closely spaced zinc fingers that recognize the core DNA element (T/A)

Table 3.1 Transcription factor genes mutated in human platelet disorders

Gene	Chromosome	Disease	Inheritance pattern	Megakaryocyte/platelet phenotype	Other phenotype
ETV6	12p13	ETV6-related thrombocytopenia	AD	Moderate thrombocytopenia; normal mean platelet volume; possible elongated platelet alpha-granules; small megakaryocytes with low cytoplasm to nuclear ratio; and hypolobulated nuclei	Elevated red blood cell MCV; increased hematologic cancer risk; possible autoimmunity
FLI-1	11q23	Paris-Trousseau/Jacobsen syndrome (heterozygous deletion)	AD	Macrothrombocytopenia; dimorphic population of megakaryocytes; platelets may contain giant alpha-granules	Developmental delay; dysmorphic facial and hand features; congenital cardiac defects
		FLI-1-related thrombocytopenia (point mutations and/or small deletions within FLI-1 gene)	AD/AR	Mild thrombocytopenia with dense granule secretion defects; may also have fused alpha-granules in platelets; and impaired collagen-induced platelet aggregation	
GATA-1	Xp11	X-linked macrothrombocytopenia −/+ dyserythropoietic anemia	X-linked	Macrothrombocytopenia; "gray" platelets; "platelets within platelets"	Dyserythropoiesis; mild beta-thalassemia; congenital erythropoietic porphyria
		Down syndrome transient myeloproliferative disorder and acute megakaryoblastic leukemia			
GATA-2	3q21	MonoMAC, Emberger syndrome	AD	Megakaryocytes with separated nuclei	Bone marrow failure/MDS; immunodeficiency; lymphedema

(continued)

Table 3.1 (continued)

Gene	Chromosome	Disease	Inheritance pattern	Megakaryocyte/platelet phenotype	Other phenotype
GFI-1b	9q34	Gray platelet syndrome	AD	"Gray" platelets (reduced alpha granules); functional platelet defects; dysplastic-appearing megakaryocytes	Red blood cell anisopoikilocytosis
HOXA11	7p15	Congenital thrombocytopenia with radioulnar synostosis (CTRUS)	AD	Impaired megakaryocyte differentiation	Proximal fusion of radius and ulna bones; possible bone marrow failure risk
RUNX-1	21q22	Familial platelet disorder with propensity to develop leukemia	AD	Moderate thrombocytopenia; normal mean platelet volume; abnormal platelet structure; impaired platelet function; reduced epinephrine and arachidonic acid-induced platelet aggregation	Increased risk for myelodysplastic syndrome and leukemia

AD autosomal dominant, *AR* autosomal recessive

GATA(A/G). GATA family transcription factors have been identified in organisms as diverse as fungi, nematodes, amphibians, insects, and mammals. Six GATA transcription factor genes exist in mammals. GATA-1, GATA-2, and GATA-3 play roles within hematopoietic tissues, whereas GATA-4, GATA-5, and GATA-6 play roles in extra-hematopoietic tissues, although there are exceptions. Megakaryocytes express both GATA-1 and GATA-2.

3.2.1.1 GATA-1

GATA-1 was first identified as a protein that binds to key *cis*-regulatory elements within the beta-globin locus [40, 153]. GATA-1 is expressed at high levels in erythroid and megakaryocytic cells and at lower levels in eosinophils, mast cells, and multipotent progenitor cells [182]. Knockout of the *GATA-1* gene in mice results in embryonic lethality between ~ embryonic day 10.5 and 11.5 (e10.5–11.5) days post-coitus (d.p.c.) due to severe anemia [45]. The first evidence for a functional role of GATA-1 in megakaryopoiesis came from studies of mice containing deletion of a megakaryocyte-specific *cis*-acting regulatory element upstream of the *GATA-1* gene ($GATA\text{-}1^{low}$ mice) [132]. These mice escape the embryonic lethality from anemia seen in full knockout mice. However, they are severely thrombocytopenic having platelet counts about 15 % of wild-type littermates. Megakaryocytes from these mice have impaired cytoplasmic and nuclear maturation, small size, reduced modal ploidy, abnormal ultrastructural features, and decreased expression of many terminal megakaryocyte marker genes including GPIbα, GPIbβ, PF4, c-mpl, and p45 NF-E2 [132, 160]. Platelets derived from these mice have ultrastructural defects similar to the megakaryocytes they are derived from and have impaired in vitro platelet aggregation in response to adenosine diphosphate (ADP) and epinephrine [160]. One of the most remarkable features of GATA-1-deficient megakaryocytes is their dramatic hyperproliferation compared to wild type when cultured from fetal liver [160]. Adult $GATA\text{-}1^{low}$ mice also develop myelofibrosis as they age [158].

The far majority of megakaryocyte-specific genes that have been studied contain functionally important GATA binding sites in their promoters and/or enhancers. Chromatin immunoprecipitation followed by deep sequencing (ChIP-seq) has identified 4,722 GATA-1 chromatin occupancy peaks (614 within promoters) in primary human megakaryocytes [150]. In combination with the finding of *GATA-1* germline mutations in human platelet disorders (see Sect. 3.2.1.4), these data indicate that GATA-1 is a central factor in megakaryopoiesis.

3.2.1.2 GATA-2

GATA-2 plays roles in early hematopoietic compartments, including hematopoietic stem and multipotential progenitor cells [152]. It is also expressed during early stages of committed erythroid and megakaryocytic precursor cells. It plays overlapping roles with GATA-1 during early megakaryopoiesis, but is dispensable for late megakaryopoiesis [28]. GATA-1 exchanges for GATA-2 at many loci throughout the genome during megakaryocyte maturation [36], similar to what is observed during erythropoiesis (reviewed in [16]). During this process, GATA-1 represses many early progenitor genes activated by GATA-2.

3.2.1.3 ZFPM1 (Previously Called Friend of GATA-1 (FOG1))

Zfpm1 was first identified in a yeast two-hybrid screen for GATA-1 interacting proteins in mouse erythroid cells [155] (for review, see [25]). It is a large molecule that contains nine multitype zinc fingers, four of which can independently bind to GATA-1 (via GATA-1's N-terminal zinc finger) [24, 42]. Zfpm1 is highly expressed in erythroid and megakaryocytic cells. It does not bind DNA directly, but acts as an essential cofactor for GATA-1. Its mechanism of action remains incompletely understood. It enhances GATA-1 chromatin occupancy at some, but not all, sites [83, 115]. It acts as either a coactivator or corepressor in a context-dependent manner. Part of its repressor activity depends upon its binding to the NuRD chromatin remodeling complex via an amino-terminal domain [58, 97]. Gene context-dependent activities of Zfpm1 in megakaryocyte versus erythroid cells depend on the combinatorial presence of ETS factors in megakaryocytic cells [164].

Zfpm1 antagonizes the cell fate decision of erythroid-megakaryocytic-mast cell multipotent progenitor cells for the mast cell lineage [23]. Loss of zfpm1 leads to marked mast cell-like outgrowth, whereas ectopic zfpm1 expression in developing mast cells (which normally downregulate Zfpm1 expression) reprograms them into the erythroid and megakaryocytic lineages [23]. This activity involves NuRD recruitment [97].

Mice containing targeted disruption of the Zfpm1 gene have a complete failure of megakaryopoiesis in addition to impaired erythropoiesis [154]. Whereas loss of other megakaryocytic transcription factor generally leads to impaired megakaryocyte maturation, loss of zfpm1 is relatively unique in that it causes a complete developmental failure of the entire megakaryocyte lineage. The discrepancy in the relatively late megakaryocyte phenotype of GATA-1-deficient megakaryocytes (hyperproliferation and impaired late megakaryocyte maturation) and early phenotype of Zfpm1 null megakaryocytes (failure to develop the megakaryocyte lineage) is explained by overlapping Zfpm1-dependent roles of GATA-2 (early) and GATA-1 (late) during megakaryopoiesis [28].

3.2.1.4 GATA Factor Mutations in Human Congenital Thrombocytopenias

Given the findings of GATA-1's central role in megakaryopoiesis based on mouse genetic experiments and human megakaryocytic gene promoter analysis, it was not surprising to find germline *GATA-1* gene mutations linked to human thrombopoiesis disorders. The first description was reported in 2000 and involved a family with X-linked macrothrombocytopenia and dyserythropoietic anemia (the *GATA-1* gene is located on the X-chromosome in humans and mice) [106]. Affected members harbored a missense mutation that substitutes methionine for valine at codon 205. This amino acid resides within the amino-terminal zinc finger of GATA-1 and mediates physical interaction with Zfpm1 [33]. There was no significant effect on DNA binding affinity based on electromobility shift assays. The female carrier had chronic thrombocytopenia, which worsened during pregnancy.

Since this initial description, a large number of other germline *GATA-1* mutations have been reported. These predominantly lead to X-linked macrothrombocytopenia and varying degrees of dyserythropoietic anemia, thalassemia, and/or congenital erythropoietic porphyria (for updated online review, see [30]). Substitutions of amino acids affecting Zfpm1 interaction (G208S, D218G, and D218Y) invariably lead to thrombocytopenia, the severity of which correlates with the degree of Zfpm1 binding loss [43, 44, 95]. The milder alleles have dyserythropoiesis, but no anemia. This suggests that the megakaryocytic lineage is more sensitive to changes in GATA-1/Zfpm1 interaction compared to the erythroid lineage.

Mutations that affect GATA-1 DNA binding are more likely to produce both thrombocytopenia and erythroid defects. GATA-1^{R216Q}, which interferes with DNA binding at palindromic GATA sites, but does not affect Zfpm1 interaction, leads to an X-linked beta-thalassemia-like disorder with imbalance of beta- and alpha-globin chains, hemolysis, and reticulocytosis, as well as mild-moderate thrombocytopenia [8, 121, 175]. GATA-1^{R216W} has a similar phenotype but, in addition, has features of congenital erythropoietic porphyria [117].

The GATA-1^{R216Q} mutation was reported to cause an X-linked form of gray platelet syndrome with affected individuals having large hypogranular platelets containing abnormal vacuoles [156]. Detailed ultrastructural studies of platelets from patients with the GATA-1^{G208S} mutation also showed hypogranular macrothrombocytes with reduced number of definitive alpha granules, although some platelets had normal number of alpha granules [167, 168]. In addition to classical gray platelet syndrome, the GATA-1 mutant platelets had a number of bizarre ultrastructural features including masses of dense tubular system channels, dense double membranes, "platelets within platelets," and even "platelets within platelets within platelets." It has been hypothesized that part of the defect may involve impaired separation of individual platelets from proplatelet processes. The dense granule number and morphology appeared to be normal.

Cis-regulatory mutations involving GATA-1 binding sites have also been reported in human thrombopoietic disease. A germline mutation that disrupts a GATA-1 binding site in the promoter of the platelet glycoprotein Ibβ gene results in Bernard-Soulier syndrome [87].

3.2.1.5 GATA-1 Mutations in Down Syndrome Transient Myeloproliferative Disorder and Acute Megakaryoblastic Leukemia

About 10 % of infants with Down syndrome are born with a clinically apparent transient myeloproliferative disorder with erythro-megakaryocytic features (previously called DS-TMD; now called Down syndrome transient abnormal myelopoiesis (DS-TAM) as per the latest nomenclature from the World Health Organization (WHO)) (for full reviews, see [22, 81, 91, 181]). Clinical manifestations are quite variable but typically reveal leukocytosis with excess circulating blasts, thrombocytopenia or thrombocytosis, and nucleated red blood cells on the peripheral blood smear. Hepatomegaly is common, and biopsy or autopsy frequently reveals hepatic megakaryocytic infiltration and fibrosis. In severe cases, there can be fulminant liver failure, jaundice, ascites, and respiratory failure [102]. Remarkably, DS-TAM resolves

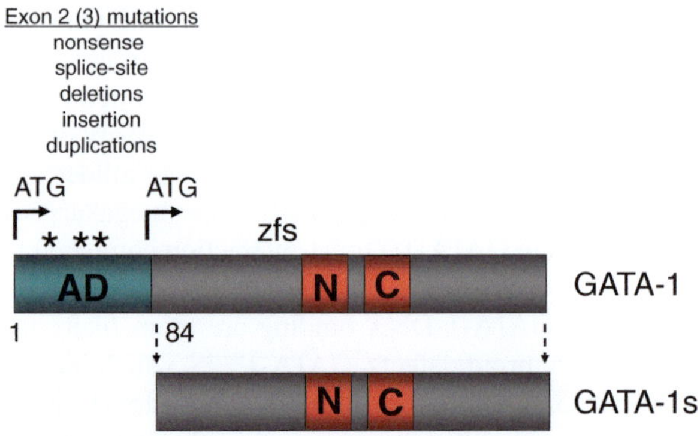

Fig. 3.1 Schematic diagram of truncating GATA-1 mutations in DS-TAM and ML-DS. Upper panel shows full-length GATA-1. Mutations in exon 2 (or rarely exon 3), indicated schematically by the *asterisks*, lead to the introduction of premature stop codons. In these cases, translation initiation occurs at a downstream ATG at codon 84. This generates a truncated protein that lacks the amino-terminal 83 amino acids. Low levels of this isoform, called "GATA-1s," are generated under normal conditions. *AD* activation domain, *C* carboxyl terminal, *N* amino terminal, *zfs* zinc fingers (Reproduced with permission from Cantor [22])

spontaneously by about 3–4 months of age [46, 73, 94]. However, ~20–25 % of infants with clinically apparent DS-TAM who survive the newborn period subsequently develop acute megakaryoblastic leukemia (previously called DS-AMKL; now called myeloid leukemia associated with Down syndrome (ML-DS)) [46, 73, 94, 104]. The mean age of diagnosis of ML-DS is 20 months (range 6–38 months). Overall, children with Down syndrome have about a 500-fold increased risk of megakaryoblastic leukemia compared to the general population [81].

In 2002, Crispino and colleagues reported that ML-DS cells harbor somatic mutations in *GATA-1* [165]. This was quickly confirmed by a number of other research groups and was also shown to occur in DS-TAM cells [48, 55, 101, 120, 172]. A wide variety of mutations have been described, but all involve exon 2 or rarely exon 3 [2]. These include missense, insertions, deletions, and splice-site mutations, and all lead to the introduction of a premature stop codon. However, translation initiates from a downstream in-frame "ATG" located at codon 84, generating a truncated GATA-1 protein product that lacks the first 83 amino acids (called "GATA-1s") (Fig. 3.1). Low levels of GATA-1s are produced normally during development and decrease relative to full-length GATA-1 during erythroid maturation [21, 88]. However, in DS-TAM and ML-DS cells, GATA-1s is produced exclusively since the *GATA-1* gene is located on the X-chromosome.

GATA-1s-producing mutations have been detected as early as 21 weeks of gestation [148]. In some cases, multiple GATA-1-generating mutations have been detected in the same individual at birth indicating an oligoclonal selection process [1, 124, 173]. Somatic GATA-1-producing mutations have only been found in Down syndrome individuals, with rare exceptions. Collectively, these findings indicate that *GATA-1* mutations are early in utero initiating events for ML-DS and are strongly selected for in a trisomy 21 genetic background. The basis for this selection is currently unknown.

How GATA-1 mutations contribute to ML-DS remains incompletely understood. Loss of the first 83 amino acids of GATA-1 results in reduced transcriptional activation capability in reporter assays. However, GATA-1s is able to rescue terminal erythroid maturation of a GATA-1 null erythroid cell line [166]. Mice engineered to exclusively produce GATA-1s in the germline have developmental stage-specific alterations in megakaryocyte proliferative control [84]. Megakaryocyte progenitor cells from these animals markedly hyperproliferate during late yolk sac and early fetal liver stages of hematopoietic development but proliferate close to normal during late gestation and postnatal stages. The stage-specific effects of GATA-1s on megakaryocyte proliferative control may involve developmental differences in insulin-like growth factor (IGF), mTOR, and/or type I interferon signaling [74, 170].

GATA-1s generating mutations by themselves appear to be insufficient to cause leukemia. This is supported by the GATA-1s mouse knock-in studies and the descriptions of several families with germline GATA-1s-producing mutations (and no trisomy 21) whose affected members have not developed leukemia or had a clinically apparent neonatal myeloproliferative disorder [57, 84, 88]. Trisomy 21 itself perturbs human hematopoiesis and likely contributes to the leukemogenesis beyond simply providing selective pressure for GATA-1s-containing clones [31, 90, 125, 126, 157]. Data supporting specific roles for the chromosome 21 genes *ERG* [105, 136, 137] and *ETS2* [47, 136] and the microRNA miR-125b [75] have been presented.

Analysis of *GATA-1* mutations in DS-TAM and ML-DS samples from the same patient universally reveals identical nucleotide changes in the ML-DS cells and in the DS-TAM cells (or at least a subpopulation of the DS-TAM cells) supporting a clonal evolution model of ML-DS. Indeed, a number of secondary mutations have been identified in the ML-DS cells and include chromosomal aberrations and copy number alterations such as trisomy 8, loss of material on chromosomes 5 and 7, gain of chromosome 21, dup(1q), del(16q), and other rarer alterations [12, 13]. Acquired missense mutations have been described in *JAK1* [12], *JAK2* [12], *JAK3* [12, 92, 162], *TP53* [93], *FLT3*, and *MPL*. Application of genomic profiling and whole exome/genome sequencing has revealed a high proportion of acquired mutations in components of the cohesin complex and other epigenetic regulators [173]. Additional studies will be needed to fully elucidate how these mutations may cooperate with GATA-1s to promote ML-DS.

3.2.1.6 *GATA-2* Mutations in Familial Myelodysplastic Syndrome and Acute Myelogenous Leukemia

Recently, germline heterozygous loss-of-function and/or dominant negative acting *GATA-2* mutations have been identified as a cause of familial myelodysplastic syndrome/leukemia predisposition syndrome [35, 50, 60, 71, 111]. These patients can present with thrombocytopenia. Megakaryocytes from these patients can have an unusual morphology characterized by widely separated nuclei. Some affected individuals also have immunodeficiency and/or lymphedema (called MonoMAC or Emberger syndromes; for review, see [135]). Somatic-activating *GATA-2* mutations have been associated with transformation of chronic myelogenous leukemia to acute myeloid leukemia [178]. Thus, it is clear that tight regulation of GATA transcription factor activity levels is critical for normal hematopoiesis.

3.2.2 ETS Family Transcription Factors

The E26 transformation-specific (ETS) transcription factors comprise a large family of transcriptional regulators that are defined by the presence of an "ETS" DNA binding domain that recognizes the core motif GGAA. As described earlier, tandem binding sites for ETS and GATA factors occur frequently in the *cis*-regulatory elements of megakaryocyte-specific genes [82]. There are over 30 different ETS family members, and at least ten of them (FLI-1, ERG, ETS1, ETS2, GABPbα, ETV6/TEL, PU.1, ELF1, ELF2, and ELK4) are expressed in megakaryocytes. Family members with the greatest evidence for roles in megakaryopoiesis and/or those mutated in human thrombopoietic disorders will be discussed in more detail.

3.2.2.1 FLI-1

FLI-1 is a member of the pointed domain subfamily of ETS factors. It plays a prominent role in terminal megakaryocyte maturation, as well as in vascular development. FLI-$1^{-/-}$ mice die around e11.5 d.p.c. of embryogenesis due to severe hemorrhage that is likely due to both vascular and thrombopoiesis defects [52]. Megakaryocytes cultured in collagen-based semisolid medium from the FLI-$1^{-/-}$ embryos have overall small size, high nuclear to cytoplasm ratio, hypolobulated nuclei, reduced number of α-granules, and disorganization of the platelet demarcation system. The embryos also have an increased number of megakaryocyte progenitors compared to wild type and heterozygous littermates, suggesting impaired megakaryocyte maturation. Inducible hematopoietic conditional knockout of FLI-1 in adult mice using the Mx1-Cre system results in a mild thrombocytopenia with a decrease in the number of mature megakaryocytes [139]. These mice also have an increase in erythrocytic cells (see Sect. 3.6 for additional discussion). Interestingly, FLI-1 appears to be expressed monoallelically during early megakaryocyte maturation ($CD41^{+}CD42b^{-}$ cells) and biallelically at later stages ($CD41^{+}CD42b^{+}$ cells) [122].

Jacobsen syndrome (OMIM 147791) and Paris-Trousseau disorder (OMIM 188925) are overlapping contiguous gene deletion syndromes that are due to constitutional heterozygous loss of a segment of human chromosome 11q23. These disorders include macrothrombocytopenia with abnormal megakaryocyte morphology and other variable features such as congenital cardiac disease, trigonocephaly, developmental delay, and/or dysmorphic hands and feet depending on the extent of the chromosomal deletion. Bone marrow examination reveals a dimorphic population of megakaryocytes with some showing marked dysmegakaryopoiesis including micromegakaryocytes and dying megakaryocytes [17]. Platelet survival is normal in these patients, indicating a predominant production defect. Platelets from these individuals can have giant α-granules, which are thought to arise from aberrant α-granule fusion during prolonged megakaryocyte bone marrow residence time.

Both the *FLI-1* and *ETS-1* genes reside within the minimally deleted region. Favier and colleagues demonstrated that lentiviral expression of FLI-1 in CD34+ cells from patients with Paris-Trousseau disorder rescued their megakaryocytic defects in vitro, providing strong evidence that FLI-1 haploinsufficiency contributes to the platelet phenotype of this disorder [122]. They also proposed that monoallelic expression of FLI-1 in early megakaryopoiesis may account for the dimorphic megakaryocyte populations in these patients, since a subpopulation of the megakaryocytes would have no FLI-1 expression due to deletion on one allele and silencing of the other. Recently, *FLI-1* gene point mutations resulting in alterations in the DNA binding domain (p.Arg 337Trp, p.Tyr343Cys, and p.Asn331Thrfs⁻4) were identified in some individuals with mild thrombocytopenia and dense granule secretion defects [142]. An autosomal recessive platelet disorder with features of Paris-Trousseau was also recently described in a consanguineous family carrying a germline DNA binding domain *FLI-1* germline mutation (p. Arg324Trp) [141]. Collectively, these findings implicate FLI-1 as the likely causative genetic defect in the 11q23 deletion interval found in Jacobsen and Paris-Trousseau syndromes.

3.2.2.2 ERG

ERG is in the same ETS subfamily as FLI-1. It plays overlapping functional roles with FLI-1 during normal murine megakaryopoiesis [76]. The human *ERG* gene is located on chromosome 21. Overexpression of ERG, as well as FLI-1, immortalizes murine fetal liver megakaryocytic progenitor cells harboring GATA-1s mutations, but not wild-type GATA-1 [136]. Trisomy of ERG is required for the myelomegakaryocytic hyperproliferation in the Ts65Dn mouse model of human trisomy 21 and leads to lineage priming of multilineage and megakaryocyte-erythroid progenitors [104, 105]. Based on these data, it has been proposed that ERG overexpression plays a role in the pathophysiology of DS-TAM and ML-DS.

3.2.2.3 ETV-6 (TEL)

ETV-6 is another member of the pointed domain containing ETS family transcription factors. It heterodimerizes with FLI-1 [78, 79]. ETV-6 plays a predominant role in gene repression. Conditional loss of ETV-6 in murine bone marrow impairs hematopoietic stem cell survival and specifically impairs megakaryocyte maturation [56]. Recently, three groups identified germline heterozygous *ETV-6* mutations in families with an autosomal dominant thrombocytopenia and leukemia predisposition syndrome [107, 151, 177]. The mutations reported to date involve the ETS DNA binding domain or a linker region between the pointed and DNA binding domain. The mutations typically have loss-of-function activity when measured in reporter assays. Megakaryocytes from these individuals are typically small and have hypolobulated nuclei and low amounts of cytoplasm. In addition to thrombocytopenia and increased risk of myelodysplastic syndrome and leukemia, these patients can also have myopathy, developmental delay, gastrointestinal motility difficulties, and possibly increased risk of inflammatory bowel disease.

3.2.2.4 ETS-1

A number of studies have implicated the ETS factor ETS-1 in megakaryopoiesis. ETS-1 cooperates with GATA-1 in the transactivation of megakaryocyte-specific genes in reporter assays [34, 82, 86]. Overexpression of ETS-1 in human primary hematopoietic progenitor cells promotes megakaryopoiesis and blocks erythropoiesis [89]. ETS-1 binding contributes to GATA factor site selection in megakaryocytic cells and correlates with histone H3 acetylation and gene activation [36].

3.2.2.5 GABPα

GABPα is another member of the pointed domain subfamily of ETS factors. It preferentially occupies ETS binding sites during early stages of megakaryopoiesis and controls expression of early megakaryocyte genes [116]. GABPα-deficient murine fetal liver-derived megakaryocytes have reduced expression of early megakaryocyte genes (e.g., c-mpl and αIIb) in contrast to FLI-1-deficient megakaryocytes, which have reduced expression of late megakaryocyte genes (e.g., PF4, GPIbα, and GPIX). The relative ratio of GABPα to FLI-1 decreases during megakaryocyte maturation. Based on these findings, it has been proposed that different ETS family members play stage-specific roles during megakaryocyte maturation [63, 116].

3.2.3 Core Binding Factors

The core binding transcription factors consist of heterodimers of DNA binding alpha (CBFα) subunits (now called RUNX-1, RUNX-2, and RUNX-3) and a common non-DNA binding CBFβ subunit. In the absence of CBFβ, the CBFα subunits have relatively low DNA binding affinity. Interaction with CBFβ induces conformation changes that markedly enhance DNA binding affinity [64] and increases CBFα subunit protein stability [119].

3.2.3.1 RUNX-1 (Previously Called CBFα2, AML-1, and PEBPA2B)

RUNX-1 is one of the hematopoietic expressed RUNX factors. RUNX-1 conventional knockout mice die between embryonic days 11.5 and 12.5 due to extensive central nervous system hemorrhage [163]. This is thought to be due to roles of RUNX-1 in vascular development. These mice also have complete failure of the definitive wave of hematopoiesis [163]. This is due to RUNX-1's requirement for the transition of hemogenic endothelial cells to the first definitive hematopoietic stem cells (HSCs) in the aorta-gonad-mesonephros (AGM) region during embryogenesis [29, 80]. RUNX-1's role in long-term hematopoietic stem cell maintenance remains controversial. Some studies have shown little consequences of RUNX-1 loss on adult LT-HSC function, whereas others have demonstrated accelerated HSC exhaustion during aging [20, 67].

Conditional RUNX-1 loss in the hematopoietic system of mice leads to marked dysmegakaryopoiesis and thrombocytopenia [49, 65]. These mice have reduced numbers of megakaryocytes in their bone marrow. In addition, the knockout megakaryocytes have hypolobulated nuclei, low DNA ploidy, and low cytoplasm-nuclear ratio, indicative of impaired megakaryocyte maturation. Heterozygous Vav-1-Cre RUNX-1 conditional knockout mice have mild thrombocytopenia [62]. Similar megakaryocyte defects are seen in mice containing CBFβ deficiency [145].

3.2.3.2 Familial Platelet Disorder with Propensity to Develop Leukemia

In humans, germline heterozygous mutations/deletions in the *RUNX-1* gene cause the autosomal dominant disorder called "familial platelet disorder with propensity to develop leukemia (FPD/AML)" [OMIM 601399] [134]. Affected individuals have lifelong thrombocytopenia. They also have bleeding symptoms that are typically out of proportion to their platelet count. This is due to functional platelet defects that include impaired in vitro aggregation to epinephrine and arachidonic acid, as well as impaired GPIIb-IIIa activation and pleckstrin phosphorylation [144]. Platelets from affected individuals can have a number of morphologic abnormalities including the presence of giant granules, inclusion of numerous vacuoles, unseparated proplatelets, and thin platelets [15]. The megakaryocytes are small and contain low cytoplasmic-nuclear ratio, hypolobulated nuclei, low DNA ploidy, and impaired proplatelet formation [15]. A number of direct RUNX-1 target genes have been identified including MYH9 [15], MYH10 [4], MYL9 [143], PKCtheta [144], platelet 12-lipoxygenase (ALOX12) [70], c-mpl [53, 63], and NR4A3 [14], among others.

Individuals with germline RUNX-1 mutations are predisposed to developing leukemia [112, 134]. Overall, they have ~35–50 % lifetime risk of developing leukemia. The leukemia is frequently preceded by a period of myelodysplastic syndrome (MDS). MDS/leukemia development occurs over a wide age range (8–72 years, reported). Both myeloid and lymphoid leukemias have been described. A wide range of acquired cytogenetic abnormalities has been described in the leukemias. These typically include common recurrent mutations found in human leukemias. A predominance of acquired *CDC25* mutations was reported in one study [174].

3.2.3.3 Sporadic RUNX-1 Mutations in Myelodysplastic Syndrome

Somatic *RUNX-1* mutations also occur in de novo MDS, particularly radiation and therapy-related cases (40–50 % of cases) [51, 180]. These mutations are associated with high incidence of leukemia transformation and poor survival [9, 32, 103, 147]. Clonal architecture studies in human AML indicate that *RUNX-1* mutations often occur early in the disease process [98]. These observations and the findings of germline *RUNX-1* mutations in FPD/AML provide strong evidence that RUNX-1 mutations can serve as early "driver" roles in human leukemogenesis. This may be related to recent work uncovering a requirement for RUNX-1 in the activation of DNA damage response pathways [5, 19, 66, 113, 114, 130, 171].

Interestingly, chromosomal translocations leading to the generation of a novel fusion molecule involving RUNX-1 and ETV-6 occur in the most common form of childhood pre-B-cell acute lymphoblastic leukemia. These translocations can occur in utero and, like the *RUNX-1* and *GATA-1* mutations, are early initiating events in leukemogenesis [41].

3.2.4 TAL-1 (Also Called SCL)

TAL-1 is a member of the basic helix-loop-helix transcription factor family. It is highly expressed in megakaryocytic and erythroid cells. It forms obligate heterodimeric complexes with ubiquitously expressed E-proteins (such as E2A, E12, and E47) and binds to E-box motifs (sequence CANNTG). It forms a pentameric complex along with E47, GATA-1, LDB-1, and LMO-2 in erythroid cells that form a bridge between GATA and E-box motifs spaced about nine base pairs apart [161]. It also interacts with the corepressor ETO-2 in erythroid and megakaryocytic cells [131]. TAL-1 is required during embryogenesis for the development of all hematopoietic lineages, likely through its role in specification of the ventral mesoderm in blood cell fate [118]. Loss of TAL-1 within the megakaryocytic lineage in mice impairs megakaryocyte proliferation, polyploidization, cytoplasmic maturation, and platelet release [27]. These effects are in part due to TAL-1's direct role in repressing the cell cycle regulator p21. TAL-1, like some other megakaryocytic transcription factors, plays additional roles in vascular development [159]. Chromosomal translocations leading to ectopic expression of TAL-1 in the T-cell lineage cause T-cell acute lymphoblastic leukemia in humans [108].

3.2.5 NF-E2

NF-E2 is a heterodimeric transcription factor complex composed of two basic region-leucine zipper (bZip) domain containing proteins: a hematopoietic specific NF-E2 p45 protein and a ubiquitously expressed p18 small Maf family member (comprised of MafF, MafG, and MafK). It is a major regulator of late megakaryocytic maturation. NF-E2 p45 null mice fail to produce platelets and succumb to hemorrhage in the neonatal period [133]. Similarly, compound MafK:MafG knockout mice have profound thrombocytopenia with impaired proplatelet formation [110]. A number of direct NF-E2 target genes have been identified in megakaryocytes. These include thromboxane synthase, β1-tubulin, 3β-hydroxysteroid dehydrogenase (3β-HSD), caspase 12, and Rab27b. β1-tubulin is a megakaryocyte-specific tubulin gene that encodes a structural component of the platelet marginal band structure, which is responsible for maintaining platelet discoid shape. 3β-HSD catalyzes autocrine estradiol biosynthesis within megakaryocytes and is important for proplatelet formation.

3.2.6 GFI-1b

GFI-1b is a member of the "growth factor independent" family of zinc finger transcriptional repressors that were originally identified as proto-oncogenes through insertional mutagenesis screens. They function, in part, by recruiting the corepressors LSD1 (KDM1) and coREST to chromatin via their SNAG domain [128]. GFI-1b is highly expressed in megakaryocyte and erythroid cells. GFI-1b null mouse embryos die from hemorrhage at ~ e15 d.p.c [127]. Megakaryocytes cultured from fetal liver of the animals are small and have reduced expression of NF-E2 p45, von Willebrand factor, c-mpl, and glycoprotein IIb, indicating a block in maturation. Definitive erythropoiesis is also marked impaired. In humans, germline dominant negative acting mutations in *GFI-1b* cause a form of autosomal dominant gray platelet syndrome [3, 99, 140]. Affected members have macrothrombocytopenia, platelet dysfunction, and red blood cell anisopoikilocytosis. The megakaryocytes appear dysplastic with some having hypolobulated nuclei and others with multiple separated nuclei.

3.2.7 Homeodomain Transcription Factors

Several homeodomain-related transcription factors have specialized roles in megakaryopoiesis and are mutated in human thrombocytopenia disorders and leukemia.

3.2.7.1 MEIS1

MEIS1 is a member of the three amino acid loop extension (TALE) family of homeodomain containing transcription factors and is highly expressed in megakaryocytes. It is closely related to the PBX subfamily of homeodomain transcription factors and heterodimerizes with other HOX proteins such as HOXA9. MEIS1$^{-/-}$ mice die by e14.5 d.p.c. due to extensive hemorrhage [7, 54]. These animals have a complete failure of megakaryopoiesis at an early progenitor cell stage. MEIS/PBX complexes synergize with GATA-1 and ETS factors to directly regulate the platelet factor 4 (PF4) gene [109]. Retroviral insertion or chromosomal translocations involving MEIS1 or HOXA9 contribute to murine and human leukemogenesis.

3.2.7.2 HOXA11

Mutations in the *HOXA11* gene cause congenital thrombocytopenia with radioulnar synostosis (CTRUS; OMIM # 605432), a rare form of human autosomal dominant thrombocytopenia associated with proximal fusion of the radius and ulna bones [149]. The mutations lead to impaired megakaryocyte differentiation in vitro [59]. Affected individuals may be at increased risk of bone marrow failure. HOXA11 is also involved in chromosomal translocations in human myeloid leukemias.

3.3 Megakaryocyte Enhanceosome Complex

A number of the transcription factors and cofactors discussed in the preceding section physically and functionally interact during megakaryopoiesis. GATA-1 interacts with RUNX-1 [38] and FLI-1 [37], in addition to its interactions with ZFPM-1 and TAL-1 discussed earlier. RUNX-1 and FLI-1 directly interact and synergistically activate megakaryocyte-specific genes [63]. GATA-2 interacts with RUNX-1 and ERG [169]. Genome-wide chromatin occupancy analysis in human CD34+ cell ex vivo differentiated megakaryocytes reveals marked enrichment for simultaneous binding by GATA-1/GATA-2, TAL-1, RUNX-1, and FLI-1 at many hematopoietic gene regulatory *cis*-elements [150]. In fact, common occupancy by all five factors occurred much more frequently than binding of individual or subsets of factors. Collectively, these findings suggest that a large "enhanceosome" complex exists that includes GATA-1/GATA-2, ZFPM-1, TAL-1, RUNX-1, and FLI-1 (and possibly additional factors), which is responsible for driving megakaryocytic cell identity and differentiation (Fig. 3.2). Whether other key megakaryocytic transcription factors, such as NF-E2, GFI-1b, MEIS1, and HOXA11, participate in all or some of these complexes or form completely independent complexes remains to be clearly defined. At least in human undifferentiated CD34+ stem/progenitor cells, GFI-1b and MEIS-1 appear to participate in separate complexes based on chromatin occupancy analysis [169].

There is some evidence that megakaryocytic transcription factor complexes change during megakaryocyte differentiation. As discussed earlier, GABPα predominantly controls genes involved in early megakaryopoiesis, whereas the related ETS factor FLI-1 mostly regulates terminal maturation genes [116]. The author's research group demonstrated physical interaction between RUNX-1 and FLI-1 that was enhanced during terminal megakaryocyte maturation [63]. Gel filtration chromatography studies documented large changes in the size of native complexes containing RUNX-1, GATA-1, Zfpm-1, and FLI-1 during megakaryocyte maturation using a mouse cell line model [63]. Thus, it seems likely that megakaryocytic transcription factor complexes change dynamically during megakaryocyte commitment and terminal maturation. Additional studies will be needed to further define these alterations and to determine their functional significance.

3.4 Cell Signaling Control of Megakaryocytic Transcription Factors

A major gap in understanding megakaryopoietic regulation involves how cell signaling pathways intersect with the major megakaryocytic transcription factors described in Sect. 3.2. Some insights are beginning to emerge (Fig. 3.2). RUNX-1 has been shown to be activated via direct extracellular signal-related kinases (ERK)-mediated serine phosphorylation [146]. The author's research group provided evidence that RUNX-1 is also negatively modulated by src-family kinase (SFK)-mediated direct tyrosine phosphorylation and that this is counteracted by the non-receptor tyrosine phosphatase Shp2 (also called PTPN11) [62]. We have also shown that a serine dephosphorylation step on FLI-1 enhances its interaction and cooperativity with RUNX-1 [63]. RUNX-1 activity is also enhanced by cyclin-dependent kinases (CDK6 and CDK1) serine phosphorylation [176] and PRMT1-mediated arginine methylation [179]. The RAS-p38 MAP kinase pathway has been recently shown to positively regulate GATA-2 activity [69]. Thrombopoietin signaling enhances HOXA9 nuclear transport during megakaryopoiesis [72]. Additional studies will be needed to fully integrate key cell signaling events with transcriptional control of megakaryopoiesis.

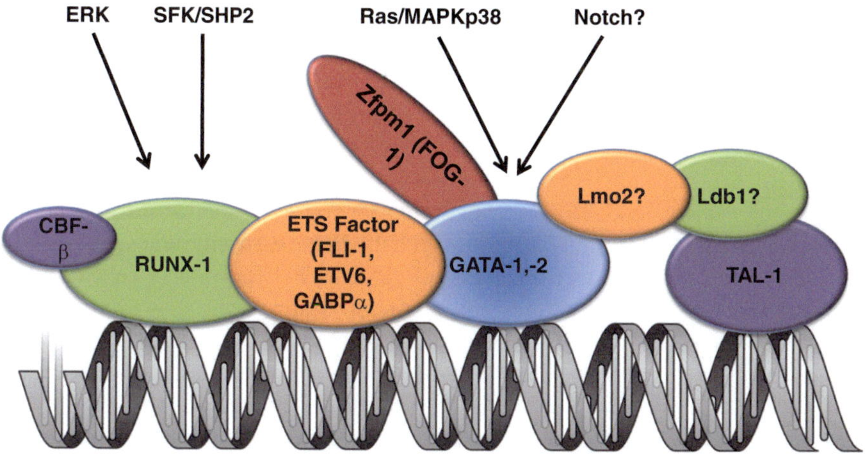

Fig. 3.2 Schematic diagram of megakaryocytic transcription factor "enhanceosome" complex. The GATA transcription factors (GATA-1, GATA-2) are shown in complex with ETS transcription factors (FLI-1, ETV6, and/or GABPα) and RUNX-1/CBFβ. These all likely involve direct protein-protein interactions. TAL-1 may interact with GATA factors indirectly via Ldb1 and Lmo2 based on studies in erythroid cells [161]. Signaling pathways known to modulate the indicated transcription factors are indicated. *ERK* extracellular signal-regulated kinase, *SFK* src-family tyrosine kinase, *SHP2* (*PTPN11*) non-receptor tyrosine phosphatase, *MAPK* mitogen-activated protein kinase

3.5 Control of Terminal Megakaryopoiesis via Bone Marrow Niche Signals

Terminal megakaryocyte maturation is spatially coordinated within the bone marrow. Megakaryocytic precursor cells migrate to vascular sinusoidal niches where they make contact with the abluminal surface of specialized endothelial cells and undergo the final platelet-forming stages of maturation [6, 68]. It is tempting to speculate that key signaling events occur via this interaction that triggers these final steps. Whether this occurs via direct modulation of megakaryocytic transcription factors remains to be determined. However, the fact that deletion of several transcription factors impairs proplatelet formation suggests this as a possibility. An intriguing prospect is the involvement of Notch signaling pathways. Notch signaling occurs via direct cell-cell contacts and results in release of the Notch receptor intracellular cytoplasmic domain (ICD), which then translocates to the nucleus and acts as transcriptional cofactor. Mercher et al. showed that Notch signaling positively regulates megakaryopoiesis [96]. Several Notch ligands, including DLL1, DLL4, Jagged-1, and Jagged-2, are expressed on bone marrow vascular endothelial cells [18]. Notch signaling has been shown to modulate GATA-2 expression [123]. Further work will be needed to test the role of Notch signaling, as well as other signaling pathways, in terminal megakaryocyte maturation in situ.

3.6 Transcriptional Control of Megakaryocytic-Erythroid Bipotential Progenitor Cell Fate Decision

Megakaryocytes and erythroid cells are derived from a common bipotential progenitor cell (MEP, megakaryocyte-erythroid progenitor). The mechanisms underlying the cell fate decision of MEPs for the megakaryocyte versus erythroid lineages are not fully understood. However, there is growing evidence for involvement of gene regulatory networks involving some of the lineage-specific transcription factors discussed previously as well as microRNAs. KLF-1 (previously called EKLF) is a major erythroid-specific transcription factor. Starck et al. showed that FLI-1 physically and functionally cross antagonizes KLF-1 during erythroid-megakaryocytic lineage choices [138]. More recently, Kuvardina et al. showed that RUNX-1, which cooperates with FLI-1 [63], directly represses expression of the *KLF-1* gene during MEP commitment to the megakaryocytic lineage [77]. Thus, there is an interconnected regulatory loop involving the major megakaryocytic transcription factors FLI-1 and RUNX-1 and the major erythroid transcription factor KLF-1 contributing to fate choice of MEPs. Interestingly, dysregulated expression of FLI-1 in the erythroid lineage via integration of the Friend murine leukemia virus (F-MuLV) leads to erythroleukemia in mice [10, 11].

The proto-oncogene transcription factor c-Myb plays a negative regulatory role during megakaryopoiesis and a positive role during erythroid differentiation [26, 39, 100]. This activity requires interaction between c-Myb and the chromatin remodeling factor p300 [129]. Lu et al. showed that the microRNA miR-150, which targets the 3′UTR of the c-Myb transcript, is markedly upregulated during megakaryocyte differentiation from human $CD34^+$ cells [85]. However, it remains at low levels during erythroid differentiation. They also showed that enhanced miR-150 expression promotes megakaryopoiesis at the expense of erythropoiesis, whereas reduced miR-150 expression had the opposite effect. Thus, miR-150 via c-Myb, and possibly other target transcripts, contributes to megakaryocyte-erythroid cell fate choice.

3.7 Megakaryocytic Transcription Factors and Hematopoietic Stem Cells

Many of the key megakaryocytic transcription factors discussed in this chapter also play important roles in hematopoietic stem cell (HSC) ontogeny and maintenance (reviewed in [61]). This includes RUNX-1, ETV6, FLI-1, GATA-2, TAL-1, HOXA9, and MEIS1. Megakaryocyte and HSCs also share common cell signaling pathways (TPO signaling), cell surface markers (e.g., CD41 on embryonic HSCs), and vascular sinusoidal niche localization. The reason for the significant overlap between HSCs and megakaryocyte features is not clear, but may reflect their recently identified close hierarchical relationship. Interestingly, many of the same transcription factors shared between megakaryocytes and HSCs are recurrent targets of mutations or chromosomal translocations in human leukemia and leukemia predisposition syndromes.

3.8 Summary and Future Directions

Considerable progress has been made over the past few decades in identifying the key transcription factors involved in megakaryocyte lineage specification and maturation. Demonstration of physical and functional interaction among a number of these core factors, notably GATA, RUNX, and ETS family members, supports the concept of a megakaryocytic-specific enhanceosome-type complex. Network interactions between these factors and key erythroid regulators are emerging as an underlying principle in fate determination of bipotent megakaryocyte-erythroid progenitor cells. A surprisingly large proportion of transcription factors involved in normal megakaryopoiesis are mutated or involved in chromosomal translocations in human leukemias, and these frequently represent early initiating steps. Whether this is related to the relatively close relationship between megakaryocytes and HSCs will require further study. Major knowledge gaps in megakaryocyte transcription control that remain include how the assembly of megakaryocytic transcription factor complexes is controlled, how cell signaling pathways modulate megakaryocytic transcription factor activity, and how bone marrow spatial cues influence megakaryocyte transcriptional control. Answers to these questions are likely to enhance our knowledge of human thrombopoiesis disorders. They will also likely lead to practical advances in generating platelets from human embryonic stem (ES) and induced pluripotent stem (IPS) cells for clinical use. The current rapid development of new tools to address these questions in vivo promises that new insights are likely on the horizon.

Acknowledgments The author would like to thank the many colleagues in the field and members of the laboratory who have contributed ideas and insights to the material discussed in this chapter. I apologize for any work from others that was omitted due to space limitations. A.B.C. is supported by grants from the NIH (R01 DK098448 and R01 HL130793).

References

1. Ahmed M, Sternberg A, Hall G, Thomas A, Smith O, O'Marcaigh A, Wynn R, Stevens R, Addison M, King D, Stewart B, Gibson B, Roberts I, Vyas P (2004) Natural history of GATA1 mutations in Down syndrome. Blood 103(7):2480–2489
2. Alford KA, Reinhardt K, Garnett C, Norton A, Bohmer K, von Neuhoff C, Kolenova A, Marchi E, Klusmann JH, Roberts I, Hasle H, Reinhardt D, Vyas P (2011) Analysis of GATA1 mutations in Down syndrome transient myeloproliferative disorder and myeloid leukemia. Blood 118(8):2222–2238. doi:10.1182/blood-2011-03-342774
3. Aminkeng F (2014) GFI1B mutation causes autosomal dominant gray platelet syndrome. Clin Genet 85(6):534–535. doi:10.1111/cge.12380
4. Antony-Debre I, Bluteau D, Itzykson R, Baccini V, Renneville A, Boehlen F, Morabito M, Droin N, Deswarte C, Chang Y, Leverger G, Solary E, Vainchenker W, Favier R, Raslova H (2012) MYH10 protein expression in platelets as a biomarker of RUNX1 and FLI1 alterations. Blood 120(13):2719–2722. doi:10.1182/blood-2012-04-422352
5. Antony-Debre I, Manchev VT, Balayn N, Bluteau D, Tomowiak C, Legrand C, Langlois T, Bawa O, Tosca L, Tachdjian G, Leheup B, Debili N, Plo I, Mills JA, French DL, Weiss MJ, Solary E, Favier R, Vainchenker W, Raslova H (2015) Level of RUNX1 activity is critical for

leukemic predisposition but not for thrombocytopenia. Blood 125(6):930–940. doi:10.1182/blood-2014-06-585513
6. Avecilla ST, Hattori K, Heissig B, Tejada R, Liao F, Shido K, Jin DK, Dias S, Zhang F, Hartman TE, Hackett NR, Crystal RG, Witte L, Hicklin DJ, Bohlen P, Eaton D, Lyden D, de Sauvage F, Rafii S (2004) Chemokine-mediated interaction of hematopoietic progenitors with the bone marrow vascular niche is required for thrombopoiesis. Nat Med 10(1):64–71
7. Azcoitia V, Aracil M, Martinez AC, Torres M (2005) The homeodomain protein Meis1 is essential for definitive hematopoiesis and vascular patterning in the mouse embryo. Dev Biol 280(2):307–320
8. Balduini CL, Pecci A, Loffredo G, Izzo P, Noris P, Grosso M, Bergamaschi G, Rosti V, Magrini U, Ceresa IF, Conti V, Poggi V, Savoia A (2004) Effects of the R216Q mutation of GATA-1 on erythropoiesis and megakaryocytopoiesis. Thromb Haemost 91(1):129–140. doi:10.1267/THRO04010129
9. Bejar R, Stevenson K, Abdel-Wahab O, Galili N, Nilsson B, Garcia-Manero G, Kantarjian H, Raza A, Levine RL, Neuberg D, Ebert BL (2011) Clinical effect of point mutations in myelodysplastic syndromes. N Engl J Med 364(26):2496–2506. doi:10.1056/NEJMoa1013343
10. Ben-David Y, Giddens EB, Bernstein A (1990) Identification and mapping of a common proviral integration site Fli-1 in erythroleukemia cells induced by Friend murine leukemia virus. Proc Natl Acad Sci U S A 87(4):1332–1336
11. Ben-David Y, Giddens EB, Letwin K, Bernstein A (1991) Erythroleukemia induction by Friend murine leukemia virus: insertional activation of a new member of the ets gene family, Fli-1, closely linked to c-ets-1. Genes Dev 5(6):908–918
12. Blink M, Buitenkamp TD, van den Heuvel-Eibrink MM, Danen-van Oorschot AA, de Haas V, Reinhardt D, Klusmann JH, Zimmermann M, Devidas M, Carroll AJ, Basso G, Pession A, Hasle H, Pieters R, Rabin KR, Izraeli S, Zwaan CM (2011) Frequency and prognostic implications of JAK 1–3 aberrations in Down syndrome acute lymphoblastic and myeloid leukemia. Leukemia 25(8):1365–1368. doi:10.1038/leu.2011.86
13. Blink M, van den Heuvel-Eibrink MM, Aalbers AM, Balgobind BV, Hollink IH, Meijerink JP, van der Velden VH, Beverloo BH, de Haas V, Hasle H, Reinhardt D, Klusmann JH, Pieters R, Calado RT, Zwaan CM (2012) High frequency of copy number alterations in myeloid leukaemia of Down syndrome. Br J Haematol 158(6):800–803. doi:10.1111/j.1365-2141.2012.09224.x
14. Bluteau D, Gilles L, Hilpert M, Antony-Debre I, James C, Debili N, Camara-Clayette V, Wagner-Ballon O, Cordette-Lagarde V, Robert T, Ripoche H, Gonin P, Swierczek S, Prchal J, Vainchenker W, Favier R, Raslova H (2011) Down-regulation of the RUNX1-target gene NR4A3 contributes to hematopoiesis deregulation in familial platelet disorder/acute myelogenous leukemia. Blood 118(24):6310–6320. doi:10.1182/blood-2010-12-325555
15. Bluteau D, Glembotsky AC, Raimbault A, Balayn N, Gilles L, Rameau P, Nurden P, Alessi MC, Debili N, Vainchenker W, Heller PG, Favier R, Raslova H (2012) Dysmegakaryopoiesis of FPD/AML pedigrees with constitutional RUNX1 mutations is linked to myosin II deregulated expression. Blood 120(13):2708–2718. doi:10.1182/blood-2012-04-422337
16. Bresnick EH, Lee HY, Fujiwara T, Johnson KD, Keles S (2010) GATA switches as developmental drivers. J Biol Chem 285:31087–31093
17. Breton-Gorius J, Favier R, Guichard J, Cherif D, Berger R, Debili N, Vainchenker W, Douay L (1995) A new congenital dysmegakaryopoietic thrombocytopenia (Paris-Trousseau) associated with giant platelet alpha-granules and chromosome 11 deletion at 11q23. Blood 85(7):1805–1814
18. Butler JM, Nolan DJ, Vertes EL, Varnum-Finney B, Kobayashi H, Hooper AT, Seandel M, Shido K, White IA, Kobayashi M, Witte L, May C, Shawber C, Kimura Y, Kitajewski J, Rosenwaks Z, Bernstein ID, Rafii S (2010) Endothelial cells are essential for the self-renewal and repopulation of Notch-dependent hematopoietic stem cells. Cell Stem Cell 6(3):251–264. doi:10.1016/j.stem.2010.02.001
19. Cai X, Gao L, Teng L, Ge J, Oo ZM, Kumar AR, Gilliland DG, Mason PJ, Tan K, Speck NA (2015) Runx1 deficiency decreases ribosome biogenesis and confers stress resistance to hematopoietic stem and progenitor cells. Cell Stem Cell 17(2):165–177. doi:10.1016/j.stem.2015.06.002

20. Cai X, Gaudet JJ, Mangan JK, Chen MJ, De Obaldia ME, Oo Z, Ernst P, Speck NA (2011) Runx1 loss minimally impacts long-term hematopoietic stem cells. PLoS ONE 6(12), e28430. doi:10.1371/journal.pone.0028430
21. Calligaris R, Bottardi S, Cogoi S, Apezteguia I, Santoro C (1995) Alternative translation initiation site usage results in two functionally distinct forms of the GATA-1 transcription factor. Proc Natl Acad Sci U S A 92(25):11598–11602
22. Cantor AB (2015) Myeloid proliferations associated with Down syndrome. J Hematop 8:169–176
23. Cantor AB, Iwasaki H, Arinobu Y, Moran TB, Shigematsu H, Sullivan MR, Akashi K, Orkin SH (2008) Antagonism of FOG-1 and GATA factors in fate choice for the mast cell lineage. J Exp Med 205(3):611–624
24. Cantor AB, Katz SG, Orkin SH (2002) Distinct domains of the GATA-1 cofactor FOG-1 differentially influence erythroid versus megakaryocytic maturation. Mol Cell Biol 22(12): 4268–4279
25. Cantor AB, Orkin SH (2005) Coregulation of GATA factors by the Friend of GATA (FOG) family of multitype zinc finger proteins. Semin Cell Dev Biol 16(1):117–128
26. Carpinelli MR, Hilton DJ, Metcalf D, Antonchuk JL, Hyland CD, Mifsud SL, Di Rago L, Hilton AA, Willson TA, Roberts AW, Ramsay RG, Nicola NA, Alexander WS (2004) Suppressor screen in Mpl-/- mice: c-Myb mutation causes supraphysiological production of platelets in the absence of thrombopoietin signaling. Proc Natl Acad Sci U S A 101(17):6553–6558. doi:10.1073/pnas.0401496101
27. Chagraoui H, Kassouf M, Banerjee S, Goardon N, Clark K, Atzberger A, Pearce AC, Skoda RC, Ferguson DJ, Watson SP, Vyas P, Porcher C (2011) SCL-mediated regulation of the cell-cycle regulator p21 is critical for murine megakaryopoiesis. Blood 118(3):723–735. doi:10.1182/blood-2011-01-328765
28. Chang AN, Cantor AB, Fujiwara Y, Lodish MB, Droho S, Crispino JD, Orkin SH (2002) GATA-factor dependence of the multitype zinc-finger protein FOG-1 for its essential role in megakaryopoiesis. Proc Natl Acad Sci U S A 99(14):9237–9242
29. Chen MJ, Yokomizo T, Zeigler BM, Dzierzak E, Speck NA (2009) Runx1 is required for the endothelial to haematopoietic cell transition but not thereafter. Nature 457(7231):887–891
30. Chou ST, Kacena MA, Weiss MJ, Rasking WH (2014) GATA1-related X-linked cytopenia. http://www.ncbi.nlm.nih.gov/books/NBK1364/
31. Chou ST, Opalinska JB, Yao Y, Fernandes MA, Kalota A, Brooks JS, Choi JK, Gewirtz AM, Danet-Desnoyers GA, Nemiroff RL, Weiss MJ (2008) Trisomy 21 enhances human fetal erythro-megakaryocytic development. Blood 112(12):4503–4506
32. Christiansen DH, Andersen MK, Pedersen-Bjergaard J (2004) Mutations of AML1 are common in therapy-related myelodysplasia following therapy with alkylating agents and are significantly associated with deletion or loss of chromosome arm 7q and with subsequent leukemic transformation. Blood 104(5):1474–1481
33. Crispino JD, Lodish MB, MacKay JP, Orkin SH (1999) Use of altered specificity mutants to probe a specific protein-protein interaction in differentiation: the GATA-1:FOG complex. Mol Cell 3(2):219–228
34. Deveaux S, Filipe A, Lemarchandel V, Ghysdael J, Romeo PH, Mignotte V (1996) Analysis of the thrombopoietin receptor (MPL) promoter implicates GATA and Ets proteins in the coregulation of megakaryocyte-specific genes. Blood 87(11):4678–4685
35. Dickinson RE, Griffin H, Bigley V, Reynard LN, Hussain R, Haniffa M, Lakey JH, Rahman T, Wang XN, McGovern N, Pagan S, Cookson S, McDonald D, Chua I, Wallis J, Cant A, Wright M, Keavney B, Chinnery PF, Loughlin J, Hambleton S, Santibanez-Koref M, Collin M (2011) Exome sequencing identifies GATA-2 mutation as the cause of dendritic cell, monocyte, B and NK lymphoid deficiency. Blood 118(10):2656–2658. doi:10.1182/blood-2011-06-360313
36. Dore LC, Chlon TM, Brown CD, White KP, Crispino JD (2012) Chromatin occupancy analysis reveals genome-wide GATA factor switching during hematopoiesis. Blood 119(16):3724–3733. doi:10.1182/blood-2011-09-380634

37. Eisbacher M, Holmes ML, Newton A, Hogg PJ, Khachigian LM, Crossley M, Chong BH (2003) Protein-protein interaction between Fli-1 and GATA-1 mediates synergistic expression of megakaryocyte-specific genes through cooperative DNA binding. Mol Cell Biol 23(10):3427–3441
38. Elagib KE, Racke FK, Mogass M, Khetawat R, Delehanty LL, Goldfarb AN (2003) RUNX1 and GATA-1 coexpression and cooperation in megakaryocytic differentiation. Blood 101(11):4333–4341
39. Emambokus N, Vegiopoulos A, Harman B, Jenkinson E, Anderson G, Frampton J (2003) Progression through key stages of haemopoiesis is dependent on distinct threshold levels of c-Myb. EMBO J 22(17):4478–4488. doi:10.1093/emboj/cdg434
40. Evans T, Reitman M, Felsenfeld G (1988) An erythrocyte-specific DNA-binding factor recognizes a regulatory sequence common to all chicken globin genes. Proc Natl Acad Sci U S A 85(16):5976–5980
41. Ford AM, Bennett CA, Price CM, Bruin MC, Van Wering ER, Greaves M (1998) Fetal origins of the TEL-AML1 fusion gene in identical twins with leukemia. Proc Natl Acad Sci U S A 95(8):4584–4588
42. Fox AH, Liew C, Holmes M, Kowalski K, Mackay J, Crossley M (1999) Transcriptional cofactors of the FOG family interact with GATA proteins by means of multiple zinc fingers. Embo J 18(10):2812–2822
43. Freson K, Devriendt K, Matthijs G, Van Hoof A, De Vos R, Thys C, Minner K, Hoylaerts MF, Vermylen J, Van Geet C (2001) Platelet characteristics in patients with X-linked macrothrombocytopenia because of a novel GATA1 mutation. Blood 98(1):85–92
44. Freson K, Matthijs G, Thys C, Marien P, Hoylaerts MF, Vermylen J, Van Geet C (2002) Different substitutions at residue D218 of the X-linked transcription factor GATA1 lead to altered clinical severity of macrothrombocytopenia and anemia and are associated with variable skewed X inactivation. Hum Mol Genet 11(2):147–152
45. Fujiwara Y, Browne CP, Cunniff K, Goff SC, Orkin SH (1996) Arrested development of embryonic red cell precursors in mouse embryos lacking transcription factor GATA-1. Proc Natl Acad Sci U S A 93(22):12355–12358
46. Gamis AS, Alonzo TA, Gerbing RB, Hilden JM, Sorrell AD, Sharma M, Loew TW, Arceci RJ, Barnard D, Doyle J, Massey G, Perentesis J, Ravindranath Y, Taub J, Smith FO (2011) Natural history of transient myeloproliferative disorder clinically diagnosed in Down syndrome neonates: a report from the Children's Oncology Group Study A2971. Blood 118(26):6752–6759. doi:10.1182/blood-2011-04-350017; quiz 6996
47. Ge Y, LaFiura KM, Dombkowski AA, Chen Q, Payton SG, Buck SA, Salagrama S, Diakiw AE, Matherly LH, Taub JW (2008) The role of the proto-oncogene ETS2 in acute megakaryocytic leukemia biology and therapy. Leukemia 22(3):521–529. doi:10.1038/sj.leu.2405066
48. Greene ME, Mundschau G, Wechsler J, McDevitt M, Gamis A, Karp J, Gurbuxani S, Arceci R, Crispino JD (2003) Mutations in GATA1 in both transient myeloproliferative disorder and acute megakaryoblastic leukemia of Down syndrome. Blood Cells Mol Dis 31(3):351–356
49. Growney JD, Shigematsu H, Li Z, Lee BH, Adelsperger J, Rowan R, Curley DP, Kutok JL, Akashi K, Williams IR, Speck NA, Gilliland DG (2005) Loss of Runx1 perturbs adult hematopoiesis and is associated with a myeloproliferative phenotype. Blood 106(2):494–504
50. Hahn CN, Chong CE, Carmichael CL, Wilkins EJ, Brautigan PJ, Li XC, Babic M, Lin M, Carmagnac A, Lee YK, Kok CH, Gagliardi L, Friend KL, Ekert PG, Butcher CM, Brown AL, Lewis ID, To LB, Timms AE, Storek J, Moore S, Altree M, Escher R, Bardy PG, Suthers GK, D'Andrea RJ, Horwitz MS, Scott HS (2011) Heritable GATA2 mutations associated with familial myelodysplastic syndrome and acute myeloid leukemia. Nat Genet 43(10):1012–1017. doi:10.1038/ng.913
51. Harada H, Harada Y, Tanaka H, Kimura A, Inaba T (2003) Implications of somatic mutations in the AML1 gene in radiation-associated and therapy-related myelodysplastic syndrome/acute myeloid leukemia. Blood 101(2):673–680
52. Hart A, Melet F, Grossfeld P, Chien K, Jones C, Tunnacliffe A, Favier R, Bernstein A (2000) Fli-1 is required for murine vascular and megakaryocytic development and is hemizygously deleted in patients with thrombocytopenia. Immunity 13(2):167–177

53. Heller PG, Glembotsky AC, Gandhi MJ, Cummings CL, Pirola CJ, Marta RF, Kornblihtt LI, Drachman JG, Molinas FC (2005) Low Mpl receptor expression in a pedigree with familial platelet disorder with predisposition to acute myelogenous leukemia and a novel AML1 mutation. Blood 105(12):4664–4670
54. Hisa T, Spence SE, Rachel RA, Fujita M, Nakamura T, Ward JM, Devor-Henneman DE, Saiki Y, Kutsuna H, Tessarollo L, Jenkins NA, Copeland NG (2004) Hematopoietic, angiogenic and eye defects in Meis1 mutant animals. Embo J 23(2):450–459
55. Hitzler JK, Cheung J, Li Y, Scherer SW, Zipursky A (2003) GATA1 mutations in transient leukemia and acute megakaryoblastic leukemia of Down syndrome. Blood 101(11):4301–4304
56. Hock H, Meade E, Medeiros S, Schindler JW, Valk PJ, Fujiwara Y, Orkin SH (2004) Tel/Etv6 is an essential and selective regulator of adult hematopoietic stem cell survival. Genes Dev 18(19):2336–2341
57. Hollanda LM, Lima CS, Cunha AF, Albuquerque DM, Vassallo J, Ozelo MC, Joazeiro PP, Saad ST, Costa FF (2006) An inherited mutation leading to production of only the short isoform of GATA-1 is associated with impaired erythropoiesis. Nat Genet 38(7):807–812
58. Hong W, Nakazawa M, Chen YY, Kori R, Vakoc CR, Rakowski C, Blobel GA (2005) FOG-1 recruits the NuRD repressor complex to mediate transcriptional repression by GATA-1. EMBO J 24(13):2367–2378. doi:10.1038/sj.emboj.7600703
59. Horvat-Switzer RD, Thompson AA (2006) HOXA11 mutation in amegakaryocytic thrombocytopenia with radio-ulnar synostosis syndrome inhibits megakaryocytic differentiation in vitro. Blood Cells Mol Dis 37(1):55–63. doi:10.1016/j.bcmd.2006.04.001
60. Hsu AP, Sampaio EP, Khan J, Calvo KR, Lemieux JE, Patel SY, Frucht DM, Vinh DC, Auth RD, Freeman AF, Olivier KN, Uzel G, Zerbe CS, Spalding C, Pittaluga S, Raffeld M, Kuhns DB, Ding L, Paulson ML, Marciano BE, Gea-Banacloche JC, Orange JS, Cuellar-Rodriguez J, Hickstein DD, Holland SM (2011) Mutations in GATA2 are associated with the autosomal dominant and sporadic monocytopenia and mycobacterial infection (MonoMAC) syndrome. Blood 118(10):2653–2655. doi:10.1182/blood-2011-05-356352
61. Huang H, Cantor AB (2009) Common features of megakaryocytes and hematopoietic stem cells: what's the connection? J Cell Biochem 107(5):857–864
62. Huang H, Woo AJ, Waldon Z, Schindler Y, Moran TB, Zhu HH, Feng G, Steen H, Cantor AB (2012) A Src family kinase-Shp2 axis controls RUNX1 activity in megakaryocyte and T lymphocyte differentiation. Genes Dev 26:1587–1601, Jul 3. [Epub ahead of print] PMID: 22759635
63. Huang H, Yu M, Akie TE, Moran TB, Woo AJ, Tu N, Waldon Z, Lin YY, Steen H, Cantor AB (2009) Differentiation-dependent interactions between RUNX-1 and FLI-1 during megakaryocyte development. Mol Cell Biol 29(15):4103–4115. doi:10.1128/MCB.00090-09, doi:MCB.00090-09 [pii]
64. Huang X, Peng JW, Speck NA, Bushweller JH (1999) Solution structure of core binding factor beta and map of the CBF alpha binding site. Nat Struct Biol 6(7):624–627
65. Ichikawa M, Asai T, Saito T, Yamamoto G, Seo S, Yamazaki I, Yamagata T, Mitani K, Chiba S, Hirai H, Ogawa S, Kurokawa M (2004) AML-1 is required for megakaryocytic maturation and lymphocytic differentiation, but not for maintenance of hematopoietic stem cells in adult hematopoiesis. Nat Med 10(3):299–304
66. Ito Y, Bae SC, Chuang LS (2015) The RUNX family: developmental regulators in cancer. Nat Rev Cancer 15(2):81–95. doi:10.1038/nrc3877
67. Jacob B, Osato M, Yamashita N, Wang CQ, Taniuchi I, Littman DR, Asou N, Ito Y (2010) Stem cell exhaustion due to Runx1 deficiency is prevented by Evi5 activation in leukemogenesis. Blood 115(8):1610–1620
68. Junt T, Schulze H, Chen Z, Massberg S, Goerge T, Krueger A, Wagner DD, Graf T, Italiano JE Jr, Shivdasani RA, von Andrian UH (2007) Dynamic visualization of thrombopoiesis within bone marrow. Science 317(5845):1767–1770
69. Katsumura KR, Yang C, Boyer ME, Li L, Bresnick EH (2014) Molecular basis of crosstalk between oncogenic Ras and the master regulator of hematopoiesis GATA-2. EMBO Rep 15(9):938–947. doi:10.15252/embr.201438808

70. Kaur G, Jalagadugula G, Mao G, Rao AK (2010) RUNX1/core binding factor A2 regulates platelet 12-lipoxygenase gene (ALOX12): studies in human RUNX1 haplodeficiency. Blood 115(15):3128–3135
71. Kazenwadel J, Secker GA, Liu YJ, Rosenfeld JA, Wildin RS, Cuellar-Rodriguez J, Hsu AP, Dyack S, Fernandez CV, Chong CE, Babic M, Bardy PG, Shimamura A, Zhang MY, Walsh T, Holland SM, Hickstein DD, Horwitz MS, Hahn CN, Scott HS, Harvey NL (2012) Loss-of-function germline GATA2 mutations in patients with MDS/AML or MonoMAC syndrome and primary lymphedema reveal a key role for GATA2 in the lymphatic vasculature. Blood 119(5):1283–1291. doi:10.1182/blood-2011-08-374363
72. Kirito K, Fox N, Kaushansky K (2004) Thrombopoietin induces HOXA9 nuclear transport in immature hematopoietic cells: potential mechanism by which the hormone favorably affects hematopoietic stem cells. Mol Cell Biol 24(15):6751–6762
73. Klusmann JH, Creutzig U, Zimmermann M, Dworzak M, Jorch N, Langebrake C, Pekrun A, Macakova-Reinhardt K, Reinhardt D (2008) Treatment and prognostic impact of transient leukemia in neonates with Down syndrome. Blood 111(6):2991–2998
74. Klusmann JH, Godinho FJ, Heitmann K, Maroz A, Koch ML, Reinhardt D, Orkin SH, Li Z (2010) Developmental stage-specific interplay of GATA1 and IGF signaling in fetal megakaryopoiesis and leukemogenesis. Genes Dev 24(15):1659–1672. doi:10.1101/gad.1903410
75. Klusmann JH, Li Z, Bohmer K, Maroz A, Koch ML, Emmrich S, Godinho FJ, Orkin SH, Reinhardt D (2010) miR-125b-2 is a potential oncomiR on human chromosome 21 in megakaryoblastic leukemia. Genes Dev 24(5):478–490. doi:10.1101/gad.1856210
76. Kruse EA, Loughran SJ, Baldwin TM, Josefsson EC, Ellis S, Watson DK, Nurden P, Metcalf D, Hilton DJ, Alexander WS, Kile BT (2009) Dual requirement for the ETS transcription factors Fli-1 and Erg in hematopoietic stem cells and the megakaryocyte lineage. Proc Natl Acad Sci U S A 106(33):13814–13819. doi:10.1073/pnas.0906556106
77. Kuvardina ON, Herglotz J, Kolodziej S, Kohrs N, Herkt S, Wojcik B, Oellerich T, Corso J, Behrens K, Kumar A, Hussong H, Urlaub H, Koch J, Serve H, Bonig H, Stocking C, Rieger MA, Lausen J (2015) RUNX1 represses the erythroid gene expression program during megakaryocytic differentiation. Blood 125(23):3570–3579. doi:10.1182/blood-2014-11-610519
78. Kwiatkowski BA, Bastian LS, Bauer TR Jr, Tsai S, Zielinska-Kwiatkowska AG, Hickstein DD (1998) The ets family member Tel binds to the Fli-1 oncoprotein and inhibits its transcriptional activity. J Biol Chem 273(28):17525–17530
79. Kwiatkowski BA, Zielinska-Kwiatkowska AG, Bauer TR Jr, Hickstein DD (2000) The ETS family member Tel antagonizes the Fli-1 phenotype in hematopoietic cells. Blood Cells Mol Dis 26(1):84–90. doi:10.1006/bcmd.2000.0282
80. Lancrin C, Sroczynska P, Stephenson C, Allen T, Kouskoff V, Lacaud G (2009) The haemangioblast generates haematopoietic cells through a haemogenic endothelium stage. Nature 457(7231):892–895
81. Lange B (2000) The management of neoplastic disorders of haematopoiesis in children with Down's syndrome. Br J Haematol 110(3):512–524
82. Lemarchandel V, Ghysdael J, Mignotte V, Rahuel C, Romeo PH (1993) GATA and Ets cis-acting sequences mediate megakaryocyte-specific expression. Mol Cell Biol 13(1):668–676
83. Letting DL, Chen YY, Rakowski C, Reedy S, Blobel GA (2004) Context-dependent regulation of GATA-1 by friend of GATA-1. Proc Natl Acad Sci U S A 101(2):476–481
84. Li Z, Godinho FJ, Klusmann JH, Garriga-Canut M, Yu C, Orkin SH (2005) Developmental stage-selective effect of somatically mutated leukemogenic transcription factor GATA1. Nat Genet 37(6):613–619
85. Lu J, Guo S, Ebert BL, Zhang H, Peng X, Bosco J, Pretz J, Schlanger R, Wang JY, Mak RH, Dombkowski DM, Preffer FI, Scadden DT, Golub TR (2008) MicroRNA-mediated control of cell fate in megakaryocyte-erythrocyte progenitors. Dev Cell 14(6):843–853. doi:10.1016/j.devcel.2008.03.012
86. Lu J, Pazin MJ, Ravid K (2004) Properties of ets-1 binding to chromatin and its effect on platelet factor 4 gene expression. Mol Cell Biol 24(1):428–441

87. Ludlow LB, Schick BP, Budarf ML, Driscoll DA, Zackai EH, Cohen A, Konkle BA (1996) Identification of a mutation in a GATA binding site of the platelet glycoprotein Ibbeta promoter resulting in the Bernard-Soulier syndrome. J Biol Chem 271(36):22076–22080
88. Ludwig LS, Gazda HT, Eng JC, Eichhorn SW, Thiru P, Ghazvinian R, George TI, Gotlib JR, Beggs AH, Sieff CA, Lodish HF, Lander ES, Sankaran VG (2014) Altered translation of GATA1 in Diamond-Blackfan anemia. Nat Med 20(7):748–753. doi:10.1038/nm.3557
89. Lulli V, Romania P, Morsilli O, Gabbianelli M, Pagliuca A, Mazzeo S, Testa U, Peschle C, Marziali G (2006) Overexpression of Ets-1 in human hematopoietic progenitor cells blocks erythroid and promotes megakaryocytic differentiation. Cell Death Differ 13(7):1064–1074. doi:10.1038/sj.cdd.4401811
90. Maclean GA, Menne TF, Guo G, Sanchez DJ, Park IH, Daley GQ, Orkin SH (2012) Altered hematopoiesis in trisomy 21 as revealed through in vitro differentiation of isogenic human pluripotent cells. Proc Natl Acad Sci U S A 109(43):17567–17572. doi:10.1073/pnas.1215468109
91. Malinge S, Izraeli S, Crispino JD (2009) Insights into the manifestations, outcomes, and mechanisms of leukemogenesis in Down syndrome. Blood 113(12):2619–2628
92. Malinge S, Ragu C, Della-Valle V, Pisani D, Constantinescu SN, Perez C, Villeval JL, Reinhardt D, Landman-Parker J, Michaux L, Dastugue N, Baruchel A, Vainchenker W, Bourquin JP, Penard-Lacronique V, Bernard OA (2008) Activating mutations in human acute megakaryoblastic leukemia. Blood 112(10):4220–4226. doi:10.1182/blood-2008-01-136366
93. Malkin D, Brown EJ, Zipursky A (2000) The role of p53 in megakaryocyte differentiation and the megakaryocytic leukemias of Down syndrome. Cancer Genet Cytogenet 116(1):1–5
94. Massey GV, Zipursky A, Chang MN, Doyle JJ, Nasim S, Taub JW, Ravindranath Y, Dahl G, Weinstein HJ (2006) A prospective study of the natural history of transient leukemia (TL) in neonates with Down syndrome (DS): Children's Oncology Group (COG) study POG-9481. Blood 107(12):4606–4613
95. Mehaffey MG, Newton AL, Gandhi MJ, Crossley M, Drachman JG (2001) X-linked thrombocytopenia caused by a novel mutation of GATA-1. Blood 98(9):2681–2688
96. Mercher T, Cornejo MG, Sears C, Kindler T, Moore SA, Maillard I, Pear WS, Aster JC, Gilliland DG (2008) Notch signaling specifies megakaryocyte development from hematopoietic stem cells. Cell Stem Cell 3(3):314–326
97. Miccio A, Wang Y, Hong W, Gregory GD, Wang H, Yu X, Choi JK, Shelat S, Tong W, Poncz M, Blobel GA (2010) NuRD mediates activating and repressive functions of GATA-1 and FOG-1 during blood development. Embo J 29(2):442–456
98. Miller CA, Wilson RK, Ley TJ (2013) Genomic landscapes and clonality of de novo AML. N Engl J Med 369:1473
99. Monteferrario D, Bolar NA, Marneth AE, Hebeda KM, Bergevoet SM, Veenstra H, Laros-van Gorkom BA, MacKenzie MA, Khandanpour C, Botezatu L, Fransen E, Van Camp G, Duijnhouwer AL, Salemink S, Willemsen B, Huls G, Preijers F, Van Heerde W, Jansen JH, Kempers MJ, Loeys BL, Van Laer L, Van der Reijden BA (2014) A dominant-negative GFI1B mutation in the gray platelet syndrome. N Engl J Med 370(3):245–253. doi:10.1056/NEJMoa1308130
100. Mukai HY, Motohashi H, Ohneda O, Suzuki N, Nagano M, Yamamoto M (2006) Transgene insertion in proximity to the c-myb gene disrupts erythroid-megakaryocytic lineage bifurcation. Mol Cell Biol 26(21):7953–7965. doi:10.1128/MCB.00718-06
101. Mundschau G, Gurbuxani S, Gamis AS, Greene ME, Arceci RJ, Crispino JD (2003) Mutagenesis of GATA1 is an initiating event in Down syndrome leukemogenesis. Blood 101(11):4298–4300
102. Muramatsu H, Kato K, Watanabe N, Matsumoto K, Nakamura T, Horikoshi Y, Mimaya J, Suzuki C, Hayakawa M, Kojima S (2008) Risk factors for early death in neonates with Down syndrome and transient leukaemia. Br J Haematol 142(4):610–615. doi:10.1111/j.1365-2141.2008.07231.x
103. Nakao M, Horiike S, Fukushima-Nakase Y, Nishimura M, Fujita Y, Taniwaki M, Okuda T (2004) Novel loss-of-function mutations of the haematopoiesis-related transcription factor,

acute myeloid leukaemia 1/runt-related transcription factor 1, detected in acute myeloblastic leukaemia and myelodysplastic syndrome. Br J Haematol 125(6):709–719
104. Ng AP, Hu Y, Metcalf D, Hyland CD, Ierino H, Phipson B, Wu D, Baldwin TM, Kauppi M, Kiu H, Di Rago L, Hilton DJ, Smyth GK, Alexander WS (2015) Early lineage priming by trisomy of erg leads to myeloproliferation in a down syndrome model. PLoS Genet 11(5), e1005211. doi:10.1371/journal.pgen.1005211
105. Ng AP, Hyland CD, Metcalf D, Carmichael CL, Loughran SJ, Di Rago L, Kile BT, Alexander WS (2010) Trisomy of Erg is required for myeloproliferation in a mouse model of Down syndrome. Blood 115(19):3966–3969. doi:10.1182/blood-2009-09-242107
106. Nichols KE, Crispino JD, Poncz M, White JG, Orkin SH, Maris JM, Weiss MJ (2000) Familial dyserythropoietic anaemia and thrombocytopenia due to an inherited mutation in GATA1. Nat Genet 24(3):266–270
107. Noetzli L, Lo RW, Lee-Sherick AB, Callaghan M, Noris P, Savoia A, Rajpurkar M, Jones K, Gowan K, Balduini CL, Pecci A, Gnan C, De Rocco D, Doubek M, Li L, Lu L, Leung R, Landolt-Marticorena C, Hunger S, Heller P, Gutierrez-Hartmann A, Xiayuan L, Pluthero FG, Rowley JW, Weyrich AS, Kahr WH, Porter CC, Di Paola J (2015) Germline mutations in ETV6 are associated with thrombocytopenia, red cell macrocytosis and predisposition to lymphoblastic leukemia. Nat Genet. doi:10.1038/ng.3253
108. O'Neil J, Look AT (2007) Mechanisms of transcription factor deregulation in lymphoid cell transformation. Oncogene 26(47):6838–6849. doi:10.1038/sj.onc.1210766
109. Okada Y, Nagai R, Sato T, Matsuura E, Minami T, Morita I, Doi T (2003) Homeodomain proteins MEIS1 and PBXs regulate the lineage-specific transcription of the platelet factor 4 gene. Blood 101(12):4748–4756. doi:10.1182/blood-2002-02-0380
110. Onodera K, Shavit JA, Motohashi H, Yamamoto M, Engel JD (2000) Perinatal synthetic lethality and hematopoietic defects in compound mafG::mafK mutant mice. Embo J 19(6): 1335–1345
111. Ostergaard P, Simpson MA, Connell FC, Steward CG, Brice G, Woollard WJ, Dafou D, Kilo T, Smithson S, Lunt P, Murday VA, Hodgson S, Keenan R, Pilz DT, Martinez-Corral I, Makinen T, Mortimer PS, Jeffery S, Trembath RC, Mansour S (2011) Mutations in GATA2 cause primary lymphedema associated with a predisposition to acute myeloid leukemia (Emberger syndrome). Nat Genet 43(10):929–931. doi:10.1038/ng.923
112. Owen CJ, Toze CL, Koochin A, Forrest DL, Smith CA, Stevens JM, Jackson SC, Poon MC, Sinclair GD, Leber B, Johnson PR, Macheta A, Yin JA, Barnett MJ, Lister TA, Fitzgibbon J (2008) Five new pedigrees with inherited RUNX1 mutations causing familial platelet disorder with propensity to myeloid malignancy (FPD/AML). Blood 112:4639–4645
113. Ozaki T, Nakagawara A, Nagase H (2013) RUNX family participates in the regulation of p53-dependent DNA damage response. Int J Genom 2013:271347. doi:10.1155/2013/271347
114. Ozaki T, Wu D, Sugimoto H, Nagase H, Nakagawara A (2013) Runt-related transcription factor 2 (RUNX2) inhibits p53-dependent apoptosis through the collaboration with HDAC6 in response to DNA damage. Cell Death Dis 4, e610. doi:10.1038/cddis.2013.127
115. Pal S, Cantor AB, Johnson KD, Moran TB, Boyer ME, Orkin SH, Bresnick EH (2004) Coregulator-dependent facilitation of chromatin occupancy by GATA-1. Proc Natl Acad Sci U S A 101(4):980–985
116. Pang L, Xue HH, Szalai G, Wang X, Wang Y, Watson DK, Leonard WJ, Blobel GA, Poncz M (2006) Maturation stage-specific regulation of megakaryopoiesis by pointed-domain Ets proteins. Blood 108(7):2198–2206
117. Phillips JD, Steensma DP, Pulsipher MA, Spangrude GJ, Kushner JP (2007) Congenital erythropoietic porphyria due to a mutation in GATA-1: the first trans-acting mutation causative for a human porphyria. Blood 109(6):2618–2621
118. Porcher C, Swat W, Rockwell K, Fujiwara Y, Alt FW, Orkin SH (1996) The T cell leukemia oncoprotein SCL/tal-1 is essential for development of all hematopoietic lineages. Cell 86(1):47–57
119. Qin X, Jiang Q, Matsuo Y, Kawane T, Komori H, Moriishi T, Taniuchi I, Ito K, Kawai Y, Rokutanda S, Izumi S, Komori T (2015) Cbfb regulates bone development by stabilizing Runx family proteins. J Bone Miner Res 30(4):706–714. doi:10.1002/jbmr.2379

120. Rainis L, Bercovich D, Strehl S, Teigler-Schlegel A, Stark B, Trka J, Amariglio N, Biondi A, Muler I, Rechavi G, Kempski H, Haas OA, Izraeli S (2003) Mutations in exon 2 of GATA1 are early events in megakaryocytic malignancies associated with trisomy 21. Blood 102(3):981–986
121. Raskind WH, Niakan KK, Wolff J, Matsushita M, Vaughan T, Stamatoyannopoulos G, Watanabe C, Rios J, Ochs HD (2000) Mapping of a syndrome of X-linked thrombocytopenia with Thalassemia to band Xp11-12: further evidence of genetic heterogeneity of X-linked thrombocytopenia. Blood 95(7):2262–2268
122. Raslova H, Komura E, Le Couedic JP, Larbret F, Debili N, Feunteun J, Danos O, Albagli O, Vainchenker W, Favier R (2004) FLI1 monoallelic expression combined with its hemizygous loss underlies Paris-Trousseau/Jacobsen thrombopenia. J Clin Invest 114(1):77–84
123. Robert-Moreno A, Espinosa L, de la Pompa JL, Bigas A (2005) RBPjkappa-dependent Notch function regulates Gata2 and is essential for the formation of intra-embryonic hematopoietic cells. Development 132(5):1117–1126
124. Roberts I, Alford K, Hall G, Juban G, Richmond H, Norton A, Vallance G, Perkins K, Marchi E, McGowan S, Roy A, Cowan G, Anthony M, Gupta A, Ho J, Uthaya S, Curley A, Rasiah SV, Watts T, Nicholl R, Bedford-Russell A, Blumberg R, Thomas A, Gibson B, Halsey C, Lee PW, Godambe S, Sweeney C, Bhatnagar N, Goriely A, Campbell P, Vyas P (2013) GATA1-mutant clones are frequent and often unsuspected in babies with Down syndrome: identification of a population at risk of leukemia. Blood 122(24):3908–3917. doi:10.1182/blood-2013-07-515148
125. Roberts I, O'Connor D, Roy A, Cowan G, Vyas P (2013) The impact of trisomy 21 on foetal haematopoiesis. Blood Cells Mol Dis 51(4):277–281. doi:10.1016/j.bcmd.2013.07.008
126. Roy A, Cowan G, Mead AJ, Filippi S, Bohn G, Chaidos A, Tunstall O, Chan JK, Choolani M, Bennett P, Kumar S, Atkinson D, Wyatt-Ashmead J, Hu M, Stumpf MP, Goudevenou K, O'Connor D, Chou ST, Weiss MJ, Karadimitris A, Jacobsen SE, Vyas P, Roberts I (2012) Perturbation of fetal liver hematopoietic stem and progenitor cell development by trisomy 21. Proc Natl Acad Sci U S A 109(43):17579–17584. doi:10.1073/pnas.1211405109
127. Saleque S, Cameron S, Orkin SH (2002) The zinc-finger proto-oncogene Gfi-1b is essential for development of the erythroid and megakaryocytic lineages. Genes Dev 16(3):301–306
128. Saleque S, Kim J, Rooke HM, Orkin SH (2007) Epigenetic regulation of hematopoietic differentiation by Gfi-1 and Gfi-1b is mediated by the cofactors CoREST and LSD1. Mol Cell 27(4):562–572
129. Sandberg ML, Sutton SE, Pletcher MT, Wiltshire T, Tarantino LM, Hogenesch JB, Cooke MP (2005) c-Myb and p300 regulate hematopoietic stem cell proliferation and differentiation. Dev Cell 8(2):153–166. doi:10.1016/j.devcel.2004.12.015
130. Satoh Y, Matsumura I, Tanaka H, Harada H, Harada Y, Matsui K, Shibata M, Mizuki M, Kanakura Y (2012) C-terminal mutation of RUNX1 attenuates the DNA-damage repair response in hematopoietic stem cells. Leukemia 26(2):303–311. doi:10.1038/leu.2011.202
131. Schuh AH, Tipping AJ, Clark AJ, Hamlett I, Guyot B, Iborra FJ, Rodriguez P, Strouboulis J, Enver T, Vyas P, Porcher C (2005) ETO-2 associates with SCL in erythroid cells and megakaryocytes and provides repressor functions in erythropoiesis. Mol Cell Biol 25(23):10235–10250. doi:10.1128/MCB.25.23.10235-10250.2005
132. Shivdasani RA, Fujiwara Y, McDevitt MA, Orkin SH (1997) A lineage-selective knockout establishes the critical role of transcription factor GATA-1 in megakaryocyte growth and platelet development. Embo J 16(13):3965–3973
133. Shivdasani RA, Rosenblatt MF, Zucker-Franklin D, Jackson CW, Hunt P, Saris CJ, Orkin SH (1995) Transcription factor NF-E2 is required for platelet formation independent of the actions of thrombopoietin/MGDF in megakaryocyte development. Cell 81(5):695–704
134. Song WJ, Sullivan MG, Legare RD, Hutchings S, Tan X, Kufrin D, Ratajczak J, Resende IC, Haworth C, Hock R, Loh M, Felix C, Roy DC, Busque L, Kurnit D, Willman C, Gewirtz AM, Speck NA, Bushweller JH, Li FP, Gardiner K, Poncz M, Maris JM, Gilliland DG (1999) Haploinsufficiency of CBFA2 causes familial thrombocytopenia with propensity to develop acute myelogenous leukaemia. Nat Genet 23(2):166–175

135. Spinner MA, Sanchez LA, Hsu AP, Shaw PA, Zerbe CS, Calvo KR, Arthur DC, Gu W, Gould CM, Brewer CC, Cowen EW, Freeman AF, Olivier KN, Uzel G, Zelazny AM, Daub JR, Spalding CD, Claypool RJ, Giri NK, Alter BP, Mace EM, Orange JS, Cuellar-Rodriguez J, Hickstein DD, Holland SM (2014) GATA2 deficiency: a protean disorder of hematopoiesis, lymphatics, and immunity. Blood 123(6):809–821. doi:10.1182/blood-2013-07-515528
136. Stankiewicz MJ, Crispino JD (2009) ETS2 and ERG promote megakaryopoiesis and synergize with alterations in GATA-1 to immortalize hematopoietic progenitor cells. Blood 113(14):3337–3347. doi:10.1182/blood-2008-08-174813
137. Stankiewicz MJ, Crispino JD (2013) AKT collaborates with ERG and Gata1s to dysregulate megakaryopoiesis and promote AMKL. Leukemia 27(6):1339–1347. doi:10.1038/leu.2013.33
138. Starck J, Cohet N, Gonnet C, Sarrazin S, Doubeikovskaia Z, Doubeikovski A, Verger A, Duterque-Coquillaud M, Morle F (2003) Functional cross-antagonism between transcription factors FLI-1 and EKLF. Mol Cell Biol 23(4):1390–1402
139. Starck J, Weiss-Gayet M, Gonnet C, Guyot B, Vicat JM, Morle F (2010) Inducible Fli-1 gene deletion in adult mice modifies several myeloid lineage commitment decisions and accelerates proliferation arrest and terminal erythrocytic differentiation. Blood 116(23):4795–4805. doi:10.1182/blood-2010-02-270405
140. Stevenson WS, Morel-Kopp MC, Chen Q, Liang HP, Bromhead CJ, Wright S, Turakulov R, Ng AP, Roberts AW, Bahlo M, Ward CM (2013) GFI1B mutation causes a bleeding disorder with abnormal platelet function. J Thromb Haemost JTH 11(11):2039–2047. doi:10.1111/jth.12368
141. Stevenson WS, Rabbolini DJ, Beutler L, Chen Q, Gabrielli S, Mackay JP, Brighton TA, Ward CM, Morel-Kopp MC (2015) Paris-Trousseau thrombocytopenia is phenocopied by the autosomal recessive inheritance of a DNA-binding domain mutation in FLI1. Blood 126(17):2027–2030. doi:10.1182/blood-2015-06-650887
142. Stockley J, Morgan NV, Bem D, Lowe GC, Lordkipanidze M, Dawood B, Simpson MA, Macfarlane K, Horner K, Leo VC, Talks K, Motwani J, Wilde JT, Collins PW, Makris M, Watson SP, Daly ME (2013) Enrichment of FLI1 and RUNX1 mutations in families with excessive bleeding and platelet dense granule secretion defects. Blood 122(25):4090–4093. doi:10.1182/blood-2013-06-506873
143. Sun L, Gorospe JR, Hoffman EP, Rao AK (2007) Decreased platelet expression of myosin regulatory light chain polypeptide (MYL9) and other genes with platelet dysfunction and CBFA2/RUNX1 mutation: insights from platelet expression profiling. J Thromb Haemost 5(1):146–154
144. Sun L, Mao G, Rao AK (2004) Association of CBFA2 mutation with decreased platelet PKC-theta and impaired receptor-mediated activation of GPIIb-IIIa and pleckstrin phosphorylation: proteins regulated by CBFA2 play a role in GPIIb-IIIa activation. Blood 103(3):948–954
145. Talebian L, Li Z, Guo Y, Gaudet J, Speck ME, Sugiyama D, Kaur P, Pear WS, Maillard I, Speck NA (2007) T-lymphoid, megakaryocyte, and granulocyte development are sensitive to decreases in CBFbeta dosage. Blood 109(1):11–21
146. Tanaka T, Kurokawa M, Ueki K, Tanaka K, Imai Y, Mitani K, Okazaki K, Sagata N, Yazaki Y, Shibata Y, Kadowaki T, Hirai H (1996) The extracellular signal-regulated kinase pathway phosphorylates AML1, an acute myeloid leukemia gene product, and potentially regulates its transactivation ability. Mol Cell Biol 16(7):3967–3979
147. Tang JL, Hou HA, Chen CY, Liu CY, Chou WC, Tseng MH, Huang CF, Lee FY, Liu MC, Yao M, Huang SY, Ko BS, Hsu SC, Wu SJ, Tsay W, Chen YC, Lin LI, Tien HF (2009) AML1/RUNX1 mutations in 470 adult patients with de novo acute myeloid leukemia: prognostic implication and interaction with other gene alterations. Blood 114:5352–5361
148. Taub JW, Mundschau G, Ge Y, Poulik JM, Qureshi F, Jensen T, James SJ, Matherly LH, Wechsler J, Crispino JD (2004) Prenatal origin of GATA1 mutations may be an initiating step in the development of megakaryocytic leukemia in Down syndrome. Blood 104(5):1588–1589. doi:10.1182/blood-2004-04-1563
149. Thompson AA, Nguyen LT (2000) Amegakaryocytic thrombocytopenia and radio-ulnar synostosis are associated with HOXA11 mutation. Nat Genet 26(4):397–398. doi:10.1038/82511

150. Tijssen MR, Cvejic A, Joshi A, Hannah RL, Ferreira R, Forrai A, Bellissimo DC, Oram SH, Smethurst PA, Wilson NK, Wang X, Ottersbach K, Stemple DL, Green AR, Ouwehand WH, Gottgens B (2011) Genome-wide analysis of simultaneous GATA1/2, RUNX1, FLI1, and SCL binding in megakaryocytes identifies hematopoietic regulators. Dev Cell 20(5):597–609. doi:10.1016/j.devcel.2011.04.008
151. Topka S, Vijai J, Walsh MF, Jacobs L, Maria A, Villano D, Gaddam P, Wu G, McGee RB, Quinn E, Inaba H, Hartford C, Pui CH, Pappo A, Edmonson M, Zhang MY, Stepensky P, Steinherz P, Schrader K, Lincoln A, Bussel J, Lipkin SM, Goldgur Y, Harit M, Stadler ZK, Mullighan C, Weintraub M, Shimamura A, Zhang J, Downing JR, Nichols KE, Offit K (2015) Germline ETV6 mutations confer susceptibility to acute lymphoblastic leukemia and thrombocytopenia. PLoS Genet 11(6), e1005262. doi:10.1371/journal.pgen.1005262
152. Tsai FY, Keller G, Kuo FC, Weiss M, Chen J, Rosenblatt M, Alt FW, Orkin SH (1994) An early haematopoietic defect in mice lacking the transcription factor GATA-2. Nature 371(6494):221–226
153. Tsai SF, Martin DI, Zon LI, D'Andrea AD, Wong GG, Orkin SH (1989) Cloning of cDNA for the major DNA-binding protein of the erythroid lineage through expression in mammalian cells. Nature 339(6224):446–451
154. Tsang AP, Fujiwara Y, Hom DB, Orkin SH (1998) Failure of megakaryopoiesis and arrested erythropoiesis in mice lacking the GATA-1 transcriptional cofactor FOG. Genes Dev 12(8):1176–1188
155. Tsang AP, Visvader JE, Turner CA, Fujiwara Y, Yu C, Weiss MJ, Crossley M, Orkin SH (1997) FOG, a multitype zinc finger protein, acts as a cofactor for transcription factor GATA-1 in erythroid and megakaryocytic differentiation. Cell 90(1):109–119
156. Tubman VN, Levine JE, Campagna DR, Monahan-Earley R, Dvorak AM, Neufeld EJ, Fleming MD (2007) X-linked gray platelet syndrome due to a GATA1 Arg216Gln mutation. Blood 109(8):3297–3299. doi:10.1182/blood-2006-02-004101
157. Tunstall-Pedoe O, Roy A, Karadimitris A, de la Fuente J, Fisk NM, Bennett P, Norton A, Vyas P, Roberts I (2008) Abnormalities in the myeloid progenitor compartment in Down syndrome fetal liver precede acquisition of GATA1 mutations. Blood 112(12):4507–4511
158. Vannucchi AM, Bianchi L, Cellai C, Paoletti F, Rana RA, Lorenzini R, Migliaccio G, Migliaccio AR (2002) Development of myelofibrosis in mice genetically impaired for GATA-1 expression (GATA-1(low) mice). Blood 100(4):1123–1132
159. Visvader JE, Fujiwara Y, Orkin SH (1998) Unsuspected role for the T-cell leukemia protein SCL/tal-1 in vascular development. Genes Dev 12(4):473–479
160. Vyas P, Ault K, Jackson CW, Orkin SH, Shivdasani RA (1999) Consequences of GATA-1 deficiency in megakaryocytes and platelets. Blood 93(9):2867–2875
161. Wadman IA, Osada H, Grutz GG, Agulnick AD, Westphal H, Forster A, Rabbitts TH (1997) The LIM-only protein Lmo2 is a bridging molecule assembling an erythroid, DNA-binding complex which includes the TAL1, E47, GATA-1 and Ldb1/NLI proteins. Embo J 16(11):3145–3157
162. Walters DK, Mercher T, Gu TL, O'Hare T, Tyner JW, Loriaux M, Goss VL, Lee KA, Eide CA, Wong MJ, Stoffregen EP, McGreevey L, Nardone J, Moore SA, Crispino J, Boggon TJ, Heinrich MC, Deininger MW, Polakiewicz RD, Gilliland DG, Druker BJ (2006) Activating alleles of JAK3 in acute megakaryoblastic leukemia. Cancer Cell 10(1):65–75. doi:10.1016/j.ccr.2006.06.002
163. Wang Q, Stacy T, Binder M, Marin-Padilla M, Sharpe AH, Speck NA (1996) Disruption of the Cbfa2 gene causes necrosis and hemorrhaging in the central nervous system and blocks definitive hematopoiesis. Proc Natl Acad Sci U S A 93(8):3444–3449
164. Wang X, Crispino JD, Letting DL, Nakazawa M, Poncz M, Blobel GA (2002) Control of megakaryocyte-specific gene expression by GATA-1 and FOG-1: role of Ets transcription factors. Embo J 21(19):5225–5234
165. Wechsler J, Greene M, McDevitt MA, Anastasi J, Karp JE, Le Beau MM, Crispino JD (2002) Acquired mutations in GATA1 in the megakaryoblastic leukemia of Down syndrome. Nat Genet 32(1):148–152
166. Weiss MJ, Yu C, Orkin SH (1997) Erythroid-cell-specific properties of transcription factor GATA-1 revealed by phenotypic rescue of a gene-targeted cell line. Mol Cell Biol 17(3):1642–1651

167. White JG, Nichols WL, Steensma DP (2007) Platelet pathology in sex-linked GATA-1 dyserythropoietic macrothrombocytopenia I ultrastructure. Platelets 18(4):273–283. doi:10.1080/09537100601065825
168. White JG, Nichols WL, Steensma DP (2007) Platelet pathology in sex-linked GATA-1 dyserythropoietic macrothrombocytopenia II. Cytochemistry. Platelets 18(6):436–450. doi:10.1080/09537100701280662
169. Wilson NK, Foster SD, Wang X, Knezevic K, Schutte J, Kaimakis P, Chilarska PM, Kinston S, Ouwehand WH, Dzierzak E, Pimanda JE, de Bruijn MF, Gottgens B (2010) Combinatorial transcriptional control in blood stem/progenitor cells: genome-wide analysis of ten major transcriptional regulators. Cell Stem Cell 7(4):532–544. doi:10.1016/j.stem.2010.07.016
170. Woo AJ, Wieland K, Huang H, Akie TE, Piers T, Kim J, Cantor AB (2013) Developmental differences in IFN signaling affect GATA1s-induced megakaryocyte hyperproliferation. J Clin Invest 123:3292–3304. doi:10.1172/JCI40609
171. Wu D, Ozaki T, Yoshihara Y, Kubo N, Nakagawara A (2013) Runt-related transcription factor 1 (RUNX1) stimulates tumor suppressor p53 protein in response to DNA damage through complex formation and acetylation. J Biol Chem 288(2):1353–1364. doi:10.1074/jbc.M112.402594
172. Xu G, Nagano M, Kanezaki R, Toki T, Hayashi Y, Taketani T, Taki T, Mitui T, Koike K, Kato K, Imaizumi M, Sekine I, Ikeda Y, Hanada R, Sako M, Kudo K, Kojima S, Ohneda O, Yamamoto M, Ito E (2003) Frequent mutations in the GATA-1 gene in the transient myeloproliferative disorder of Down syndrome. Blood 102(8):2960–2968
173. Yoshida K, Toki T, Okuno Y, Kanezaki R, Shiraishi Y, Sato-Otsubo A, Sanada M, Park MJ, Terui K, Suzuki H, Kon A, Nagata Y, Sato Y, Wang R, Shiba N, Chiba K, Tanaka H, Hama A, Muramatsu H, Hasegawa D, Nakamura K, Kanegane H, Tsukamoto K, Adachi S, Kawakami K, Kato K, Nishimura R, Izraeli S, Hayashi Y, Miyano S, Kojima S, Ito E, Ogawa S (2013) The landscape of somatic mutations in Down syndrome-related myeloid disorders. Nat Genet 45(11):1293–1299. doi:10.1038/ng.2759
174. Yoshimi A, Toya T, Kawazu M, Ueno T, Tsukamoto A, Iizuka H, Nakagawa M, Nannya Y, Arai S, Harada H, Usuki K, Hayashi Y, Ito E, Kirito K, Nakajima H, Ichikawa M, Mano H, Kurokawa M (2014) Recurrent CDC25C mutations drive malignant transformation in FPD/AML. Nat Commun 5:4770. doi:10.1038/ncomms5770
175. Yu C, Niakan KK, Matsushita M, Stamatoyannopoulos G, Orkin SH, Raskind WH (2002) X-linked thrombocytopenia with thalassemia from a mutation in the amino finger of GATA-1 affecting DNA binding rather than FOG-1 interaction. Blood 100(6):2040–2045
176. Zhang L, Fried FB, Guo H, Friedman AD (2008) Cyclin-dependent kinase phosphorylation of RUNX1/AML1 on 3 sites increases transactivation potency and stimulates cell proliferation. Blood 111(3):1193–1200. doi:10.1182/blood-2007-08-109702
177. Zhang MY, Churpek JE, Keel SB, Walsh T, Lee MK, Loeb KR, Gulsuner S, Pritchard CC, Sanchez-Bonilla M, Delrow JJ, Basom RS, Forouhar M, Gyurkocza B, Schwartz BS, Neistadt B, Marquez R, Mariani CJ, Coats SA, Hofmann I, Lindsley RC, Williams DA, Abkowitz JL, Horwitz MS, King MC, Godley LA, Shimamura A (2015) Germline ETV6 mutations in familial thrombocytopenia and hematologic malignancy. Nat Genet 47(2):180–185. doi:10.1038/ng.3177
178. Zhang SJ, Ma LY, Huang QH, Li G, Gu BW, Gao XD, Shi JY, Wang YY, Gao L, Cai X, Ren RB, Zhu J, Chen Z, Chen SJ (2008) Gain-of-function mutation of GATA-2 in acute myeloid transformation of chronic myeloid leukemia. Proc Natl Acad Sci U S A 105(6):2076–2081. doi:10.1073/pnas.0711824105
179. Zhao X, Jankovic V, Gural A, Huang G, Pardanani A, Menendez S, Zhang J, Dunne R, Xiao A, Erdjument-Bromage H, Allis CD, Tempst P, Nimer SD (2008) Methylation of RUNX1 by PRMT1 abrogates SIN3A binding and potentiates its transcriptional activity. Genes Dev 22(5):640–653
180. Zharlyganova D, Harada H, Harada Y, Shinkarev S, Zhumadilov Z, Zhunusova A, Tchaizhunusova NJ, Apsalikov KN, Kemaikin V, Zhumadilov K, Kawano N, Kimura A, Hoshi

M (2008) High frequency of AML1/RUNX1 point mutations in radiation-associated myelodysplastic syndrome around Semipalatinsk nuclear test site. J Radiat Res 49(5):549–555
181. Zipursky A (2003) Transient leukaemia – a benign form of leukaemia in newborn infants with trisomy 21. Br J Haematol 120(6):930–938
182. Zon LI, Yamaguchi Y, Yee K, Albee EA, Kimura A, Bennett JC, Orkin SH, Ackerman SJ (1993) Expression of mRNA for the GATA-binding proteins in human eosinophils and basophils: potential role in gene transcription. Blood 81(12):3234–3241

Impact of the Megakaryocytic Vascular Niche on Platelet Biogenesis

4

Harald Schulze, Daniela Semeniak, and Alessandra Balduini

Abstract

The biogenesis of blood platelets is a process referred to as thrombopoiesis. Platelets are derived from large precursor cells residing in the bone marrow and are designated megakaryocytes. Mature megakaryocytes are mostly located juxtaposed to bone marrow sinusoids where they release long protrusions across the endothelial barrier into the blood stream where the final platelet biogenesis and sizing occurs. So far, it has remained unclear how bone marrow matrix proteins and cell-cell contacts at the endothelial influence the maturation, migration and terminal differentiation of megakaryocytes and prevent premature platelet release into the bone marrow cavity. Here, we provide the current concepts of a vascular niche for megakaryocytes and discuss relevant components of the microenvironment for platelet biogenesis. This includes matrix proteins fibronectin, vitronectin, collagens, and laminins and their receptors. In addition, we discuss the impact of chemokines, diffusable gases and lipids as well as altered conditions in extramedullary thrombopoiesis and malignancy. Finally, we discuss how platelet bioreactors can mimick this process to study the underlying mechanisms ex vivo.

H. Schulze (✉) • D. Semeniak
Experimental Biomedicine, University Hospital Würzburg, Josef-Schneider-Str. 2, D15, 97080 Würzburg, Germany
e-mail: harald.schulze@uni-wuerzburg.de

A. Balduini
Department of Molecular Medicine, University of Pavia, via Forlanini 6, 27100 Pavia, Italy

Department of Biomedical Engineering, Tufts University, 4 Colby Street, Medford, MA 02155, Boston, USA

H. Schulze, J. Italiano (eds.), *Molecular and Cellular Biology of Platelet Formation*, DOI 10.1007/978-3-319-39562-3_4

4.1 Introduction: Megakaryocytes and Niches

The biogenesis of blood platelets is a process referred to as thrombopoiesis. For more than 100 years, it has been known that they are derived from large cells residing in the bone marrow which have been coined megakaryocytes [117]. Despite being a rare population among the nucleated cells, they are readily recognized by their size, and their maturation stages have been identified by staining with specific dyes. Bone marrow biopsies provide a cell suspension that can be subjected to smears allowing quantitative and qualitative, mostly morphological, evaluation. However, the exact localization of megakaryocytes within the bone marrow, especially in respect to their maturation from hematopoietic stem and progenitor cells (HSC), their migration to the sinusoids, and their terminal release of platelets across the endothelial layer are still poorly understood.

With an increase in understanding that stem cells reside at specific compartments within the bone marrow, this microenvironment has been named "niche," originally in a seminal paper by Schofield [100] and since then evolved over more than three decades [64]. While this term has immediate connotative implications of an architectonic element, in respect to a biological specimen, "niche" can readily be understood as an "anatomic" structure. For the stem cell population, this has often been recognized as an "osteoblastic" or "endosteal" niche, emphasizing that at least one subpopulation of HSC resides at this structure, while other subpopulations are found perivascular or at the arteriolar niche [51, 55]. In general, almost all HSC are in contact with perivascular cells that produce CXCL12 and are thus referred to as *CXCL12-abundant reticular* (CAR) cells [105]. However, the microenvironment for stem cells might also be rather dynamic, allowing a quick response to damage or an increased demand of progenitor cells. Work in axolotl has clearly shown that limb amputation can generate de novo sites of (tissue) stem cells, induced or generated at the site of injury. "Niche" might therefore also mean that a certain surrounding of cell-cell, cell-matrix, and humoral factors (in addition to factors like interstitial pressure or oxygen pressure) provides the setting for any specific niche. This does clearly apply to the spatiotemporal changes within the developing organism.

For many cell types, characterization of such a "niche" might be of interest to better understand their natural habitat. For the liquid organ "blood," the blood-bone marrow barrier is the fundamental frontier between the circulating blood and the compartment generating these cells from hematopoietic precursors. The outer lining of sinusoids in bone marrow, an overall discontinuous endothelium, has been recognized as the primary site for mature blood cells before they enter the blood stream. This site embedded by endothelial cells forming the sinusoids and its subendothelial matrix composition, comprising collagens (mostly type IV), laminin, and cross-linking compounds like nidogen or perlecan, provides the basis with an overall normoxic oxygen condition.

Most mature megakaryocytes (MKs) line up to this site and are thus present at this "vascular niche." A recent review provides an extended overview about factors that influence MK maturation [65]. This chapter summarizes information on how MK maturation and platelet biogenesis occurs within the bone marrow. Here, we aim to provide current concepts that direct our thinking about this unidirectional maturation and that leads to the generation of hundreds of virtually identical blood platelets from a single MK.

4.2 Hematopoietic Niches

Osteoblastic cells are present at the lining of the endosteal surface and found to produce and release a variety of cytokines that might influence hematopoietic stem cells. In addition, transplantation experiments have implied that labeled HSCs tend to home at the sites near the endosteum, thus concluding that this site might be a niche for HSC. The importance of osteoblasts in regulating HSCs has been shown in elegant studies [52] (and references within), suggesting rather HSC maintenance than their formation. Osteoblasts express osteopontin, a glycoprotein that limits the amount of HSC in the niche and angiopoietin 1, that acts on its receptor Tie2 on HSC to support the long-term repopulating capacity. Ca^{2+} ions released by bone-degrading osteoclasts and sensed by G-protein-coupled receptors maintain HSC at the endosteal bone lining [3]. At least a subset of HSC express c-Mpl, the receptor for thrombopoietin which is required to bring quiescent stem cells into cycling, in addition to their well-known function as a lineage-specific cytokine for the megakaryocytic lineage. Patients lacking c-Mpl or having compound mutations that abrogate its function are typically born with a severe isolated thrombocytopenia (referred to as congenital amegakarocytic thrombocytopenia, CAMT) [7], but later develop into pancytopenia due to an exhausted stem cell pool that definitely requires hematopoietic stem cell transplantation as the only curative therapy [29] in addition to gene therapy (--> Chap. 20) [113].

However, a subset of HSCs must be associated with blood vessels in the bone marrow, as cytokines can induce a rapid transition of HSCs into the circulation [38, 58], these sinusoidal endothelial cells could maintain a HSC niche in extra-medullary organs (like spleen, liver, lung) where there are no osteoblasts [111]. More recent evidence comes from studies showing that mobilized HSCs originate from both bone and vascular niches [13]. In humans, mobilization of HSC is mostly induced by G-CSF prior to apheresis of CD34-positive cells [33]. In addition, CXCR4 antagonists are applied to support the mobilization, indicating that CXCL12 (SDF-1) plays a pivotal role in keeping the HSC at their niche within the marrow [20, 114].

4.2.1 Components of the Microenvironment

In a multicellular organism with organs, specialized tissues, and cells, any cells of interest are somehow embedded in a variety of factors that ultimately might define their microenvironment. They comprise cell-cell and cell-matrix interactions, the presence of humoral factors (like cytokines or chemokines), or lipids. Finally, the oxygen concentration and the interstitial pressure influence this environment.

For HSCs (or differentiating MKs), this is obviously the extracellular matrix which provides the physical scaffolding. Structural proteins include collagens, laminins, fibronectin, as well as nidogen or perlecan. The most relevant for MK maturation and platelet biogenesis are described in more detail below. At the endosteum, HSCs reside in the osteoblastic niche where osteopontin is considered a marker protein. There has been evidence that MKs can influence HSCs indirectly, as these home preferentially to MKs in irradiated mice and their engraftment is impaired when MKs are inhibited [37, 84]. More recently, two studies have shown that MKs have also a direct influence on HSC proliferation [14, 122]: Loss of CXCL4 (PF4) expressing MKs results in increased stem cell cycling and numbers, while FGF1 produced by MKs enhances the recovery of HSC in mice after myeloablation. These findings imply that under steady-state conditions, MKs produce factors that fine-tune HSC proliferation.

At the vascular niche, the subendothelial layer has to be remodeled, and collagen filaments are digested by matrix metalloproteases MMP2 and MMP9 [57]. These enzymes are specifically activated downstream of concentrated, specialized actin structures, referred to as podosomes and nodules [90, 96]. These structures are mainly found in vitro, when MKs are plated on matrix components, but they might indeed be challenging to visualize in situ during the transendothelial passage. This polarization is essential to prevent that proplatelet formation occurs within the marrow cavity. The lipid composition of the membrane seems to contribute to this process: the efflux of cholesterol in MKs is a negative regulator for platelet biogenesis [77].

4.2.2 Cell-Cell Contacts

Megakaryocytic cells are in direct contact with other cells, either by direct cell-cell contact or by indirect, paracrine mechanisms. Some of the guidance and repulsion proteins have been identified: Adiponectin is considered a negative regulator, possibly acting only on very early progenitor cells [106]. Semaphorin 7A (Sem7A) acts also negatively on the differentiation of CD34-positive HSC and early progenitors to MKs, and colony formation is reduced in the presence of Sem7A. This effect is mediated by interaction with integrin β1 and also of clinical importance, as Sem7A is upregulated in about half of patients undergoing chemotherapy, which might explain the late engraftment of megakaryocytic progenitors [41]. In addition, the involvement of Wnt and Notch signaling has been shown [64]. When murine MKs are plated on the stromal cell line OP9 expressing the Notch ligand Delta-1, proplatelet formation is highly upregulated [72]. During final thrombopoiesis, MKs will reach the endothelial layers of the bone marrow sinusoids where they interact with VE-cadherins [5].

MKs also express receptor activator of nuclear factor kappa B ligand (RANKL) and modulate the osteoblast and osteoclast proliferation and differentiation. MKs thus have a direct, but complex impact on the metabolism of bone structure and density [44–46].

4.3 The Term "Vascular Niche"

As indicated above, MKs and their progenitors are distributed throughout the hematopoietic bone marrow tissues. Any approach to define a population of MKs requires referral to anatomic structures, histological positioning, or immunohistochemical stainings in a complex tissue. More recent advances have come from novel cutting techniques of bones, protocols that do not require decalcification, as well as substantially improved microscopy including secondary antibodies conjugated to next-generation fluorophores. The structure in mouse femur including the visualization of niches within the bone marrow has recently been published in an extensive study [107]. Analysis of single (or at its best a series of consecutive) sections of one bone further restricts any option to analyze dynamic processes but solely allows us to quantify MK numbers or distances to other cells or vessels, always assuming that preparation and staining artifacts will not contribute significantly to skew results. The concept and the localization of a HSC niche are subject of many reviews [24, 49, 64, 74] and are not further outlined here. Microanatomic locations have been described for bone or endosteal niche [25] and are often set in opposite to a "vascular niche" which is somehow close to vessels [50, 105] in [13]. This term has thus been used to describe cells that are associated with the vasculature, typically with sinusoids. Although this approach seems overall feasible, there is still debate how the relevant vessels need to be identified. Endoglin (CD105) is a coreceptor for TGF-beta and considered a bona fide marker for sinusoid vessels [108], where it is expressed in endothelial cells. However, it is also present in small cells within the marrow and might mark a subset of mesenchymal stem cells [9]. CD105 is absent from MKs and colocalization found in one study might be due to staining artifacts [89]. VEGFR3 has also been used to stain sinusoids [66, 92]. Best evidence might come from costaining of any marker with PECAM (CD31) or laminins that should lead to a substantial and meaningful overlap of signals. The megakaryocytic "vascular niche" as used in this chapter refers to the microenvironment that is juxtaposed or in close vicinity to the extraluminal site of these vessels. As other markers that might further refine this niche are up to now not yet identified, MKs that show direct contact to a vessel might be considered at this vascular niche. As the average distance between two sinusoids in femur sections is about 40–50 μm with MKs being 20–30 μm in diameter, MKs that are without vessel contact might still be in this niche, while we will refer to them as "cavity" or interluminal MKs.

4.3.1 Considering a "Malignant Niche" or a "Vascular Niche" Post Irradiation

Many medical (pathological as well as iatrogenic) conditions have a major impact on the bone marrow. Chemotherapy (i.e., drugs like 5-fluorouracil) as well as total body irradiation alter the bone marrow cellularity, the extracellular matrix protein composition, as well as the microvasculature [30, 32]. Reports describe that MKs can secrete matrix proteins and this secretion might be upregulated in response to irradiation or post injury [60, 66, 79, 122]. Typically, bone marrow sinusoids start to dilate immediately after lethal or sublethal irradiation [84]. These conditions are complex and discussed elsewhere [91].

4.4 Migration of Megakaryocytes: Gradients in the Bone Marrow

Hematopoietic stem and progenitor cells reside in specific microenvironments where they contribute to constitutive hematopoiesis as well as on the production of specific cell lines on demand. In addition, these cells are capable of leaving their tissue and migrate to other tissues. These have been shown by Massberg and colleagues demonstrating that hematopoietic stem cells contribute to immunosurveillance and follow certain routes including the lymphatic vessels [43]. This circulation throughout the body clearly requires that cells migrate within the bone marrow to enter the blood stream somewhere across the endothelial lining of the blood-bone marrow barrier. Migration of cells is readily visualized by lymphocytes in lymph nodes or in the bone marrow where they crawl in and out of vessels. This is typically performed by intravital microscopy and has been facilitated by two- or multiphoton microscopy. Migration of other stem and progenitor cells is more difficult to address and typically only deduced indirectly. Often cell numbers in two distinct compartments are counted at two different time points, and absolute or relative numbers are determined and the difference translated into a number of "migrating cells." MKs are overall sessile cells in the marrow when compared to lymphocytes [43], and migration of MKs has been analyzed from a proliferative osteoblastic niche to the capillary-rich vascular niche [20, 86]. Seminal work by Rafii's group has shown this in the context of the humoral factors FGF-4 and CXCL12 that drive MKs toward the vascular niche and across the sinusoids into the blood stream [5]. Their major finding was that these drivers of migration are active even in the absence of the thrombopoietin (TPO)-c-Mpl axis and that coinjection of both chemokines can transiently restore platelet counts to normal in mice lacking either TPO or its receptor. CXCL12 also dynamically mediates MK niche occupancy and thrombopoiesis [83]. This chemokine is expressed in highly levels by specialized CAR cells (CXCL12-abundant reticular cells) located at the sinusoids [87].

As mentioned above, the bone marrow endothelial barrier is discontinuous, but the endothelial cells still have interjunctional adhesion molecules, of which the VE-cadherins are prominent. Application of antibodies to VE-cadherins in myeloablative mice disrupted the vascular niche and reduced the number of megakaryocytic cells and interfered with the CXCL12/FGF4-mediated thrombopoiesis [5, 22].

There are many more cells that are now recognized to interact directly or indirectly with MKs. One of the first reports was the cell-cell interaction of MKs with plasma B-cells suggesting that MKs support their niche [115]. The impact of the B-cell lineage was further corroborated by demonstration that B-cell-derived growth factor (BAFF, APRIL) expression within the megakaryocytic lineage increases during their differentiation and mutually support thriving of both lineages [11, 12, 62, 116].

4.4.1 Humoral Factors and Gradients

The maturation of HSCs to MKs and platelets is predominantly regulated by thrombopoietin, a cytokine that primarily expands the stem cell pool and leads to differentiation into the megakaryocytic lineage. Other cytokines, including interleukin (IL)-3, IL-6, or IL-11, can partly contribute, but they are overall much weaker, even when combined. MKs have several chemokine receptors on their surface, among which CXCR4 is expressed abundantly. CXCL12, the main ligand for CXCR4, shares many signaling events with TPO but has also distinct functions [34]. While fibroblast-growth factor 4 (FGF-4) pushes the migration of megakaryocytic precursor cells to the endothelial layer, CXCL12 supports their transendothelial passage. The final step of platelet biogenesis might be directed by a sphingosine phosphate, a lipid metabolite whose expression is low in the bone marrow but enriched in the blood stream. Sphingosine phosphates are described to be global directors for the egression of lymphocytes [31], and recently it has been shown that MKs express the corresponding sphingosine receptors SPR1 and SPR2. Two studies from Steffen Massberg's group provided elegant evidence that this gradient contributes to both the release of preplatelets and proplatelets into the circulation as well as their disintegration to platelets in the blood stream [55, 121], as postulated before [10].

Finally, hypoxic conditions curb the maturation of MKs and reduce proplatelet formation, suggesting that normoxic conditions at the vascular niche support the final release of proplatelets [76]. The mechanisms and receptors mediating this process are still mostly unknown, as the impact of distinct reactive oxygen species (ROS) on MK maturation and platelet biogenesis leads to activation of PI3K, ROS-p38 MAPK, and the FOXO3 transcription factor [16, 23].

4.5 Matrix Proteins and Megakaryocytes

HSCs have the capacity to proliferate and differentiate into all blood cells. The process of differentiation, however, is still ill-defined. A lack of certain cell types is typically associated with an elevated level of a lineage-specific growth factor (G-SCF, erythropoietin, thrombopoietin) that acts upon stem and precursor cells, most likely by transducing a variety of signals. These might include antiapoptotic signals, stimulate proliferation, and induce differentiation. The maturing cell has to leave the stem cell niche, sense the location of the vasculature, and finally egress into the circulation. Typically, very few immature cells are found in blood, like reticulocytes or band neutrophils, suggesting that final maturation is required for transition into the blood stream. The bone marrow compartment is full of matrix proteins that form a complex, three-dimensional network that supports the directed migration toward the endothelial barrier. Some of the main matrix components and their effect on MKs are described in the following section.

4.5.1 Fibronectin

Fibronectin (FN) is a multi-domain dimeric glycoprotein of approximately 220 kDa produced by many types of cells that forms ECM when secreted [40]. It is composed of an array of repeated modular structures: 12 FN type I repeat (FNI), 2 type II repeats (FNII), 15 constitutively expressed and 2 alternatively spliced (EIIIA and EIIIB domains) type III repeats (FNIII), and a nonhomologous variable (V) or type III connecting segment (IIICS) region [8, 42, 93]. Three alternative splicing sites (termed EDA, EDB, and IIICS in human or EIIIA, EIIIB, and V in rats) in FN mRNA allow 20 different isoforms of FN mRNA. EIIIA and EIIIB exons are included or excluded from the FN mRNA by exon skipping [54, 101]. Plasma FN (pFN) lacks both EIIIA and EIIIB segments and is a soluble form secreted by hepatocytes. It circulates in blood plasma at 300–400 μg/ml in a soluble, compact, and inactive form, with only one subunit containing a V domain. Cellular FN (cFN) is synthesized by many cell types, including fibroblasts, endothelial cells, chondrocytes, and myocytes. It contains variable proportions of EIIIA and EIIIB segments and is found as fibrils in the ECM [75]. The specific domains of FN can interact with multiple binding partners, including other extracellular matrix components (i.e., type I collagen, heparin, and chondroitin sulfate) and cell-surface receptors [88]. Differently from fibrillar collagens, FN in solution alone does not polymerize or form a three-dimensional matrix in the absence of cells. Both pFN and cFN are secreted in a soluble, compact form, which is maintained by intramolecular electrostatic interactions between the FNI_{1-5}, $FNIII_{1-2}$, $FNIII_{2-3}$, and $FNIII_{12-14}$ domains. FN matrix assembly is a cell-mediated process. FN binding to cell-surface receptors induces cytoskeletal reorganization at focal contacts on the cell periphery, which are rich in paxillin, vinculin, and integrins. Myosin II-dependent contractility results in receptor clustering and activation, causing the tethered FN molecules to become unfolded and exposing FN binding sites that allow FN-FN intermolecular interactions to occur [68]. Both platelets and MKs are involved in the process of FN-fibrillogenesis. Platelets assemble soluble FN when adsorbed on a polymerized fibrin matrix but not onto unprocessed fibrinogen [17]. Similarly, human MKs can polymerize FN upon adhesion onto type I collagen, but not onto fibrinogen-coated surface [67] (Fig. 4.1).

Fibronectin (FN) forms a reticulated network throughout the bone marrow, comparable to fibrinogen (Fig. 4.2a, b). Overall these studies demonstrate the importance of cytoskeleton and cell tensions in generating traction forces that induce conformational changes of the cell-surface-bound FN [103]. In the bone marrow, FN may modulate the migration and homing of cells to specific regions that promote maturation and the release of mature blood cells into the circulation [35]. Staining of mouse femur sections with an antibody recognizing FN demonstrates that this ECM is distributed throughout the marrow cavity, forming septa that may provide anchorage for migrating hematopoietic progenitor cells [82]. α4β1 (VLA4, very late antigen-4) and α5β1 (VLA5, very late antigen-4) are the major integrins involved in the binding of fibronectin in hematopoietic cells. In addition, MKs can bind fibronectin through functional αIIbβ3 and αvβ3 integrins [2]. A role for FN in MK maturation and platelet production is suggested by the expression of FN in fetal

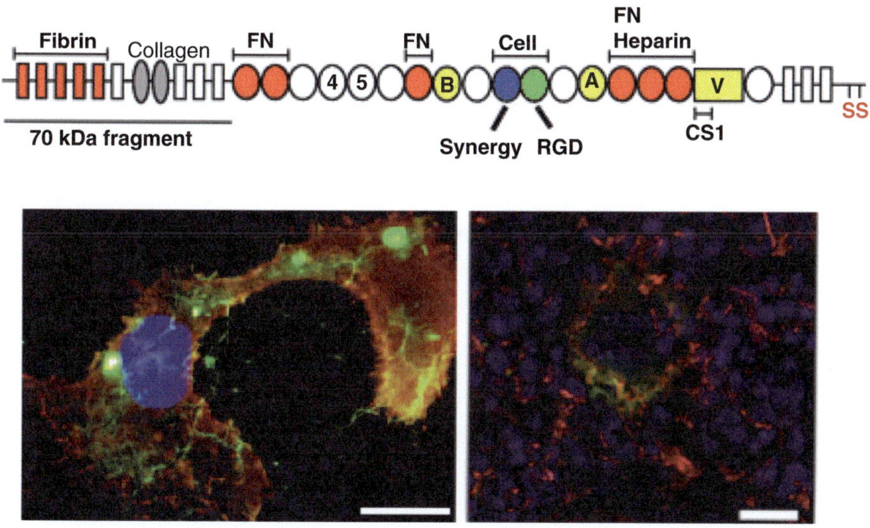

Fig. 4.1 Fibronectin-megakaryocyte interaction in the bone marrow microenvironment. (**a**) Domain structure of fibronectin. Fibronectin consists of type I (*rectangles*), type II (*ovals*), and type III (*circles*) repeats. Sets of repeats constitute binding domains for fibrin, collagen, cells, and heparin as indicated. Three alternatively spliced segments, EIIIA, EIIIB, and V (or IIICS), are in *yellow* (Image adapted from *J Cell Sci*. 2003; 116:3269–76). (**b**) Assembly of FITC-labeled plasma fibronectin (*green*) by CD61 (*red*)-positive megakaryocyte adherent onto immobilized type I collagen. Scale bar represents 5 μm (Image adapted from *Blood* 2011; 117:2476–83). (**c**) Confocal microscopy analysis of ex vivo MK-ECM interaction within the bone marrow shows that MK (CD41+, *green*) are surrounded by a fibrillar pericellular matrix mainly composed of fibronectin (*red*). Scale bar represents 10 μm (Adapted from *Stem Cells* 2014;32:926–937)

liver-derived MKs before its expression in fetal hepatocytes [112]. Mature MKs derived from guinea pigs, mice, and humans synthesize FN with a level 7.5-fold greater than other bone marrow hematopoietic cells [18, 67, 98]. Interestingly, FN of megakaryocytic origin contains both EIIIA and EIIIB domains [67, 98]. FN is reported to play key roles in the proliferation, differentiation, and platelet release of megakaryocytic cells through FN receptors. FN sustains viability and expansion of $CD34^+$ cells and MK-progenitors from human cord blood in the presence of cytokines [19, 36] and enhances the expansion and engraftibility of primitive murine hematopoietic stem cells [94]. Moreover, adhesion to fibronectin induces megakaryocytic differentiation of JAS-REN cells, a megakaryocytic-erythroid cell line [119]. Engagement of α4β1 integrin enhanced MK growth in the presence of TPO [27]. FN was demonstrated to support, but not to enhance PPF in human cord blood-derived MKs [6], while increased PPF was observed in phorbol-12-myristate-13-acetate (PMA)-differentiated CHRF-288 cells when adhering to FN in vitro. Accelerated PPF was accompanied by increased extracellular signal-regulated protein kinase 1 (ERK 1/ERK 2) and prevented using anti β1 integrin monoclonal antibodies, indicating that adhesive interaction with FN via VLA-4 and VLA-5 was required for PPF [69]. Similarly, addition of soluble FN to cultured megakaryocytic UT-7/TPO cells increased PPF through ERK 1/ERK 2 and phosphoinositide 3-kinase (PI3K)

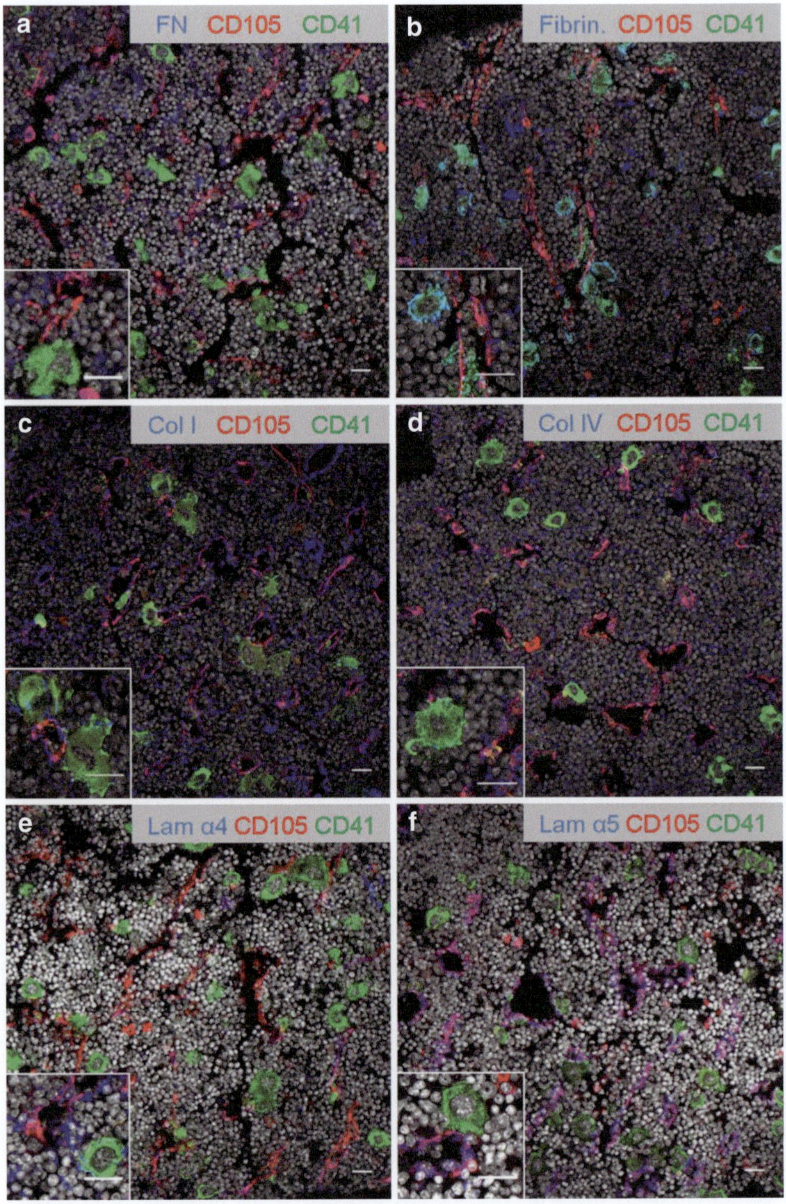

Fig. 4.2 Localization of matrix proteins in the bone marrow. Megakaryocytes, stained with CD41 (*green*) and sinusoids (CD105, *red*), are in close contact with distinct extracellular matrix proteins. However, the distribution pattern of those matrix proteins is diverse. (**a**) Fibronectin (FN, *blue*) is expressed to a lesser extent and shows a dotted-like pattern. (**b**) In contrast fibrinogen (fibrin, *blue*) is detectable in patches within the bone marrow but also colocalizes with MKs. (**c**) Collagen type I (*blue*) and (**d**) collagen type IV (*blue*) are abundantly expressed throughout the whole bone marrow and colocalize with sinusoids (*red*). Bone marrow endothelial sinusoids are negative for (**e**) laminin-α4 (*blue*), whereas they are positive for (**f**) laminin-α5 (*blue*). Scale bars represent 20 μm

activation. Anti β1 antibody and the RGD (arginine-glycine-aspartate) peptide were demonstrated to inhibit FN-induced PPF [47].

Bone marrow (BM) fibrosis is the consequence of a reactive environment in several hematologic diseases. In general, fibrosis associated with hematological malignancies is characterized by fibroblast proliferation, increased angiogenesis, and abnormal MK accumulation. The latter event is an uncontrolled deposition of ECM components with excessive fibrous tissue formation in the BM. Two recent works [65, 99] demonstrated cell-intrinsic alterations in the mesenchymal stromal cells (MSCs) derived from primary myelofibrosis (PMF) patient BM. Both researches demonstrated that, among the different components, FN expression and deposition significantly increased in PMF MSCs with respect to controls.

4.5.2 Vitronectin

Another member of the nectin superfamily is vitronectin which is structurally similar to FN. Vitronectin is a glycoprotein present in bone marrow, but also in plasma. MKs and platelets express αV-β3 integrin. This integrin is also found on endothelial cells where it contributes to platelet adhesion [28], most likely via kindlin-2 [61]. Interestingly, antibodies against αV-β3 did not inhibit PPF of guinea pig MKs when plated on collagen type I [59].

4.5.3 Collagens

Collagens comprise the largest group of structural proteins. They form filaments composed of three homologous subunits and differ by their composition. Collagen type I fibers are present in the bone cortex as well as throughout the marrow in an elaborated, reticulated network (Fig. 4.2c). The characteristic motif is a tripeptide of glycine, followed by two proline residues of which the second is substrate to proline hydroxylases. Specific hydrogen bonds stabilize the trimeric structure and bring the melting temperature above 37 °C. This GPO motif comprises up to 10 % of all amino acids in both type I and type III collagens. The latter often accompanies type I fibers, but it is absent from bone cortex. Type IV has been described to be associated with the subendothelial basal lamina, but it is also present in the marrow cavity and can also partially colocalize with other collagens (Fig. 4.2d). It is less fiber forming and lacks the GPO motif present in type I and III. Collagens are widely used as agonists, but its use is hampered by two issues: first, each preparation is somewhat different, depending on the source and the purity and second, the solubility in physiological buffers is limited. Therefore, artificial collagen-related peptides (CRP) have been synthesized that concentrate the GPO motif, flanked by a potential cross-linker. Even here, typically every new preparation and lot needs to be tested and compared to a standard, as the degree of cross-linking can have an important impact on its reactivity. The GKO-(GPO)10-GKOG-motif is often used to mimic collagen, a sequence motif derived from types I or III. In contrast, a

GFOGER polypeptide is used to mimic domains present in collagen IV, and these peptides are typically recognized by α2-β1 integrins rather than by GPVI, which is the main receptor for CRP.

Several distinct collagen receptors are expressed throughout the megakaryocytic lineage up to platelets. The discoidin domain receptor-1 (DDR1) has been detected on early megakaryocytes and is believed to transmit negative signals [1]. Leukocyte-associated immunoglobulin receptor-1 (LAIR1) is found during MK maturation, but also absent from platelets. In contrast, glycoprotein VI (GPVI) is the major collagen receptor on platelets and responsible to transduce the activatory signal to a variety of stimulatory responses that ultimately lead to full platelet activation and aggregation [81]. In addition, the integrin heterodimer α2-β1 also binds to collagen. Mice lacking the α2-specific subunit show a delayed response to collagen, whereas high concentrations and collagen agonists like CRP or the snake venom convulxin can transmit full answers even in the absent of this receptor. Binding studies suggest that adhesive binding is rather mediated by the integrin, whereas the signaling occurs via GPVI. Both receptors are present on bone marrow MKs, but the signaling is substantially altered. They still can respond to CRP or convulxin, but the responses are overall weak. Indeed, in such assays MKs are incubated with these agonists in stirring conditions resulting in partial phosphorylation of Syk kinase or causing calcium influx [71]. Collagen types I and IV have been attributed diametral effects on MK proplatelet formation (PPF). While collagen type IV, enriched at sinusoids, was reported to support PPF, type I can suppress or attenuate PPF. Using a combined approach of mice lacking specific collagen receptors or using wild-type MKs where the receptors have been blocked by antibodies, there is evidence that GPVI is the main receptor to transduce an inhibitory signal by collagen type I, while type IV competes for binding but does not inhibit PPF. Interestingly, although this inhibition is selectively dependent on the presence and expression of GPVI, it seems that the downstream signaling is distinct, as we failed to find phosphorylation of Syk in this context. Moreover, increasing concentrations of CRP could abrogate the inhibitory effect of type I, suggesting that different binding to GPVI or downstream signaling occurs during this final step of platelet biogenesis [102].

Finally, several collagen isotypes that are expressed at low, rather modulating concentrations that might fine-tune the network and several cross-linking proteins, including perlecan and nidogen, are present, but their role in platelet function is still poorly characterized.

4.5.4 Laminins

Laminins are critical components of basal membranes typically found in the subendothelial layer. Laminins are heterotrimers, each comprising one out of five α, four β, and three distinct γ chains, and this composition determines the nomenclature. So far, 18 distinct laminins have been detected, most of them during embryogenesis [120]. The tissue specificity is mostly determined by expression of the specific α-chain. In the bone marrow, expression of α4 and α5 chains has been detected.

α4 is found at many larger vessels throughout the bone marrow, while α5 becomes expressed postnatally at rather specific vessel types. Thus, in mouse femur both types are readily found in smaller arterioles, whereas the mouse bone marrow sinusoids stain positive for laminin α5 (Fig. 4.2e, f).

There are mouse models that lack specific laminin genes: LAMA5-null animals are embryonic lethal [73]. In contrast, LAMA4-null mice are viable but show a deteriorated microvasculature during development, mice present with severe bleedings during the embryonic and postnatal period. Platelet counts are reported to be reduced by half [110], suggesting that thrombopoiesis is markedly abrogated in the absence of α4-chain. Laminin α5 is overall upregulated throughout the vascular tree in these mice, showing an increased extravasation of T lymphocytes. Laminin α5 might thus provide a stop signal for T lymphocyte extravasation through laminin α4 [39, 48]. The role of selective laminin isoform depletion needs to be studied in conditional animals, i.e., when conditional knockout mice are interbred with Cre mice that lead to gene ablation in endothelial cells [53].

Several surface receptors present on MKs and platelets have been reported to bind to laminin, but so far the integrin α6-β1 is considered the main receptor for laminin α4 and α5 (reviewed in [120]). Mice lacking the ITGA6 gene, which encodes the α6-integrin subunit, clearly show a marked reduction in arterial thrombosis in three different injury models, whereas the hemostatic function including the tail bleeding time was unaltered [97]. A blocking antibody against α6-integrin in tumorigenic mice improved the overall survival due to new bone formation [56]. Recent evidence suggests that the α5-β1 integrin (which typically is referred to as the FN receptor) and αV-β3 (the vitronectin receptor) that are both expressed in the megakaryocytic lineage can specifically bind to laminin α5 isoforms [95]. More research is required to elucidate the role of laminins in megakaryopoiesis and platelet biogenesis.

4.5.5 Other Receptors: PECAM1

The platelet and endothelial cell adhesion molecule (PECAM, CD31) is expressed during the megakaryocytic differentiation [80]. As the name indicates, it is also present on endothelial cells and can be used as a marker for blood vessels. Mice lacking PECAM show an increased reactivity to collagen, suggesting that it acts as a negative regulator for megakaryocyte differentiation and maturation with the potential to fine-tune megakaryopoiesis [20, 118].

4.6 Extramedullary Hematopoiesis

A first pool of MKs is present during fetal hematopoiesis in the liver when hematopoiesis shifts from primitive to definitive. Fetal MKs are capable of producing platelets that circulate in the developing fetus and share most features with platelets in the adult organism, although some significant differences have been identified [63]. Obviously, the microenvironment in fetal liver is different from bone marrow and platelet biogenesis with the consequent modification of two critical parameters: megakaryocytic cells derived from fetal HSC might be different in the first place and are in addition in a distinct compartment. This has been become obvious when platelets derived from preterm newborns are analyzed. These platelets show an overall reduced reactivity to agonists and are often referred to as "hyporeactive." This becomes the more prominent the "younger" the preterm is [4]. However, this altered reactivity might rather be their physiological role at this gestational age, and so transfusion of platelets from adults (as provided in any standard platelet concentrate) into preterm newborns has been considered problematic with respect to number and the "improved" reactivity.

In malignant conditions, the capacity of the bone marrow environment might be limited for a full replenishment of all hematopoietic cells. This can happen in response to an advanced myelofibrosis as found in a subset of myeloproliferative disorders like essential thrombocythemia, polycythemia vera, or primary myelofibrosis. These are often grouped as "myeloproliferative disorders" with underlying mutations in Jak2, c-Mpl, or calreticulin that are mostly acquired can become clonal and finally develop into leukemic conditions [104]. Excessive presence of MKs is known to contribute to myelofibrosis by several mechanisms, including the release of TGF-β [15]. During leukemia, transformed blasts proliferate within or migrate into the bone marrow and restrict the space (and niches) required for steady-state hematopoiesis. Peripheral blood counts might thus reduce, and the progenitor cells might have found new environments outside the bone marrow, i.e., in the spleen, liver, or lung. Here again, MKs are in contact to an altered microenvironment and platelet function might be altered.

4.7 Rebuilding the "Vascular Niche" with Platelet Bioreactors

Engineering the vascular niche has proven to be an effective strategy for studying in vitro megakaryopoiesis and platelet production, thanks to the resulting physiologic environment which is more effective than traditional culture systems (Fig. 4.3). Major advancements in the field have been achieved recently by three different three-dimensional (3D) models, two employing polydimethylsiloxane (PDMS), a cell-inert silicon-based organic polymer [26, 78, 109], and the other silk fibroin, a naturally derived protein extracted from *Bombyx mori* silkworm cocoons ([85]; Di Buduo et al.). The system by Nakagawa et al. was made of poly-L-lactide-*co*-epsilon-caprolactone (PLCL), with a membrane scaffold composed of PDMS to support platelet release from MKs. The scaffolds allowed the use of two different media flows: one to promote MK adhesion on the supporting membrane, and the second to allow platelets recover. The microfluidic bioreactor, devised by Thon et al., consists of transparent PDMS bonded to glass slides to ensure efficient gas exchange and to support high-resolution live-cell imaging. The device consists of an upper channel, filled with alginates and matrigel for MK seeding, and a lower channel for platelet collection. The channels were separated by a series of columns

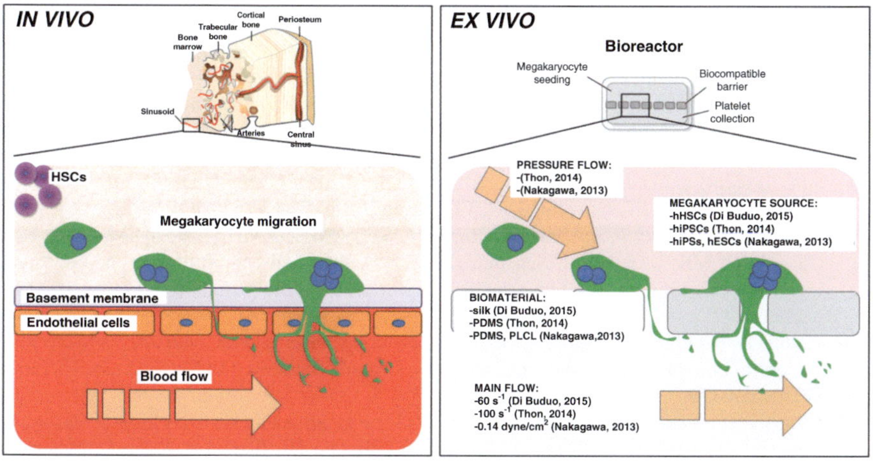

Fig. 4.3 Schematic representation of in vivo vs ex vivo thrombopoiesis. In vivo: maturing megakaryocytes migrate toward bone marrow sinusoids prior to extend proplatelets. Proplatelets traverse the basement membrane and extend, through the vascular endothelium, into the lumen of sinusoidal vessels where platelet particles, stemming from their terminal ends, are finally released. Ex vivo: all bioengineered tissue models mimicking the bone marrow environment are made of a biocompatible porous barrier supporting mature megakaryocyte function. Once in contact with the biomaterial, megakaryocytes extend proplatelets through the pores. Platelet release is finally prompted by media flow, mimicking the bloodstream, obtained by controlled electronic pumps. (*hHSC* human hematopoietic stem cell, *hiPSC* human-induced pluripotent stem cell, *hESC* human embryonic stem cell, *PDMS* polydimethylsiloxane, *PLCL* poly-L-lactide-co-epsilon-caprolactone)

to prevent MKs entering the upper channel from crossing into the lower channel upon fluid withdrawal while allowing proplatelet extension [109]. Pallotta et al. proposed a 3D model, in which the vascular niche environment was composed of organic silk-based vascular micro-tubes, coated with fibrinogen and von Willebrand factor embedded in a gel matrix of collagen to house MKs. Once seeded into the gel matrix, megakaryocytes were able to migrate toward the vascular micro-tubes in response to SDF-1α, extend proplatelets, and release platelets [85]. Advanced features of the silk have been exploited in the latest 3D human bone marrow niche tissue models for functional platelet production [21]. In this system Di Buduo et al. showed how silk vascular tubes can be embedded in a silk sponge mimicking the "spongy" structure of the bone marrow and support MK growth and platelet release into the perfused vascular tube lumen. Functionalization of silk with different ECM components present in the vascular niche in vivo supported MK function without loss of bioactivity [21]. Platelets released in the perfused silk vascular tube were morphologically similar to peripheral blood platelets and functional under specific physiologic activating stimuli. Finally, MK coculture with endothelial cells resulted in a significant increase in platelet production. Independently of the features and the architecture of the vascular niche model, all these approaches proved the pivotal role of vascular shear stress for proper platelet production [70].

References

1. Abbonante V, Gruppi C, Rubel D, Gross O, Moratti R, Balduini A (2013) Discoidin domain receptor 1 protein is a novel modulator of megakaryocyte-collagen interactions. J Biol Chem 288(23):16738–16746. doi:10.1074/jbc.M112.431528
2. Abrams CS, Ruggeri ZM, Taub R, Hoxie JA, Nagaswami C, Weisel JW, Shattil SJ (1992) Anti-idiotypic antibodies against an antibody to the platelet glycoprotein (GP) IIb-IIIa complex mimic GP IIb-IIIa by recognizing fibrinogen. J Biol Chem 267(4):2775–2785
3. Adams GB, Chabner KT, Alley IR, Olson DP, Szczepiorkowski ZM, Poznansky MC, Kos CH, Pollak MR, Brown EM, Scadden DT (2006) Stem cell engraftment at the endosteal niche is specified by the calcium-sensing receptor. Nature 439(7076):599–603. doi:10.1038/nature04247
4. Andres O, Schulze H, Speer CP (2015) Platelets in neonates: central mediators in haemostasis, antimicrobial defence and inflammation. Thromb Haemost 113(1):3–12. doi:10.1160/TH14-05-0476
5. Avecilla ST, Hattori K, Heissig B, Tejada R, Liao F, Shido K, Jin DK, Dias S, Zhang F, Hartman TE, Hackett NR, Crystal RG, Witte L, Hicklin DJ, Bohlen P, Eaton D, Lyden D, de Sauvage F, Rafii S (2004) Chemokine-mediated interaction of hematopoietic progenitors with the bone marrow vascular niche is required for thrombopoiesis. Nat Med 10(1):64–71
6. Balduini A, Pallotta I, Malara A, Lova P, Pecci A, Viarengo G, Balduini CL, Torti M (2008) Adhesive receptors, extracellular proteins and myosin IIA orchestrate proplatelet formation by human megakaryocytes. J Thromb Haemost 6(11):1900–1907. doi:10.1111/j.1538-7836.2008.03132.x
7. Ballmaier M, Germeshausen M, Schulze H, Cherkaoui K, Lang S, Gaudig A, Krukemeier S, Eilers M, Strauss G, Welte K (2001) c-mpl mutations are the cause of congenital amegakaryocytic thrombocytopenia. Blood 97(1):139–146
8. Baneyx G, Baugh L, Vogel V (2002) Fibronectin extension and unfolding within cell matrix fibrils controlled by cytoskeletal tension. Proc Natl Acad Sci U S A 99(8):5139–5143. doi:10.1073/pnas.072650799
9. Barry FP, Boynton RE, Haynesworth S, Murphy JM, Zaia J (1999) The monoclonal antibody SH-2, raised against human mesenchymal stem cells, recognizes an epitope on endoglin (CD105). Biochem Biophys Res Commun 265(1):134–139. doi:10.1006/bbrc.1999.1620
10. Behnke O, Forer A (1998) From megakaryocytes to platelets: platelet morphogenesis takes place in the bloodstream. Eur J Haematol 61:3–23
11. Belnoue E, Pihlgren M, McGaha TL, Tougne C, Rochat AF, Bossen C, Schneider P, Huard B, Lambert PH, Siegrist CA (2008) APRIL is critical for plasmablast survival in the bone marrow and poorly expressed by early-life bone marrow stromal cells. Blood 111(5):2755–2764. doi:10.1182/blood-2007-09-110858
12. Bonci D, Hahne M, Felli N, Peschle C, De Maria R (2004) Potential role of APRIL as autocrine growth factor for megakaryocytopoiesis. Blood 104(10):3169–3172. doi:10.1182/blood-2003-11-3861
13. Bonig H, Papayannopoulou T (2013) Hematopoietic stem cell mobilization: updated conceptual renditions. Leukemia 27(1):24–31. doi:10.1038/leu.2012.254
14. Bruns I, Lucas D, Pinho S, Ahmed J, Lambert MP, Kunisaki Y, Scheiermann C, Schiff L, Poncz M, Bergman A, Frenette PS (2014) Megakaryocytes regulate hematopoietic stem cell quiescence through CXCL4 secretion. Nat Med 20(11):1315–1320. doi:10.1038/nm.3707
15. Chagraoui H, Komura E, Tulliez M, Giraudier S, Vainchenker W, Wendling F (2002) Prominent role of TGF-beta 1 in thrombopoietin-induced myelofibrosis in mice. Blood 100(10):3495–3503
16. Chen S, Su Y, Wang J (2013) ROS-mediated platelet generation: a microenvironment-dependent manner for megakaryocyte proliferation, differentiation, and maturation. Cell Death Dis 4, e722. doi:10.1038/cddis.2013.253
17. Cho J, Degen JL, Coller BS, Mosher DF (2005) Fibrin but not adsorbed fibrinogen supports fibronectin assembly by spread platelets. Effects of the interaction of alphaIIb beta3 with the

C terminus of the fibrinogen gamma-chain. J Biol Chem 280(42):35490–35498. doi:10.1074/jbc.M506289200
18. Courtney MA, Stoler MH, Marder VJ, Haidaris PJ (1991) Developmental expression of mRNAs encoding platelet proteins in rat megakaryocytes. Blood 77(3):560–568
19. Deutsch V, Hubel E, Kay S, Ohayon T, Katz BZ, Many A, Zander A, Naparstek E, Grisaru D (2010) Mimicking the haematopoietic niche microenvironment provides a novel strategy for expansion of haematopoietic and megakaryocyte-progenitor cells from cord blood. Br J Haematol 149(1):137–149. doi:10.1111/j.1365-2141.2009.08041.x
20. Dhanjal TS, Pendaries C, Ross EA, Larson MK, Protty MB, Buckley CD, Watson SP (2007) A novel role for PECAM-1 in megakaryocytokinesis and recovery of platelet counts in thrombocytopenic mice. Blood 109(10):4237–4244
21. Di Buduo CA, Wray LS, Tozzi L, Malara A, Chen Y, Ghezzi CE, Smoot D, Sfara C, Antonelli A, Spedden E, Bruni G, Staii C, De Marco L, Magnani M, Kaplan DL, Balduini A (2015) Programmable 3D silk bone marrow niche for platelet generation ex vivo and modeling of megakaryopoiesis pathologies. Blood 125(14):2254–2264. doi:10.1182/blood-2014-08-595561
22. Ding BS, Nolan DJ, Butler JM, James D, Babazadeh AO, Rosenwaks Z, Mittal V, Kobayashi H, Shido K, Lyden D, Sato TN, Rabbany SY, Rafii S (2010) Inductive angiocrine signals from sinusoidal endothelium are required for liver regeneration. Nature 468(7321):310–315. doi:10.1038/nature09493
23. Eliades A, Matsuura S, Ravid K (2012) Oxidases and reactive oxygen species during hematopoiesis: a focus on megakaryocytes. J Cell Physiol 227(10):3355–3362. doi:10.1002/jcp.24071
24. Eliasson P, Jonsson JI (2010) The hematopoietic stem cell niche: low in oxygen but a nice place to be. J Cell Physiol 222(1):17–22. doi:10.1002/jcp.21908
25. Ellis SL, Nilsson SK (2012) The location and cellular composition of the hemopoietic stem cell niche. Cytotherapy 14(2):135–143. doi:10.3109/14653249.2011.630729
26. Feng Q, Shabrani N, Thon JN, Huo H, Thiel A, Machlus KR, Kim K, Brooks J, Li F, Luo C, Kimbrel EA, Wang J, Kim KS, Italiano J, Cho J, Lu SJ, Lanza R (2014) Scalable generation of universal platelets from human induced pluripotent stem cells. Stem Cell Rep 3(5):817–831. doi:10.1016/j.stemcr.2014.09.010
27. Fox NE, Kaushansky K (2005) Engagement of integrin alpha4beta1 enhances thrombopoietin-induced megakaryopoiesis. Exp Hematol 33(1):94–99. doi:10.1016/j.exphem.2004.10.002
28. Gawaz M, Neumann FJ, Dickfeld T, Reininger A, Adelsberger H, Gebhardt A, Schomig A (1997) Vitronectin receptor (alpha(v)beta3) mediates platelet adhesion to the luminal aspect of endothelial cells: implications for reperfusion in acute myocardial infarction. Circulation 96(6):1809–1818
29. Germeshausen M, Ballmaier M, Welte K (2006) MPL mutations in 23 patients suffering from congenital amegakaryocytic thrombocytopenia: the type of mutation predicts the course of the disease. Hum Mutat 27(3):296
30. Giannopoulou E, Katsoris P, Hatziapostolou M, Kardamakis D, Kotsaki E, Polytarchou C, Parthymou A, Papaioannou S, Papadimitriou E (2001) X-rays modulate extracellular matrix in vivo. Int J Cancer 94(5):690–698
31. Golfier S, Kondo S, Schulze T, Takeuchi T, Vassileva G, Achtman AH, Graler MH, Abbondanzo SJ, Wiekowski M, Kremmer E, Endo Y, Lira SA, Bacon KB, Lipp M (2010) Shaping of terminal megakaryocyte differentiation and proplatelet development by sphingosine-1-phosphate receptor S1P4. FASEB J 24(12):4701–4710. doi:10.1096/fj.09-141473
32. Green DE, Adler BJ, Chan ME, Lennon JJ, Acerbo AS, Miller LM, Rubin CT (2013) Altered composition of bone as triggered by irradiation facilitates the rapid erosion of the matrix by both cellular and physicochemical processes. PLoS ONE 8(5), e64952. doi:10.1371/journal.pone.0064952
33. Greenbaum AM, Link DC (2011) Mechanisms of G-CSF-mediated hematopoietic stem and progenitor mobilization. Leukemia 25(2):211–217. doi:10.1038/leu.2010.248

34. Hamada T, Mohle R, Hesselgesser J, Hoxie J, Nachman RL, Moore MA, Rafii S (1998) Transendothelial migration of megakaryocytes in response to stromal cell-derived factor 1 (SDF-1) enhances platelet formation. J Exp Med 188(3):539–548
35. Hamilton R, Campbell FR (1991) Immunochemical localization of extracellular materials in bone marrow of rats. Anat Rec 231(2):218–224. doi:10.1002/ar.1092310210
36. Han P, Guo XH, Story CJ (2002) Enhanced expansion and maturation of megakaryocytic progenitors by fibronectin. Cytotherapy 4(3):277–283. doi:10.1080/146532402320219790
37. Heazlewood SY, Neaves RJ, Williams B, Haylock DN, Adams TE, Nilsson SK (2013) Megakaryocytes co-localise with hemopoietic stem cells and release cytokines that up-regulate stem cell proliferation. Stem Cell Res 11(2):782–792. doi:10.1016/j.scr.2013.05.007
38. Heissig B, Hattori K, Dias S, Friedrich M, Ferris B, Hackett NR, Crystal RG, Besmer P, Lyden D, Moore MA, Werb Z, Rafii S (2002) Recruitment of stem and progenitor cells from the bone marrow niche requires MMP-9 mediated release of kit-ligand. Cell 109(5):625–637
39. Hooper AT, Butler JM, Nolan DJ, Kranz A, Iida K, Kobayashi M, Kopp HG, Shido K, Petit I, Yanger K, James D, Witte L, Zhu Z, Wu Y, Pytowski B, Rosenwaks Z, Mittal V, Sato TN, Rafii S (2009) Engraftment and reconstitution of hematopoiesis is dependent on VEGFR2-mediated regeneration of sinusoidal endothelial cells. Cell Stem Cell 4(3):263–274. doi:10.1016/j.stem.2009.01.006
40. Hynes R (1985) Molecular biology of fibronectin. Annu Rev Cell Biol 1:67–90. doi:10.1146/annurev.cb.01.110185.000435
41. Jaimes Y, Gras C, Goudeva L, Buchholz S, Eiz-Vesper B, Seltsam A, Immenschuh S, Blasczyk R, Figueiredo C (2012) Semaphorin 7A inhibits platelet production from CD34+ progenitor cells. J Thromb Haemost 10(6):1100–1108. doi:10.1111/j.1538-7836.2012.04708.x
42. Johnson KJ, Sage H, Briscoe G, Erickson HP (1999) The compact conformation of fibronectin is determined by intramolecular ionic interactions. J Biol Chem 274(22):15473–15479
43. Junt T, Schulze H, Chen Z, Massberg S, Goerge T, Krueger A, Wagner DD, Graf T, Italiano JE Jr, Shivdasani RA, von Andrian UH (2007) Dynamic visualization of thrombopoiesis within bone marrow. Science 317(5845):1767–1770
44. Kacena MA, Ciovacco WA (2010) Megakaryocyte-bone cell interactions. Adv Exp Med Biol 658:31–41. doi:10.1007/978-1-4419-1050-9_4
45. Kacena MA, Gundberg CM, Nelson T, Horowitz MC (2005) Loss of the transcription factor p45 NF-E2 results in a developmental arrest of megakaryocyte differentiation and the onset of a high bone mass phenotype. Bone 36(2):215–223. doi:10.1016/j.bone.2004.09.024
46. Kacena MA, Shivdasani RA, Wilson K, Xi Y, Troiano N, Nazarian A, Gundberg CM, Bouxsein ML, Lorenzo JA, Horowitz MC (2004) Megakaryocyte-osteoblast interaction revealed in mice deficient in transcription factors GATA-1 and NF-E2. J Bone Miner Res 19(4):652–660
47. Kawaguchi T, Hatano R, Yamaguchi K, Nawa K, Hashimoto R, Yokota H (2012) Fibronectin promotes proplatelet formation in the human megakaryocytic cell line UT-7/TPO. Cell Biol Int 36(1):39–45. doi:10.1042/CBI20110383
48. Kenne E, Soehnlein O, Genove G, Rotzius P, Eriksson EE, Lindbom L (2010) Immune cell recruitment to inflammatory loci is impaired in mice deficient in basement membrane protein laminin alpha4. J Leukoc Biol 88(3):523–528. doi:10.1189/jlb.0110043
49. Kfoury Y, Mercier F, Scadden DT (2014) SnapShot: the hematopoietic stem cell niche. Cell 158(1):228–228.e221. doi:10.1016/j.cell.2014.06.019
50. Kiel MJ, Morrison SJ (2006) Maintaining hematopoietic stem cells in the vascular niche. Immunity 25(6):862–864. doi:10.1016/j.immuni.2006.11.005
51. Kiel MJ, Morrison SJ (2008) Uncertainty in the niches that maintain haematopoietic stem cells. Nat Rev Immunol 8(4):290–301. doi:10.1038/nri2279
52. Kiel MJ, Yilmaz OH, Iwashita T, Yilmaz OH, Terhorst C, Morrison SJ (2005) SLAM family receptors distinguish hematopoietic stem and progenitor cells and reveal endothelial niches for stem cells. Cell 121(7):1109–1121. doi:10.1016/j.cell.2005.05.026
53. Kisanuki YY, Hammer RE, Miyazaki J, Williams SC, Richardson JA, Yanagisawa M (2001) Tie2-Cre transgenic mice: a new model for endothelial cell-lineage analysis in vivo. Dev Biol 230(2):230–242. doi:10.1006/dbio.2000.0106

54. Kornblihtt AR, Vibe-Pedersen K, Baralle FE (1984) Human fibronectin: molecular cloning evidence for two mRNA species differing by an internal segment coding for a structural domain. Embo J 3(1):221–226
55. Kunisaki Y, Bruns I, Scheiermann C, Ahmed J, Pinho S, Zhang D, Mizoguchi T, Wei Q, Lucas D, Ito K, Mar JC, Bergman A, Frenette PS (2013) Arteriolar niches maintain haematopoietic stem cell quiescence. Nature 502(7473):637–643. doi:10.1038/nature12612
56. Landowski TH, Gard J, Pond E, Pond GD, Nagle RB, Geffre CP, Cress AE (2014) Targeting integrin alpha6 stimulates curative-type bone metastasis lesions in a xenograft model. Mol Cancer Ther 13(6):1558–1566. doi:10.1158/1535-7163.MCT-13-0962
57. Lane WJ, Dias S, Hattori K, Heissig B, Choy M, Rabbany SY, Wood J, Moore MA, Rafii S (2000) Stromal-derived factor 1-induced megakaryocyte migration and platelet production is dependent on matrix metalloproteinases. Blood 96(13):4152–4159
58. Laterveer L, Lindley IJ, Hamilton MS, Willemze R, Fibbe WE (1995) Interleukin-8 induces rapid mobilization of hematopoietic stem cells with radioprotective capacity and long-term myelolymphoid repopulating ability. Blood 85(8):2269–2275
59. Leven RM, Tablin F (1992) Extracellular matrix stimulation of guinea pig megakaryocyte proplatelet formation in vitro is mediated through the vitronectin receptor. Exp Hematol 20(11):1316–1322
60. Li G, Li XG, Jia DZ, Wang DH, Su YH, Zhang QL (2007) Functional and structural recovery of injured spinal cord following delayed X-irradiation in rats. Chin Med Sci J 22(1):58–61
61. Liao Z, Kato H, Pandey M, Cantor JM, Ablooglu AJ, Ginsberg MH, Shattil SJ (2015) Interaction of kindlin-2 with integrin beta3 promotes outside-in signaling responses by the alphaVbeta3 vitronectin receptor. Blood 125(12):1995–2004. doi:10.1182/blood-2014-09-603035
62. Liu CC, Wang SC, Kao CW, Hsieh RK, Chang MC, Chang YF, Lim KH, Chen CG (2014) B cells facilitate platelet production mediated by cytokines in patients with essential thrombocythaemia. Thromb Haemost 112(3):537–550. doi:10.1160/TH13-11-0949
63. Liu ZJ, Italiano J Jr, Ferrer-Marin F, Gutti R, Bailey M, Poterjoy B, Rimsza L, Sola-Visner M (2011) Developmental differences in megakaryocytopoiesis are associated with up-regulated TPO signaling through mTOR and elevated GATA-1 levels in neonatal megakaryocytes. Blood 117(15):4106–4117
64. Lymperi S, Ferraro F, Scadden DT (2010) The HSC niche concept has turned 31. Has our knowledge matured? Ann N Y Acad Sci 1192:12–18. doi:10.1111/j.1749-6632.2009.05223.x
65. Malara A, Abbonante V, Di Buduo CA, Tozzi L, Currao M, Balduini A (2015) The secret life of a megakaryocyte: emerging roles in bone marrow homeostasis control. Cell Mol Life Sci 72(8):1517–1536. doi:10.1007/s00018-014-1813-y
66. Malara A, Currao M, Gruppi C, Celesti G, Viarengo G, Buracchi C, Laghi L, Kaplan DL, Balduini A (2014) Megakaryocytes contribute to the bone marrow-matrix environment by expressing fibronectin, type IV collagen, and laminin. Stem Cells 32(4):926–937. doi:10.1002/stem.1626
67. Malara A, Gruppi C, Rebuzzini P, Visai L, Perotti C, Moratti R, Balduini C, Tira ME, Balduini A (2011) Megakaryocyte-matrix interaction within bone marrow: new roles for fibronectin and factor XIII-A. Blood 117(8):2476–2483. doi:10.1182/blood-2010-06-288795
68. Mao Y, Schwarzbauer JE (2005) Fibronectin fibrillogenesis, a cell-mediated matrix assembly process. Matrix Biol 24(6):389–399. doi:10.1016/j.matbio.2005.06.008
69. Matsunaga T, Fukai F, Kameda T, Shide K, Shimoda H, Torii E, Kamiunten A, Sekine M, Yamamoto S, Hidaka T, Kubuki Y, Yokokura S, Uemura M, Matsuoka A, Waki F, Matsumoto K, Kanaji N, Ishii T, Imataki O, Dobashi H, Bandoh S, Shimoda K (2012) Potentiated activation of VLA-4 and VLA-5 accelerates proplatelet-like formation. Ann Hematol 91(10):1633–1643. doi:10.1007/s00277-012-1498-y
70. Mazo IB, von Andrian UH (1999) Adhesion and homing of blood-borne cells in bone marrow microvessels. J Leukoc Biol 66(1):25–32
71. Melford SK, Turner M, Briddon SJ, Tybulewicz VL, Watson SP (1997) Syk and Fyn are required by mouse megakaryocytes for the rise in intracellular calcium induced by a collagen-related peptide. J Biol Chem 272(44):27539–27542

72. Mercher T, Cornejo MG, Sears C, Kindler T, Moore SA, Maillard I, Pear WS, Aster JC, Gilliland DG (2008) Notch signaling specifies megakaryocyte development from hematopoietic stem cells. Cell Stem Cell 3(3):314–326
73. Miner JH, Cunningham J, Sanes JR (1998) Roles for laminin in embryogenesis: exencephaly, syndactyly, and placentopathy in mice lacking the laminin alpha5 chain. J Cell Biol 143(6):1713–1723
74. Moore KA (2004) Recent advances in defining the hematopoietic stem cell niche. Curr Opin Hematol 11(2):107–111
75. Moretti FA, Chauhan AK, Iaconcig A, Porro F, Baralle FE, Muro AF (2007) A major fraction of fibronectin present in the extracellular matrix of tissues is plasma-derived. J Biol Chem 282(38):28057–28062. doi:10.1074/jbc.M611315200
76. Mostafa SS, Papoutsakis ET, Miller WM (2001) Oxygen tension modulates the expression of cytokine receptors, transcription factors, and lineage-specific markers in cultured human megakaryocytes. Exp Hematol 29(7):873–883
77. Murphy AJ, Bijl N, Yvan-Charvet L, Welch CB, Bhagwat N, Reheman A, Wang Y, Shaw JA, Levine RL, Ni H, Tall AR, Wang N (2013) Cholesterol efflux in megakaryocyte progenitors suppresses platelet production and thrombocytosis. Nat Med 19(5):586–594. doi:10.1038/nm.3150
78. Nakagawa Y, Nakamura S, Nakajima M, Endo H, Dohda T, Takayama N, Nakauchi H, Arai F, Fukuda T, Eto K (2013) Two differential flows in a bioreactor promoted platelet generation from human pluripotent stem cell-derived megakaryocytes. Exp Hematol 41(8):742–748. doi:10.1016/j.exphem.2013.04.007
79. Narayan K, Juneja S, Garcia C (1994) Effects of 5-fluorouracil or total-body irradiation on murine bone marrow microvasculature. Exp Hematol 22(2):142–148
80. Newman PJ, Newman DK (2003) Signal transduction pathways mediated by PECAM-1: new roles for an old molecule in platelet and vascular cell biology. Arterioscler Thromb Vasc Biol 23(6):953–964. doi:10.1161/01.ATV.0000071347.69358.D9
81. Nieswandt B, Watson SP (2003) Platelet-collagen interaction: is GPVI the central receptor? Blood 102(2):449–461. doi:10.1182/blood-2002-12-3882
82. Nilsson SK, Debatis ME, Dooner MS, Madri JA, Quesenberry PJ, Becker PS (1998) Immunofluorescence characterization of key extracellular matrix proteins in murine bone marrow in situ. J Histochem Cytochem 46(3):371–377
83. Niswander LM, Fegan KH, Kingsley PD, McGrath KE, Palis J (2014) SDF-1 dynamically mediates megakaryocyte niche occupancy and thrombopoiesis at steady state and following radiation injury. Blood 124(2):277–286. doi:10.1182/blood-2014-01-547638
84. Olson TS, Caselli A, Otsuru S, Hofmann TJ, Williams R, Paolucci P, Dominici M, Horwitz EM (2013) Megakaryocytes promote murine osteoblastic HSC niche expansion and stem cell engraftment after radioablative conditioning. Blood 121(26):5238–5249. doi:10.1182/blood-2012-10-463414
85. Pallotta I, Lovett M, Kaplan DL, Balduini A (2011) Three-dimensional system for the in vitro study of megakaryocytes and functional platelet production using silk-based vascular tubes. Tissue Eng Part C Methods 17(12):1223–1232. doi:10.1089/ten.tec.2011.0134
86. Pallotta I, Lovett M, Rice W, Kaplan DL, Balduini A (2009) Bone marrow osteoblastic niche: a new model to study physiological regulation of megakaryopoiesis. PLoS ONE 4(12), e8359. doi:10.1371/journal.pone.0008359
87. Petit I, Jin D, Rafii S (2007) The SDF-1-CXCR4 signaling pathway: a molecular hub modulating neo-angiogenesis. Trends Immunol 28(7):299–307. doi:10.1016/j.it.2007.05.007
88. Pickford A, Campbell I (2004) NMR studies of modular protein structures and their interactions. Chem Rev 104(8):3557–3566
89. Ponzoni M, Savage DG, Ferreri AJ, Pruneri G, Viale G, Servida P, Bertolini F, Orazi A (2004) Chronic idiopathic myelofibrosis: independent prognostic importance of bone marrow microvascular density evaluated by CD105 (endoglin) immunostaining. Mod Pathol 17(12):1513–1520. doi:10.1038/modpathol.3800224

90. Poulter NS, Pollitt AY, Davies A, Malinova D, Nash GB, Hannon MJ, Pikramenou Z, Rappoport JZ, Hartwig JH, Owen DM, Thrasher AJ, Watson SP, Thomas SG (2015) Platelet actin nodules are podosome-like structures dependent on Wiskott-Aldrich syndrome protein and ARP2/3 complex. Nat Commun 6:7254. doi:10.1038/ncomms8254
91. Psaila B, Lyden D, Roberts I (2012) Megakaryocytes, malignancy and bone marrow vascular niches. J Thromb Haemost 10(2):177–188. doi:10.1111/j.1538-7836.2011.04571.x
92. Remedio L, Carvalho T, Caiado F, Bastos-Carvalho A, Martins D, Duarte A, Yagita H, Dias S (2012) Context- and cell-dependent effects of Delta-like 4 targeting in the bone marrow microenvironment. PLoS ONE 7(12), e52450. doi:10.1371/journal.pone.0052450
93. Rocco M, Infusini E, Daga MG, Gogioso L, Cuniberti C (1987) Models of fibronectin. Embo J 6(8):2343–2349
94. Sagar BM, Rentala S, Gopal PN, Sharma S, Mukhopadhyay A (2006) Fibronectin and laminin enhance engraftibility of cultured hematopoietic stem cells. Biochem Biophys Res Commun 350(4):1000–1005. doi:10.1016/j.bbrc.2006.09.140
95. Sasaki T, Timpl R (2001) Domain IVa of laminin alpha5 chain is cell-adhesive and binds beta1 and alphaVbeta3 integrins through Arg-Gly-Asp. FEBS Lett 509(2):181–185
96. Schachtner H, Calaminus SD, Sinclair A, Monypenny J, Blundell MP, Leon C, Holyoake TL, Thrasher AJ, Michie AM, Vukovic M, Gachet C, Jones GE, Thomas SG, Watson SP, Machesky LM (2013) Megakaryocytes assemble podosomes that degrade matrix and protrude through basement membrane. Blood 121(13):2542–2552. doi:10.1182/blood-2012-07-443457
97. Schaff M, Tang C, Maurer E, Bourdon C, Receveur N, Eckly A, Hechler B, Arnold C, de Arcangelis A, Nieswandt B, Denis CV, Lefebvre O, Georges-Labouesse E, Gachet C, Lanza F, Mangin PH (2013) Integrin alpha6beta1 is the main receptor for vascular laminins and plays a role in platelet adhesion, activation, and arterial thrombosis. Circulation 128(5):541–552. doi:10.1161/CIRCULATIONAHA.112.000799
98. Schick PK, Wojensk CM, Bennett V, Denisova L (1996) Fibronectin isoforms in megakaryocytes. Stem Cells 14(Suppl 1):212–219. doi:10.1002/stem.5530140728
99. Schneider RK, Ziegler S, Leisten I, Ferreira MS, Schumacher A, Rath B, Fahrenkamp D, Muller-Newen G, Crysandt M, Wilop S, Jost E, Koschmieder S, Knuchel R, Brummendorf TH, Ziegler P (2014) Activated fibronectin-secretory phenotype of mesenchymal stromal cells in pre-fibrotic myeloproliferative neoplasms. J Hematol Oncol 7:92. doi:10.1186/s13045-014-0092-2
100. Schofield R (1978) The relationship between the spleen colony-forming cell and the haemopoietic stem cell. Blood Cells 4(1–2):7–25
101. Schwarzbauer JE, Tamkun JW, Lemischka IR, Hynes RO (1983) Three different fibronectin mRNAs arise by alternative splicing within the coding region. Cell 35(2 Pt 1):421–431
102. Semeniak D, Kulawig R, Stegner D, Meyer I, Schwiebert S, Bösing H, Eckes B, Nieswandt B, Schulze H (2016) Proplatelet formation is selectively inhibited by collagen type I via Syk-independent GPVI signaling, J Cell Sci, 129(18):3473–3484
103. Sottile J, Hocking DC (2002) Fibronectin polymerization regulates the composition and stability of extracellular matrix fibrils and cell-matrix adhesions. Mol Biol Cell 13(10):3546–3559. doi:10.1091/mbc.E02-01-0048
104. Stein BL, Gotlib J, Arcasoy M, Nguyen MH, Shah N, Moliterno A, Jamieson C, Pollyea DA, Scott B, Wadleigh M, Levine R, Komrokji R, Klisovic R, Gundabolu K, Kropf P, Wetzler M, Oh ST, Ribeiro R, Paschal R, Mohan S, Podoltsev N, Prchal J, Talpaz M, Snyder D, Verstovsek S, Mesa RA (2015) Historical views, conventional approaches, and evolving management strategies for myeloproliferative neoplasms. J Natl Compr Canc Netw 13(4):424–434
105. Sugiyama T, Kohara H, Noda M, Nagasawa T (2006) Maintenance of the hematopoietic stem cell pool by CXCL12-CXCR4 chemokine signaling in bone marrow stromal cell niches. Immunity 25(6):977–988. doi:10.1016/j.immuni.2006.10.016
106. Sun S, Wang W, Latchman Y, Gao D, Aronow B, Reems JA (2013) Expression of plasma membrane receptor genes during megakaryocyte development. Physiol Genomics 45(6):217–227. doi:10.1152/physiolgenomics.00056.2012

107. Takaku T, Malide D, Chen J, Calado RT, Kajigaya S, Young NS (2010) Hematopoiesis in 3 dimensions: human and murine bone marrow architecture visualized by confocal microscopy. Blood 116(15):e41–e55. doi:10.1182/blood-2010-02-268466
108. Tanaka F, Otake Y, Yanagihara K, Kawano Y, Miyahara R, Li M, Yamada T, Hanaoka N, Inui K, Wada H (2001) Evaluation of angiogenesis in non-small cell lung cancer: comparison between anti-CD34 antibody and anti-CD105 antibody. Clin Cancer Res 7(11):3410–3415
109. Thon JN, Mazutis L, Wu S, Sylman JL, Ehrlicher A, Machlus KR, Feng Q, Lu S, Lanza R, Neeves KB, Weitz DA, Italiano JE Jr (2014) Platelet bioreactor-on-a-chip. Blood 124(12):1857–1867
110. Thyboll J, Kortesmaa J, Cao R, Soininen R, Wang L, Iivanainen A, Sorokin L, Risling M, Cao Y, Tryggvason K (2002) Deletion of the laminin alpha4 chain leads to impaired microvessel maturation. Mol Cell Biol 22(4):1194–1202
111. Truss M, Swat M, Kielbasa SM, Schafer R, Herzel H, Hagemeier C (2005) HuSiDa – the human siRNA database: an open-access database for published functional siRNA sequences and technical details of efficient transfer into recipient cells. Nucleic Acids Res 33(Database issue):D108–D111
112. Weiss RE, Reddi AH (1981) Appearance of fibronectin during the differentiation of cartilage, bone, and bone marrow. J Cell Biol 88(3):630–636
113. Wicke DC, Meyer J, Buesche G, Heckl D, Kreipe H, Li Z, Welte KH, Ballmaier M, Baum C, Modlich U (2010) Gene therapy of MPL deficiency: challenging balance between leukemia and pancytopenia. Mol Ther J Am Soc Gene Ther 18(2):343–352. doi:10.1038/mt.2009.233
114. Winkler IG, Pettit AR, Raggatt LJ, Jacobsen RN, Forristal CE, Barbier V, Nowlan B, Cisterne A, Bendall LJ, Sims NA, Levesque JP (2012) Hematopoietic stem cell mobilizing agents G-CSF, cyclophosphamide or AMD3100 have distinct mechanisms of action on bone marrow HSC niches and bone formation. Leukemia 26(7):1594–1601. doi:10.1038/leu.2012.17
115. Winter O, Moser K, Mohr E, Zotos D, Kaminski H, Szyska M, Roth K, Wong DM, Dame C, Tarlinton DM, Schulze H, MacLennan IC, Manz RA (2010) Megakaryocytes constitute a functional component of a plasma cell niche in the bone marrow. Blood 116(11):1867–1875. doi:10.1182/blood-2009-12-259457
116. Wong D, Winter O, Hartig C, Siebels S, Szyska M, Tiburzy B, Meng L, Kulkarni U, Fahnrich A, Bommert K, Bargou R, Berek C, Chu VT, Bogen B, Jundt F, Manz RA (2014) Eosinophils and megakaryocytes support the early growth of murine MOPC315 myeloma cells in their bone marrow niches. PLoS ONE 9(10), e109018. doi:10.1371/journal.pone.0109018
117. Wright J (1906) The origin and nature of blood platelets. Boston Med Surg J 154:643–645
118. Wu Y, Welte T, Michaud M, Madri JA (2007) PECAM-1: a multifaceted regulator of megakaryocytopoiesis. Blood 110(3):851–859
119. Yamada H, Sekikawa T, Agawa M, Iwase S, Suzuki H, Horiguchi-Yamada J (2008) Adhesion to fibronectin induces megakaryocytic differentiation of JAS-REN cells. Anticancer Res 28(1A):261–266
120. Yousif LF, Di Russo J, Sorokin L (2013) Laminin isoforms in endothelial and perivascular basement membranes. Cell Adh Migr 7(1):101–110. doi:10.4161/cam.22680
121. Zhang L, Orban M, Lorenz M, Barocke V, Braun D, Urtz N, Schulz C, von Bruhl ML, Tirniceriu A, Gaertner F, Proia RL, Graf T, Bolz SS, Montanez E, Prinz M, Muller A, von Baumgarten L, Billich A, Sixt M, Fassler R, von Andrian UH, Junt T, Massberg S (2012) A novel role of sphingosine 1-phosphate receptor S1pr1 in mouse thrombopoiesis. J Exp Med 209(12):2165–2181. doi:10.1084/jem.20121090
122. Zhao M, Perry JM, Marshall H, Venkatraman A, Qian P, He XC, Ahamed J, Li L (2014) Megakaryocytes maintain homeostatic quiescence and promote post-injury regeneration of hematopoietic stem cells. Nat Med 20(11):1321–1326. doi:10.1038/nm.3706

Morphogenesis of Platelets in the Circulation

5

Matthew T. Rondina and Hansjörg Schwertz

Abstract

Platelets are defined as terminally differentiated cells incapable of more complex cell processes like cell division or fission processes. Therefore, megakaryocytes are viewed as the only parent cell producing mature platelets. Nevertheless, the cellular fragments, termed proplatelet extension, released by the megakaryocytes exceed platelet dimension, leaving the question if further maturation of platelets in the blood stream is a potential concept of platelet formation. In this chapter we will review established and emerging findings relevant to platelet morphogenesis in the circulation. Furthermore, some of the known regulatory mechanisms involved will be discussed and useful methods and tool to study platelet morphogenesis will be highlighted.

M.T. Rondina
Molecular Medicine Program, University of Utah, Salt Lake City, UT, USA

Department of Internal Medicine, University of Utah, Salt Lake City, UT, USA

Department of Internal Medicine, George E. Wahlen Salt Lake City VAMC, Salt Lake City, UT, USA

H. Schwertz, MD/PhD (✉)
Molecular Medicine Program, University of Utah, Salt Lake City, UT, USA

Department of Surgery, University of Utah, Salt Lake City, UT, USA

Department of Immunology and Transfusion Medicine, University of Greifswald, Greifswald, Germany

Lichtenberg-Professor for Experimental Hemostasis, University of Greifswald, Greifswald, Germany

Department of Immunology and Transfusion Medicine, University of Greifswald, Sauerbruchstr, Greifswald 17475, Germany

Division of Vascular Surgery, Program in Molecular Medicine, University of Utah, 15 North 2030 East, Bldg. 533, Rm. 4225, Salt Lake City, UT 84112, USA
e-mail: hansjoerg.schwertz@uni-greifswald.de; hansjorg.schwertz@u2m2.utah.edu

H. Schulze, J. Italiano (eds.), *Molecular and Cellular Biology of Platelet Formation*,
DOI 10.1007/978-3-319-39562-3_5

5.1 Introduction

Platelets have been traditionally viewed as terminally differentiated cells incapable of undergoing cell division or fission processes. In this manner, platelet production has classically been thought of occurring solely in the megakaryocyte, the platelet parent cell. The megakaryocyte, during the final stages of proliferation and differentiation, extends pseudopodia-like cytoplasmic protrusions referred to as proplatelet extensions [1, 2]. These cellular extensions are comprised of multiple cell body-like swellings being connected by thin, fusiform shafts. As demonstrated in both in vitro model systems [1] and in vivo live tracking approaches [3–5], proplatelet extensions then fragment into shorter sections or as single platelets that enter the circulation. It is important to note that most of the released fragments exceed platelet dimensions, suggesting that these fragments, also termed preplatelets [6], may retain the potential to undergo maturation in the circulation. Recent studies have also described the release of two different types of platelet progenitors into circulation. These platelet precursors are released via distinct intravascular protrusions that depend, in part, on physiologic or pathophysiologic conditions in the organism [4]. The morphogenesis of platelet precursors into mature platelets in the circulation was initially proposed in 1998 [7], and recent reports have contributed new findings to this field [6, 8], although many questions remain. In this chapter, we will review established and emerging findings relevant to platelet morphogenesis in the circulation, including the trafficking of megakaryocytes to the lung and the contribution of the pulmonary circulation to platelet production and release. In addition, we will briefly discuss some of the known regulatory mechanisms involved and highlight several methods and tools currently available and utilized to fill knowledge gaps in this field.

5.2 A Historical Review of Platelet Morphogenesis

In the past, distinct morphologic variations of normal platelets – not resembling the ordinary discoid phenotype but rather an elongated, fusiform appearance – have been observed in freshly prepared platelet-rich plasma from rats [7, 9–11], guinea pigs [12], and dogs [13]. Whether these elongated, fusiform cell bodies were platelets that were directly shed from the cytoplasm of megakaryocytes or derived from existing platelets in the circulation, however, was unknown. The following paragraph will highlight some of these publications, providing a more detailed insight into morphologies and hypothesis describing potential mechanisms regulating the formation of fusiform platelets and their fate in the circulation.

In 1990, Smith and colleagues described their finding that 15 % of platelets isolated by cardiac puncture from the circulatory system of guinea pigs were elongated, resembling beaded or cylindrical phenotypes [12]. These studies were performed under steady-state conditions of thrombopoiesis and provided initial, and intriguing, evidence to support the hypothesis that these elongated, fusiform platelets may be of significant importance to platelet production and physiology. These investigators also found that they could aspirate individual fusiform platelets into micropipettes, suggesting these elongated platelets were of a single cellular body.

Moreover, elongated platelets did not fragment into smaller "mature" platelets during aspiration, even under conditions mimicking physiologic shear, suggesting that the process by which this elongated, fusiform platelet morphed into single discoid platelets was a more complex and regulated process rather than being exclusively a shear-stress-mediated event.

Subsequently, in 1993, Behnke described his observations of the transformation of elongated, fusiform proplatelets to discoid platelets in the rat circulation using a variety of light, vital, and electron microscopic approaches [9]. Behnke described two very distinct and disparate platelet morphologic events. The first observation was made using live microscopy techniques to study isolated cells. In this manner, Behnke identified a roughly cylindrical platelet cell body that he measured to be approximately 12 μm long. During a 22-min observation period, this cell extended to almost three times its original length, became thinner, and formed cytoplasmic concentrations like pearls on a string. This same phenotypic transformation, where a cell developed pointed tips and then elongated until resembling a beaded and slender phenotype, was also noted in later studies in rat and human cells [7]. At this time, Behnke speculated, based on other observations made decades earlier, that megakaryocytes trapped in lung capillaries might pinch these elongated cellular processes, leading to the release of proplatelets that developed into mature platelets. We will discuss the concept of platelet formation in the pulmonary vasculature niche in more detail below.

Behnke also put forth a model of platelet formation that was based on his identification of these elongated structures having varying degrees of curvatures. Some of the elongated, spindle-like platelets that Behnke observed appeared to curve sufficiently to allow their ends to visually appear to touch. This platelet phenotype has also been observed in our laboratory by SEM (Fig. 5.1a). Behnke also observed crescent-shaped and ring-shaped platelet forms with elongated, beaded cell bodies, similar to morphologies we have detected in platelet preparations from human healthy donors (Fig. 5.1b, c). Based on his observations, Behnke postulated that these ring-shaped platelets may form when tips of the crescent-shaped cells fuse together. With further elongation, he speculated that the ring could subsequently be closed by veil-like membrane structures. This platelet morphology can also be detected in human platelets (Fig. 5.1d) and is visually distinct from quiescent discoid platelets (Fig. 5.1e). Based on Behnke's model, the proposed succession of platelet developmental steps would progress from fusiform to ring and, finally, to discoid platelets, while logically, this sequence of events could obviously occur in the reverse manner, specifically that a discoid platelet first undergoes thinning of the central cell body with shifting of its content toward the outer perimeter. Under this alternative hypothesis, platelet morphological changes would result in a beaded ring formation (Fig. 5.1b, c) and ultimately the formation of a dumbbell-shaped cell (Fig. 5.1f). The precise order of this process, however, remains purely speculation, and additional single-cell experiments are needed to further understand these various platelet forms, the order of their formation, and the mechanisms regulating their formation.

Using transmission electron microscopy (TEM), Behnke went on to demonstrate that both fusiform proplatelets and mature platelets contain a microtubular cytoskeleton that is oriented in the long axis of the cell. Based on these and other

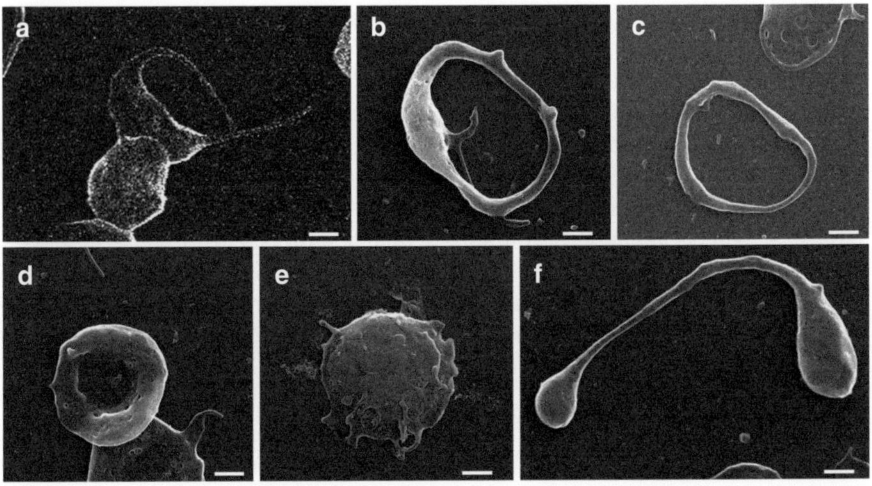

Fig. 5.1 Variations of platelet morphology. Platelets were isolated from healthy donors and either fixed immediately (**e**) or incubated for 6 h at 37 °C, fixed and subsequently analyzed by scanning electron microscopy (*SEM*). (**a**) Cell body with curving fusiform extension. (**b**, **c**) Complete rings with multiple cell body swellings. (**d**) Cell body with thin inner membrane veil. (**e**) Discoid platelet with some protrusions due to the preparation process. (**f**) Platelet dumbbell showing two cell bodies connected by a thin elongated shaft. Scale bars = 1 μm

observations, investigators went on to hypothesize that the bundling and bending of the microtubules would lead to the formation of the marginal band, one of the hallmark structures of mature circulating platelets [1, 14, 15]. Currently, the concept of proplatelet formation and platelet maturation being microtubule dependent is generally accepted [6, 16]. Subsequent studies have gone on to demonstrate that microtubule polymerization and sliding drive proplatelet elongation [17]. The importance of the platelet cytoskeleton and microtubule network will be discussed in more detail below.

5.3 Preplatelet and Platelet Progeny Formation

We recently reported that isolated discoid human platelets from healthy human volunteers generate progeny. This occurs through a morphological process that involves platelet elongation and extension to beaded isoforms that resemble dumbbells or long proplatelet-extension-like processes [8]. At baseline, these platelets had typical discoid morphology, the typical marginal band formed by β-tubulin fibers, and a homogenous cell size distribution containing only very few large platelets (Fig. 5.2). During conditions where platelets were incubated for various times, however, the appearance of barbell-shaped platelets that contained two or more cell bodies was observed. Moreover, when confined in microfluidic-based media droplets, platelets underwent fission to generate daughter cells, a process we have termed progeny formation (Fig. 5.3 and [8]). Platelet counts also increased in transfusion bags during storage, suggesting that under these settings, platelets undergo fission and

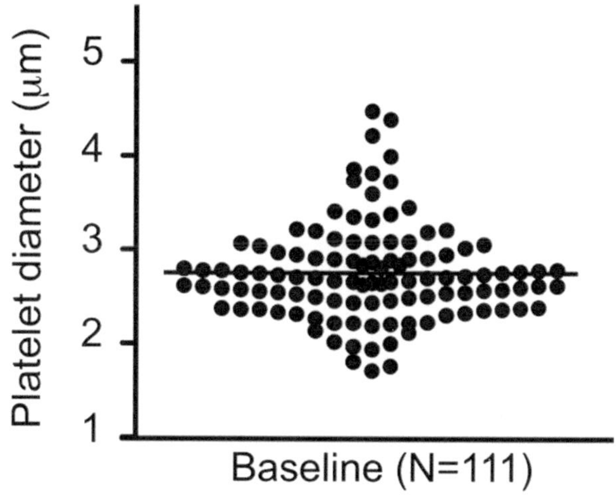

Fig. 5.2 Large platelets are rare in freshly isolated human cell preparations. The dot blot indicates *TEM* diameter measurements along the longest axis of freshly isolated platelets (baseline). The closed circles show a homogenous cell population that contains very few large platelets

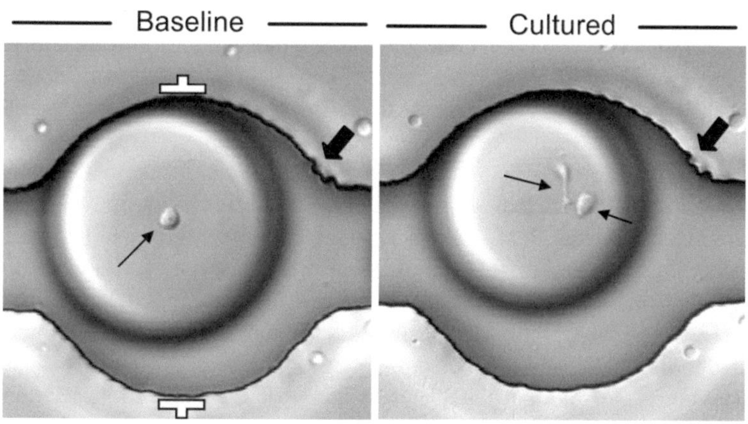

Fig. 5.3 Individual platelets confined in media droplets produce daughter cells. Platelets were loaded into microdrops generated by microfluidics technique and examined at baseline or after incubation (cultured). The *thin arrows* point to a single platelet (baseline) or the same platelet that formed two new cell bodies (cultured). The *thick arrows* point to unique landmarks for characteristic of this position on the microfluidic device

progeny formation, thus increasing their number. Finally, when incubating freshly isolated platelets at low concentrations representative of thrombocytopenic conditions in humans, the percentage of dumbbell or beaded platelets increased (Schwertz, unpublished observations).

At approximately the same time, in an elegant set of murine studies, Thon and colleagues discovered a new intermediate stage in platelet production, which they termed the preplatelet [6]. They defined the preplatelet as a circulating discoid cell that is larger than normal platelets (the normal size of circulating murine platelets is between 0.5 and 2 μm [18]) but which retains the capacity to elongate and extend into barbell-shaped proplatelets. Their finding that these larger cellular processes

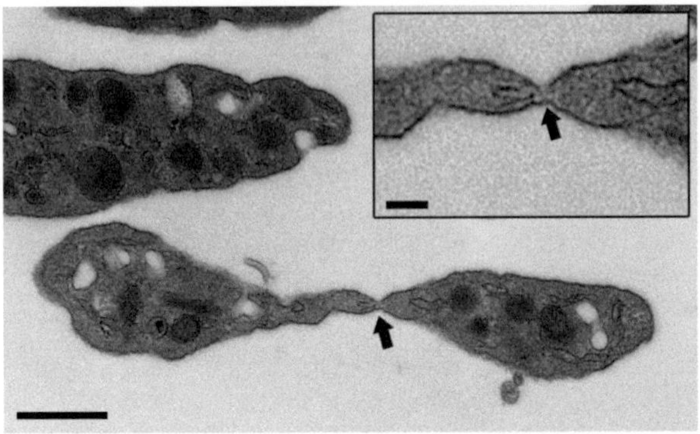

Fig. 5.4 Cell bodies of extended platelets are separated by segmented constrictions. A constricted region partially missing a well-defined cell membrane and resembling a cleavage furrow (*arrow*) is noted along the shaft of a cultured platelet. Scale bar = 500 nm (insert scale bar = 100 nm)

circulate and are megakaryocyte derived is supported by the observation that bone marrow megakaryocytes extend and release fragments exceeding platelet dimensions [3]. Using isolations of released proplatelets/preplatelets from mouse fetal liver-derived megakaryocyte cultures, these investigators found that preplatelets only developed an extended cell phenotype if their diameter was between 2 and 10 μm, a larger size than mature murine platelets, suggesting that specific size restrictions might influence the formation of this platelet phenotype.

Thon and colleagues then isolated proplatelets, which are platelet precursors, from mice. Following fluorescent labeling of these proplatelets, they were subsequently transfused back into the mice. By recovering mature platelets from the circulation post-infusion, they demonstrated that the infused labeled precursors generated individual, discoid, labeled platelets.

In both murine and human platelets, small, segmented constrictions located in the thin shafts connecting newly formed platelet cell bodies have been identified (Fig. 5.4 and [6, 8]). These structures are consistent with cellular pinching actions and resemble constriction furrows in dividing nucleated cells, suggesting perhaps that platelet cell body separation is regulated by more complex processes than shear stress alone. The observations by both groups may also be an explanation for the sub-megakaryocyte size fragments described previously in human platelet-rich plasma [7].

Taken together, these two studies highlight the identification of new platelet morphologies that may play a role in platelet production and/or maturation although many additional questions remain to be answered. For example, the size restriction for the development of extended platelets in a murine system may not be conserved across human platelets as ex vivo, human platelets of normal size (e.g., approximately 2–3 μm) are capable of forming new platelet extensions and beaded cell bodies [8]. In addition, large human platelets (>4 μm), which are thought to be younger, are infrequently observed in platelet preparations (Fig. 5.2 and Schwertz et al., unpublished observations), suggesting that the process of extension and new cell body formation might not be exclusively restricted to very young platelets.

Finally, while it remains unknown as to whether preplatelets develop into proplatelets in vivo, limitations in demonstrating this may be the result of the fragility of interconnecting shafts between platelet bodies, which when exposed to shear forces in the bloodstream might be severed. Additional studies will hopefully fill current knowledge gaps on the processes and mechanisms regulating these stages of platelet development.

5.4 Endogenous Factors That Influence Platelet Morphogenesis

It is a well-accepted concept that components of the cytoskeleton, namely, microtubules (especially β-tubulin) as well as the actin cytoskeleton, are an integral part of proplatelet formation and thrombopoiesis [1, 14, 19, 20]. Below, we will briefly review these published studies highlighting the importance of the cytoskeleton for platelet morphogenesis.

5.4.1 Role of the Cytoskeleton in Platelet Morphology

The microtubular cytoskeleton is known to play an important role in megakaryocyte maturation, proplatelet formation, and platelet function [1, 14, 19, 20]. In the aforementioned publication from 1993 [9], Behnke set the stage by asking how a fusiform platelet might be transformed into a discoid platelet with its defining marginal band of β-tubulin fibers. In fusiform proplatelet-like structures, the microtubules were observed to be arranged along the long cell axis [21]. Therefore, Behnke [9] and others [22] suggested that bending of the microtubular cytoskeleton could allow the tips of the platelet cell body to lie in close proximity, resulting in fusion and subsequent ring formation. It was further speculated that microtubule associated proteins might constrain the microtubules as a bundled structure together. Since these initial reports, ongoing work has filled in some of the intricate details related to the function of microtubules in proplatelet formation and platelet biology [6, 8, 16, 17, 20].

For example, Thon and colleagues [6] incubated proplatelets at either 37 °C or 4 °C to induce tubulin depolymerization and, in some experiments, again at 37 °C to induce tubulin repolymerization. Depolymerization of tubulin resulted in a shift in the morphology of the proplatelet population toward discoid phenotypes. However, in conditions where tubulin repolymerization was induced, this shift was reversed completely. Changes in microtubule organization were also directly visualized by expressing GFP-labeled β1-tubulin in isolated released mouse proplatelet extensions. Using this technique, the authors were able to demonstrate that the oval marginal band of a preplatelet ($\approx 7\,\mu$m in diameter) twisted around its center, leading to the formation of a barbell-shaped extended cell. The importance of the microtubular cytoskeleton for preplatelet morphogenesis was further confirmed using pharmacological manipulations of microtubules [6].

In experiments with either human or murine platelets [6, 8], incubated in the presence of either taxol (a microtubule-stabilizing agent) or nocodazole (a tubulin depolymerizing substance), platelets and proplatelets were prevented from undergoing morphologic shifts to larger beaded cell bodies or preplatelets, respectively. Intriguingly, in the presence of these tubulin-destabilizing agents, platelets were observed with a phenotype described as "teardrop-like," suggesting that these ordered cellular events may have been interrupted during the fission and morphogenesis process [8]. The results from these studies highlight that the dramatic reorganization of the tubulin network during the conversion from discoid cell to dumbbell-shaped cell plays an essential role in platelet morphogenesis.

Actin is also an important regulator of both platelet spreading and filopodia formation [15]. In ongoing studies, we have examined the effects of inhibiting actin polymerization on platelet progeny formation. In the presence of cytochalasin D, an inhibitor of actin polymerization, platelet progeny formation is blocked (Schwertz, unpublished observations), demonstrating the importance of actin and associated proteins in these processes. While our understanding of the mechanisms and cytoskeletal components regulating proplatelet and preplatelet development and platelet progeny formation is increasing, detailed mechanistic studies are still needed to better understand these processes in platelets and megakaryocytes.

5.5 Megakaryocyte Niches Outside the Bone Marrow: Implications for Extra-marrow Thrombopoiesis and Platelet Morphogenesis

The adult bone marrow is recognized traditionally as a major reservoir for megakaryocytes where they develop, like other cells from a master stem cell, and the site where platelet production may primarily occur [23]. Nevertheless, many of the control checkpoints and signaling events regulating thrombopoiesis are not contingent upon the bone marrow environment. In addition to the bone marrow niche, a growing body of human and experimental data has identified that megakaryocytes are found in mammalian lungs, which represents the largest capillary bed. Moreover, thrombopoiesis can occur under certain circumstances and settings in the lungs of human and nonhuman mammals, leading to platelet release [24]. Based on some historic studies of retrograde perfused rat lungs, it also seems reasonable to assume that circulating megakaryocytes can be trapped in the lung capillaries producing proplatelet extensions, leading to circulating platelet precursors [25].

The first description of megakaryocytes within the vascular niche of the lung has been attributed to Aschoff in 1893, where he suggested that megakaryocytes from the bone marrow migrate into the systemic circulation, lodge within the capillary bed of the lung due in part to their size, and then may release platelets [26] (and reviewed in [27]). Consistent with this initial finding, images of megakaryocytes migrating through openings in the endothelium that are 3–6 μm in diameter have been recorded [28]. While retention of megakaryocytes in the microvessels of the pulmonary circulation due to the size of these large cells may be expected, other morphological and structural aspects of megakaryocytes may also contribute. For example, the cytoskeletal rigidity of megakaryocytes as well as megakaryocyte nuclear size and complexity may limit the deformability of these cells, promoting their conservation in the pulmonary vessels [24, 27]. Furthermore, platelets being morphed into beaded extensions can reach up to 100 μm in length (Schwertz et al., unpublished observation), which might lead to a deposition in the lung capillary bed. In addition, the expression of adhesion molecules on the surface of megakaryocytes and platelet precursors may contribute to their adhesion to the pulmonary endothelium and subsequent retention within the lungs, although the evidence for this remains limited. These mechanisms could also be attributed to preplatelets or barbell-shaped platelet morphing into mature discoid platelets. In settings of human disease where inflammatory insults or vascular injury activates the endothelium, expression of these molecules may be upregulated, promoting increased adhesion of megakaryocytes to the pulmonary endothelium, megakaryocyte maturation, and platelet release [24, 29–31]. Nevertheless, the precise morphological and structural features and signaling pathways controlling megakaryocyte retention, or even morphogenesis of preplatelets and platelets in the lung, remain incompletely understood.

Driven by Aschoff's observations and other lines of experimental evidence, subsequent investigations in humans and experimental model systems in vivo have consistently identified megakaryocytes within the pulmonary vasculature. For example, Howell and colleagues, in an early set of experiments, performed comparative counting of platelets in the arterial and venous circulation of anesthetized dogs and cats. These investigators concluded that new platelets are added to the capillary circulation of the lungs, pointing toward thrombopoiesis taking place in the lung. In addition, through histological examination, they identified "giant cells" within lung sections, morphological findings consistent with an interpretation that these cells were megakaryocytes [32]. These cells were also noted to have retained cytoplasm. Subsequent studies identified megakaryocytes within the lungs of other nonhuman mammals, including rabbits, rats, and mice (reviewed in [24, 27, 33]).

Megakaryocyte numbers in the pulmonary vasculature are remarkably high with as many as 250,000 megakaryocytes traveling through the lung every hour [27]. Moreover, in experimental in vivo settings where thrombocytopenia is induced in rabbits, there is a three- to sixfold increase in the number of megakaryocytes detected histologically in the lungs (reviewed in [33]). While investigations utilizing immunostaining or hematoxylin and eosin staining (combined with light microscopy) concluded that megakaryocytes were rare in murine lungs [33], more recent ultrastructural-based studies identified intact megakaryocytes present under basal conditions in CB6FL/J mice [34], suggesting that demonstration of megakaryocytes may be influenced by detection techniques. In human lungs, megakaryocytes can be identified in necropsy studies and during routine diagnostic lung biopsies [30]. During human disease settings, including ALI/ARDS, disseminated intravascular coagulation, burns, and thrombosis, megakaryocyte numbers increase in the lung [24, 30, 35]. In addition, some microscopy-based investigations have noted cells

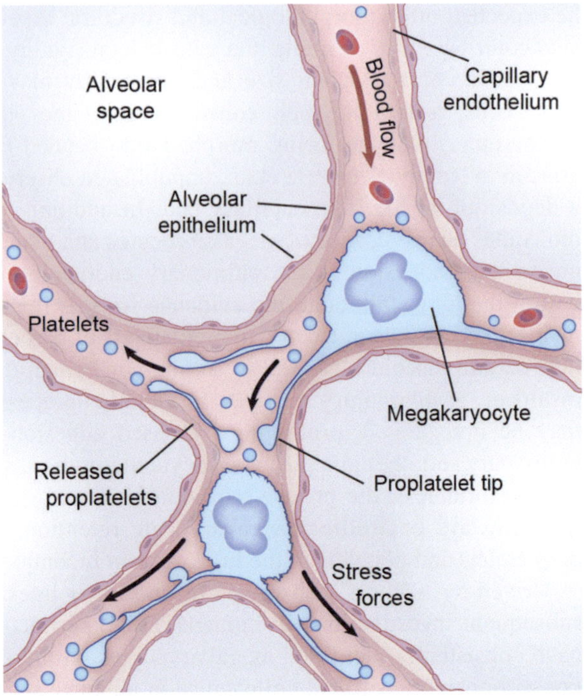

Fig. 5.5 Megakaryocytes traffic to, and are retained within, human lung microvessels. In these settings, megakaryocytes may release platelets, contributing to thrombopoiesis, although this process has not been formally demonstrated in the lungs of human subjects. In some human disease settings, such as ALI/ARDS, the number of megakaryocytes may increase, further leading to increased platelet production. The process of platelet genesis from megakaryocytes residing within the lung, however, remains incompletely understood

identified as megakaryocytes that have visible cytoplasmic extensions characteristic of proplatelet formation [36, 37].

Taken together, these established and emerging studies identify that megakaryocytes circulating within the pulmonary vasculature may be retained in substantial numbers in the microvessels of the lung and may increase in number during disease settings (Fig. 5.5). In addition, this mechanism might also be true for proplatelet processes, preplatelets, or even dumbbell-shaped platelets undergoing morphogenesis, which have an increased size.

Nevertheless, important questions remain regarding how effectively platelet production occurs by megakaryocytes retained in the lung and the relative contribute of this thrombopoiesis or platelet morphogenetic mechanisms to the circulating platelet pool. In human subjects where blood was sampled simultaneously from both pulmonary arterial vessels and the aorta, megakaryocytes are approximately tenfold more abundant in pulmonary arterial blood, with many of the megakaryocytes having an intact cytoplasm (reviewed in [24]. Intriguingly, some of these megakaryocytes identified and examined microscopically had cytoplasmic extensions that morphologically were consistent with proplatelet formation [36–38]. Approximately 90 % of intact megakaryocytes that are found within blood samples taken from the pulmonary artery are retained within the lung vasculature, with only a fraction passing through into the systemic circulation. Advancements in imaging techniques and resolution and ongoing investigations of thrombopoiesis in extra-marrow niches should help shed additional light on this subject.

Bierman measured platelets in humans during settings of cross-transfusion (where blood is removed from the right ventricle and then reintroduced into the pulmonary artery) and in the presence of epinephrine [39]. They found that both cross-transfusion and the administration of epinephrine led to immediate and large increases in platelets from the lung, leading them to conclude that the lungs are a significant reservoir for platelets and are involved both in their production and storage. Intriguingly, calculations suggest that the average production of platelets each day could occur in human lungs, based on the volume of cytoplasm of trapped megakaryocytes in the pulmonary circulation [40]. Even more, computational modeling suggested that physical fragmentation processes within the cytoplasm induce megakaryocytes to release all platelets into the circulation [41]. Earlier studies proposed that this fragmentation process occurred along demarcation membrane system (DMS) fracture lines and involved an extensive internal membrane reorganization process [42–44]. Subsequent investigations have cast some uncertainty on the DMS model for platelet production, suggesting instead that the DMS may function primarily to provide plasma membrane for proplatelet formation. Currently, thrombopoiesis is thought to primarily occur through proplatelet formation, a process recently discovered to be mediated in part by shear forces (reviewed in [27]).

The force applied by circulating blood and the cross-sectional area of the vessel through which the blood is moving determines shear stress in the vasculature niche. In the human circulatory system, shear stresses may vary markedly based on flow rates and vascular diameters. For example, shear rates in bone marrow sinusoids are typically low, while in the circulation (including the pulmonary vasculature), shear rates may be up to fivefold higher and may increase further during human diseases or experimentally induced pathophysiologic conditions [45–47]. High shear rates, in vitro, accelerate platelet release from megakaryocytes [6, 48]. Thus, in the pulmonary circulation, as compared to bone marrow sinusoids, the increased shear may stimulate platelet release from resident megakaryocytes and in addition accelerate the morphologic processes implied in pre- and proplatelet formation. Nevertheless, the precise molecular events involved in platelet release in response to shear, particularly in human lungs, remain incompletely understood. Alterations in this process in human diseases are even less well examined yet likely play important roles in both adaptive and maladaptive responses and influences clinical outcomes.

5.6 Toolsets and Techniques in the Evaluation of Platelet Morphogenesis

Many of the regulatory mechanisms involved in regulating these various stages of platelet morphogenesis remain poorly understood. Some of these limitations in knowledge may reflect the only recent development of more sophisticated tools and techniques for advances in this field.

The careful handling and isolation of platelets and proplatelets is requisite for reliable and valid experimental results as pre-analytical techniques may influence experimental outcomes. For example, the inadvertent activation of platelets or the introduction of shear forces, either during blood sampling or isolation, may markedly impair the detection of these various platelet morphologies. Best results may be achieved when passive, gravitational-force-based separation of platelets from platelet-rich plasma is utilized. In addition, as there has not yet been any identification of surface or intracellular markers unique to these platelet precursors, it may be necessary, at times, to perform experiments using single-cell approaches. Traditional tools in single-cell systems are often either cell-culture coated or made from non-inert surfaces (i.e., glass coverslips). These properties can result in inadvertent, ex vivo platelet activation preventing morphogenesis (Schwertz, unpublished observations). New experimental techniques, such as the use of microfluidic techniques, may help overcome some of these obstacles [8]. One innovative technique is the use of controlled ultra low-flow conditions where single platelets can be confined in media drops enclosed by an oil-phase, preventing surface contact of the platelet cell bodies [8]. The drops can then be parked in a custom microfluidic device and analyzed using a confocal laser-scanning microscope setup with an environmental control chamber, to ensure constant temperature, CO_2-concentration, and humidity. Thon and colleagues [49] recently described another very promising microfluidic-based technique. In this application, a microfluidic-based bioreactor mimics the bone marrow environment to produce proplatelet extensions and functional platelets from either mouse fetal liver-derived megakaryocytes or human-induced pluripotent stem cell (hiPSC)-derived megakaryocytes. This technique allows for production of large numbers of uniform platelets in vitro under carefully controlled conditions that may be manipulated.

Other emerging toolsets and techniques include the genetic manipulation of megakaryocytes and experiments investigating the role of protein synthesis, degradation, and turnover in platelet morphogenesis. By targeting candidate genes involved in platelet morphogenesis in highly selected fashion, the molecular mechanisms and pathways controlling platelet progeny formation and platelet morphogenesis may be further dissected. In addition, a deeper understanding of protein synthetic events in platelets and megakaryocytes will likely be integral to deciphering the events regulating platelet morphogenesis. Investigations into whether platelets synthesize membrane lipids and subsequently new double lipid layer membranes or if they reutilize the abundant membrane reservoir of the open canalicular system (OCS) during elongation and fission events will likely yield crucial new insights into these regulatory pathways. Platelets have been shown to possess requisite machinery for lipid metabolism and readily synthesize fatty acids [50, 51]. While initial seminal experiments in this area were performed approximately 45 years ago, current methods will enable investigators to get a more detailed picture of the lipid synthetic repertoire of human platelets.

In addition, it remains to be clarified if and how platelets can synthesize new organelles, such as mitochondria, α-granules, δ-granules, and even ribosomes. Live super-resolution microscopy techniques in conjunction with organelle tracking need to be developed to gain further knowledge about these cellular events. In particular, the accurate live staining of organelles might only be achievable when combining the techniques of in vitro platelet generation using bioreactors [49, 52, 53], with a robust and reliable transfection technique of hiPSCs or megakaryocytes inducing labeled and organelle-specific proteins. More detailed functional tests to evaluate the responsiveness of newly formed platelets are needed to determine if these platelets have altered functional responses. Studies using NOD-SCID mouse models [53, 54] treated with human platelets, megakaryocytes, or hiPSCs may help determine if newly formed platelets represent a distinct population from those produced by megakaryocytes.

5.7 Summary

Studies performed over the last two decades have led to exciting new developments, demonstrating the complex dynamic interplay between proplatelets, preplatelets, platelets, and platelet progeny formation. Emerging evidence also implicates the role of extra-marrow thrombopoiesis in some settings. Nevertheless, many questions remain surrounding the precise sequence of events in these processes, the regulatory mechanisms and pathways involved, and how platelet morphogenesis and production may be altered during human disease settings. Evolving tools and techniques will likely help advance this field further, filling in current knowledge gaps and bringing important new insights to the processes of platelet production events.

Acknowledgments We thank Ms. Diana Lim for her excellent figure preparation and Dr. Guy Zimmerman for his thoughtful suggestions and insights. We appreciate Ms. Nancy Chandler's assistance with preparing SEM images and we thank Dr. Walter H. Kahr for his collaboration with TEM pictures. This work was supported by the NIH (HL126547 and HL112311) and the NIA (AG048022). Dr. Hansjörg Schwertz was supported by a Lichtenberg-Professorship from the Volkswagen Foundation.

References

1. Italiano JE Jr, Lecine P, Shivdasani RA, Hartwig JH (1999) Blood platelets are assembled principally at the ends of proplatelet processes produced by differentiated megakaryocytes. J Cell Biol 147(6):1299–1312
2. Geddis AE, Kaushansky K (2007) Immunology. The root of platelet production. Science 317(5845):1689–1691
3. Junt T, Schulze H, Chen Z et al (2007) Dynamic visualization of thrombopoiesis within bone marrow. Science 317(5845):1767–1770
4. Kowata S, Isogai S, Murai K et al (2014) Platelet demand modulates the type of intravascular protrusion of megakaryocytes in bone marrow. Thromb Haemost 112(4):743–756
5. Zhang L, Orban M, Lorenz M et al (2012) A novel role of sphingosine 1-phosphate receptor S1pr1 in mouse thrombopoiesis. J Exp Med 209(12):2165–2181
6. Thon JN, Montalvo A, Patel-Hett S et al (2010) Cytoskeletal mechanics of proplatelet maturation and platelet release. J Cell Biol 191(4):861–874
7. Behnke O, Forer A (1998) From megakaryocytes to platelets: platelet morphogenesis takes place in the bloodstream. Eur J Haematol Suppl 61:3–23
8. Schwertz H, Koster S, Kahr WH et al (2010) Anucleate platelets generate progeny. Blood 115(18):3801–3809

9. Behnke O (1993) The Formation of Fusiform Proplatelets and their Transformation to Discoid Platelets. Platelets 4(5):262–267
10. Handagama PJ, Feldman BF, Jain NC, Farver TB, Kono CS (1987) Circulating proplatelets: isolation and quantitation in healthy rats and in rats with induced acute blood loss. Am J Vet Res 48(6):962–965
11. Tong M, Seth P, Penington DG (1987) Proplatelets and stress platelets. Blood 69(2):522–528
12. Smith CM 2nd, Burris SM, White JG (1990) High frequency of elongated platelet forms in guinea pig blood: ultrastructure and resistance to micropipette aspiration. J Lab Clin Med 115(6):729–737
13. Handagama P, Jain NC, Kono CS, Feldman BF (1986) Scanning electron microscopic studies of megakaryocytes and platelet formation in the dog and rat. Am J Vet Res 47(11): 2454–2460
14. Italiano JE Jr, Shivdasani RA (2003) Megakaryocytes and beyond: the birth of platelets. J Thromb Haemost 1(6):1174–1182
15. Hartwig JH (2006) The platelet: form and function. Semin Hematol 43(1 Suppl 1):S94–S100
16. Thon JN, Macleod H, Begonja AJ et al (2012) Microtubule and cortical forces determine platelet size during vascular platelet production. Nat Commun 3:852
17. Bender M, Thon JN, Ehrlicher AJ et al (2015) Microtubule sliding drives proplatelet elongation and is dependent on cytoplasmic dynein. Blood 125(5):860–8
18. Schmitt A, Guichard J, Masse JM, Debili N, Cramer EM (2001) Of mice and men: comparison of the ultrastructure of megakaryocytes and platelets. Exp Hematol 29(11):1295–1302
19. Italiano J Jr, Hartwig JH (2007) Megakaryocyte development and platelet formation. In: Michelson AD (ed) Platelets. Elsevier, San Diego, pp 23–44
20. Italiano JE Jr, Patel-Hett S, Hartwig JH (2007) Mechanics of proplatelet elaboration. J Thromb Haemost 5(Suppl 1):18–23
21. Stenberg PE, Levin J (1989) Mechanisms of platelet production. Blood Cells 15(1):23–47
22. Radley JM, Hartshorn MA (1987) Megakaryocyte fragments and the microtubule coil. Blood Cells 12(3):603–614
23. Ogawa M (1993) Differentiation and proliferation of hematopoietic stem cells. Blood 81(11):2844–2853
24. Weyrich AS, Zimmerman GA (2013) Platelets in lung biology. Annu Rev Physiol 75:569–591
25. Pedersen NT (1974) The pulmonary vessels as a filter for circulating megakaryocytes in rats. Scand J Haematol 13(3):225–231
26. Aschoff L (1893) Ueber capillare Embolien von niesenkernhaltigen Zellen. Arch Pathol Anat Phys 134:11–14
27. Italiano JE, Hartwig JH (2013) Megakaryocyte Development and Platelet Formation, 3rd edn. Academic Press, London/Waltham
28. Tavassoli M, Aoki M (1981) Migration of entire megakaryocytes through the marrow–blood barrier. Br J Haematol 48(1):25–29
29. Avraham H, Cowley S, Chi SY, Jiang S, Groopman JE (1993) Characterization of adhesive interactions between human endothelial cells and megakaryocytes. J Clin Invest 91(6):2378–2384
30. Bozza FA, Shah AM, Weyrich AS, Zimmerman GA (2009) Amicus or adversary: platelets in lung biology, acute injury, and inflammation. Am J Respir Cell Mol Biol 40(2):123–134
31. Nicolls MR, Tamosiuniene T, Babu AN, Voelkel NF (2009) Interactions of pulmonary endothelial cells with immune cells and platelets: implications for disease pathogenesis. Wiley, Chichester/Hoboken
32. Howell WH, Donahue DD (1937) The production of blood platelets in the lungs. J Exp Med 65(2):177–203
33. Davis RE, Stenberg PE, Levin J, Beckstead JH (1997) Localization of megakaryocytes in normal mice and following administration of platelet antiserum, 5-fluorouracil, or radiostrontium: evidence for the site of platelet production. Exp Hematol 25(7):638–648

34. Zucker-Franklin D, Philipp CS (2000) Platelet production in the pulmonary capillary bed: new ultrastructural evidence for an old concept. Am J Pathol 157(1):69–74
35. Mandal RV, Mark EJ, Kradin RL (2007) Megakaryocytes and platelet homeostasis in diffuse alveolar damage. Exp Mol Pathol 83(3):327–331
36. Kaufman RM, Airo R, Pollack S, Crosby WH (1965) Circulating megakaryocytes and platelet release in the lung. Blood 26(6):720–731
37. Pedersen NT (1978) Occurrence of megakaryocytes in various vessels and their retention in the pulmonary capillaries in man. Scand J Haematol 21(5):369–375
38. Kallinikos-Maniatis A (1969) Megakaryocytes and platelets in central venous and arterial blood. Acta Haematol 42(6):330–335
39. Bierman HR (1955) The hematologic role of the lung in man. Am J Surg 89(1):130–140
40. Levine RF, Eldor A, Shoff PK, Kirwin S, Tenza D, Cramer EM (1993) Circulating megakaryocytes: delivery of large numbers of intact, mature megakaryocytes to the lungs. Eur J Haematol 51(4):233–246
41. Trowbridge EA, Martin JF, Slater DN (1982) Evidence for a theory of physical fragmentation of megakaryocytes, implying that all platelets are produced in the pulmonary circulation. Thromb Res 28(4):461–475
42. Kosaki G (2005) In vivo platelet production from mature megakaryocytes: does platelet release occur via proplatelets? Int J Hematol 81(3):208–219
43. Shaklai M, Tavassoli M (1978) Demarcation membrane system in rat megakaryocyte and the mechanism of platelet formation: a membrane reorganization process. J Ultrastruct Res 62(3):270–285
44. Yamada E (1957) The fine structure of the megakaryocyte in the mouse spleen. Acta Anat (Basel) 29(3):267–290
45. Kroll MH, Afshar-Kharghan V (2012) Platelets in pulmonary vascular physiology and pathology. Pulm Circ 2(3):291–308
46. Machlus KR, Italiano JE Jr (2013) The incredible journey: from megakaryocyte development to platelet formation. J Cell Biol 201(6):785–796
47. Weinbaum S, Cowin SC, Zeng Y (1994) A model for the excitation of osteocytes by mechanical loading-induced bone fluid shear stresses. J Biomech 27(3):339–360
48. Dunois-Larde C, Capron C, Fichelson S, Bauer T, Cramer-Borde E, Baruch D (2009) Exposure of human megakaryocytes to high shear rates accelerates platelet production. Blood 114(9):1875–1883
49. Thon JN, Mazutis L, Wu S et al (2014) Platelet bioreactor-on-a-chip. Blood 124(12): 1857–1867
50. Lewis N, Majerus PW (1969) Lipid metabolism in human platelets. II. De novo phospholipid synthesis and the effect of thrombin on the pattern of synthesis. J Clin Invest 48(11): 2114–2123
51. Majerus PW, Smith MB, Clamon GH (1969) Lipid metabolism in human platelets. I. Evidence for a complete fatty acid synthesizing system. J Clin Invest 48(1):156–164
52. Nakagawa Y, Nakamura S, Nakajima M et al (2013) Two differential flows in a bioreactor promoted platelet generation from human pluripotent stem cell-derived megakaryocytes. Exp Hematol 41(8):742–748
53. Nakamura S, Takayama N, Hirata S et al (2014) Expandable megakaryocyte cell lines enable clinically applicable generation of platelets from human induced pluripotent stem cells. Cell Stem Cell 14(4):535–548
54. Feng Q, Shabrani N, Thon JN et al (2014) Scalable generation of universal platelets from human induced pluripotent stem cells. Stem Cell Reports 3(5):817–831

Part II

Platelet Biology: Signals and Functions

Platelet Contents

6

Robert Flaumenhaft and Secil Koseoglu

Abstract

Following their release from megakaryocytes, platelets circulate as tiny sentinels of the vasculature. In this capacity, they are equipped with organelles and membrane systems that contribute to many aspects of host defense including hemostasis and thrombosis, inflammation, angiogenesis, and wound healing [1–5]. Prominent in their armory are granules, self-contained vesicular structures that store high concentrations of bioactive cargos. Granule cargos are released from activated platelets in a variety of physiological and pathophysiological settings and function to maintain the integrity of blood vessels [6–9]. The most prominent platelet organelles, dense granules and α-granules, are unique to platelets. However, platelets also contain a number of organelles such as lysosomes and mitochondria that are common to all eukaryotic cells. Although these organelles primarily serve housekeeping functions, there is evidence that they also participate more directly in maintenance of vascular integrity. In addition, as a consequence of its unusual mode of biogenesis, the platelet is endowed with unique membrane systems including the open canalicular system (OCS) and dense tubular system (DTS) that course throughout its interior [10, 11]. These membrane systems also contribute to the function of platelets in host defense. This chapter will provide an overview of platelet contents with a focus on platelet granules.

R. Flaumenhaft, MD, PhD (✉)
Division of Hemostasis and Thrombosis, Department of Medicine,
Beth Israel Deaconess Medical Center, Harvard Medical School, Boston, MA 02215, USA

Center for Life Science, Rm 905, Beth Israel Deaconess Medical Center,
3 Blackfan Circle, Boston, MA 02215, USA
e-mail: rflaumen@bidmc.harvard.edu

S. Koseoglu
Division of Hemostasis and Thrombosis, Department of Medicine, Beth Israel Deaconess Medical Center, Harvard Medical School, Boston, MA 02215, USA

H. Schulze, J. Italiano (eds.), *Molecular and Cellular Biology of Platelet Formation*,
DOI 10.1007/978-3-319-39562-3_6

6.1 Introduction

Following their release from megakaryocytes, platelets circulate as tiny sentinels of the vasculature. In this capacity, they are equipped with organelles and membrane systems that contribute to many aspects of host defense including hemostasis and thrombosis, inflammation, angiogenesis, and wound healing [1–5]. Prominent in their armory are granules, self-contained vesicular structures that store high concentrations of bioactive cargos. Granule cargos are released from activated platelets in a variety of physiological and pathophysiological settings and function to maintain the integrity of blood vessels [6–9]. The most prominent platelet organelles, dense granules and α-granules, are unique to platelets. However, platelets also contain a number of organelles such as lysosomes and mitochondria that are common to all eukaryotic cells. Although these organelles primarily serve housekeeping functions, there is evidence that they also participate more directly in maintenance of vascular integrity. In addition, as a consequence of its unusual mode of biogenesis, the platelet is endowed with unique membrane systems including the open canalicular system (OCS) and dense tubular system (DTS) that course throughout its interior [10, 11]. These membrane systems also contribute to the function of platelets in host defense. This chapter will provide an overview of platelet contents with a focus on platelet granules (Fig. 6.1).

6.2 Platelet Granules

Granules were first observed in platelets using light microscopy in the late nineteenth century [12]. However, it was not until the application of electron microscopy (EM) in the 1960s that platelets were imaged with adequate resolution to demonstrate the diversity of their granules [13]. The first granules to be identified by electron microscopy were dense granules [13], which were easily detected on the basis of their intrinsic electron density. α-Granules were identified a year later as being distinct from dense granules and from typical lysosomes, which had been identified in many nucleated cells [14]. However, the distinction between α-granules and lysosomes remained controversial until the early 1980s, when electron microscopy coupled with cytochemistry was used to distinguish these granules types [15]. Thus, by convention, we think of platelets as having three categories of granules: dense granules, α-granules, and lysosomes.

6.2.1 Platelet Dense Granules

Platelet dense granules are lysosome-related organelles that are unique to platelets [16]. Platelets contain an average of three to eight dense granules [17, 18]. These granules are approximately 150 nm in diameter and are so electron dense that they can be detected by whole mount electron microscopy in the absence of staining (Fig. 6.2). When chemically fixed and viewed by transmission electron microscopy (TEM), dense granules can give the appearance of a "bull's-eye," which represents the retraction of the electron-dense core from the granule

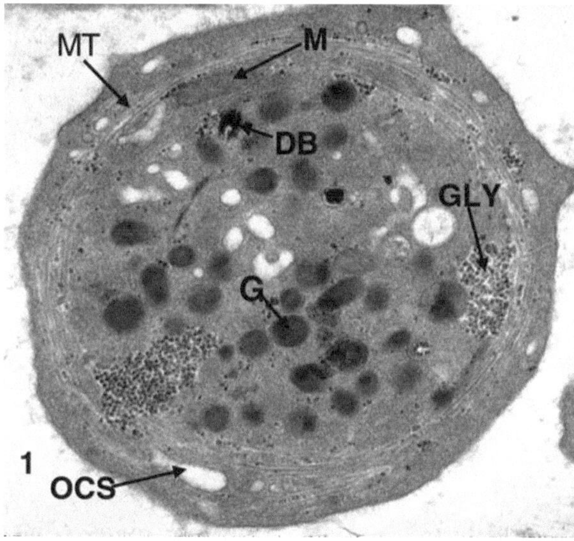

Fig. 6.1 Platelet structure. TEM image of a human platelet (×23,000). Platelet α-granules (*G*), microtubules (*MT*), mitochondria (*M*), open canalicular system (*OCS*), dense bodies (*DB*), and glycogen (*GLY*) are identified (Adapted with permission from Ref. [10])

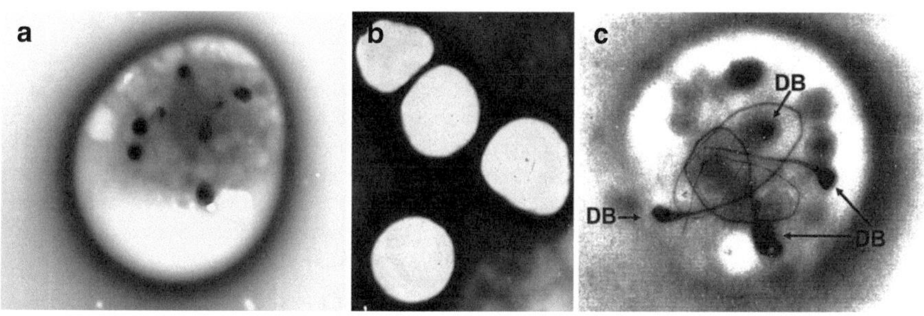

Fig. 6.2 Whole-mounted platelets viewed by TEM. (**a**) Dense bodies observed in platelets from healthy donors. (**b**) Platelets from HPS patient lacking dense bodies. (**c**) Dense bodies with one or two tails are observed in platelets from healthy donor (Figures were adapted from Ref. [19], with permission)

membrane during processing [17–19]. Less typical dense granule appearances include ringed forms on whole mount or TEM and dense granules with a "whip-like" tail observed on whole mount EM [20]. More complex dense granule morphologies have also been observed.

The luminal pH of dense granules is ~5.4, compared to a cytosolic pH of 7.2 [21]. The pH gradient across the dense granule membrane is maintained by a vesicular H^+-ATPase proton pump (Fig. 6.3). This proton pump is essential for generating a luminal positive electrochemical gradient ($\Delta\Psi$) of 30–40 mV [22]. The electron gradient powers transporters (described below) responsible for the substantial concentration of cytosolic nucleotides and bioactive amines observed in dense granules (Table 6.1). In addition, to a high H^+ content, platelet dense granules contain an extremely high calcium concentration (2.2 M), which renders these granules electron dense [23, 24]. Platelet dense granules also contain high concentrations of magnesium and potassium, which can be free or complexed to anions. These divalent cations have well-established functions in hemostasis and thrombosis. For example,

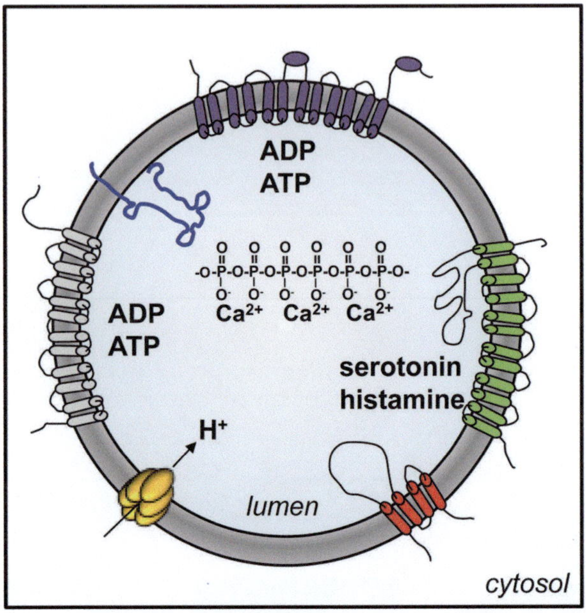

Fig. 6.3 Schematic representation of dense granule contents. Platelet dense granules are the storage unit for the ionic species (Ca^{2+}, polyphosphate), adenine nucleotides (ADP and ATP), and bioactive amines including serotonin and histamine. A vesicular H^+-ATPase proton pump (*yellow*) maintains a low intraluminal pH and a transmembrane electrochemical gradient. ADP and ATP are concentrated in dense granules via MRP4 (*purple*) and VNUT (*gray*). VMAT2 (*green*) is responsible for the transfer of serotonin from cytosol into dense granules. Platelet dense granules also express several membrane proteins including LAMP-2 (*blue*) and CD63 (*red*)

Table 6.1 Platelet dense granule content

Type	Examples
Ionic species	Ca^{2+}, Mg^{2+}, K^+, polyphosphate, pyrophosphate
Membrane proteins	CD63 (granulophysin), LAMP-2, GPIb, αIIbβ3
Nucleotides	ATP, GTP, ADP, GDP
Bioactive amines	Serotonin, histamine
Transporter proteins	MRP4, VNUT, VMAT2

calcium and magnesium participate in the binding of platelet adhesion receptors to their substrates [25–27]. Calcium and magnesium are also cofactors for several coagulation proteins. Dense granule cations can be free or complexed to anions.

Abundant anions in platelet dense granules include adenine nucleotides, which are concentrated at ~653 mM ADP and ~436 mM ATP [24, 28–30]. Adenine nucleotides are thought to be transported into platelets via specific transporters that have been identified on dense granules including multidrug resistance protein 4 (MRP4) and vesicular nucleotide transport (VNUT or SCL17A9) (Fig. 6.3). MRP4 is a 1325-amino acid member of the ATP-binding cassette transporter family [31–33]. These transporters include 12 transmembrane domains that form a channel, which facilitates transport of nucleotides across the plasma membrane [34]. VNUT is a 430-amino acid membrane protein that contains 12 transmembrane domains [35]. It is found on ATP storage organelles in several cell types including chromaffin large dense-core vesicles and synaptic vesicles [36, 37]. VNUT was first predicted to reside on platelet dense granules based on their similarity to the large dense-core vesicles of chromaffin cells [38] and later found to reside on platelet dense granules [39]. An inhibitor of VNUT (glyoxylate) blocked ATP uptake into these megakaryocyte-like cells [39]. The relative contributions of VNUT and MRP4 in loading adenosine nucleotides into dense granules remain to be determined. ADP secreted from dense granules serves a critical role in stimulating platelet autocrine activation during hemostasis and thrombosis via stimulation of P2Y12, P2Y1, and P2X1 receptors on platelets [40–42].

Polyphosphate and pyrophosphates are present in dense granules in millimolar concentrations (in terms of Pi residues) and can complex calcium (Fig. 6.3). Although the length of polyphosphate chains within platelet dense granules varies, they are typically ~60–100 phosphate units long [43, 44]. How these polyphosphate chains come to reside in dense granules is unclear. It is likely that they are synthesized within dense granules rather than being transported. Although the mechanism by which polyphosphates are synthesized is not known, mice lacking inositol hexakisphosphate kinase 1 have reduced levels of polyphosphate [45], implicating inositol hexakisphosphate kinase 1 in their synthesis. The intraluminal source of phosphate for synthesizing polyphosphate chains has not been established; however, it is worth noting that although cytosol contains high concentrations of ATP and relatively little ADP, both ATP and ADP accumulate at high concentrations into dense granules. One possibility is that VNUT and/or MRP4 transport primarily ATP into dense granules, and ATP is hydrolyzed to ADP intraluminally, generating phosphate substrates for synthesis of polyphosphate [39]. Platelet polyphosphates may potentiate activation of factor V and factor XI activation of thrombin during thrombosis [43, 46].

Platelet dense granules also contain several bioactive amines. Serotonin is present in platelet dense granules at a concentration of ~0.5 M [47, 48]. Uptake of serotonin into platelet granules is mediated by vesicular monoamine transporter 2 (VMAT2) [49] (Fig. 6.3). Serotonin is first pumped into the cytosol from plasma by a serotonin transporter (SERT), and the electrochemical proton gradient across the dense granule membrane drives VMAT2-mediated uptake of serotonin from cytosol into platelet granules. VMAT2 also appears to mediate the uptake of histamine into dense granules [24, 50]. Platelet serotonin is an important mediator of vascular tone at sites of injury [48, 51–54].

The fact that the packing of nucleotides and bioactive amines to such high concentrations is dependent on transporters raises the question of how these transporters are targeted to dense granule membranes. Many transporters contain targeting motifs within their primary amino acid sequence. For example, VMAT2 contains signals in its C-terminus for directing vesicular transport to dense granules [55]. The C-terminal dileucine motif is the best characterized of these signals, and site-directed mutagenesis studies have confirmed that the dileucine motif of VMAT2 is essential for its incorporation into the dense granule membrane [55]. In addition, the

C-terminus of VMAT2 contains an acidic motif including two serines that is required for retention in granule membranes [56]. Sorting signals also direct the trafficking of other integral membrane proteins to dense granules.

In addition to transporters, platelet dense granules contain several other membrane proteins. Dense granule membranes express markers such as the tetraspanin CD63 (granulophysin) and lysosomal-associated membrane protein-2 (LAMP-2, CD107b) (Fig. 6.3). CD63 is a 238-amino acid member of the tetraspanin family, which is characterized by four membrane-spanning domains and extensive glycosylation of its luminal/extracellular domains. CD63 functions in membrane trafficking and assists in the delivery of membrane proteins to vesicles. It is expressed on the platelet surface following activation where it associates with αIIbβ3 [57, 58]. CD63 also participates in lipid raft formation on platelets [59]. LAMP-2 is a 410-amino acid, extensively glycosylated protein that contains a single transmembrane domain. LAMP-2 functions in maintaining the integrity of lysosomal and presumably dense granule membranes as well as in trafficking. Site-directed mutagenesis of a tyrosine-based sorting signal on LAMP-2 resulted in mistargeting of the protein to the plasma membrane [55]. Adhesive receptors such as GPIb and $\alpha_{IIb}\beta_3$ have also been reported to reside on platelet dense granule membranes [60], although the contribution of this adhesion receptor pool to platelet function is unproven.

6.2.2 Platelet α-Granules

α-Granules constitute the largest granule population in platelets with 50–80 granules present in a single platelet (Fig. 6.4). α-Granules have a diameter of 200–500 nm, comprise 10% of platelet volume, and have a calculated surface area of 14 μm^2/platelet [61, 62]. The classic platelet α-granule as described by TEM possesses several features including (1) the peripheral membrane of the granule, (2) an electron-dense nucleoid that contains chemokines and proteoglycan, (3) a less electron-dense area adjacent to the nucleoid that contains fibrinogen, and (4) a peripheral electronlucent zone that contains von Willebrand factor (vWf) [61]. However, more recent imaging techniques have called into question whether or not this classic α-granule is representative of all α-granules. Electron tomography with three-dimensional reconstruction of the platelet interior demonstrates morphologically distinct α-granule populations [20]. Superresolution fluorescence microscopy has identified α-granules that appear not to have all the constituents of the classical α-granule [63]. These observations of α-granule heterogeneity have raised fundamental questions regarding the identity of α-granules. Thus, while granules with a classical appearance on electron microscopy and those that are stained with antibodies to P-selectin, vWf, and fibrinogen on immunofluorescence can be positively identified as α-granules, the identification of other vesicular structures with alternative protein cargos is less easily categorized.

Platelet α-granules, like dense granules, are considered lysosome-related organelles and share some similarities with dense granules. They have an acidic interior generated by a Ca^{2+}- or $Mg2^{+}$-sensitive H^+-pumping ATPase. This pump is

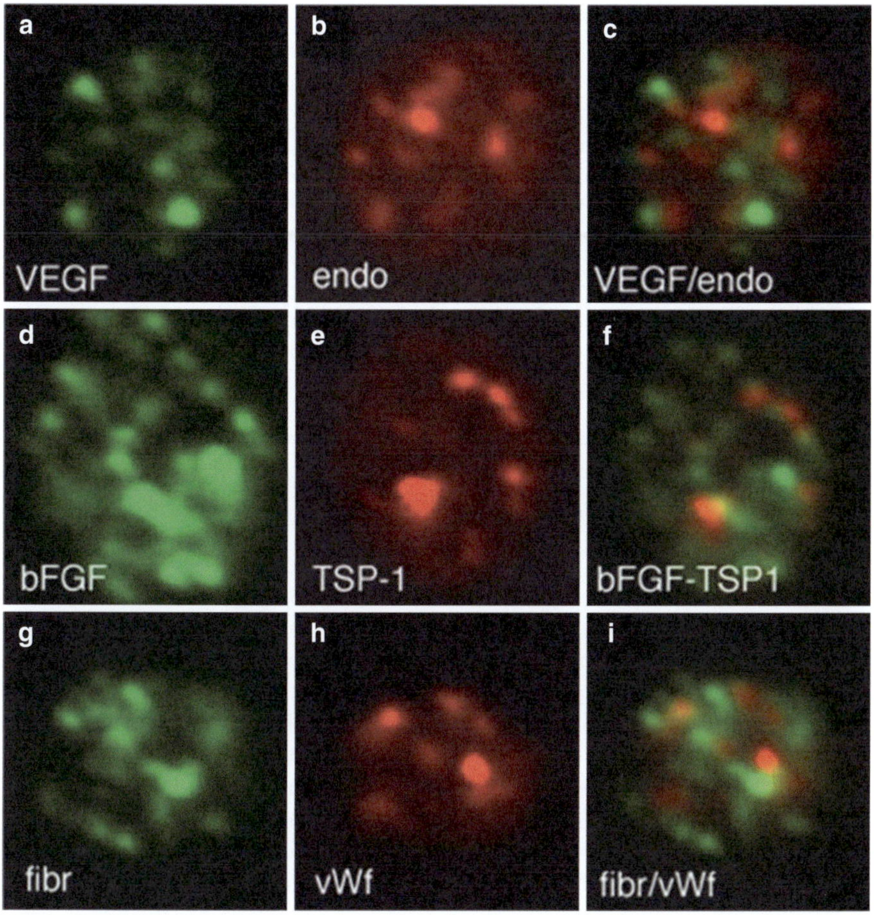

Fig. 6.4 Double immunofluorescence microscopy of VEGF (**a**), endostatin (**b**), and an overlay (**c**) shows that VEGF and endostatin localize in different α-granules. Cargo separation is also observed using antibodies against bFGF (**d**), TSP-1, (**e**) and an overlay (**f**) as well as with antibodies against fibrinogen (**g**), von Willebrand factor (**h**), and an overlay (**i**) demonstrates the colocolization of these proteins (With permission from Ref. [81])

responsible for generating an electrochemical gradient across the α-granule membrane [64–66]. Platelet α-granules have been reported to contain high concentrations of cations (14 mM Ca^{2+} and >60 mM Mg^{2+}) [66], although not as high as dense granules. However, while dense granules are known for their high content of small molecules, α-granules are known for their abundant protein content (Table 6.2).

6.2.2.1 α-Granule Protein Content

The protein content of α-granules is diverse. The platelet α-granule proteome consists of over 300 proteins including both membrane-bound proteins and soluble proteins that are secreted into the extracellular space [67, 68]. Although the majority

Table 6.2 Examples of platelet alpha-granule contents

Type	Examples
Integral membrane proteins	αIIbβ3, GPIbα-IX-V, GPVI, TLT-1, P-selectin, fibrocystin, CD109, Scamp2
Coagulants, anticoagulants, and fibrinolytic proteins	Factor V, factor IX, factor XIII, antithrombin, protein S, tissue factor pathway inhibitor, plasminogen, plasminogen activator inhibitor 1, α_2-macroglobulin
Adhesion proteins	Fibrinogen, von Willebrand factor, thrombospondin, LAMA5, ESAM
Chemokines	CXCL1 (GRO-α), CXCL4 (PF4), CXCL5 (ENA–78), CXCL7 (PBP, β-TG, CTAP-III, NAP-2), CXCL8 (IL-8), CXCL12 (SDF-1α), CCL2 (MCP-1), CCL3 (MIP-1α), and CCL5 (RANTES)
Growth factors	VEGF, platelet-derived growth factor, fibroblast growth factor, epidermal growth factor, hepatocyte growth factor, and insulin-like growth factor, transforming growth factor β
Microbicidal proteins	Thymosin-β4, thrombocidins 1 and 2 (from NAP-2)
Immune mediators	Complement C3 precursor, complement C4 precursor, β1H Globulin, factor D, factor H, C1 inhibitor, IgG

of α-granule proteins are synthesized in megakaryocytes and trafficked into α-granules via the regulated secretory pathway (see Chap. 1), some proteins such as fibrinogen, IgG, and albumin are taken up by the receptor-mediated endocytosis. Platelet α-granules are capable of storing a large variety of proteins at very high concentrations and still maintaining the osmolarity of the granule equivalent to the cytosol. This concentration may be achieved via aggregation and multimerization (e.g., vWf, multimerin) of proteins during protein sorting [9]. Recent developments in proteomic technologies and immunodetection methods have enabled not only identification of the platelet α-granule cargos but also quantification of granule proteins. These studies have shown that the most abundant proteins found in platelet α-granules are platelet factor 4 (PF4, CXCL4) (706 k copies/platelet), beta-thromboglobulin (479 k copies/platelet), thrombospondin 1 (101 k copies/platelet), fibrinogen (89 k copies/platelet), and factor XIII (83 k copies/platelet). However, α-granule cargo content is not static, and changes in α-granule cargos associated with disease states have been characterized [69–71].

6.2.2.2 Function of α-Granule Cargos

α-Granule proteins participate in a broad array of physiological functions including hemostasis and thrombosis, inflammation, angiogenesis, wound repair, and antimicrobial host defense. The supposition that α-granules serve an essential role in hemostasis and thrombosis was initially supported by the observation that patients with gray platelet syndrome, who lack normal α-granules, have a bleeding diathesis [72–77]. More recently, *NBEAL2*-deficient mice, whose platelets lack α-granules, were shown to have defects in hemostasis and thrombosis [78, 79]. Platelet proteins that contribute to hemostasis and thrombosis include prothrombotic proteins such as vWf, fibrinogen, factor V, factor IX, and factor XIII. α-Granules also contain anticoagulant proteins such as antithrombin, protein S, and tissue factor pathway inhibitor as well as components of the fibrinolytic system such as plasminogen and plasminogen activator inhibitor. Angiogenic factors in α-granules include vascular endothelial growth factor, platelet-derived growth factor, fibroblast growth factor, and transforming growth factor-beta among others. α-Granules also contain antiangiogenic factors such as endostatin and platelet factor 4 (PF4). The observation that α-granules contain both prothrombotic and antithrombotic as well as proangiogenic and antiangiogenic proteins raises the question of how cargos with opposing activities can be organized and released so as to effectively promote physiological processes. One possibility is that these cargos are physically separated into different α-granule subpopulations that respond to distinct platelet agonists [63, 80, 81] (Fig. 6.4). A second hypothesis is that cargos are spatially separated within α-granules and the kinetics of their release differs [82]. These issues regarding the structural and functional organization of platelet α-granule cargos remain to be resolved.

As a role for platelets in inflammation and innate immunity has become increasingly evident [83–86], the function of platelet α-granule cargos serve in this capacity has become more prominent. Platelets are immune cells, and their α-granules contain many mediators of inflammation. Among the most abundant α-granule contents are chemokines, which include CXCL1 (GRO-α), CXCL4 (PF4), CXCL5 (ENA–78), CXCL7 (PBP, β-TG, CTAP-III, NAP-2), CXCL8 (IL-8), CXCL12 (SDF-1α), CCL2 (MCP-1), CCL3 (MIP-1α), and CCL5 (RANTES) [87–89]. Several of these chemokines function in the development of atherosclerosis, in which platelets have been shown to serve an early role [86]. Platelets α-granules contain many mediators of the complement cascade such as complement C3, complement C4 precursor, C1 inhibitor, factor H, and factor D [90]. Cargos that are directly microbicidal are also secreted from α-granules. Such microbicidal proteins include thymosin-β4 and thrombocidins 1 and 2, which are derived from NAP-2 [84]. PF4 is essential for the ability of platelets to kill *Plasmodium falciparum* during malarial infection [91–93].

Despite the large number of α-granule proteins and the diversity of their proposed functions, few cargos are specific to platelets. In the vast majority of cases, α-granule cargo proteins are also constitutes of plasma or secreted by other cells. There are, however, a few proteins that are either specific for α-granules or that have α-granule-specific isoforms with special properties.

PF4 is found almost exclusively in platelets and is among the most abundant proteins in α-granules (16–25 % of the α-granule proteins, 706 k copies/platelet) [70, 94]. PF4 is a 70-amino acid protein derived from a 101-amino acid precursor that contains a CXCL motif and binds CXCR3. It is synthesized in megakaryocytes and subsequently transported into platelet granules by virtue of a Leu-Lys-Asn-Gly motif at the 45–48 position [95]. Although PF4 is mostly known for its strong affinity for heparin (KD = 10–30 nM) and role in coagulation, recent studies demonstrated a role for PF4 in a variety of physiological processes including control of angiogenesis, inflammation, and protection from infection [94, 96–99].

Von Willebrand factor has long been a marker of platelet α-granules. vWf is a 2813-amino acid protein that circulates in plasma in multimers ranging in size from 500 kDa to more than 10,000 kDa. vWf can be synthesized by both endothelial cells and megakaryocytes. vWf stored in platelet α-granules constitutes 20 % of total vWf and is enriched in high-molecular-weight multimers [100–105]. It has been shown that vWf originating from platelets is structurally and functionally distinct from endothelial-derived vWf. In addition to being enriched in high-molecular-weight multimers, platelet-derived vWf has a different glycosylation pattern that increases its resistance to the protease ADAMTS13 and contributes to its hemostatic function in platelet-rich thrombi [105]. Studies in which normal bone marrow was transplanted into pigs with severe von Willebrand disease demonstrated that platelet vWf could partially compensate for lack of plasma vWf [106]. Conversely, normal thrombus formation was observed following gene transfer of vWf, resulting in its ectopic expression in the liver of mice with severe von Willebrand disease [107]. Taken together, these results suggest that α-granule vWf can contribute to, but is not necessary for, hemostasis and thrombus formation.

Factor V (FV) is an important coagulation factor whose activation to FVa plays a major role in the conversion of prothrombin to thrombin [108]. Approximately 25 % of the total FV resides in platelets. Platelet FV is either synthesized by megakaryocytes or taken up from the plasma. In platelet α-granules FV strongly associates with a high-molecular-weight protein termed multimerin via disulfide linkages and other noncovalent interactions [109–112]. An average of 2 ng/10^6 platelets FV is found in α-granules interacting with about 17.1–19.9 ng/10^6 platelets multimerin. Upon platelet activation, the FV and multimerin complex slowly dissociates into the extracellular matrix. Although multimerin influences the packing of FV in α-granules, it is not crucial for FV storage as individuals with multimerin deficiency have normal FV storage [111–114]. FV acts as a cofactor in the prothrombinase complex. Compared to plasma FV, platelet FV is more heterogeneous and more resistant to activated protein C (APC) [112].

6.2.2.3 α-Granule Transmembrane Proteins

α-Granule transmembrane proteins also serve important functions in platelet biology. P-selectin, TMEMs, and integrins reside on the platelet α-granule surface and translocate to the platelet plasma membrane surface upon platelet activation. These proteins contribute to platelet-platelet, platelet-leukocyte, and platelet-endothelial cell interactions. Most α-granule transmembrane receptors, such αIIbβ3 and GPIb, are also present on the plasma membrane of the resting platelet, and receptors from α-granules are thought to constitute a reserve pool. Other receptors, such as P-selectin CD109 and fibrocystin L, are sequestered in α-granules in resting platelets, and their translocation to the plasma membrane serves as a sensitive indicator of platelet activation.

P-Selectin (CD62P) is perhaps the best-characterized protein found in the α-granule membrane. Tissue distribution of P-selectin is limited to platelets and endothelial cells. This 140-kDa molecule is composed of an N-terminal C-type lectin domain, an epidermal growth factor-like motif, a series of complement repeats, a transmembrane domain, and a C-terminal cytoplasmic tail [115, 116]. In resting platelets, the protein resides intraluminally but translocates to the extracellular surface upon platelet activation. Once exposed on the platelet surface, the lectin domain of P-selectin interacts with PSGL-1 on immune cells, recruiting them to sites of inflammation [117]. Selective knockdown of P-selectin in the hematopoietic cells compartment inhibits neutrophil recruitment to sites of inflammation, proving that platelet P-selectin is critical for this function [118]. Platelet P-selectin also functions in atherosclerosis and leukocyte recruitment during thrombus formation [119–121].

6.3 Platelet Lysosomes

A platelet typically contains 0–3 lysosomes, which are 200–250 nm in diameter and can be identified by cytochemical stains such as acid phosphatase or arylsulfatase [15]. Like lysosomes in other cell types, platelet lysosomes contain enzymes involved in degradation of proteins, carbohydrates, and lipids. The most abundant class is the acid hydrolases. Platelet lysosome contents are released during platelet activation in vivo following vascular injury [122, 123]. Lysosomal membranes contain LAMP-1, LAMP-2, and CD63 [15, 123]. These proteins are highly glycosylated, decorating the luminal lysosome surface and contributing to the protective function of the lysosomal membrane. The role of platelet lysosomes in hemostasis and thrombosis is not known, although some have suggested that enzymes released by lysosomes may serve a role in digestion and resolution of thrombi [124].

6.4 Platelet Membrane Systems

The platelet contains internal membranes that are distinct entities based on their ultrastructure, their function, and/or their origin in the megakaryocyte. This section will focus on the two most well-studied systems of internal membranes, the open canalicular system and the dense tubular system. Other internal membranes include Golgi complexes and rough endoplasmic reticulum, which are rare platelet vestiges of the nucleated megakaryocyte and are found primarily in the setting of pathological, rapid platelet turnover.

6.4.1 Open Canalicular System

The open canalicular system (OCS) is a membranous network of tunneling invaginations of the platelet membrane. This network is highly interconnected within the platelet, and openings of the OCS onto the plasma membrane are dispersed throughout the platelet surface [125–128]. The OCS can serve as a conduit for the release of cargos that are discharged into from granules [127]. Immunostaining has shown that the OCS is lined with syntaxin and SNAP-23, t-SNAREs that facilitate the fusion of granule membranes with membranes of the OCS [129]. In addition, the OCS can serve as a vehicle for the incorporation of substances into the platelet [127]. Perhaps the most obvious example of the role of the OCS in host defense is its ability to form engulfment vacuoles around bacteria [130]. α-Granules

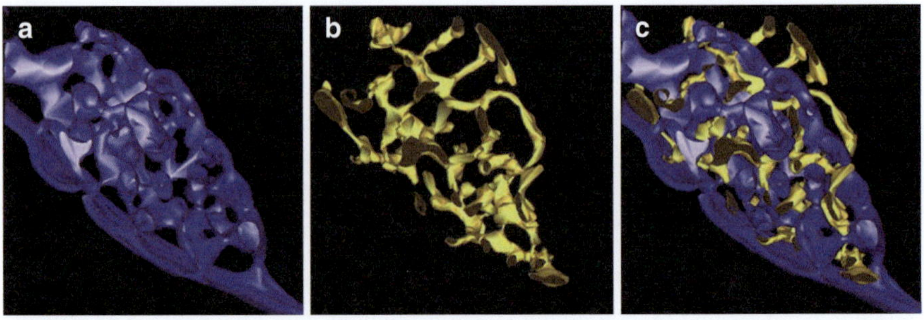

Fig. 6.5 The DTS and OCS are highly intertwined membrane complexes. Shown are three-dimensional reconstructions of the OCS (*dark blue*) and DTS (*yellow*) from resting vitrified platelets as imaged using cryoelectron microscopy (Adopted with permission from Ref. [20])

subsequently release bactericidal cargo into the vacuole. Investigators have also shown that the OCS acts as a membrane reservoir, contributing membrane to the periphery of the platelet during spreading [126]. How this intricately interconnected system of tunnels disassembles in order to evaginate and incorporate into the growing plasma membrane of spread platelets remains speculative.

6.4.2 Dense Tubular System

The dense tubular system (DTS) is the major calcium-sequestering organelle in the platelet. It is thought to derive from ribosome-depleted endoplasmic reticulum from the megakaryocyte and consists of thin elongated membranes. The DTS is similar to the OCS in that it is a complex membrane system that courses through the platelet interior. However, lumens of the DTS and OCS are not interconnected. Nonetheless, the separate DTS and OCS membrane systems closely associate with one another in membrane complexes (Fig. 6.5). Unlike the OCS, the DTS is not open to the extracellular space. In addition, the DTS is distinct on EM by virtue of its more electron-dense lumen (the OCS lumen is clear). This electron density is secondary to the high concentration of Ca^{2+} in the DTS. In response to platelet agonists that stimulate increased $[Ca^{2+}]i$, the DTS changes morphology from thin elongated membranes to more vesicular structures [131, 132].

In the resting platelet, cytosolic calcium concentrations are maintained in the nanomolar range owing to the activity of pumps promoting calcium efflux from the cytosol. Sarco/endoplasmic reticulum Ca^{2+}-ATPases (SERCA) located in the DTS serve a prominent role in maintaining this low level of cytosolic calcium. Following activation, the Ca^{2+}-ATPases pump Ca^{2+} into the platelet cytosol following platelet activation [133], contributing to the rise in intracellular calcium. In this manner, the DTS serves a critical function in platelet activation.

T granules are related to the electron-dense tubular system. Although the T granule is not a well-established granule type, it appears to be distinct from α-granules in that it contains unique cargo, which is secreted upon platelet activation. The first T granule cargo to be identified was toll-like receptor 9 (TLR9) [140]. Immunogold

EM has shown that protein disulfide isomerase (PDI), an ER-resident protein that is secreted upon platelet activation, is stored in and secreted from T granules [147, 148]. Further studies are required to identify additional T granule cargos and clarify the relationship of this granule to the dense tubular system.

6.5 Other Membranous Structures in Platelets

6.5.1 Mitochondria

Platelets contain two to eight mitochondria per cell [134–137]. Mitochondria are dynamic organelles in shape, number, and size. Proteins of the dynamin-related protein 1 family, which function in mitochondrial scission and perhaps fusion, are enriched in platelets [138]. As in most cells, mitochondria within platelets function in energy metabolism, forming ATP primarily from fatty acids, palmitate, glucose, and glutamine. The transport of electrons across the inner membrane of the mitochondria creates a proton-motive force, providing energy for ATP production. However, platelet mitochondria have additional roles that are more specialized in platelets. Exposure to high concentrations of agonists results in activation of mitochondrial permeability transition (MPT) with $\Delta\Psi_m$ collapse and increased production of reactive oxygen species [135, 139–143]. ROS production contributes to platelet activation and to mitochondria-dependent phosphatidylserine exposure [135].

Platelet mitochondria can also function in apoptosis. Prosurvival proteins such as BCL-X_L and BCL2 keep pro-apoptotic proteins such as BAK1 and BAX in check, thereby preventing apoptosis. When prosurvival pathways are overwhelmed, pro-apoptotic proteins translocate to the mitochondria, oligomerize and permeabilize the mitochondrial membrane. These events lead to collapse of $\Delta\Psi_m$ and cytochrome c release with induction of apoptosis [144, 145]. Apoptosis has been invoked in controlling both megakaryopoiesis and platelet lifespan, underscoring the importance of mitochondria in platelet physiology.

6.5.2 Peroxisomes

Platelets also contain peroxisomes as indicated by staining of catalase [146]. Peroxisomes typically function in the breakdown of long chain and branched fatty acids. However, the significance of this organelle in platelet function is undetermined.

6.6 Perspective

It is not remarkable that given the distinct and extraordinary process of megakaryopoiesis, the platelet interior is so unusual: lacking a nucleus, filled with unique membrane systems, and replete with granules found in no other cell. Yet despite this bizarre process of formation, platelet contents are remarkably uniform among the

10^{11} platelets that are formed daily. While some mechanistic insight into how these organelles come to reside in platelets has been provided by live cell imaging [149], much remains to be learned about how organelles form in and are distributed by megakaryocytes. Basic questions regarding platelet granule contents and membrane systems remain to be definitively answered. Although much has been learnt about the mechanisms of membrane fusion leading to granule exocytosis [150–155], much remains to be learnt. Do α-granules represent a homogenous granule population, or are they divided into subpopulations that possess different cargos? How are dense granule contents filled with adenosine nucleotides and polyphosphates? How does the OCS form during megakaryopoiesis and deconvolute to contribute to platelet spreading? Technical advances in electron microscopy and superresolution microscopy with three-dimensional reconstructions of the platelet interior and live cell imaging will propel our understanding of the organization of the platelet and thereby enable improved knowledge of platelet function.

Acknowledgments This work was supported by HL112809, HL125275, and HL112302 from NHLBI. The author would like to acknowledge the many significant contributions to the field of platelet granule secretion that were not cited in this chapter owing to space limitations.

Conflict of Interest The authors have no conflict of interest to declare.

References

1. Boilard E, Nigrovic PA, Larabee K et al (2010) Platelets amplify inflammation in arthritis via collagen-dependent microparticle production. Science 327:580–583
2. Gawaz M (2006) Platelets in the onset of atherosclerosis. Blood Cells Mol Dis 36:206–210
3. Gawaz M, Langer H, May AE (2005) Platelets in inflammation and atherogenesis. J Clin Invest 115:3378–3384
4. Wagner DD, Burger PC (2003) Platelets in inflammation and thrombosis. Arter Thromb Vasc Biol 23:2131–2137
5. Frojmovic MM, Milton JG (1982) Human platelet size, shape, and related functions in health and disease. Physiol Rev 62:185–261
6. Flaumenhaft R (2003) Molecular basis of platelet granule secretion. Arterioscler Thromb Vasc Biol 23:1152–1160
7. Whiteheart SW (2011) Platelet granules: surprise packages. Blood 118:1190–1191
8. Lemons PP, Chen D, Bernstein AM et al (1997) Regulated secretion in platelets: identification of elements of the platelet exocytosis machinery. Blood 90:1490–1500
9. King SM, Reed GL (2002) Development of platelet secretory granules. Semin Cell Dev Biol 13:293–302
10. White JG (2004) Medich giant platelet disorder: a unique alpha granule deficiency I. Structural abnormalities. Platelets 15:345–353
11. White JG, Key NS, King R a, Vercellotti GM (2005) A "touch" of the white platelet syndrome. Platelets 16:346–361
12. Wright JH (1910) The histogenesis of the blood platelets. Publ Mass Gen Hosp 3:1–16
13. Tranzer JP, Da Prada M, Pletscher A (1966) Ultrastructural localization of 5-hydroxytryptamine in blood platelets. Nature 212:1574–1575
14. Siegel A, Luscher EF (1967) Non-identity of the alpha-granules of human blood platelets with typical lysosomes. Nature 215:745–747

15. Bentfeld-Barker ME, Bainton DF (1982) Identification of primary lysosomes in human megakaryocytes and platelets. Blood 59:472–481
16. Raposo G, Marks MS, Cutler DF (2007) Lysosome-related organelles: driving post-Golgi compartments into specialisation. Curr Opin Cell Biol 19:394–401
17. Gerrard JM, Rao GH, White JG (1977) The influence of reserpine and ethylenediaminetetraacetic acid (EDTA) on serotonin storage organelles of blood platelets. Am J Pathol 87:633–646
18. White JG, Gerrard JM (1978) Recent advances in platelet structural physiology. Suppl Thromb Haemost 63:49–60
19. Amesse LS, Pfaff-Amesse T, Gunning WT et al (2013) Clinical and laboratory characteristics of adolescents with platelet function disorders and heavy menstrual bleeding. Exp Hematol Oncol 2:3–12
20. Van Nispen tot Pannerden H, de Haas F, Geerts W et al (2010) The platelet interior revisited: electron tomography reveals tubular alpha-granule subtypes. Blood 116:1147–1156
21. Dean GE, Fishkes H, Nelson PJ, Rudnick G (1984) The hydrogen ion-pumping adenosine triphosphatase of platelet dense granule membrane. Differences from F1F0- and phosphoenzyme-type ATPases. J Biol Chem 259:9569–9574
22. Carty SE, Johnson RG, Scarpa A (1981) Serotonin transport in isolated platelet granules. Coupling to the electrochemical proton gradient. J Biol Chem 256:11244–11250
23. Ruiz FA, Lea CR, Oldfield E, Docampo R (2004) Human platelet dense granules contain polyphosphate and are similar to acidocalcisomes of bacteria and unicellular eukaryotes. J Biol Chem 279:44250–44257
24. Fukami MH, Holmsen H, Ugurbil K (1984) Histamine uptake in pig platelets and isolated dense granules. Biochem Pharmacol 33:3869–3874
25. Coller BS, Beer JH, Scudder LE, Steinberg MH (1989) Collagen-platelet interactions: evidence for a direct interaction of collagen with platelet GPIa/IIa and an indirect interaction with platelet GPIIb/IIIa mediated by adhesive proteins. Blood 74:182–192
26. Farndale RW, Siljander PR, Onley DJ et al (2003) Collagen-platelet interactions: recognition and signalling. Biochem Soc Symp 70:81–94
27. Zhu J, Choi W-S, McCoy JG et al (2012) Structure-guided design of a high-affinity platelet integrin IIb 3 receptor antagonist that disrupts Mg2+ binding to the MIDAS. Sci Transl Med 4:125ra32–125ra32
28. Lages B, Weiss HJ (1999) Secreted dense granule adenine nucleotides promote calcium influx and the maintenance of elevated cytosolic calcium levels in stimulated human platelets. Thromb Haemost 81:286–292
29. Holmsen H, Dangelmaier CA (1989) Measurement of secretion of serotonin. Methods Enzymol 169:195–205
30. Holmsen H, Weiss HJ (1979) Secretable storage pools in platelets. Annu Rev Med 30:119–134
31. Jedlitschky G, Tirschmann K, Lubenow LE et al (2004) The nucleotide transporter MRP4 (ABCC4) is highly expressed in human platelets and present in dense granules, indicating a role in mediator storage. Blood 104:3603–3610
32. Ritter CA, Jedlitschky G, Meyer zu Schwabedissen H et al (2005) Cellular export of drugs and signaling molecules by the ATP-binding cassette transporters MRP4 (ABCC4) and MRP5 (ABCC5). Drug Metab Rev 37:253–278
33. Oevermann L, Scheitz J, Starke K et al (2009) Hematopoietic stem cell differentiation affects expression and function of MRP4 (ABCC4), a transport protein for signaling molecules and drugs. Int J Cancer 124:2303–2311
34. Moro MA, Russel RJ, Cellek S et al (1996) cGMP mediates the vascular and platelet actions of nitric oxide: confirmation using an inhibitor of the soluble guanylyl cyclase. Proc Natl Acad Sci U S A 93:1480–1485
35. Van Liefferinge J, Massie A, Portelli J et al (2013) Are vesicular neurotransmitter transporters potential treatment targets for temporal lobe epilepsy? Front Cell Neurosci 7:139–163
36. Sawada K, Echigo N, Juge N et al (2008) Identification of a vesicular nucleotide transporter. Proc Natl Acad Sci U S A 105:5683–5686

37. Rose SD, Lejen T, Casaletti L et al (2002) Molecular motors involved in chromaffin cell secretion. Ann N Y Acad Sci 971:222–231
38. Fitch-Tewfik JL, Flaumenhaft R (2013) Platelet granule exocytosis: a comparison with chromaffin cells. Front Endocrinol (Lausanne) 4:77–86
39. Hiasa M, Togawa N, Miyaji T et al (2014) Essential role of vesicular nucleotide transporter in vesicular storage and release of nucleotides in platelets. Physiol Rep 2:1–14
40. Andre P, Delaney SM, LaRocca T et al (2003) P2Y12 regulates platelet adhesion/activation, thrombus growth, and thrombus stability in injured arteries. J Clin Invest 112:398–406
41. Turner NA, Moake JL, McIntire LV (2001) Blockade of adenosine diphosphate receptors P2Y(12) and P2Y(1) is required to inhibit platelet aggregation in whole blood under flow. Blood 98:3340–3345
42. Foster CJ, Prosser DM, Agans JM et al (2001) Molecular identification and characterization of the platelet ADP receptor targeted by thienopyridine antithrombotic drugs. J Clin Invest 107:1591–1598
43. Morrissey JH, Choi SH, Smith SA (2012) Polyphosphate: an ancient molecule that links platelets, coagulation, and inflammation. Blood 119:5972–5979
44. Muller F, Mutch NJ, Schenk WA et al (2009) Platelet polyphosphates are proinflammatory and procoagulant mediators in vivo. Cell 139:1143–1156
45. Ghosh S, Shukla D, Suman K et al (2013) Inositol hexakisphosphate kinase 1 maintains hemostasis in mice by regulating platelet polyphosphate levels. Blood 122:1478–1486
46. Smith SA, Mutch NJ, Baskar D et al (2006) Polyphosphate modulates blood coagulation and fibrinolysis. Proc Natl Acad Sci U S A 103:903–908
47. Ge S, Wittenberg NJ, Haynes CL (2008) Quantitative and real-time detection of secretion of chemical messengers from individual platelets. Biochemistry 47:7020–7024
48. Ge S, White JG, Haynes CL (2009) Quantal release of serotonin from platelets. Anal Chem 81:2935–2943
49. Leitner B, Lovisetti-Scamihorn P, Heilmann J et al (1999) Subcellular localization of chromogranins, calcium channels, amine carriers, and proteins of the exocytotic machinery in bovine splenic nerve. J Neurochem 72:1110–1116
50. Tharmapathy P, Fukami MH, Holmsen H (2000) The stimulatory effects of cationic amphiphilic drugs on human platelets treated with thrombin. Biochem Pharmacol 60:1267–1277
51. Holland JM (1976) Serotonin deficiency and prolonged bleeding in beige mice. Proc Soc Exp Biol Med 151:32–39
52. Malyszko J, Malyszko JS, Pawlak D et al (1996) Hemostasis, platelet function and serotonin in acute and chronic renal failure. Thromb Res 83:351–361
53. Sauer WH, Berlin JA, Kimmel SE (2003) Effect of antidepressants and their relative affinity for the serotonin transporter on the risk of myocardial infarction. Circulation 108:32–36
54. Golino P, Crea F, Willerson JT (2002) How to study the effects of platelet aggregation and thrombosis on coronary vasomotion and their clinical relevance. Ital Heart J 3:220–225
55. Ambrosio AL, Boyle JA, Di Pietro SM (2012) Mechanism of platelet dense granule biogenesis: study of cargo transport and function of Rab32 and Rab38 in a model system. Blood 120:4072–4081
56. Waites CL, Mehta A, Tan PK et al (2001) An acidic motif retains vesicular monoamine transporter 2 on large dense core vesicles. J Cell Biol 152:1159–1168
57. Van der Zee PM, Biro E, Ko Y et al (2006) P-selectin- and CD63-exposing platelet microparticles reflect platelet activation in peripheral arterial disease and myocardial infarction. Clin Chem 52:657–664
58. Israels SJ, McMillan-Ward EM, Easton J et al (2001) CD63 associates with the alphaIIb beta3 integrin-CD9 complex on the surface of activated platelets. Thromb Haemost 85:134–141
59. Israels SJ, McMillan-Ward EM (2007) Platelet tetraspanin complexes and their association with lipid rafts. Thromb Haemost 98:1081–1087
60. Youssefian T, Masse JM, Rendu F et al (1997) Platelet and megakaryocyte dense granules contain glycoproteins Ib and IIb-IIIa. Blood 89:4047–4057

61. Blair P, Flaumenhaft R (2009) Platelet alpha-granules: basic biology and clinical correlates. Blood Rev 23:177–189
62. Harrison P, Cramer EM (1993) Platelet alpha-granules. Blood Rev 7:52–62
63. Sehgal S, Storrie B (2007) Evidence that differential packaging of the major platelet granule proteins von Willebrand factor and fibrinogen can support their differential release. J Thromb Haemost 5:2009–2016
64. Grinstein S, Furuya W (1983) The electrochemical H+ gradient of platelet secretory alpha-granules. Contribution of a H+ pump and a Donnan potential. J Biol Chem 258: 7876–7882
65. Grinstein S, Vander Meulen J, Furuya W (1982) Possible role of H+--alkali cation countertransport in secretory granule swelling during exocytosis. FEBS Lett 148:1–4
66. Grinstein S, Furuya W, Vander Meulen J, Hancock RG (1983) The total and free concentrations of Ca2+ and Mg2+ inside platelet secretory granules. Measurements employing a novel double null point technique. J Biol Chem 258:14774–14777
67. Coppinger JA, Cagney G, Toomey S et al (2004) Characterization of the proteins released from activated platelets leads to localization of novel platelet proteins in human atherosclerotic lesions. Blood 103:2096–2104
68. Maynard DM, Heijnen HFG, Horne MK et al (2007) Proteomic analysis of platelet alpha-granules using mass spectrometry. J Thromb Haemost 5:1945–1955
69. Maynard DM, Heijnen HF, Gahl WA, Gunay-Aygun M (2010) The alpha granule proteome: novel proteins in normal and ghost granules in gray platelet syndrome. J Thromb Haemost 8:1786–1796
70. Burkhart JM, Gambaryan S, Watson SP et al (2014) What can proteomics tell us about platelets? Circ Res 114:1204–1219
71. Peterson JE, Zurakowski D, Italiano JE et al (2010) Normal ranges of angiogenesis regulatory proteins in human platelets. Am J Hematol 85:487–493
72. Gunay-Aygun M, Zivony-Elboum Y, Gumruk F et al (2010) Gray platelet syndrome: natural history of a large patient cohort and locus assignment to chromosome 3p. Blood 116: 4990–5001
73. Gunay-Aygun M, Falik-Zaccai TC, Vilboux T et al (2011) NBEAL2 is mutated in gray platelet syndrome and is required for biogenesis of platelet alpha-granules. Nat Genet 43:732–734
74. Levy-Toledano S, Caen JP, Breton-Gorius J et al (1981) Gray platelet syndrome: alpha-granule deficiency. Its influence on platelet function. J Lab Clin Med 98:831–848
75. Swank RT, Reddington M, Novak EK (1996) Inherited prolonged bleeding time and platelet storage pool deficiency in the subtle gray (sut) mouse. Lab Anim Sci 46:56–60
76. Gerrard JM, Phillips DR, Rao GH et al (1980) Biochemical studies of two patients with the gray platelet syndrome. Selective deficiency of platelet alpha granules. J Clin Invest 66:102–109
77. Nurden AT, Nurden P (2007) The gray platelet syndrome: clinical spectrum of the disease. Blood Rev 21:21–36
78. Deppermann C, Cherpokova D, Nurden P et al (2013) Gray platelet syndrome and defective thrombo-inflammation in Nbeal2-deficient mice. J Clin Invest 123:3331–3342
79. Kahr WHA, Lo RW, Li L et al (2013) Abnormal megakaryocyte development and platelet function in Nbeal2-/- mice. Blood 122:3349–3358
80. Italiano JE, Battinelli EM (2009) Selective sorting of alpha-granule proteins. J Thromb Haemost 7:173–176
81. Italiano JE, Richardson JL, Patel-Hett S et al (2008) Angiogenesis is regulated by a novel mechanism: pro- and antiangiogenic proteins are organized into separate platelet granules and differentially released. Blood 111:1227–1233
82. Jonnalagadda D, Izu LT, Whiteheart SW (2012) Platelet secretion is kinetically heterogeneous in an agonist-responsive manner. Blood 120:5209–5216
83. Semple JW, Freedman J (2010) Platelets and innate immunity. Cell Mol Life Sci 67:499–511
84. Semple JW, Italiano JE, Freedman J (2011) Platelets and the immune continuum. Nat Rev Immunol 11:264–274

85. Von Hundelshausen P, Koenen RR, Weber C (2009) Platelet-mediated enhancement of leukocyte adhesion. Microcirculation 16:84–96
86. Von Hundelshausen P, Weber C (2007) Platelets as immune cells: bridging inflammation and cardiovascular disease. Circ Res 100:27–40
87. Schenk BI, Petersen F, Flad HD, Brandt E (2002) Platelet-derived chemokines CXC chemokine ligand (CXCL)7, connective tissue-activating peptide III, and CXCL4 differentially affect and cross-regulate neutrophil adhesion and transendothelial migration. J Immunol 169:2602–2610
88. Massberg S, Konrad I, Schurzinger K et al (2006) Platelets secrete stromal cell-derived factor 1alpha and recruit bone marrow-derived progenitor cells to arterial thrombi in vivo. J Exp Med 203:1221–1233
89. Koenen RR, von Hundelshausen P, Nesmelova IV et al (2009) Disrupting functional interactions between platelet chemokines inhibits atherosclerosis in hyperlipidemic mice. Nat Med 15:97–103
90. Peerschke EIB, Yin W, Ghebrehiwet B (2008) Platelet mediated complement activation. Adv Exp Med Biol 632:81–91
91. McMorran BJ, Marshall VM, de Graaf C et al (2009) Platelets kill intraerythrocytic malarial parasites and mediate survival to infection. Science 323:797–800
92. Derby B (2012) Printing and prototyping of tissues and scaffolds. Science 338:921–926
93. Love MS, Millholland MG, Mishra S et al (2012) Platelet factor 4 activity against P. Falciparum and its translation to nonpeptidic mimics as antimalarials. Cell Host Microbe 12:815–823
94. Slungaard A (2005) Platelet factor 4: a chemokine enigma. Int J Biochem Cell Biol 37:1162–1167
95. El Golli N, Issertial O, Rosa JP, Briquet-Laugier V (2005) Evidence for a granule targeting sequence within platelet factor 4. J Biol Chem 280:30329–30335
96. Bikfalvi A (2004) Platelet factor 4: an inhibitor of angiogenesis. Semin Thromb Hemost 30:379–385
97. Peterson JE, Zurakowski D, Italiano JE et al (2012) VEGF, PF4 and PDGF are elevated in platelets of colorectal cancer patients. Angiogenesis 15:265–273
98. Krauel K, Weber C, Brandt S et al (2012) Platelet factor 4 binding to lipid A of Gram-negative bacteria exposes PF4/heparin-like epitopes. Blood 120:3345–3352
99. Vandercappellen J, Van Damme J, Struyf S (2011) The role of the CXC chemokines platelet factor-4 (CXCL4/PF-4) and its variant (CXCL4L1/PF-4var) in inflammation, angiogenesis and cancer. Cytokine Growth Factor Rev 22:1–18
100. Gralnick HR, Williams SB, McKeown LP et al (1985) Platelet von Willebrand factor: comparison with plasma von Willebrand factor. Thromb Res 38:623–633
101. Denis C (2002) Molecular and cellular biology of von Willebrand factor. Int J Hematol 75:3–8
102. Wagner DD, Saffaripour S, Bonfanti R et al (1991) Induction of specific storage organelles by von Willebrand factor propolypeptide. Cell 64:403–413
103. Titani K, Kumar S, Takio K et al (1986) Amino acid sequence of human von Willebrand factor. Biochemistry 25:3171–3184
104. Kanaji S, Fahs S a, Shi Q et al (2012) Contribution of platelet vs. endothelial VWF to platelet adhesion and hemostasis. J Thromb Haemost 10:1646–1652
105. McGrath RT, van den Biggelaar M, Byrne B et al (2013) Altered glycosylation of platelet-derived von Willebrand factor confers resistance to ADAMTS13 proteolysis. Blood 122:4107–4110
106. Bowie EJ, Solberg LA Jr, Fass DN et al (1986) Transplantation of normal bone marrow into a pig with severe von Willebrand's disease. J Clin Invest 78:26–30
107. De Meyer SF, Vandeputte N, Pareyn I et al (2008) Restoration of plasma von Willebrand factor deficiency is sufficient to correct thrombus formation after gene therapy for severe von Willebrand disease. Arter Thromb Vasc Biol 28:1621–1626
108. Mann KG, Kalafatis M (2003) Factor V: a combination of Dr Jekyll and Mr Hyde. Blood 101:20–30
109. Jeimy SB, Tasneem S, Cramer EM, Hayward CP (2008) Multimerin 1. Platelets 19:83–95

110. Jeimy SB, Fuller N, Tasneem S et al (2008) Multimerin 1 binds factor V and activated factor V with high affinity and inhibits thrombin generation. Thromb Haemost 100: 1058–1067
111. Hayward CP, Furmaniak-Kazmierczak E, Cieutat AM et al (1995) Factor V is complexed with multimerin in resting platelet lysates and colocalizes with multimerin in platelet alpha-granules. J Biol Chem 270:19217–19224
112. Hayward CPM, Fuller N, Zheng S et al (2004) Human platelets contain forms of factor V in disulfide-linkage with multimerin. Thromb Haemost 92:1349–1357
113. Hayward CPM, Weiss HJ, Lages B et al (2001) The storage defects in grey platelet syndrome and αδ-storage pool deficiency affect α-granule factor V and multimerin storage without altering their proteolytic processing. Br J Haematol 113:871–877
114. Hayward CPM, Warkentin TE, Horsewood P, Kelton JG (2015) Multimerin: a series of large disulfide-linked multimeric proteins within platelets. Blood 77:2556–2560
115. Koedam J a, Cramer EM, Briend E et al (1992) P-selectin, a granule membrane protein of platelets and endothelial cells, follows the regulated secretory pathway in AtT-20 cells. J Cell Biol 116:617–625
116. Furie B, Furie BC, Flaumenhaft R (2001) A journey with platelet P-selectin: the molecular basis of granule secretion, signalling and cell adhesion. Thromb Haemost 86:214–221
117. Hrachovinova I, Cambien B, Hafezi-Moghadam A et al (2003) Interaction of P-selectin and PSGL-1 generates microparticles that correct hemostasis in a mouse model of hemophilia A. Nat Med 9:1020–1025
118. Singbartl K, Forlow SB, Ley K (2001) Platelet, but not endothelial, P-selectin is critical for neutrophil-mediated acute postischemic renal failure. FASEB J 15:2337–2344
119. Palabrica T, Lobb R, Furie BC et al (1992) Leukocyte accumulation promoting fibrin deposition is mediated in vivo by P-selectin on adherent platelets. Nature 359:848–851
120. Yokoyama S, Ikeda H, Haramaki N et al (2005) Platelet P-selectin plays an important role in arterial thrombogenesis by forming large stable platelet-leukocyte aggregates. J Am Coll Cardiol 45:1280–1286
121. Mayadas TN, Johnson RC, Rayburn H et al (1993) Leukocyte rolling and extravasation are severely compromised in P selectin-deficient mice. Cell 74:541–554
122. Holmsen H, Day HJ (1968) Thrombin-induced platelet release reaction and platelet lysosomes. Nature 219:760–761
123. Ciferri S, Emiliani C, Guglielmini G et al (2000) Platelets release their lysosomal content in vivo in humans upon activation. Thromb Haemost 83:157–164
124. Rendu F, Brohard-Bohn B (2001) The platelet release reaction: granules' constituents, secretion and functions. Platelets 12:261–273
125. White JG, Clawson CC (1980) The surface-connected canalicular system of blood platelets – a fenestrated membrane system. Am J Pathol 101:353–364
126. Escolar G, Leistikow E, White JG (1989) The fate of the open canalicular system in surface and suspension-activated platelets. Blood e74:1983–1988
127. Escolar G, White JG (1991) The platelet open canalicular system: a final common pathway. Blood Cells 17:467–495
128. Choi W, Karim ZA, Whiteheart SW (2010) Protein expression in platelets from six species that differ in their open canalicular system. Platelets 21:167–175
129. Flaumenhaft R, Rozenvayn N, Feng D, Dvorak AM (2007) SNAP-23 and syntaxin-2 localize to the extracellular surface of the platelet plasma membrane. Blood 110:1492–1501
130. White JG (2005) Platelets are covercytes, not phagocytes: uptake of bacteria involves channels of the open canalicular system. Platelets 16:121–131
131. Ebbeling L, Robertson C, McNicol a, Gerrard JM (1992) Rapid ultrastructural changes in the dense tubular system following platelet activation. Blood 80:718–723
132. Yarom R, Lijovetzky G, Havivi Y (1981) Platelet heterogeneity and dense tubular system changes on activation. Experientia 37:96–98
133. Kovàcs T, Berger G, Corvazier E et al (1997) Immunolocalization of the multi-sarco/endoplasmic reticulum Ca2+ ATPase system in human platelets. Br J Haematol 97:192–203

134. Shuster RC, Rubenstein AJ, Wallace DC (1988) Mitochondrial DNA in anucleate human blood cells. Biochem Biophys Res Commun 155:1360–1365
135. Choo HJ, Saafir TB, Mkumba L et al (2012) Mitochondrial calcium and reactive oxygen species regulate agonist-initiated platelet phosphatidylserine exposure. Arterioscler Thromb Vasc Biol 32:2946–2955
136. Verhoeven AJ, Verhaar R, Gouwerok EG, de Korte D (2005) The mitochondrial membrane potential in human platelets: a sensitive parameter for platelet quality. Transfusion 45:82–89
137. Fukami MH, Salganicoff L (1973) Isolation and properties of human platelet mitochondria. Blood 42:913–918
138. Koseoglu S, Dilks JR, Peters CG et al (2013) Dynamin-related protein-1 controls fusion pore dynamics during platelet granule exocytosis. Arterioscler Thromb Vasc Biol 33:481–488
139. Bhandari V, Choo-Wing R, Lee CG et al (2006) Hyperoxia causes angiopoietin 2-mediated acute lung injury and necrotic cell death. Nat Med 12:1286–1293
140. Serve O, Kamiya Y, Maeno A et al (2010) Redox-dependent domain rearrangement of protein disulfide isomerase coupled with exposure of its substrate-binding hydrophobic surface. J Mol Biol 396:361–374
141. Dale GL, Remenyi G, Friese P (2005) Quantitation of microparticles released from coated-platelets. J Thromb Haemost 3:2081–2088
142. Jobe SM, Wilson KM, Leo L et al (2008) Critical role for the mitochondrial permeability transition pore and cyclophilin D in platelet activation and thrombosis. Blood 111:1257–1265
143. Garcia-Souza LF, Oliveira MF (2014) Mitochondria: biological roles in platelet physiology and pathology. Int J Biochem Cell Biol 50:156–160
144. Kile BT (2014) The role of apoptosis in megakaryocytes and platelets. Br J Haematol 165:217–226
145. Mason KD, Carpinelli MR, Fletcher JI et al (2007) Programmed anuclear cell death delimits platelet life span. Cell 128:1173–1186
146. White JG (1979) Ultrastructural studies of the gray platelet syndrome. Am J Pathol 95:445–462
147. Thon JN, Peters CG, Machlus KR et al (2012) T granules in human platelets function in TLR9 organization and signaling. J Cell Biol 198:561–574
148. Jasuja R, Passam FH, Kennedy DR et al (2012) Protein disulfide isomerase inhibitors constitute a new class of antithrombotic agents. J Clin Invest 122:2104–2113
149. Richardson JL, Shivdasani RA, Boers C et al (2005) Mechanisms of organelle transport and capture along proplatelets during platelet production. Blood 106:4066–4075
150. Ren Q, Ye S, Whiteheart SW (2008) The platelet release reaction: just when you thought platelet secretion was simple. Curr Opin Hematol 15:537–541
151. Graham GJ, Ren Q, Dilks JR et al (2009) Endobrevin/VAMP-8-dependent dense granule release mediates thrombus formation in vivo. Blood 114:1083–1090
152. Woronowicz K, Dilks JR, Rozenvayn N et al (2010) The platelet actin cytoskeleton associates with SNAREs and participates in alpha-granule secretion. Biochemistry 49:4533–4542
153. Rozenvayn N, Flaumenhaft R (2001) Phosphatidylinositol 4,5-bisphosphate mediates Ca2+-induced platelet alpha-granule secretion. Evidence for type II phosphatidylinositol 5-phosphate 4-kinase function. J Biol Chem 276:22410–22419
154. Rozenvayn N, Flaumenhaft R (2003) Protein kinase C mediates translocation of type II phosphatidylinositol 5-phosphate 4-kinase required for platelet alpha-granule secretion. J Biol Chem 278:8126–8134
155. Feng D, Flaumenhaft R, Bandeira-Melo C et al (2001) Ultrastructural localization of vesicle-associated membrane protein(s) to specialized membrane structures in human pericytes, vascular smooth muscle cells, endothelial cells, neutrophils, and eosinophils. J Histochem Cytochem 49:293–304

Signalling Pathways Regulating Platelet Biogenesis

7

Alexandra Mazharian and Yotis A. Senis

Abstract

Platelet biogenesis is a complex process controlled by a combination of cell-extrinsic and cell-intrinsic factors. Cell-extrinsic factors include cytokines and growth factors, chemokines, extracellular matrix proteins, cell-cell interactions and shear forces within the bone marrow *milieu* that act on megakaryocytes. Cell-intrinsic factors include receptors and associated signalling pathways, cytoskeletal structures, lipid and cation concentrations found within megakaryocytes. Collectively, these two sets of factors control the differentiation, proliferation and survival of megakaryocytes and produce platelets. Recent discoveries have greatly increased our understanding of the molecular mechanisms controlling platelet biogenesis; however, major gaps remain in our knowledge regarding signalling events regulating the transition from megakaryopoiesis to thrombopoiesis, the triggering of proplatelet formation and platelet release and the inhibition of activation signals within megakaryocytes during the course of platelet biogenesis. A major step will be to map all of the signalling networks within megakaryocytes and elucidate their interconnectedness. In this chapter, we describe some of the major signalling pathways that regulate platelet biogenesis, novel modes of regulation and inhibitory mechanisms that prevent uncontrolled megakaryocyte and platelet activation. We also highlight key questions that remain to be addressed and propose potential mechanisms. It is only through a comprehensive understanding of how platelet biogenesis is regulated that we will be

A. Mazharian (✉) • Y.A. Senis
Institute of Cardiovascular Sciences (ICVS), College of Medical and Dental Sciences, Institute of Biomedical Research, University of Birmingham, Birmingham B15 2TT, UK
e-mail: a.mazharian@bham.ac.uk; Y.Senis@bham.ac.uk

H. Schulze, J. Italiano (eds.), *Molecular and Cellular Biology of Platelet Formation*, DOI 10.1007/978-3-319-39562-3_7

able to identify key signalling nodes that can be targeted to modulate platelet homeostasis in health and disease.

7.1 Introduction

Megakaryocytes make up only 0.01 % of bone marrow cells, but produce approximately 100 billion platelets per day in an adult human. The normal platelet range is 150–400 × 10^9/L in humans, and 600–1,200 × 10^9/L in mice, the model of choice for studying platelet production and function *in vivo*. The risk of pathological bleeding rises substantially when platelet counts fall below 50 × 10^9/L and thrombosis when counts rise above 70 % of normal [1, 2]. Thus, platelet production must be tightly regulated and coordinated with platelet clearance in order to maintain the blood platelet concentration within the normal range.

The process of *thrombopoiesis* or platelet production takes place in the bone marrow and to a lesser extent the spleen [3]. Prior to this happening, haematopoietic stem cells (HSCs) must differentiate into megakaryocytes in a process referred to as *megakaryopoiesis* that culminates in the formation of mature polyploid megakaryocytes with the capacity to produce platelets [4]. Interestingly, megakaryocytes of any ploidy can produce platelets, although presumably the higher the ploidy, the more platelets that can ultimately be produced. An essential part of the process is for megakaryocytes to come into contact with sinusoidal blood vessels where they can release the platelets into the circulation. This can come about by megakaryocytes either migrating towards blood vessels or developing adjacent to them [5, 6]. Once in contact with the abluminal surface of the blood vessel, megakaryocytes breach the outer lining of the vessel wall and extrude fragments into the lumen in a polarized manner. These large fragments further subdivide into *preplatelets* and eventually platelets within the circulation. Preplatelets and platelets are ultimately shed from the tips of long filamentous projections, referred to as *proplatelets* that arise from the main body of the megakaryocyte and the megakaryocyte fragments that have entered the circulation [7, 8]. Again, proplatelets form in a polarized manner, away from the megakaryocyte and bone marrow, and towards the lumen of the blood vessel (Fig. 7.1). Although proplatelets can arise directly from megakaryocytes *in vitro*, it appears as though most platelets found in the circulation are derived from the large megakaryocyte fragments released into the circulation that further subdivide into platelets, rather than proplatelets that extend directly from the main body of the megakaryocyte. Further work is required to determine the proportions derived from both processes *in vivo*. Doing so has important implications for understanding the molecular basis of platelet production.

The processes of megakaryopoiesis and thrombopoiesis are regulated by cell-extrinsic and cell-intrinsic factors within sites of haematopoiesis and megakaryocytes, respectively. Many of these factors have been elucidated, but suspect many

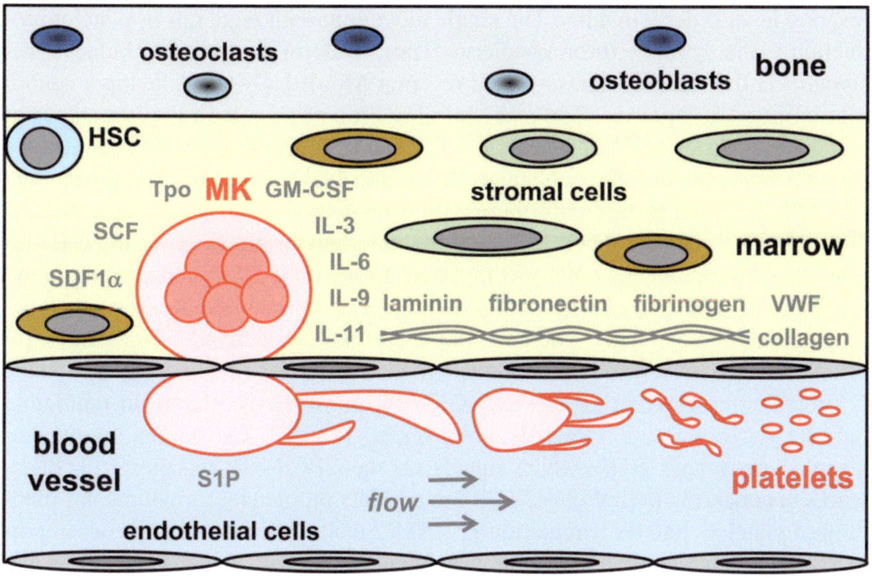

Fig. 7.1 Schematic representation of platelet biogenesis in the bone marrow. Cross section of the bone marrow and sinusoidal blood vessel. Megakaryocytes (MKs) are exposed to a variety of cell-extrinsic factors, including cytokine and growth factors (Tpo, SCF, GM-CSF, IL-3, IL-6, IL-9, IL-11), chemokines (SDF-1α), extracellular matrix proteins (collagen, laminin, fibronectin, fibrinogen and VWF), cell-cell interactions and shear forces within blood vessels. Large fragments of megakaryocytes are extruded into the blood vessels where they are exposed to shear forces, causing them to detach from the megakaryocyte body that remains in the bone marrow. Megakaryocytes form long filamentous projections called proplatelets from the tips of which preplatelets and platelets are released.

more remain undefined. Our inability to culture large numbers of platelets *in vitro* is evidence of our lack of understanding of the molecular mechanism controlling this process. Cell-extrinsic factors include any external stimuli that act on megakaryocytes and their progenitors to regulate differentiation, proliferation, survival, cellular responses and platelet biogenesis. Cell-extrinsic factors can be broadly divided into cytokines and growth factors, chemokines, extracellular matrix (ECM) proteins, cell-cell interactions and shear forces within sinusoidal blood vessels. All of these factors act directly on receptors on the megakaryocyte surface that transmit biochemical signals within megakaryocytes. Cell-intrinsic factors include all receptors and associated signalling pathways; cytoskeletal proteins and structures that regulate the formation of signalling complexes and cellular responses; and lipid composition of membranes and cation concentrations within megakaryocytes. These two sets of factors are interrelated and work in conjunction to coordinate megakaryocyte development and function.

Megakaryocytes are continuously exposed to a myriad of stimuli within the bone marrow and spleen, but are somehow able to integrate these stimuli and respond in an orderly manner. The single most important regulator of platelet production is the cytokine thrombopoietin (Tpo), made in the liver and kidneys that signals via the tyrosine kinase-linked receptor Mpl [9]. Despite playing a central role in megakaryopoiesis, Tpo and Mpl play little or no role in the late stages of platelet biogenesis, and mice lacking either Tpo or Mpl still make megakaryocytes and platelets, albeit far fewer platelets than control mice (10–15 % of normal) [10–12]. Other cytokines that contribute to the regulation of megakaryopoiesis and thrombopoiesis are stem cell factor (SCF), Interleukin (IL)-3, IL-6, IL-9, IL-11, granulocyte-macrophage colony-stimulating factor (GM-CSF) and erythropoietin (Epo), all of which signal in a similar manner, but are required at different stages of megakaryopoiesis [13]. The chemokine stromal cell-derived factor (SDF)-1α, also known as C-X-C motif chemokine ligand 12 (CXCL12), which signals via the G protein-coupled receptor (GPCR) CXCR4, plays a critical role in regulating megakaryocyte migration towards blood vessels [14–16]. The sphingolipid sphingosine-1-phosphate (S1P), which signals via the GPCR S1P receptor 1 (S1PR1), works in concert with SDF-1α-CXCR4 to regulate proplatelet formation and shedding of platelets into the circulation [17, 18]. Finally, shear forces play an important role in regulating the rate of proplatelet extension, platelet shedding and release into the circulation [19]. However, it remains to be proven whether shear forces activate biochemical signalling pathways within megakaryocytes or whether the shear force itself is sufficient to induce mechanical shedding of platelets.

In the remainder of this chapter, we discuss four key aspects of cell-intrinsic factors that regulate platelet biogenesis. First, we discuss our current understanding of Tpo signal transduction and how this regulates megakaryopoiesis. We also introduce the concept of developmental-functional switching that occurs when megakaryocytes stop developing and start producing platelets, which requires a different set of signalling events. Second, we discuss the importance of the Src family kinase (SFK)-Syk-PLCγ2 signalling axis to megakaryocyte development and platelet production (Fig. 7.2a). This signalling pathway is utilized by integrins and immunoreceptor tyrosine-based activation motif (ITAM)-containing receptors involved in regulating cellular responses to ECM proteins. This signalling pathway is central to platelet activation, but it remains unclear how it is regulated in megakaryocytes, which are constantly in contact with highly thrombogenic ECMs in the bone marrow. Third, we discuss a relatively new and under-investigated area of megakaryocyte signal transduction, namely the role of protein-tyrosine phosphatases (PTPs) in regulating megakaryopoiesis and thrombopoiesis. We focus on the receptor-like PTP CD148 that plays a critical role in regulating SFKs in conjunction with C-terminal Src kinase (Csk) in platelets and presumably megakaryocytes [20] (Fig. 7.2b) and the immunoreceptor tyrosine-based inhibition motif (ITIM)-containing receptor G6b-B, which signals via and regulates the non-transmembrane PTPs Shp1 and Shp2 (Fig. 7.3). Finally, we discuss signalling events regulating the terminal stages of platelet biogenesis, including megakaryocyte migration, proplatelet formation and platelet release, focusing mainly on the GPCRs CXCR4 and S1PR1 that have been implicated in regulating these processes.

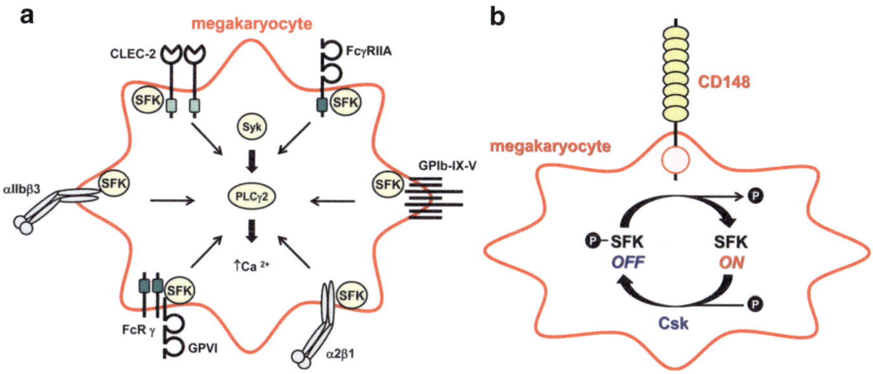

Fig. 7.2 The Src family kinase (SFK)-Syk-PLCγ2-Ca^{2+} signalling pathway in megakaryocytes. (**a**) Many of the main primary activation receptors are expressed in platelets signalling via the Src family kinases (SFKs)-Syk-PLCγ2-Ca^{2+}; however, it remains unclear how signals from these receptors are inhibited in megakaryocytes within the bone marrow. Dark green box in the cytoplasmic tail of the FcR γ-chain represents an immunoreceptor tyrosine-based activation motif (ITAM), and the light green box in the cytoplasmic tail of CLEC-2 a hem-ITAM domain. (**b**) A growing body of evidence has implicated the phosphatase-kinase pair CD148-Csk as a major regulator of SFKs in platelets and megakaryocytes.

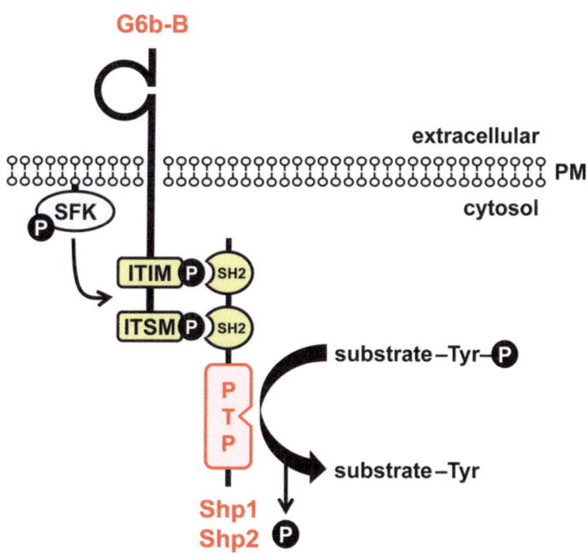

Fig. 7.3 The G6b-B-Shp1/2 signalling complex is a critical regulator of platelet homeostasis. Phosphorylation of conserved tyrosine residues within the immunoreceptor tyrosine-based inhibition/switch motif (ITIM, ITSM)-containing receptor G6b-B by Src family kinases (SFKs) provides a docking site for the structurally related tandem SH2 domain-containing protein-tyrosine phosphatases Shp1 and Shp2. Docking to G6b-B acts to compartmentalize and activate Shp1 and Shp2, allowing them to dephosphorylate key substrates at the plasma membrane (PM), thus regulating signalling from activation receptors, including the immunoreceptor tyrosine-based activation (ITAM)-containing collagen receptor complex GPVI-FcR γ-chain and the hem-ITAM-containing podoplanin receptor CLEC-2.

7.2 Tpo-Mpl Receptor Signal Transduction

Several cytokines are essential for optimal megakaryopoiesis, first and foremost being Tpo and additional ones being SCF, IL-3, IL-6, IL-11 and GM-CSF. All cytokines bind to and signal via structurally related type I cytokine receptors, which lack intrinsic catalytic activity, instead relying on signalling proteins that interact with their cytoplasmic tails. There are three main signalling pathways activated downstream of all type I cytokine receptors, including (1) the Jak-signal transducer and activator of transcription (STAT), (2) the Ras-mitogen-activated protein kinase (MAPK) and (3) the phosphatidylinositol 3-kinase (PI3K)-Akt pathways [21]. Additional signalling events include activation of suppressors of cytokine signalling (SOCS), which inhibit Jak-STAT signalling; SFKs which can be both activatory and inhibitory; the non-transmembrane PTPs Shp1 and Shp2, which both positively regulate megakaryocyte polyploidization, the latter by positively regulating the Ras-MAPK signalling [22]; and the lipid phosphatase SHIP1, which negatively regulates the PI3K-Akt pathway (Fig. 7.4).

Mpl is a typical type I cytokine receptor that is expressed throughout megakaryopoiesis [9, 23]. It is detected as early as HSCs and as late as terminally differentiated

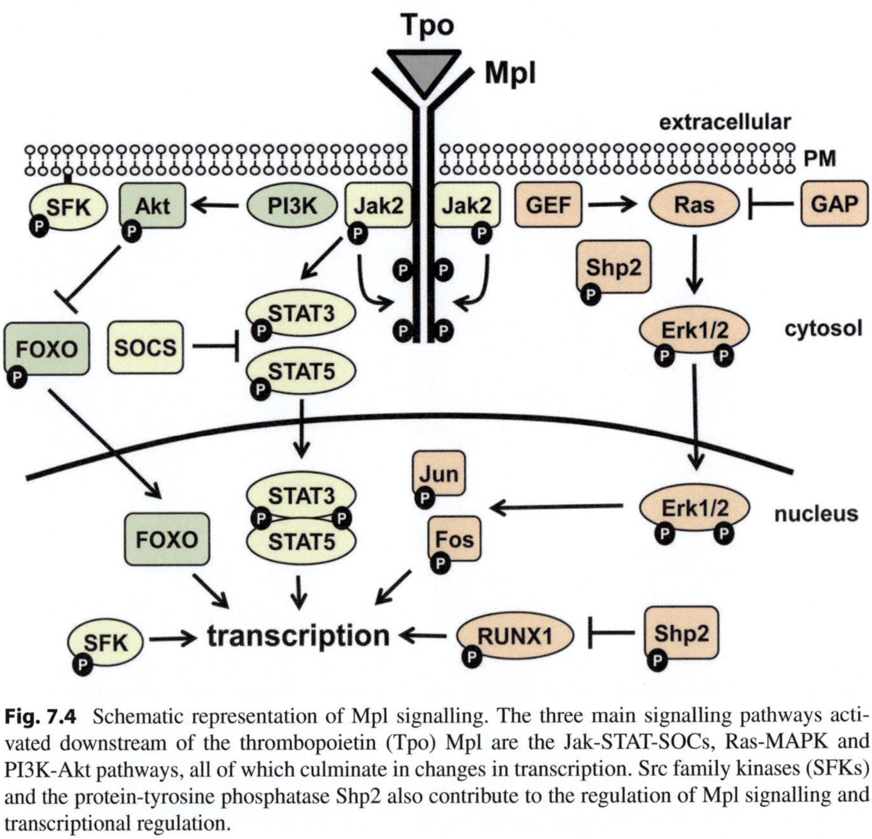

Fig. 7.4 Schematic representation of Mpl signalling. The three main signalling pathways activated downstream of the thrombopoietin (Tpo) Mpl are the Jak-STAT-SOCs, Ras-MAPK and PI3K-Akt pathways, all of which culminate in changes in transcription. Src family kinases (SFKs) and the protein-tyrosine phosphatase Shp2 also contribute to the regulation of Mpl signalling and transcriptional regulation.

platelets. Structurally, it consists of two cytokine receptor homology domains in its extracellular region; a single transmembrane domain and a relatively long cytoplasmic tail containing binding sites for various signalling proteins, including the protein-tyrosine kinase Jak2 that is constitutively associated with motifs in the juxtamembrane region and SH2 domain-containing adaptors and signalling proteins via phosphotyrosine residues in its cytoplasmic tail [24]. The receptor exists as a dimeric form on the surface of cells. Engagement of Tpo causes a conformational change in the cytoplasmic tails of adjacent Mpl monomers, bringing them into closer proximity and allowing Jak2 molecules to *trans*-phosphorylate one another and initiate signalling. Activation of Jak2 is one of the earliest and most important signalling events following ligand engagement. Downstream targets of Jak2 include tyrosine residues in the cytoplasmic tail of Mpl which act as docking sites for STAT1, STAT3 and STAT5 [25], and the adaptor protein Shc that in turn acts as a docking site for the adaptor Grb2 and the guanine nucleotide exchange factor (GEF) SOS, which activates the small G protein Ras and the downstream Raf-MAPK pathway that culminates in activation of transcription and gene expression. A recent landmark discovery was the demonstration that a gain-of-function mutation in Jak2 underlies many myeloproliferative disorders. A valine to phenylalanine substitution at position 617 (V617F) in the Jak homology 2 pseudokinase domain of Jak2 was shown to be present in the majority of patients with polycythaemia vera (PV) [26]. This mutation results in increased Mpl signalling in megakaryocytes and platelets. The disease was subsequently recapitulated in knockin mouse models, expressing the same mutation [27, 28]. Interestingly, Jak2 V617F results in increased platelet reactivity and predisposes both humans and mice to an increased risk of thrombotic complications, purportedly by priming platelets for activation by classical agonists, including collagen and thrombin.

MAPKs are divided into four structurally related subfamilies, including extracellular signal-regulated kinase (Erk1/Erk2), Erk5, c-Jun N-terminal kinase (JNK) and p38, all of which contain a distinctive T-x-Y (threonine-x-tyrosine) motif that when phosphorylated on both residues locks the activation loop in an active conformation [29]. Phosphorylated MAPKs subsequently enter the nucleus where they phosphorylate and activate key transcription factors, including Jun and Fos, key components of the AP-1 transcription factor complex, which regulates gene expression. The two most well-studied MAPKs in megakaryocytes are Erk1 and Erk2 [30]. Both play a critical role in regulating the development of human and mouse megakaryocytes *in vitro*, but this has yet to be definitively proven *in vivo*. In addition all of the studies to date were performed using inhibitors of MEK1 and MEK2, kinases upstream of Erk1 and Erk2, raising the question of non-specificity [31]. These inhibitors also do not differentiate between Erk1 and Erk2, thus the specific functions of each remain ambiguous. Knockout mouse studies demonstrate that both are required for haematopoiesis [32], but this has yet to be demonstrated in the megakaryocyte lineage. Recent findings suggest that Erk1 and Erk2 are essential for transmitting activation signals downstream of Mpl and that these signals must be inhibited in order for platelet biogenesis to take place [33]. This cannot simply be explained by Erk1/Erk2 being degraded in the final stages of megakaryocytopoiesis, because Erk1 and Erk2 are important for transmitting activation signals downstream of integrin and ITAM-containing receptors in platelets. Another MAPK activated downstream of Mpl is p38, resulting in induction of the transcription factor HoxB4 [34]. HoxA9 also translocates from the cytoplasm to the cytosol, favouring HSC expansion [34].

Another critical signalling pathway activated downstream of Mpl is the PI3K-Akt pathway [35], blocking of which inhibits Tpo-mediated cell survival and proliferation [36]. Upstream of PI3K is a complex consisting of the PTP Shp2, the scaffolding protein Gab/IRS and the p85 regulatory subunit of PI3K. Distal of Akt is the activation of the transcription factor FOXO3a, the cell cycle inhibitor p27 and glycogen synthase kinase 3β (GSK3β). PI3K isoforms α and β play important roles in transmitting activation signals downstream of integrin and ITAM-containing receptors in platelets [37], thus like MAPKs, PI3K and Akt must be inhibited at some point during platelet biogenesis and packaged into platelets in an inactivate state for later use.

Besides these three major signalling pathways, several other key signalling events are essential for optimal Mpl signalling. The SFK Lyn negatively regulates Mpl signalling by activating the lipid phosphatase SHIP-1, which inhibits signalling via the PI3K-Akt pathway. Mice lacking Lyn exhibit a complex platelet phenotype, including mild thrombocytopenia after 10 weeks of age, secondary to enhanced inflammation, increased numbers of megakaryocyte progenitors and mature megakaryocytes in the bone marrow and aberrant platelet function [38]. The phenotype can be partially explained by increased Mpl-mediated Erk1/Erk2 and Akt activation in Lyn-deficient megakaryocytes. SHIP-1 phosphorylation is concomitantly decreased, suggesting that Lyn negatively regulates the Ras/MAPK and PI3K/Akt pathways in a SHIP-1-dependent manner. More recently, Lyn was implicated as a membrane cholesterol sensor in megakaryocyte progenitors, linking platelet production with membrane cholesterol levels and atherogenesis [39]. Increased membrane cholesterol accumulation in mouse megakaryocyte progenitors led to increased Mpl expression and signalling, platelet overproduction, arterial thrombosis and atherogenesis, in a hyper-cholesterolemic mouse model. Mechanistically, Lyn was proposed to be the dominant SFK mediating downregulation of Mpl via the E3 ubiquitin ligase Cbl [40]. Excessive membrane cholesterol accumulation led to decreased Lyn kinase activity and reduced Cbl-mediated downregulation of Mpl by Tpo.

The structurally related SH2 domain-containing non-transmembrane PTPs Shp1 and Shp2 are positive regulators of megakaryocyte polyploidization [22]. This is mainly based on findings from Shp1 and Shp2 conditional knockout mice. Primary mouse megakaryocytes lacking either Shp1 or Shp2 exhibit dramatic block at the 4N–8N transition *in vitro*. These effects are synergistic in primary mouse megakaryocytes lacking both Shp1 and Shp2, which exhibit an even more severe block in the 4N–8N transition. In addition, Shp1 and Shp2 double-deficient megakaryocytes fail to proliferate and survive as well as control megakaryocytes cultured under the same conditions [22] (Fig. 7.5a). In the case of Shp2, this could be explained by an attenuation of Mpl-mediated Erk1/Erk2 activation and integrin-mediated SFK and Erk1/Erk2 activation. However, the mechanism by which Shp2 positively regulates Ras-MAPK signalling remains ambiguous. This may be by negatively regulating a negative regulator of Ras activation, such as the dephosphorylation of a docking site

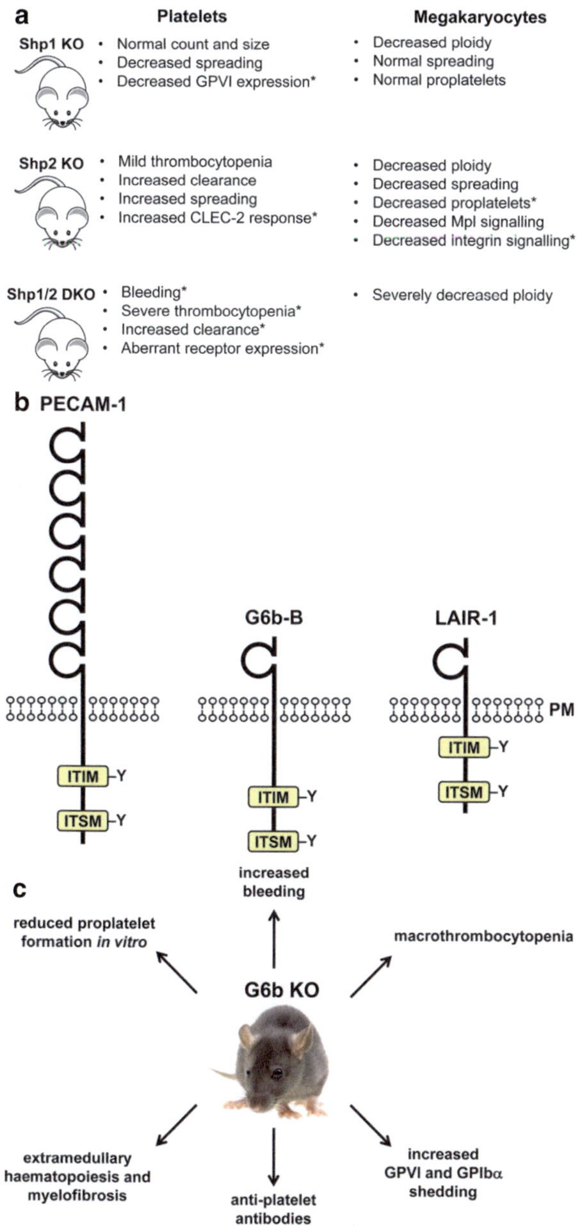

Fig. 7.5 (**a**) Distinct and overlapping features of G6b, Shp1, Shp2 and Shp1/Shp2 conditional KO and double-knockout (DKO) mice. Asterisk denotes phenocopy with G6b conditional KO mice. (**b**) Megakaryocyte-platelet immunoreceptor tyrosine-based inhibition motif (ITIM)-containing receptors: PECAM-1, G6b-B and LAIR-1. All three receptors have structural similarities, including immunoglobulin domains in their extracellular regions and consensus ITIM and immunoreceptor tyrosine-based switch motifs (ITSMs) in their cytoplasmic tails that act as docking sites for SH2 domain-containing lipid and protein-tyrosine phosphatases (SHIP-1, Shp1 and Shp2) and kinases (Csk) upon tyrosine phosphorylation by Src family kinases. (**c**) Summary of the phenotype of G6b knockout (KO) mice.

on the Mpl receptor or tyrosine residue that inhibits formation of the Shc-Grb2-SOS complex, but this remains to be proven. Less is known about the mechanism underlying how Shp1 positively regulates polyploidization. There is no evidence of aberrant Mpl or integrin signalling in Shp1-deficient megakaryocytes [22]. The current hypothesis is that Shp1 positively regulates the small G protein RhoA or some other component of the machinery controlling endomitosis in megakaryocytes.

Thus, despite the massive amount we already know about the Mpl signalling pathway, much remains to be elucidated.

7.3 The SFK-Syk-PLCγ2 Signalling Axis

Despite expressing the same repertoire of integrins and ITAM-containing receptors as platelets, megakaryocytes respond differently to ECM proteins compared with platelets. A case in point being collagen type I, which is highly abundant in the bone marrow and vessel wall, and one of the most powerful platelet agonists. Collagen engagement and cross-linking of the ITAM-containing GPVI-FcR γ-chain complex transmit a rapid and powerful activation signal within platelets that culminates in secretion, integrin activation and procoagulant activity. In addition to the GPVI receptor, platelets also express the integrin α2β1, which binds collagen with high affinity and increases the avidity to collagen. In contrast to the platelet response, collagen does not cause cultured primary mouse megakaryocytes secretion and inhibits proplatelet formation. The mechanism remains unclear, but it reported to be mediated via the integrin α2β1 [41, 42]. In addition to causing platelet activation, prolonged engagement of GPVI-FcR γ-chain by various GPVI agonists causes ectodomain shedding, internalization and degradation, but presumably this does not happen in megakaryocytes, despite being surrounded by collagen in the bone marrow. Again, the mechanism remains unknown.

In addition to the collagen receptors GPVI-FcR γ-chain and the integrin α2β1, megakaryocytes also express the integrin αIIbβ3, which is the most abundant receptor on the platelet surface and is essential for platelet aggregation and thrombus formation. Ligands of αIIbβ3 include fibrinogen, fibronectin and VWF, all of which are present in the blood and bone marrow. Megakaryocytes also express the integrins α6β1, which binds laminin, α5β1, which binds fibronectin and fibrinogen and αvβ3, which binds vitronectin and osteopontin, respectively, all of which found in the bone marrow. Interestingly, laminin has also been reported to bind to and cause activation of GPVI-FcR γ-chain in platelets [43]. Megakaryocytes also express the hem-ITAM-containing podoplanin receptor CLEC-2, which causes powerful platelet activation. Podoplanin is expressed in lymphatic endothelial cells and gets up-regulated in other cell types under inflammatory conditions, so whether or not megakaryocytes come into contact with it in the bone marrow or spleen microenvironments may depend on the pathological condition.

All of the integrin and ITAM-containing receptors described above signal via the SFK-Syk-PLCγ2 axis (Fig. 7.2). Platelets in fact express very high levels of SFKs relative to all other cell types, thus presumably megakaryocytes do as well.

Interestingly, several of the SFKs are proto-oncogenes, yet it has yet to be definitively proven how they are regulated in megakaryocytes and platelets for that matter [44].

7.4 Inhibitory ITIM-Containing Receptors and the PTPs Shp1 and Shp2

Two fundamental questions that remain unresolved regarding platelet biogenesis are as follows: (1) How are activation signals turned off in megakaryocyte fragments that enter the circulation (2) What triggers proplatelet formation? If one supposes that mature megakaryocytes express the same repertoire of surface receptors as platelets, as well as the same signalling molecules, one would expect megakaryocytes to become activated in the bone marrow *milieu* where they are constantly exposed to a variety of platelet agonists, including collagen, fibronectin, laminin and Tpo, to name a few. In addition, it remains unclear what triggers megakaryocytes to form proplatelets and undergo platelet biogenesis. Only a fraction of megakaryocytes undergo proplatelet formation *in vitro*, despite being exposed to Tpo and various other cytokines that drive megakaryopoiesis. A growing body of evidence suggest that a family of ITIM-containing receptors, including the collagen receptor LAIR-1, G6b-B and PECAM-1, as well as the non-transmembrane PTPs Shp1 and Shp2, play critical roles in regulating that activation status of megakaryocytes in the bone marrow and proplatelet formation [22, 45, 46].

ITIM-containing receptors were first reported as inhibitors of ITAM-containing receptor signalling in immune cells. Ligand-mediated co-clustering of the ITIM-containing FcγRIIb receptor with the ITAM-containing B cell receptor (BCR) was shown to inhibit signalling from the BCR. Mechanistically, this was elucidated to be caused by binding and compartmentalization of the SH2 domain-containing lipid phosphatase SHIP-1 to the plasma membrane, via the phosphorylated ITIM-containing FcγRIIb receptor, where it dephosphorylated PIP3 to PIP2 and disrupts interactions of PH domain-containing signalling proteins, such as Btk/Tec PTKs and the lipid hydrolase PLCγ2 with the plasma membrane, hence disrupting BCR signalling [47]. Several other ITIM-containing receptors have since been identified and shown to inhibit signalling from other ITAM-containing receptors, including CD19 and CD22 [48], PECAM-1, G6b-B and LAIR-1 (Fig. 7.5b). All of these receptors contain tandem ITIM domains or an ITIM and an immunoreceptor tyrosine-based switch motif (ITSM) in their cytoplasmic tails, which are separated by 12–24 amino acids. Phosphorylation of conserved tyrosine residues within the ITIM and ITSM acts as docking sites for the phosphatases SHIP1, Shp1 and Shp2 and in some instances the inhibitory PTK Csk, all of which act to inhibit signal transduction. Thus, ITIM-containing receptors act as docking sites for inhibitory signalling proteins that bring them into close proximity with activation receptors that they can subsequently inhibit [49]. However, there is a growing body of evidence that ITIM-containing receptors can also facilitate signalling from some receptors, such as integrins.

PECAM-1 was the first and is the most highly investigated ITIM-containing receptor in platelets [50]; however, little is known about its role in regulating megakaryocyte development and function. It is highly expressed in platelets and endothelial cells, but is also found in other haematopoietic and non-haematopoietic cells. There are estimated to be approximately 10,000 copies of PECAM-1 on the surface of resting human platelets. When PECAM-1 is first expressed in the megakaryocyte lineage has yet to be elucidated. Structurally, it consists of a large ectodomain, containing six immunoglobulin (Ig) domains, a single transmembrane domain and an ITIM and ITSM in its cytoplasmic tail. Phosphorylation of the ITIM and ITSM by SFKs provides a docking site for Shp1 and Shp2, although Shp2 appears to bind preferentially. The ectodomain binds homophilically to PECAM-1 expressed on other platelets and endothelial cells, which are thought to limit platelet activation under normal and pathological conditions, much the same as prostaglandin I_2 (PGI_2) and nitric oxide (NO) released by endothelial cells, albeit to a lesser degree. Antibody- or PECAM-1-mediated cross-linking of PECAM-1 on the platelet surface results in an increase in tyrosine phosphorylation of the ITIM and ITSM, association with Shp1 and Shp2 and inhibition of collagen and thrombin-mediated platelet activation and aggregation, suggesting PECAM-1 is an inhibitor of platelet activation [51]. Conversely, PECAM-1-deficient mouse platelets hyper-respond to collagen and thrombin *in vitro*, and exhibit increased thrombus formation *in vivo*, following laser-induced vessel injury [52]. Interestingly, PECAM-1-deficient platelets spread less well than control platelets on a fibrinogen-coated surface and exhibit reduced clot retraction *in vitro*, suggesting that PECAM-1 is a positive regulator of αIIbβ3-mediated function and signalling [53]. Platelet counts are normal in unchallenged PECAM-1-deficient mice, but the rate of platelet recovery is reduced following anti-GPIbα antibody-mediated thrombocytopenia [46]. Megakaryocytes from these mice developed normally *in vitro*, but exhibited reduced migration towards an SDF-1α gradient, which could be explained by a lack of compartmentalization of the SDF-1α receptor CXCR4 to the leading edge of the megakaryocyte [46]. Mechanistically, it remains unresolved how it does so. No other defects have been reported, most likely due to a lack of investigation, rather than a lack of defects.

G6b-B is much smaller than PECAM-1, containing only a single Ig-variable domain in its extracellular region, a single transmembrane domain and a shorter cytoplasmic tail, containing a proline-rich juxtamembrane region, an ITIM and an ITSM (Fig. 7.5b). G6b-B is highly expressed in mature megakaryocytes and platelets, with no detectable expression in HSCs, megakaryocyte progenitors and immature megakaryocytes [45]. There are estimated to be between 6,000–20,000 copies of G6b-B on the surface of human platelets, making it one of the most abundant platelet surface receptors. The ligand has yet to be identified. Interestingly, expression of G6b-B mirrors that of GPVI in the megakaryocyte lineage. The most well-documented function of G6b-B is as a negative regulator of signalling from the ITAM-containing collagen receptor complex GPVI-FcR γ-chain and the hem-ITAM-containing podoplanin receptor CLEC-2. Interestingly, antibody-mediated cross-linking of G6b-B inhibited CRP- and ADP-induced platelet aggregation, suggesting it plays a broader role than just inhibiting ITAM-containing receptor signalling. This hypothesis is supported by findings from G6b knockout (KO) mice, which exhibit a unique and complex platelet phenotype that provides important new insights into the molecular mechanism controlling platelet homeostasis. Mice lacking G6b are severely macrothrombocytopenic, exhibit aberrant platelet activation and function, and have elevated megakaryocyte counts in the bone marrow and spleen (Fig. 7.5c). Elevated megakaryocyte counts in sites of haematopoiesis are most likely as a consequence of elevated Tpo levels in the plasma. The macrothrombocytopenia is a consequence of an increase in platelet clearance and reduction in platelet production, resulting in a futile cycle and a net loss of platelets from the circulation [45]. Why platelets are rapidly cleared from the circulation remains to be conclusively determined. The leading hypothesis is that G6b-B-deficient platelets are in a primed, hyperactive state resulting in recognition and clearance by the mononuclear phagocyte system. This is supported by biochemical evidence demonstrating increased basal Syk activity and enhanced shedding of the ectodomains of GPVI and GPIbα from the surface of resting G6b-B-deficient platelets and cultured megakaryocytes. There was however no evidence of increased P-selectin or phosphatidylserine (PS) exposure on the surface of resting G6b-B-deficient platelets and only a marginal increase in P-selectin expression on the surface of resting G6b-B-deficient megakaryocytes *in vitro*. Interestingly, platelets from G6b KO mice had increased levels of IgG and IgM on their surface, suggesting an autoimmune defect [45]. Platelet surface immunoglobulins are recognized by macrophages in the spleen, leading to phagocytosis, but it remains unclear what the antibodies are binding to on the platelet surface, whether they are recognizing 'foreign' epitopes exposed on the surface glycoproteins, whether this represents nonspecific binding or whether this is due to upregulation of Fc receptors in platelets.

One of the most striking features of the G6b KO phenotype is the dramatic reduction in proplatelet formation of G6b-B-deficient megakaryocytes on fibrinogen- and fibronectin-coated surfaces *in vitro*. Like control megakaryocytes, G6b-B-deficient megakaryocytes failed to form proplatelets on a collagen-coated surface, demonstrating that G6b-B does not inhibit this process. These findings coincide with the reduction in platelet counts seen in G6b KO mice, suggesting that G6b-B and its putative ligand play some role in regulating this process. This is not due to a developmental defect, as G6b-B-deficient megakaryocytes differentiate and proliferate normally *in vitro*, from bone marrow-derived haematopoietic progenitors. No overt signalling defects were detected downstream of Mpl in G6b-B-deficient megakaryocytes. In addition to the dramatic reduction in proplatelet formation, G6b-B-deficient megakaryocytes failed to spread normally on fibrinogen-, fibronectin- and collagen-coated surfaces [45]. Integrin αIIbβ3 expression was normal in G6b-B-deficient megakaryocytes, whereas integrin α2β1 levels were reduced by >50 %, which may partially explain the reduced spreading on collagen. Mechanistically, SFK and Erk1/Erk2 activation were attenuated in fibrinogen-adhered G6b-B-deficient megakaryocytes, suggesting that G6b-B is a positive regulator of outside-in αIIbβ3 signalling, and underlies the fibrinogen- and fibronectin-spreading defects. It should be noted that the same phenotype was seen in G6b conventional and conditional KO mice, suggesting that G6b-B is lineage specific [22].

Interestingly, targeted deletion of Shp1 and/or Shp2 in the megakaryocyte lineage recapitulates many of the features of the G6b KO phenotype, suggesting G6b-B signals via and is an important regulator of Shp1 and Shp2 in megakaryocytes and platelets [22]. Disruption of Shp1 and Shp2 expression had different phenotypic effects, indicating that both molecules may exert their effect through distinctive mechanisms (Fig. 7.5a). Although the Shp1 conditional KO showed normal platelet count and size, platelets from these mice exhibited an abnormal response to collagen due to decreased expression of the collagen receptor GPVI. On the other hand, the Shp2 conditional KO showed low platelet counts, a finding likely attributed to defective thrombopoiesis that was perhaps due to decreased integrin and Mpl signalling and increased platelet clearance. Interestingly, the Shp1/Shp2 double KO mice expressed a much more severe phenotype, indicating that these two molecules are intrinsically involved in the proliferation, survival and normal function of megakaryocytes. Although the platelet phenotype of the G6b-B mice is strikingly similar to the Shp1/Shp2 double KO, megakaryocytes in the G6b-B mice show normal development and intact Mpl signalling, which is fundamentally affected in the Shp1/Shp2 double KO. This finding is extremely relevant for the further dissection of these pathways and their critical involvement in megakaryopoiesis.

Leukocyte-associated immunoglobulin (Ig)-like receptor-1 (LAIR-1) is a type I transmembrane glycoprotein of 287 amino acids containing a single extracellular C2-type Ig-like domain and two ITIMs in its cytoplasmic tail [54] (Fig. 7.5b). It is widely expressed on haematopoietic cells, including NK cells, T and B cells, monocytes and dendritic cells, but not on platelets or non-haematopoietic cells [54–57]. It is also present on CD34+ haematopoietic progenitor cells [58–60] and human [61] and mouse megakaryocytes. LAIR-1 is structurally related to other inhibitory immunoglobulin superfamily members localized to the leukocyte receptor complex, suggesting that these molecules have evolved from a common ancestral gene [62]. Furthermore, the gene encoding GPVI is also located within the LCR complex. The extracellular domain of LAIR-1 is structurally related to the corresponding region of LAIR-2 (86 % sequence homology) although this lacks a transmembrane domain and is absent in the mouse genome.

Cross-linking of LAIR-1 using specific mAbs to the extracellular domain has shown that LAIR-1 has a broad inhibitory function on many cell types and against several classes of receptor. LAIR-1 has been shown to inhibit target cell lysis by resting and activated NK cells [63], the cytotoxic activity of effector T cells [56, 64], prevent the proliferation and induction of apoptosis in human myeloid leukaemia cell lines and primary leukaemia cell lines [65, 66] and inhibit differentiation of peripheral blood precursors to dendritic cells [67, 68]. These effects have been shown to be mediated through tyrosine phosphorylation of the conserved tyrosine residues leading to recruitment of the tyrosine phosphatases Shp1 and Shp2 [69] and the inhibitor of Src family kinases, Csk [70]. Collagen is the only known endogenous ligand for LAIR-1 thereby revealing a novel inhibitory function of the extracellular matrix component in the immune system, possibly by setting a threshold for activation of cells that have migrated into tissue [71]. LAIR-1 binds with higher affinity to the same GPO repeat in collagen that mediates activation of GPVI [71].

The inhibitory effect of LAIR-1 is likely to be mediated through phosphorylation of the conserved tyrosines in the two ITIM sequences that are present in its cytosolic tail. Cross-linking of phosphorylated LAIR-1 to GPVI mediates inhibition by the juxtaposition of one or more of these inhibitory proteins adjacent to the FcR γ-chain ITAM. LAIR-1 is phosphorylated upon activation by pervanadate and by the GPVI ligand convulxin in immature megakaryocytes. Thus, we hypothesize that collagen preferentially stimulates LAIR-1 phosphorylation in immature megakaryocytes in the bone marrow due to its high affinity for the ITIM-containing receptor and that this prevents activation of GPVI. This therefore helps to prevent premature proplatelet formation and platelet release. Although the inhibitory function of LAIR-1 has been described in immune cells [72, 85], its physiological role in megakaryocytes and platelet formation has not been explored.

7.5 GPCR Signalling During Platelet Biogenesis

Platelets express a variety of GPCRs that are essential for regulating platelet activation and thrombus formation following vascular injury that play little or no role in regulating platelet production. Examples include the Gq-coupled thrombin receptors, protease-activated receptors (PAR) 1 and 4, the thromboxane A2 receptor (TxA2) TP, the ADP receptor P2Y1 and the Gi-coupled ADP receptor P2Y12. Targeted deletion or pharmacological inhibition of these receptors has no effect on platelet production. Conversely, there are examples of GPCRs that play important roles in regulating platelet biogenesis, but little or no role in platelet activation and thrombosis. Two of the best-studied GPCRs involved in platelet biogenesis are the Gαi-coupled SDF-1α (CXCL12) chemokine receptor CXCR4 (also referred to as fusin or CD184), which regulates megakaryocyte homing to the vascular niche [16], and the G-coupled S1P lipid receptor S1PR1 that reportedly regulates proplatelet formation and platelet release [17] (Fig. 7.6). Neither the CXCR4 nor the S1PR1 is confined to the megakaryocyte lineage and plays important roles in regulating leukocyte homing and the immune response. Indeed, all of what we know about signalling from both of these receptors has been derived from studies performed other cell types.

SDF-1α is expressed by a population of bone marrow stromal cells, which are adipo-osteogenic progenitors, and endothelial cells [73]. It has also been reported in megakaryocytes and gets packaged into α-granules in platelets making platelets a major source of SDF-1α in the vasculature. SDF-1α binds and signals via the GPCRs CXCR4 and CXCR7, the former being expressed throughout megakaryopoiesis, including platelets. Several physiological functions have been attributed to SDF-1α in megakaryocytes, including megakaryocyte chemotaxis, adhesion to endothelium and transmigration through the bone marrow endothelium [74]. It has also been implicated in regulating surface receptor expression in megakaryocytes.

Chemokine receptors such as CXCR4 are typically Gαi-coupled and act mainly by inhibiting adenylate cyclase, causing a reduction in intracellular cAMP and a reduction in PKA activation. Gαi has also been shown to bind directly to and

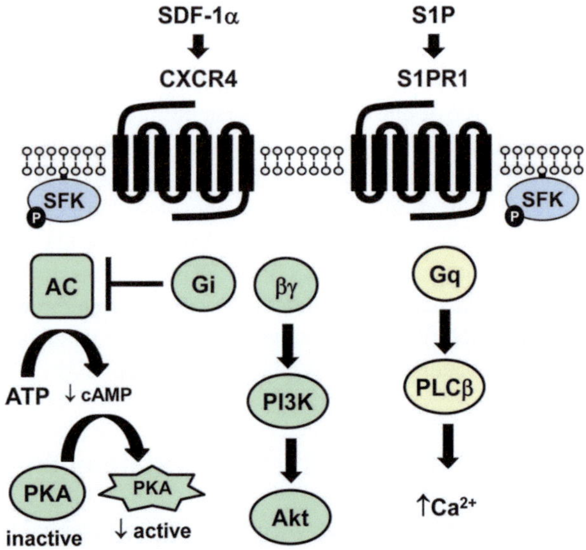

Fig. 7.6 Putative signalling pathways of the SDF-1α chemokine receptor CXCR4 and the sphingosine 1-phosphate receptor S1PR1. This is based on what is known from other cell types in which CXCR4 and S1PR1 are expressed.

activate SFKs. The Ras-MAPK signalling pathway has also been shown to get activated downstream of Gαi in some cell types. Interestingly, in the case of CXCR4, three major signalling pathways have been shown to get activated downstream of CXCR4, including Jak-STAT, PI3K-Akt and the Ras-MAPK pathway [75, 76]. However, this needs to be verified in megakaryocytes. Furthermore, all three of these signalling pathways culminate in regulation of gene expression, which presumably is not necessary for regulating megakaryocyte migration and homing to the vascular niche within the bone marrow, which are the best described functions of SDF-1α and CXCR4 in megakaryocytes [74]. Interestingly, SDF-1α also mediates calcium fluxes within cells via CXCR4, suggesting that CXCR4 may also be Gq coupled. Gq-coupled receptors signals mainly via the activation of PLCβ, the generation of DAG and IP3, leading to an increase in intracellular Ca^{2+} and activation of PKC. β-arrestin is also recruited downstream of CXCR4 to control receptor trafficking and desensitization. Thus, CXCR4 does more than just signal via the Gαi-adenylate cyclase axis. It likely also couples to Gq and leads to activation of several prominent signalling pathways that not only regulate cellular responses but also regulate gene expression, perhaps preparing the megakaryocytes for proplatelet formation and platelet release. A CXCR4 KO mouse model [77] and inhibitor (Plerixafor) [78, 79] are available that can be used to further investigate the function of CXCR4 in megakaryocyte function and platelet biogenesis. Knockout mouse models and pharmacological inhibitors are also available for many of the signalling pathways activated downstream of CXCR4 that can be employed to elucidate the main pathways regulating megakaryocyte migration and platelet production.

S1P is a blood-borne lipid mediator that is associated with lipoproteins, such as high density lipoprotein (HDL) [80]. It is generated in a variety of tissues, including megakaryocytes and platelets, through a process that involves the release of sphingosine from ceramides by ceramidase and phosphorylation by sphingosine kinase to S1P. It can be dephosphorylated by sphingosine phosphatases and returned to sphingosine and irreversibly degraded by sphingosine phosphate lyase. S1P is much less abundant in tissues compared with the blood, establishing a gradient that is important for immune cell trafficking, megakaryocyte chemotaxis and polarized platelet biogenesis into the blood. The receptor for S1P is the GPCR S1PR of which there are five subfamilies (S1PR1-5) [81]. S1PR1 is expressed in a variety of tissues, including megakaryocytes and platelets, and was recently reported to play a critical role in regulating proplatelet formation and platelet release. Like CXCR4, most of what we know about how S1PR1 signals was elucidate in other cell types, besides megakaryocytes and platelets [17, 86]. S1PR1 activation is involved in immune cell regulation and development and is also involved in immune modulation and directly involved in suppression of innate immune responses from T cells [82]. Depending on the G protein coupled with the S1PR1, diverse cellular effects are achieved: Gαi and Gαo modulate cellular survival, proliferation and motility; Gα12 and Gα13 modulate cytoskeletal remodelling and cell-shape changes and Gαq modulates several cellular effector functions. All the intracellular functions occur via the interaction with Gαi and Gαq: these two proteins recruit other proteins for downstream amplification of the signal. The main functions of S1P-S1PR1 system are (1) the PI3K-Akt signalling pathway increases the survival of lymphocytes and other immune cells by inhibiting apoptosis, and (2) PI3K and the GTPase Rac are responsible of the lymphocytes migration and their interactions with other cells or with connective tissue surfaces [8]. S1PR1-deficient thymocytes do not emigrate from the thymus, resulting in an increased numbers of mature thymocytes in the thymus and in medullary hyperplasia, and few S1PR1-deficient T cells can be detected in the blood, lymph nodes, spleen or non-lymphoid organs in these mouse models [83]. In addition, it has been shown that the proliferation of immune cells is due to S1P-mediated signals via the GTPase Ras and Erk1/Erk2 pathway [84].

7.6 Summary

Our understanding of the molecular basis regulating platelet biogenesis has progressed substantially in the last 50 years. We know that Jak2; STAT1, STAT3 and STAT5; PI3K and Akt; and Erk1/Erk2 MAPKs are involved in the primary signals that emanate from Mpl receptor. However, a number of other signalling proteins are also involved such as phosphatases Shp1 and Shp2 and the ITIM-containing receptor G6b-B. But despite the progress, many questions remain: (1) how are activation signals turned off in megakaryocyte fragments that enter the circulation and (2) what triggers proplatelet formation? The years to come will provide new insights into how platelets develop and how this process is regulated.

References

1. Thaulow E et al (1991) Blood platelet count and function are related to total and cardiovascular death in apparently healthy men. Circulation 84(2):613–617
2. Kaushansky K (2009) Determinants of platelet number and regulation of thrombopoiesis. Hematol Am Soc Hematol Educ Program 41:147–152
3. Schulze H, Shivdasani RA (2005) Mechanisms of thrombopoiesis. J Thromb Haemost 3(8):1717–1724
4. Kuter DJ (2014) Milestones in understanding platelet production: a historical overview. Br J Haematol 165(2):248–258
5. Junt T et al (2007) Dynamic visualization of thrombopoiesis within bone marrow. Science 317(5845):1767–1770
6. Mazharian A (2012) Assessment of megakaryocyte migration and chemotaxis. Methods Mol Biol 788:275–288
7. Thon JN, Italiano JE Jr (2012) Does size matter in platelet production? Blood 120(8): 1552–1561
8. Thon JN, Italiano JE (2012) Platelets: production, morphology and ultrastructure. Handb Exp Pharmacol 210:3–22
9. Kaushansky K (2005) Thrombopoietin and the hematopoietic stem cell. Ann N Y Acad Sci 1044:139–141
10. Alexander WS et al (1996) Deficiencies in progenitor cells of multiple hematopoietic lineages and defective megakaryocytopoiesis in mice lacking the thrombopoietic receptor c-Mpl. Blood 87(6):2162–2170
11. Murone M, Carpenter DA, de Sauvage FJ (1998) Hematopoietic deficiencies in c-mpl and TPO knockout mice. Stem Cells 16(1):1–6
12. Gurney AL et al (1994) Thrombocytopenia in c-mpl-deficient mice. Science 265(5177): 1445–1447
13. Kaushansky K (2005) The molecular mechanisms that control thrombopoiesis. J Clin Invest 115(12):3339–3347
14. Juarez J, Bendall L (2004) SDF-1 and CXCR4 in normal and malignant hematopoiesis. Histol Histopathol 19(1):299–309
15. Molyneaux KA et al (2003) The chemokine SDF1/CXCL12 and its receptor CXCR4 regulate mouse germ cell migration and survival. Development 130(18):4279–4286
16. Wang JF, Liu ZY, Groopman JE (1998) The alpha-chemokine receptor CXCR4 is expressed on the megakaryocytic lineage from progenitor to platelets and modulates migration and adhesion. Blood 92(3):756–764
17. Zhang L et al (2012) A novel role of sphingosine 1-phosphate receptor S1pr1 in mouse thrombopoiesis. J Exp Med 209(12):2165–2181
18. Zhang L et al (2013) Sphingosine kinase 2 (Sphk2) regulates platelet biogenesis by providing intracellular sphingosine 1-phosphate (S1P). Blood 122(5):791–802
19. Thon JN et al (2010) Cytoskeletal mechanics of proplatelet maturation and platelet release. J Cell Biol 191(4):861–874
20. Senis YA et al (2009) The tyrosine phosphatase CD148 is an essential positive regulator of platelet activation and thrombosis. Blood 113(20):4942–4954
21. Kaushansky K (2009) Molecular mechanisms of thrombopoietin signaling. J Thromb Haemost 7(Suppl 1):235–238
22. Mazharian A et al (2013) Megakaryocyte-specific deletion of the protein-tyrosine phosphatases Shp1 and Shp2 causes abnormal megakaryocyte development, platelet production, and function. Blood 121(20):4205–4220
23. Hitchcock IS, Kaushansky K (2014) Thrombopoietin from beginning to end. Br J Haematol 165(2):259–268
24. Sattler M et al (1995) The thrombopoietin receptor c-MPL activates JAK2 and TYK2 tyrosine kinases. Exp Hematol 23(9):1040–1048

25. Schulze H et al (2000) Thrombopoietin induces the generation of distinct Stat1, Stat3, Stat5a and Stat5b homo- and heterodimeric complexes with different kinetics in human platelets. Exp Hematol 28(3):294–304
26. James C et al (2005) A unique clonal JAK2 mutation leading to constitutive signalling causes polycythaemia vera. Nature 434(7037):1144–1148
27. Hobbs CM et al (2013) JAK2V617F leads to intrinsic changes in platelet formation and reactivity in a knock-in mouse model of essential thrombocythemia. Blood 122(23):3787–3797
28. Li J et al (2010) JAK2 V617F impairs hematopoietic stem cell function in a conditional knock-in mouse model of JAK2 V617F-positive essential thrombocythemia. Blood 116(9): 1528–1538
29. Pearson G et al (2001) Mitogen-activated protein (MAP) kinase pathways: regulation and physiological functions. Endocr Rev 22(2):153–183
30. Mazharian A, Watson SP, Severin S (2009) Critical role for ERK1/2 in bone marrow and fetal liver-derived primary megakaryocyte differentiation, motility, and proplatelet formation. Exp Hematol 37(10):1238–1249, e5
31. Borsch-Haubold AG, Pasquet S, Watson SP (1998) Direct inhibition of cyclooxygenase-1 and -2 by the kinase inhibitors SB 203580 and PD 98059. SB 203580 also inhibits thromboxane synthase. J Biol Chem 273(44):28766–28772
32. Chan G, Gu S, Neel BG (2013) Erk1 and Erk2 are required for maintenance of hematopoietic stem cells and adult hematopoiesis. Blood 121(18):3594–3598
33. Bluteau D et al (2014) Thrombocytopenia-associated mutations in the ANKRD26 regulatory region induce MAPK hyperactivation. J Clin Invest 124(2):580–591
34. Kirito K, Fox N, Kaushansky K (2003) Thrombopoietin stimulates Hoxb4 expression: an explanation for the favorable effects of TPO on hematopoietic stem cells. Blood 102(9):3172–3178
35. Miyakawa Y et al (2001) Thrombopoietin induces phosphoinositol 3-kinase activation through SHP2, Gab, and insulin receptor substrate proteins in BAF3 cells and primary murine megakaryocytes. J Biol Chem 276(4):2494–2502
36. Rojnuckarin P et al (2001) The roles of phosphatidylinositol 3-kinase and protein kinase Czeta for thrombopoietin-induced mitogen-activated protein kinase activation in primary murine megakaryocytes. J Biol Chem 276(44):41014–41022
37. Moroi AJ, Watson SP (2015) Impact of the PI3-kinase/Akt pathway on ITAM and hemITAM receptors: haemostasis, platelet activation and antithrombotic therapy. Biochem Pharmacol 94(3):186–194
38. Lannutti BJ et al (2006) Increased megakaryocytopoiesis in Lyn-deficient mice. Oncogene 25(23):3316–3324
39. Murphy AJ et al (2013) Cholesterol efflux in megakaryocyte progenitors suppresses platelet production and thrombocytosis. Nat Med 19(5):586–594
40. Oneyama C et al (2009) Transforming potential of Src family kinases is limited by the cholesterol-enriched membrane microdomain. Mol Cell Biol 29(24):6462–6472
41. Zou Z et al (2009) Negative regulation of activated alpha-2 integrins during thrombopoiesis. Blood 113(25):6428–6439
42. Sabri S et al (2004) Differential regulation of actin stress fiber assembly and proplatelet formation by alpha2beta1 integrin and GPVI in human megakaryocytes. Blood 104(10):3117–3125
43. Ozaki Y, Suzuki-Inoue K, Inoue O (2009) Novel interactions in platelet biology: CLEC-2/podoplanin and laminin/GPVI. J Thromb Haemost 7(Suppl 1):191–194
44. Senis YA, Mazharian A, Mori J (2014) Src family kinases: at the forefront of platelet activation. Blood 124(13):2013–2024
45. Mazharian A et al (2012) Mice lacking the ITIM-containing receptor G6b-B exhibit macrothrombocytopenia and aberrant platelet function. Sci Signal 5(248):ra78
46. Dhanjal TS et al (2007) A novel role for PECAM-1 in megakaryocytokinesis and recovery of platelet counts in thrombocytopenic mice. Blood 109(10):4237–4244
47. Kurosaki T (2002) Regulation of B-cell signal transduction by adaptor proteins. Nat Rev Immunol 2(5):354–363

48. Doody GM et al (1995) A role in B cell activation for CD22 and the protein tyrosine phosphatase SHP. Science 269(5221):242–244
49. Daeron M et al (2008) Immunoreceptor tyrosine-based inhibition motifs: a quest in the past and future. Immunol Rev 224:11–43
50. Newman PJ et al (1990) PECAM-1 (CD31) cloning and relation to adhesion molecules of the immunoglobulin gene superfamily. Science 247(4947):1219–1222
51. Patil S, Newman DK, Newman PJ (2001) Platelet endothelial cell adhesion molecule-1 serves as an inhibitory receptor that modulates platelet responses to collagen. Blood 97(6):1727–1732
52. Falati S et al (2006) Platelet PECAM-1 inhibits thrombus formation in vivo. Blood 107(2):535–541
53. Wee JL, Jackson DE (2005) The Ig-ITIM superfamily member PECAM-1 regulates the "outside-in" signaling properties of integrin alpha(IIb)beta3 in platelets. Blood 106(12):3816–3823
54. Meyaard L et al (1997) LAIR-1, a novel inhibitory receptor expressed on human mononuclear leukocytes. Immunity 7(2):283–290
55. van der Vuurst de Vries AR et al (1999) Leukocyte-associated immunoglobulin-like receptor-1 (LAIR-1) is differentially expressed during human B cell differentiation and inhibits B cell receptor-mediated signaling. Eur J Immunol 29(10):3160–3167
56. Maasho K et al (2005) The inhibitory leukocyte-associated Ig-like receptor-1 (LAIR-1) is expressed at high levels by human naive T cells and inhibits TCR mediated activation. Mol Immunol 42(12):1521–1530
57. Jansen CA et al (2007) Regulated expression of the inhibitory receptor LAIR-1 on human peripheral T cells during T cell activation and differentiation. Eur J Immunol 37(4):914–924
58. Meyaard L (2008) The inhibitory collagen receptor LAIR-1 (CD305). J Leukoc Biol 83(4):799–803
59. Xue J et al (2011) Leukocyte-associated immunoglobulin-like receptor-1 is expressed on human megakaryocytes and negatively regulates the maturation of primary megakaryocytic progenitors and cell line. Biochem Biophys Res Commun 405(1):128–133
60. Ouyang W et al (2003) 9.1C3 is identical to LAIR-1, which is expressed on hematopoietic progenitors. Biochem Biophys Res Commun 310(4):1236–1240
61. Steevels TA et al (2010) Co-expression of collagen receptors leukocyte-associated Ig-like receptor-1 and glycoprotein VI on a subset of megakaryoblasts. Haematologica 95:2005–2012
62. Martin AM et al (2002) Leukocyte Ig-like receptor complex (LRC) in mice and men. Trends Immunol 23(2):81–88
63. Poggi A et al (1997) p40 molecule regulates NK cell activation mediated by NK receptors for HLA class I antigens and TCR-mediated triggering of T lymphocytes. Int Immunol 9(9):1271–1279
64. Meyaard L et al (1999) Leukocyte-associated Ig-like receptor-1 functions as an inhibitory receptor on cytotoxic T cells. J Immunol 162(10):5800–5804
65. Poggi A et al (2000) Engagement of the leukocyte-associated Ig-like receptor-1 induces programmed cell death and prevents NF-kappaB nuclear translocation in human myeloid leukemias. Eur J Immunol 30(10):2751–2758
66. Zocchi MR et al (2001) Leukocyte-associated Ig-like receptor-1 prevents granulocyte-monocyte colony stimulating factor-dependent proliferation and Akt1/PKB alpha activation in primary acute myeloid leukemia cells. Eur J Immunol 31(12):3667–3675
67. Poggi A et al (1998) p40/LAIR-1 regulates the differentiation of peripheral blood precursors to dendritic cells induced by granulocyte-monocyte colony-stimulating factor. Eur J Immunol 28(7):2086–2091
68. Fournier N et al (2000) FDF03, a novel inhibitory receptor of the immunoglobulin superfamily, is expressed by human dendritic and myeloid cells. J Immunol 165(3):1197–1209
69. Verbrugge A et al (2003) Differential contribution of the immunoreceptor tyrosine-based inhibitory motifs of human leukocyte-associated Ig-like receptor-1 to inhibitory function and phosphatase recruitment. Int Immunol 15(11):1349–1358

70. Verbrugge A et al (2006) Leukocyte-associated Ig-like receptor-1 has SH2 domain-containing phosphatase-independent function and recruits C-terminal Src kinase. Eur J Immunol 36(1):190–198
71. Lebbink RJ et al (2006) Collagens are functional, high affinity ligands for the inhibitory immune receptor LAIR-1. J Exp Med 203(6):1419–1425
72. Tang X et al (2012) Leukocyte-associated Ig-like receptor-1-deficient mice have an altered immune cell phenotype. J Immunol 188(2):548–558
73. Yu M, Cantor AB (2012) Megakaryopoiesis and thrombopoiesis: an update on cytokines and lineage surface markers. Methods Mol Biol 788:291–303
74. Niswander LM et al (2014) SDF-1 dynamically mediates megakaryocyte niche occupancy and thrombopoiesis at steady state and following radiation injury. Blood 124(2):277–286
75. Vila-Coro AJ et al (1999) The chemokine SDF-1alpha triggers CXCR4 receptor dimerization and activates the JAK/STAT pathway. FASEB J 13(13):1699–1710
76. Vicente-Manzanares M et al (1999) Involvement of phosphatidylinositol 3-kinase in stromal cell-derived factor-1 alpha-induced lymphocyte polarization and chemotaxis. J Immunol 163(7):4001–4012
77. Nie Y, Han YC, Zou YR (2008) CXCR4 is required for the quiescence of primitive hematopoietic cells. J Exp Med 205(4):777–783
78. Davies SLSN, Bolós J, Bayés M (2007) Molecule of the month. Plerixafor hydrochloride. Drug News Perspect 20(8):543–544
79. Khan A, Greenman J, Archibald SJ (2007) Small molecule CXCR4 chemokine receptor antagonists: developing drug candidates. Curr Med Chem 14(21):2257–2277
80. Sattler K, Levkau B (2009) Sphingosine-1-phosphate as a mediator of high-density lipoprotein effects in cardiovascular protection. Cardiovasc Res 82(2):201–211
81. Hanson MA et al (2012) Crystal structure of a lipid G protein-coupled receptor. Science 335(6070):851–855
82. Sharma N, Akhade AS, Qadri A (2013) Sphingosine-1-phosphate suppresses TLR-induced CXCL8 secretion from human T cells. J Leukoc Biol 93(4):521–528
83. Allende ML et al (2004) Expression of the sphingosine 1-phosphate receptor, S1P1, on T-cells controls thymic emigration. J Biol Chem 279(15):15396–15401
84. Rosen H (2005) Chemical approaches to the lysophospholipid receptors. Prostaglandins Other Lipid Mediat 77(1–4):179–184
85. Kang X et al (2015) The ITIM-containing receptor LAIR1 is essential for acute myeloid leukaemia development. Nat Cell Biol 17(5):665–677.
86. Urtz N et al (2015) Sphingosine 1-phosphate produced by sphingosine kinase 2 intrinsically controls platelet aggregation in vitro and in vivo. Circ Res 31;117(4):376–387.

Rap GTPase Signaling in Platelets and Megakaryocytes

8

Wolfgang Bergmeier and Ellen C. O'Shaughnessy

Abstract

Platelets are essential for primary hemostasis and repair of the endothelium, but they also play a key role in the development of acute coronary syndromes and they contribute to cerebrovascular events [1, 2]. In humans, billions of platelets are produced by megakaryocytes and released into the blood stream every day. If they are not consumed in the hemostatic process, these platelets circulate for approximately 10 days before being destroyed by phagocytic cells in the spleen and the liver. While in circulation, platelets remain in a resting state even when exposed to high shear stress or small amounts of soluble agonists generated in the intact vasculature. Vascular injury, however, leads to exposure of platelets to strong agonists, rapid integrin activation, and thrombus formation.

8.1 Introduction

Platelets are essential for primary hemostasis and repair of the endothelium, but they also play a key role in the development of acute coronary syndromes and they contribute to cerebrovascular events [1, 2]. In humans, billions of platelets are produced by megakaryocytes and released into the blood stream every day. If they are not consumed in the hemostatic process, these platelets circulate for approximately 10 days before being destroyed by phagocytic cells in the spleen and the liver. While

W. Bergmeier, PhD (✉)
Department of Biochemistry and Biophysics, University of North Carolina at Chapel Hill, 3113 Genetics Medicine Building, 120 Mason Farm Road, NC 27599-7260, USA
e-mail: bergmeie@email.unc.edu

E.C. O'Shaughnessy, PhD
Department of Pharmacology, University of North Carolina at Chapel Hill, 4023 Genetics Medicine Building, 120 Mason Farm Road, NC 27599, USA
e-mail: eoshaugh@email.unc.edu

H. Schulze, J. Italiano (eds.), *Molecular and Cellular Biology of Platelet Formation*, DOI 10.1007/978-3-319-39562-3_8

in circulation, platelets remain in a resting state even when exposed to high shear stress or small amounts of soluble agonists generated in the intact vasculature. Vascular injury, however, leads to exposure of platelets to strong agonists, rapid integrin activation, and thrombus formation.

Platelet thrombus formation is a complex process that occurs in several phases [1–3]: (1) transient platelet adhesion (tethering) to the extracellular matrix (ECM), (2) agonist-induced activation of intracellular signaling cascades, and (3) firm adhesion mediated by activated integrins, in particular the fibrinogen receptor αIIbβ3. The current model suggests that tethering and pre-activation of platelets is mediated by platelet GPIbα binding to its various ligands in the ECM. Platelet activation is then triggered by collagen exposed in the vessel wall and/or locally generated thrombin. The formation of stable thrombi further depends on co-stimulatory signaling provided by the second wave agonists, thromboxane A_2 (TxA_2) and ADP, which are released from activated platelets. Thrombin, TxA_2, and ADP activate platelets through G protein-coupled receptors (GPCRs). In contrast, collagen signaling is mainly driven by GPVI, an immunoglobulin-like receptor which signals through its association with the Fc receptor γ-chain [4]. Cellular stimulation through GPCRs and GPVI leads to the activation of phospholipase C (PLC) β2 and PLCγ2, respectively, which in turn mediate the generation of the second messengers, calcium (Ca^{2+}) and diacylglycerol (DAG). DAG is critical for protein kinase C (PKC) activation, a key event in platelet granule release and integrin activation. Platelets express at least 6 PKC isoforms, which have activating as well as inhibitory roles in platelet activation [5]. Ca^{2+} is critical to various platelet responses, including integrin activation (see below).

Small GTPases of the RAP family provide a critical link between the second messengers, Ca^{2+} and DAG, and surface-expressed integrins in platelets (reviewed in [6]). RAP GTPases act as molecular switches by cycling between an inactive GDP-bound form and an active GTP-bound form. They are tightly regulated by guanine nucleotide exchange factors (GEFs), which stimulate GTP loading, and GTPase activating proteins (GAPs), which catalyze GTP hydrolysis. Upon GTP binding, RAP GTPases undergo a conformational change that enables interactions with a variety of effector proteins, thus controlling a wide range of essential biochemical pathways in eukaryotic cells. The RAP protein family includes five members that are grouped into two subfamilies, RAP1 and RAP2. The role of RAP1 in cell signaling has been studied extensively (reviewed in [7, 8]). It is a well-known regulator of several aspects of cell adhesion, including integrin-mediated cell adhesion and cadherin-mediated cell junction formation. RAP GTPases also regulate cellular differentiation and proliferation via pathways that involve the regulation of the mitogen-activated protein kinase (MAPK) cascade. RAP1 either interferes with Ras-mediated ERK MAPK activation or it activates ERK independently of Ras in a cell-context-dependent manner. Furthermore, a link between RAP GTPase signaling and actin dynamics has been established, as activated RAP1, via its interaction with specific GEFs, can induce the activation of RAC small GTPases important for cytoskeletal rearrangement [9, 10].

In this chapter, we will review existing literature on the role of RAP GTPases in platelet and megakaryocyte function, and we will briefly discuss some of the open questions regarding this important signaling pathway.

8.2 RAP Signaling in Platelets

8.2.1 Expression of RAP GTPases in Platelets

Platelets express all five RAP isoforms, RAP1A,B and RAP2A,B,C. RAP1B is the predominant isoform, both in human and mouse platelets. Early estimates suggested that RAP1B accounts for ~0.1 % of total protein and ~90 % of total RAP protein in platelets [11]. More recent studies utilizing unbiased proteome (Fig. 8.1) or transcriptome (not shown) analysis largely confirmed these findings [12–14], putting the estimated number of RAP1B molecules per platelet at ~100–200,000 copies. Thus, these proteins are expressed at copy numbers greater than that of the main platelet integrin, αIIbβ3. Interestingly, the proteome studies also identified RAP1A as a highly expressed protein, especially in human platelets. The importance of this isoform for platelet function is currently unknown.

Unbiased expression profiling also provided much needed information on the RAP-GEFs and -GAPs expressed in human and mouse platelets. In both species, CalDAG-GEFI (RASGRP2) and RASA3 ($GAP1^{IP4bp}$) are the predominant GEF and GAP, respectively. The copy numbers per platelet for other RAP-GEFs, such as PDZ-GEF1/2 and SOS1, are very low and likely not sufficient to have strong effects on the activity state of the very abundant RAP1 proteins. A similar conclusion can

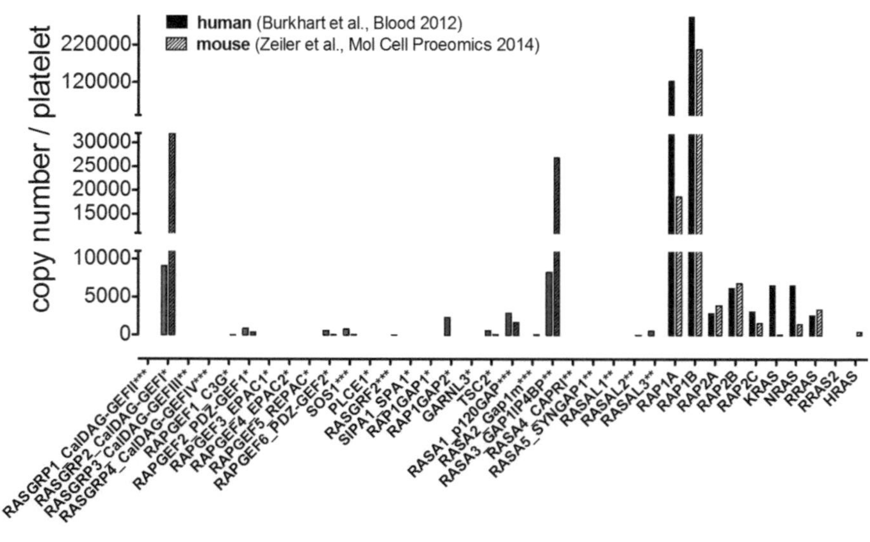

Fig. 8.1 Expression of RAP-GEFs, RAP-GAPs, and the various RAP and RAS GTPase family members in mouse and human platelets. Shown are copy numbers per platelet for the indicated small GTPases and their GEFs and GAPs, determined by proteome analysis [12, 13]

be drawn for RAP-GAPs other than RASA3, such as RAP1GAP2 and RASA1. However, these low expressed GEFs and GAPs could play an important role in the regulation of distinct pools of RAP proteins, such as RAP2, which could be important for cellular functions other than integrin activation (see below). In addition, they may provide limited backup in case CalDAG-GEFI or RASA3 expression/function is impaired.

8.2.2 RAP GTPases and Platelet Signaling

8.2.2.1 Platelet Signaling Pathways Controlling RAP1 Activity

In the late 1990s and early 2000s, several groups demonstrated that RAP1 activation in platelets depends on two main signaling pathways. A calcium-dependent first wave of RAP1 activation, followed by a second phase of activation requiring signaling by protein kinase C (PKC) and the Gαi-coupled receptor for ADP, P2Y12 [15–18]. Importantly, the kinetics of RAP1 activation closely correlated with the kinetics of integrin-mediated aggregation, suggesting a critical role for RAP1 in the inside-out activation of αIIbβ3. Studies in megakaryocytes expressing low levels of CalDAG-GEFI [19] or mutant forms of RAP1B [20] provided the first genetic evidence supporting a key role for RAP1 signaling in integrin activation. Shortly thereafter, this conclusion was confirmed by studies in mice deficient in *Caldaggef1* or *Rap1b*, which showed a marked defect in integrin activation in stimulated platelets from the respective mice [21, 22]. CalDAG-GEFI is a RAP-GEF regulated by a pair of calcium-binding EF hands and a C1-like domain with hitherto unknown function. Our group went on to demonstrate that CalDAG-GEFI in mouse platelets is critical for calcium-dependent RAP1 activation and that full RAP1 activation in platelets depends on synergistic signaling by CalDAG-GEFI and PKC/P2Y12 [23–25]. Consistent with earlier studies with BAPTA-AM-treated platelets, RAP1 activation in platelets lacking CalDAG-GEFI was delayed but sustained. In contrast, RAP1 activation in platelets lacking functional PKC or P2Y12 was fast but reversible. Together, these studies shaped a very simplified two-pathway model for RAP1 activation in stimulated platelets, where CalDAG-GEFI provides speed and sensitivity to the system, while PKC and P2Y12 are required to sustain the signal. Importantly, very similar findings were recently reported for human platelets from a patient with a loss-of-function point mutation in *Caldaggef1* [26].

While these studies identified CalDAG-GEFI as a critical GEF for RAP1, the molecular nature of the RAP1 modulator operating downstream of P2Y12 was only recently established. Using unique mouse models, we were able to demonstrate that signaling by P2Y12 leads to the inactivation of the RAP-GAP, RASA3 [27]. Early biochemical studies demonstrated that RASA3 is highly expressed in human platelets, where it is located to the plasma membrane (PM) [28–30]. This interaction with the PM is mediated by a unique PH/Btk domain, which shows high affinity for phosphoinositol 4,5-biphosphate (PIP_2) and phosphoinositol 3,4,5-triphosphate (PIP_3). Furthermore, proteome and transcriptome analyses confirmed that RASA3 is the highest expressed RAP-GAP in human and mouse platelets [12–14]. To determine how RASA3 contributes to platelet function, we generated germline and megakaryocyte-specific *Rasa3* knockout mice [27]. To our surprise, both strains of mice showed very high neonatal lethality due to almost complete thrombocytopenia, consistent with earlier findings in mice expressing a truncated version of the protein [31]. To circumvent this problem, we utilized mice with a point mutation in the C-terminus of RASA3 (H794L). This mouse strain, called *hlb381* (*Rasa3hlb*), was identified in a forward genetic screen for modifiers of the platelet count [27]. *Rasa3hlb* mice were also very thrombocytopenic but did not show any neonatal lethality. In contrast to the knockout mice, however, significant numbers of platelets were detected in circulating blood (platelet count ~3–5 % of controls). *Rasa3hlb* platelets showed signs of pre-activation, such as increased baseline activation of RAP1 and αIIbβ3 integrin, and a markedly reduced survival in circulation. Interestingly, both pre-activation and reduced survival were normalized by concomitant deletion of *Caldaggef1* in *Rasa3hlb* mice. Together, these studies suggest that RASA3 is critical to prevent unwanted platelet activation in circulating platelets, mediated by the highly sensitive CalDAG-GEFI/RAP1 signaling module. At sites of vascular injury, however, negative feedback by RASA3 would impair hemostatic plug formation. To release this "brake," adhering platelets rely on signaling via P2Y12 and phosphoinositide 3-kinase (PI3K). The exact mechanism by which PI3K signaling affects RASA3 function is currently unclear.

8.2.2.2 Platelet Responses Controlled by RAP1

Long before the studies identifying a critical role for calcium and PKC/P2Y12 signaling in RAP1 activity regulation, it was demonstrated that the same pathways control a variety of cellular responses in platelets, including integrin inside-out activation [32–35], ERK MAPK signaling and TxA_2 production [36–38], and secretion [32, 33, 39]. Together with the fact that active RAP1 is known to communicate with multiple target proteins (reviewed in [7, 8]), these findings suggested a more global role for RAP1 in platelet activation. As there were no patients with documented defects in RAP1 signaling or tools to specifically inhibit RAP1 signaling in human platelets, animal models, including knockout mice for *Caldaggef1* or *Rap1b* and various dog breeds with mutations in *Caldaggef1*, were used to define the contribution of RAP1 to platelet activation [21–25, 40–44]. A result common to all studies was a marked defect in integrin inside-out activation and hemostatic plug formation. In *Caldaggef1* mutant platelets, this integrin activation defect was most prominent when the cells were activated with ADP, collagen, TxA_2 analog, or calcium ionophore. A much milder integrin activation defect was observed when mutant platelets were stimulated with thrombin or PAR4 receptor-activating peptide, suggesting that a CalDAG-GEFI-independent pathway provides significant backup under these conditions. This alternative pathway to integrin activation depends on signaling by PKC, as (1) *Caldaggef1* mutant platelets showed a normal aggregation response to PKC stimulation with phorbol ester [21, 26, 43], and (2) inhibition of PKC abolished integrin-mediated aggregation in *Caldaggef1* knockout platelets [23]. Importantly, however, even when stimulated with high-dose thrombin or

PAR4-activating peptide, i.e., under conditions of strong PLC activation and DAG production, aggregation of *Caldaggef1* knockout platelets occurred with a delay. The delay in aggregation was also observed in platelets with impaired function in both CalDAG-GEFI and RASA3 [27]. Deficiency in RASA3, however, led to a significant increase in integrin activation, confirming its important role as an inhibitor of RAP1-dependent integrin activation.

In addition to integrin inside-out activation, CalDAG-GEFI/RAP1 signaling in platelets was shown to contribute to ERK MAPK signaling and TxA_2 production [24], granule secretion [22–24, 41–43], integrin outside-in signaling [44], spreading and clot retraction [41, 42], cAMP metabolism [43], and phosphatidylserine exposure and microvesiculation [45]. At this point, little is known about how active RAP1 can control these various cellular functions. Studies in heterologous cell systems suggested a critical role for RAP1-GTP-interacting adaptor molecule (RIAM) in RAP1-dependent αIIbβ3 activation [46, 47]. However, platelets from *Riam* knockout mice did not exhibit any integrin activation defect [48], suggesting that in platelets this process depends on a different adapter protein. Functional studies with *Caldaggef1* knockout platelets identified a role for RAP1 in ERK MAPK [24] and RAC1 GTPase signaling [41]. These findings are in agreement with studies in other cell types, in which the MAP kinase kinase kinase, B-RAF (RAF1) [49, 50], and the RAC-GEFs, TIAM1 and VAV2 [9], were identified as downstream targets of RAP1. Confirmation that B-RAF and RAC-GEFs are part of the RAP1 signaling pathway in platelets is currently not available. It is also important to note that, at least in part, the observed defects could be secondary to a defect in integrin outside-in signaling, a well-established co-stimulatory signal for primary agonist receptors in platelets [51]. Studies with blocking antibodies against αIIbβ3 suggest that this is not the case for TxA_2 production and granule release [24, 41], but genetic evidence is required to completely exclude this possibility.

8.2.2.3 RAP1 Signaling and Hemostatic Plug Formation

Given the central role of RAP1 in integrin activation and other platelet activation responses (see above), it is not surprising that hemostasis is impaired in all situations where this pathway is severely disrupted. *Rap1b* knockout mice exhibit ~five-fold increased tail bleeding times, with >50 % of the mice not being able to control bleeding throughout the entire observation period [22]. A similar defect in primary hemostasis was reported for mice, dogs, and humans with mutations in CalDAG-GEFI [21, 26, 43] or P2Y12 [52–54]. Importantly, mice lacking *Rap1b*, *Caldaggef1*, or *P2y12* were also strongly protected in several models of experimental thrombosis. To investigate how the CalDAG-GEFI and P2Y12 pathways contribute to RAP1-dependent platelet adhesion in vivo, we compared thrombus formation at sites of vascular injury in *Caldaggef1-/-* and clopidogrel-treated *wild-type (Wt)* mice [25]. Consistent with the role of P2Y12 in sustained RAP1 activation, platelet adhesion and thrombus growth occurred rapidly in *Wt*/clopidogrel mice, but forming thrombi were very unstable. In contrast, thrombi in *Caldaggef1-/-* mice grew very slowly, in particular under high shear stress conditions such as found in arterioles. Thus, while CalDAG-GEFI is important for the initiation of platelet adhesion, in particular under high shear stress conditions, P2Y12 signaling is required to stabilize the forming thrombus.

As outlined above, engagement of P2Y12 during platelet activation leads to sustained RAP1 activation due to its inhibitory effect on RASA3 function [27]. In vivo, *Rasa3* mutant platelets adhered more firmly to sites of vascular injury than *Wt* controls. Furthermore, treatment with clopidogrel to inhibit P2Y12 signaling did not affect thrombus formation in *Rasa3* mutant mice. Thus, with regard to RAP1-dependent integrin activation, *Rasa3* mutant platelets are in a state of constitutive P2Y12 signaling, i.e., these platelets do not require feedback activation by P2Y12 and are insensitive to inhibitors of this receptor.

8.2.3 Open Questions

While it is well established that RAP GTPases are critical for platelet function, many important questions still need to be addressed. For example, it is not clear if there are different pools of RAP GTPases that control the various cellular functions affected by this pathway? These pools could be defined by unique spatiotemporal activation signatures or simply by the fact that different isoforms control specific cellular responses. Studies in various mammalian cell types identified different pools of RAP GTPases, which simultaneously direct distinct cellular functions [55]. These individual pools of RAP protein seem to be regulated by specific RAP-GEFs and RAP-GAPs, which use unique anchoring mechanisms for targeting to specific cellular locations, such as the perinuclear region or the PM. However, platelets do not have a nucleus, and GEFs and GAPs other than CalDAG-GEFI and RASA3 are expressed at very low levels (see above).

It will also be important to address whether there is redundancy between the various RAP GTPase isoforms, i.e., whether the residual activation response seen in *Rap1b* knockout mice is mediated by other RAP GTPases. For example, it was shown that RAP1A is activated upon stimulation of *Rap1b*$^{-/-}$ platelets [56]. One would expect that the highly homologous RAP1A isoform can perform similar functions as RAP1B and thus that platelet activation in *Rap1a/b* double knockout mice is more severely impaired than in *Rap1b*$^{-/-}$ mice. While RAP1A and RAP1B share ~90 % sequence homology, only a ~70 % homology exists between RAP1 and the various RAP2 isoforms. Compared to RAP1, RAP2 proteins are less sensitive to GAP inactivation and can mediate cellular responses distinct from those regulated by RAP1 [57, 58]. Consistent with the insensitivity of RAP2 to GAP-mediated inactivation, we detected significant amounts of RAP2-GTP in resting mouse platelets [41]. Agonist-induced RAP2 activation, however, was also dependent on CalDAG-GEFI and P2Y12, suggesting that similar signaling pathways control the activity state of RAP1 and RAP2. Functional data on the role of RAP isoforms other than RAP1B in platelet activation is not available. The same is true for RAS GTPases that are expressed at low levels in platelets.

Much has yet to be learned about the activity regulation of CalDAG-GEFI and RASA3. For example, how does calcium binding to the EF hands affect

CalDAG-GEFI activity? What are the contributions of the C1 and C2 regulatory domains to the function of CalDAG-GEFI and RASA3, respectively? How does PI3 kinase signaling affect RASA3 function? How important are posttranslational modifications, such as phosphorylation by protein kinase A [59, 60], for CalDAG-GEFI and RASA3 activity during platelet adhesion? And how can we improve our ability to measure the expression, localization, and activity of CalDAG-GEFI and RASA3 in platelets isolated from interesting animal models or from patients with changes in platelet number or function? Answering these fundamental questions may prove vital for our ability to diagnose and treat patients with defects in platelet function or number.

Lastly, the clinical relevance of impaired RAP1 signaling should be investigated in more detail. For example, it needs to be established if there are significant inter-individual differences in the antagonistic balance between CalDAG-GEFI and RASA3. If yes, one should detect a high CalDAG-GEFI/RASA3 ratio in hyper-reactive platelets, while a low ratio should be identified in hypofunctional cells. In addition to measuring the expression of these proteins, novel assays should be developed that can easily determine the activity of CalDAG-GEFI, RASA3, or RAP1. With these tools in hand, one could screen patient populations with altered platelet function or number for defects in RAP1 signaling. The long-term benefit of this work could be a more personalized approach to thrombosis prevention, based on altered regulation of this important platelet signaling pathway.

8.3 RAP Signaling in Megakaryocytes

8.3.1 Subcellular Localization and Integrin Signaling

In light of the critical importance of RAP1 in platelet biology, it is perhaps surprising that no effect on platelet count has been seen in various animal models including mice deficient in *Rap1b* [22] or *Caldaggef1* [21] and three different breeds of dogs with mutations in *Caldaggef1* [43]. Many studies, however, have shown that RAP1 signaling mediates αIIbβ3 activation also in megakaryocytes (MKs) [19, 20, 61–63]. In human cord blood-derived MKs, RAP1 is expressed at very low levels in immature MKs and becomes much more pronounced in intermediate and mature MKs [63]. In addition, RAP1 is stored on internal, α-granule, membranes in resting cells, and upon thrombin stimulation, it is translocated to the PM in a dose-dependent manner. Using a novel confocal microscopy technique, Balduini et al. showed that active, GTP-bound RAP1 is found only on the PM. Though the authors do not address the activation of αIIbβ3 specifically, they do show that CD41 (αIIb) translocates in a manner similar to active RAP1 and is co-localized within the spatial resolution of their microscopy technique. Further, expression of constitutively active RAP1B (V12) enhanced fibrinogen binding to αIIbβ3 in primary murine megakaryocytes, while expression of RAP1GAP abrogated this effect nearly completely [20]. Disruption of the actin cytoskeleton by treatment with either cytochalasin D or latrunculin A blocked RAP1-mediated adhesion, suggesting that activation of αIIbβ3 depends on cytoskeletal regulation. Interestingly, mix-induced turbulence lead to rapid and robust activation of RAP1 in megakaryocytic cell lines, and this effect did not depend on the actin cytoskeleton or any signaling pathways currently known to regulate RAP1 [61]. Though the mechanism underlying RAP1-mediated integrin activation in MKs has not been fully elucidated, it has been suggested that the RAP1-GEF, CalDAG-GEFI, and the RAP1 effector RIAM are involved. CalDAG-GEFI expression in murine embryonic stem cell-derived MKs enhanced agonist-induced activation of RAP1 and αIIbβ3 [19] while shRNA mediated knock down of RIAM attenuated αIIbβ3 activation [62]. Interestingly, αIIbβ3 activation in platelets was not affected in *Riam*$^{-/-}$ mice [48] suggesting potential differences in integrin activation between MKs and platelets.

8.3.2 RAP GTPases and Megakaryopoiesis

8.3.2.1 MAP Kinase Signaling

While MKs have served as a genetically tractable surrogate for the study of RAP1-mediated integrin activation in platelets, the requirement for αIIbβ3 activation in megakaryocyte biology is unclear. Nonetheless, several lines of evidence suggest that RAP1 may be involved in megakaryopoiesis and platelet production. RAP1 has been shown to mediate mitogen-activated protein kinase (MAPK) signaling in megakaryocytes. Thrombopoietin, TPO, is a primary cytokine involved in megakaryocyte differentiation and proplatelet formation [64, 65]. TPO binds the Mpl (myeloproliferative virus ligand) receptor and activates a host of signaling pathways including the canonical MAPK cascade, Raf-MEK-ERK. A series of studies in leukemia cell lines [66–70] and primary MKs derived from murine bone marrow/fetal liver cells [71] and human cord blood [72–76] have documented a role for ERK activation in MK differentiation, directed migration, polyploidy, and/or proplatelet formation (reviewed in [77]).

Initial efforts to elucidate the signaling pathways responsible for ERK phosphorylation made use of human leukemia cell lines engineered to express the TPO receptor, Mpl [78–80]. In F36P-Mpl cells, exogenous expression of constitutively active RAS, a well-known activator of RAF-1 and thereby ERK, increased αIIb expression and polyploidy formation in response to TPO, while dominant negative RAS abrogated both TPO-induced effects [78]. Work in another megakaryoblastic cell line showed that, in addition to RAS, RAP1 can mediate ERK phosphorylation [79]. In TPO-differentiated UT7-Mpl cells, RAS-mediated activation of ERK via RAF-1 was required for rapid but transient MAPK activation, while RAP1 regulation of B-RAF was responsible for sustained ERK phosphorylation. Interestingly, it has been shown that direct contact with stromal cells blocks megakaryocyte differentiation in leukemic cell lines [81] and primary human CD34+ stem cell-derived MKs [82] in part through the suppression of ERK phosphorylation. It has been shown that stromal contact blocks differentiation of TPA stimulated K562 cells by preventing late-phase RAP1 activation and ERK phosphorylation with no observable effect on RAS activity [82]. These data further strengthen the evidence that RAP1 is required for the sustained ERK activation necessary for MK differentiation. Strikingly, genetic deletion of both Raf-1 and B-Raf in mouse models has shown that RAF-1 is dispensable for TPO-induced ERK phosphorylation and MK differentiation [83], while B-RAF is essential for TPO-mediated megakaryopoiesis [84]. These data suggest that RAP1 may be the predominant small GTPase responsible for ERK activation and megakaryopoiesis in vivo.

8.3.2.2 The RAP-RHO Connection

Many studies have demonstrated the importance of RHO-family GTPases in directed migration of MKs and proplatelet formation, consistent with the role of these proteins in polarization and cytoskeletal regulation [85–91]. In addition, extensive evidence from other cell types and platelets themselves has shown that RAP1 regulates the RHO-family GTPases CDC42 and RAC1 including in polarization and migration of T cells [92], axon-specification in neurons [93, 94], cell spreading and adhesion of platelets [41], directed migration of fibroblasts [95], and establishment of epithelial apical-basal polarity [96, 97]. Given the documented thrombocytopenia in RHO GTPase mutant mice [85] and the heavy integration of RAP1 and RHO-family GTPases in cytoskeletal regulation seen in a broad range of cell types (Reviewed in [98]), it is plausible and perhaps likely that such signaling networks are also critical for normal megakaryopoiesis.

8.3.2.3 Rasa3 Mutant Mice

Two studies identified the RAP-GAP, RASA3 as a major regulator of megakaryocyte and platelet biology in vivo [27, 99]. RASA3 is a member of the GAP1 subfamily with enzymatic activity toward both RAS and RAP1 [100]. Expression of a catalytically inactive Rasa3 mutant lacking exons 11 and 12 (*Rasa3*$^{\Delta}$*GAP*) is embryonic lethal at E12.5-E13.5 due to extensive bleeding [31]. To circumvent embryonic lethality, Molina-Oritz et al. used two strategies to study MKs expressing catalytically inactive RASA3: they made fetal liver cell (FLC) chimeras from wild-type and *Rasa3*$^{\Delta}$*GAP* mice in the Severe Combined Immune Deficient (SCID) background and they differentiated FLC in vitro in the presence of TPO [99]. Using FLC-derived megakaryocytes, they showed increased RAP1 activation in Rasa3$^{\Delta}$GAP cells while RAS activity was unaffected. Consistent with the role of RAP1 in integrin activation, they found enhanced αIIbβ3 activation, cell adhesion, and cell spreading in Rasa3$^{\Delta}$GAP MKs. Interestingly, increased adhesion was blocked by treatment with the RAP1 inhibitor GGTI-298. In addition to changes in integrin activation, they report large abnormalities in the actin cytoskeleton in which mutant cells fail to form stress fibers and instead exhibit extensive punctate structures. In SCID-Rasa3$^{\Delta}$GAP chimeras, they found thrombocytopenia despite a significant increase in the number of MKs in the bone marrow (BM). MKs in these mice were mislocalized within the bone marrow compared with SCID-Rasa3$^{+/+}$ chimeras, and MKs from BM explants migrated more slowly and were spread more frequently and failed to form proplatelets when compared with wild-type controls. Taken together, these data suggest that loss of RASA3 GAP activity has profound implications on MK biology in general, and this effect is mediated, at least in part, by altered RAP1 activity regulation.

Work from our lab using both genetic deletion of Rasa3 in C57BL/6J mice and expression of a mutant RASA3 with a missense mutation (H794L, $Rasa3^{hlb}$) that destabilizes the protein, resulting in a near complete loss of expression, also identified this GAP as a key regulator of platelet biology acting through RAP1 [27]. Mice lacking Rasa3 exhibited severe thrombocytopenia and significant expansion of MKs in the spleen and bone marrow, similar to what was reported by Molina-Ortiz et al. However, we observed no significant defects in $Rasa3^{hlb/hlb}$ megakaryocytes with respect to in vitro proplatelet formation, gross MK morphology, and ultrastructure or proliferation of BM-derived progenitors. These discrepancies may arise from differences in the animal models employed: FLC chimeras in the SCID background compared with genetic deletion/mutation in C57BL/6J mice or the Rasa3 mutant expressed (the H794L mutation results in a dramatic reduction in protein expression while inactivation of the catalytic domain produces a stable protein that may participate in GAP-independent functions yet to be identified). Future studies will be required to fully elucidate the role of RASA3 in MK biology; nonetheless, current work strongly suggests that increased RAP1 activation plays a role in MK differentiation, migration, and/or proplatelet formation.

8.3.3 Open Questions

There is substantial evidence supporting the hypothesis that RAP GTPases contribute to megakaryopoiesis. However, direct proof is missing, as the peripheral platelet count in Rap1b knockout mice was normal and mice lacking multiple RAP isoforms have not been reported to date. On a molecular level, it needs to be established whether and how RAP GTPases affect MAPK and RHO GTPase signaling in megakaryocytes. Recent studies on the role of RASA3 in megakaryopoiesis are somewhat contradictory and incomplete. Studies in knockout rather than mutant mouse models may be required to fully understand the impact of uncontrolled RAP signaling on platelet production. The effect of Rasa3 deficiency on MAPK and RHO GTPase signaling should be established. Aside from these mechanistic studies, it will be important to determine whether alterations in RAP signaling affect platelet production and/or platelet survival in humans.

8.4 Summary

RAP GTPases play a critical role in platelet biology (Fig. 8.2). They control various important platelet functions, such as integrin activation, TxA_2 generation, granule release, and spreading/clot retraction. RAP1 activity itself is controlled by two main pathways: (1) fast but reversible RAP1 activation by CalDAG-GEFI in response to an increase in the cytosolic calcium concentration and (2) sustained activation due to inactivation of the RAP-GAP, RASA3, mediated downstream of the platelet receptor for ADP, P2Y12. Impaired expression/function of CalDAG-GEFI or P2Y12 markedly impairs platelet function and hemostatic plug formation. RASA3

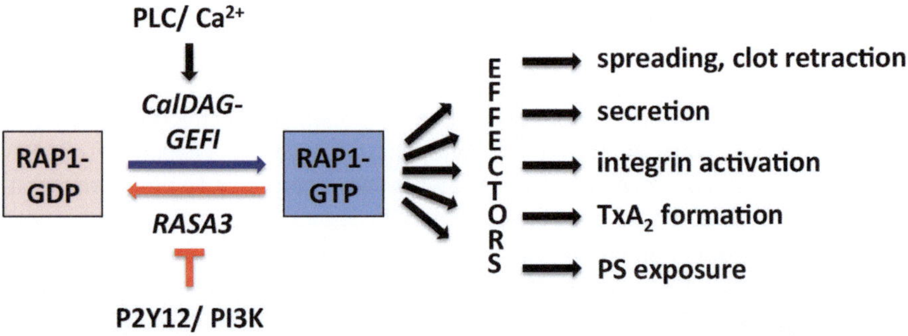

Fig. 8.2 RAP1 signaling in platelets. Main regulators of RAP1 activity in platelets are CalDAG-GEFI and RASA3. While CalDAG-GEFI is activated by an increase in the cytosolic Ca^{2+} concentration downstream of phospholipase C (PLC) signaling, RASA3 is inactivated as part of the P2Y12-PI3 kinase signaling pathway. Once activated, RAP1 can control the indicated platelet responses by communicating with hitherto unknown effector proteins

deficiency causes platelet pre-activation and severe thrombocytopenia. While critical for platelet function, the contribution of RAP GTPases to megakaryopoiesis is less well characterized. Thrombocytopenia and impaired megakaryopoiesis were shown for mice lacking the RAP1 effector B-RAF or mice expressing a mutant form of RASA3. However, direct evidence for impaired megakaryopoiesis in RAP mutant mice is missing. Many open questions regarding the importance of individual RAP isoforms and the molecular details of RAP GTPase activity regulation in platelets and megakaryocytes remain unanswered. The clinical relevance of findings in animal models has to be established.

Acknowledgment We thank Lucia Stefanini for her help with data analysis. This work was supported by the American Heart Association (14EIA18910004) and NIH grants R01 HL121650 and P01 HL120846.

References

1. Ruggeri ZM (2002) Platelets in atherothrombosis. Nat Med 8(11):1227–1234
2. Broos K, Feys HB, De Meyer SF, Vanhoorelbeke K, Deckmyn H (2011) Platelets at work in primary hemostasis. Blood Rev 25(4):155–167
3. Savage B, Almus-Jacobs F, Ruggeri ZM (1998) Specific synergy of multiple substrate-receptor interactions in platelet thrombus formation under flow. Cell 94(5):657–666
4. Nieswandt B, Watson SP (2003) Platelet-collagen interaction: is GPVI the central receptor? Blood 102(2):449–461
5. Harper MT, Poole AW (2007) Isoform-specific functions of protein kinase C: the platelet paradigm. Biochem Soc Trans 35(Pt 5):1005–1008
6. Guidetti GF, Torti M (2012) The small GTPase Rap1b: a bidirectional regulator of platelet adhesion receptors. J Signal Transduct 2012:412089
7. Hattori M (2003) Rap1 GTPase: functions, regulation, and malignancy. J Biochem 134(4):479–484

8. Bos JL (2005) Linking Rap to cell adhesion. Curr Opin Cell Biol 17(2):123–128
9. Arthur WT, Quilliam LA, Cooper JA (2004) Rap1 promotes cell spreading by localizing Rac guanine nucleotide exchange factors. J Cell Biol 167(1):111–122
10. Maillet M, Robert SJ, Cacquevel M et al (2003) Crosstalk between Rap1 and Rac regulates secretion of sAPPalpha. Nat Cell Biol 5(7):633–639
11. Torti M, Lapetina EG (1994) Structure and function of rap proteins in human platelets. Thromb Haemost 71(5):533–543
12. Burkhart JM, Vaudel M, Gambaryan S et al (2012) The first comprehensive and quantitative analysis of human platelet protein composition allows the comparative analysis of structural and functional pathways. Blood 120(15):e73–e82
13. Zeiler M, Moser M, Mann M (2014) Copy number analysis of the murine platelet proteome spanning the complete abundance range. Mol Cell Proteomics 13(12):3435–3445
14. Rowley JW, Oler AJ, Tolley ND et al (2011) Genome-wide RNA-seq analysis of human and mouse platelet transcriptomes. Blood 118(14):e101–e111
15. Franke B, Akkerman JW, Bos JL (1997) Rapid Ca2+-mediated activation of Rap1 in human platelets. EMBO J 16(2):252–259
16. Lova P, Paganini S, Hirsch E et al (2002) A selective role for phosphatidylinositol 3,4,5-trisphosphate in the Gi-dependent activation of platelet Rap1B. J Biol Chem 278(1):131–138
17. Woulfe D, Jiang H, Mortensen R, Yang J, Brass LF (2002) Activation of Rap1B by Gi family members in platelets. J Biol Chem 277(26):23382–23390
18. Larson MK, Chen H, Kahn ML et al (2003) Identification of P2Y12-dependent and -independent mechanisms of glycoprotein VI-mediated Rap1 activation in platelets. Blood 101(4):1409–1415
19. Eto K, Murphy R, Kerrigan SW et al (2002) Megakaryocytes derived from embryonic stem cells implicate CalDAG-GEFI in integrin signaling. Proc Natl Acad Sci U S A 99(20):12819–12824
20. Bertoni A, Tadokoro S, Eto K et al (2002) Relationships between Rap1b, affinity modulation of integrin alpha IIbbeta 3, and the actin cytoskeleton. J Biol Chem 277(28):25715–25721
21. Crittenden JR, Bergmeier W, Zhang Y et al (2004) CalDAG-GEFI integrates signaling for platelet aggregation and thrombus formation. Nat Med 10(9):982–986
22. Chrzanowska-Wodnicka M, Smyth SS, Schoenwaelder SM, Fischer TH (2005) White GC2. Rap1b is required for normal platelet function and hemostasis in mice. J Clin Invest 115(3):680–687
23. Cifuni SM, Wagner DD, Bergmeier W (2008) CalDAG-GEFI and protein kinase C represent alternative pathways leading to activation of integrin alphaIIbbeta3 in platelets. Blood 112(5):1696–1703
24. Stefanini L, Roden RC, Bergmeier W (2009) CalDAG-GEFI is at the nexus of calcium-dependent platelet activation. Blood 114(12):2506–2514
25. Stolla M, Stefanini L, Roden RC et al (2011) The kinetics of alphaIIbbeta3 activation determines the size and stability of thrombi in mice: implications for antiplatelet therapy. Blood 117(3):1005–1013
26. Canault M, Ghalloussi D, Grosdidier C et al (2014) Human CalDAG-GEFI gene (RASGRP2) mutation affects platelet function and causes severe bleeding. J Exp Med 211(7):1349–1362
27. Stefanini L, Paul DS, Robledo RF et al (2015) RASA3 is a critical inhibitor of RAP1-dependent platelet activation. J Clin Invest 125(4):1419–1432
28. Cullen PJ, Patel Y, Kakkar VV, Irvine RF, Authi KS (1994) Specific binding sites for inositol 1,3,4,5-tetrakisphosphate are located predominantly in the plasma membranes of human platelets. Biochem J 298(Pt 3):739–742
29. Lockyer PJ, Vanlingen S, Reynolds JS et al (1999) Tissue-specific expression and endogenous subcellular distribution of the inositol 1,3,4,5-tetrakisphosphate-binding proteins GAP1(IP4BP) and GAP1(m). Biochem Biophys Res Commun 255(2):421–426

30. El-Daher SS, Patel Y, Siddiqua A et al (2000) Distinct localization and function of (1,4,5) IP(3) receptor subtypes and the (1,3,4,5)IP(4) receptor GAP1(IP4BP) in highly purified human platelet membranes. Blood 95(11):3412–3422
31. Iwashita S, Kobayashi M, Kubo Y et al (2007) Versatile roles of R-Ras GAP in neurite formation of PC12 cells and embryonic vascular development. J Biol Chem 282(6):3413–3417
32. Worner P, Brossmer R (1975) Platelet aggregation and the release induced by inophores for divalent cations. Thromb Res 6(4):295–305
33. Feinstein MB, Fraser C (1975) Human platelet secretion and aggregation induced by calcium ionophores. Inhibition by PGE1 and dibutyryl cyclic AMP. J Gen Physiol 66(5):561–581
34. Quinton TM, Ozdener F, Dangelmaier C, Daniel JL, Kunapuli SP (2002) Glycoprotein VI-mediated platelet fibrinogen receptor activation occurs through calcium-sensitive and PKC-sensitive pathways without a requirement for secreted ADP. Blood 99(9):3228–3234
35. Quinton TM, Kim S, Dangelmaier C et al (2002) Protein kinase C- and calcium-regulated pathways independently synergize with Gi pathways in agonist-induced fibrinogen receptor activation. Biochem J 368(Pt 2):535–543
36. Nadal-Wollbold F, Pawlowski M, Levy-Toledano S et al (2002) Platelet ERK2 activation by thrombin is dependent on calcium and conventional protein kinases C but not Raf-1 or B-Raf. FEBS Lett 531(3):475–482
37. Garcia A, Quinton TM, Dorsam RT, Kunapuli SP (2005) Src family kinase-mediated and Erk-mediated thromboxane A2 generation are essential for VWF/GPIb-induced fibrinogen receptor activation in human platelets. Blood 106(10):3410–3414
38. Shankar H, Garcia A, Prabhakar J, Kim S, Kunapuli SP (2006) P2Y12 receptor-mediated potentiation of thrombin-induced thromboxane A2 generation in platelets occurs through regulation of Erk1/2 activation. J Thromb Haemost 4(3):638–647
39. Walker TR, Watson SP (1993) Synergy between Ca2+ and protein kinase C is the major factor in determining the level of secretion from human platelets. Biochem J 289(Pt 1):277–282
40. Bergmeier W, Goerge T, Wang H-W et al (2007) Mice lacking the signaling molecule CalDAG-GEFI represent a model for leukocyte adhesion deficiency type III. J Clin Invest 117(6):1699–1707
41. Stefanini L, Boulaftali Y, Ouellette TD et al (2011) Rap1-Rac1 circuits potentiate platelet activation. Arterioscler Thromb Vasc Biol 32:434–441
42. Zhang G, Xiang B, Ye S et al (2011) Distinct roles for Rap1b protein in platelet secretion and integrin αIIbβ3 outside-in signaling. J Biol Chem 286(45):39466–39477
43. Boudreaux MK, Catalfamo JL, Klok M (2007) Calcium-diacylglycerol guanine nucleotide exchange factor I gene mutations associated with loss of function in canine platelets. Transl Res 150(2):81–92
44. Bernardi B, Guidetti GF, Campus F et al (2006) The small GTPase Rap1b regulates the cross talk between platelet integrin alpha2beta1 and integrin alphaIIbbeta3. Blood 107(7):2728–2735
45. Ahmad F, Boulaftali Y, Greene TK et al (2011) Relative contributions of stromal interaction molecule 1 and CalDAG-GEFI to calcium-dependent platelet activation and thrombosis. J Thromb Haemost 9(10):2077–2086
46. Han J, Lim CJ, Watanabe N et al (2006) Reconstructing and deconstructing agonist-induced activation of integrin alphaIIbbeta3. Curr Biol 16(18):1796–1806
47. Lee H-S, Lim CJ, Puzon-McLaughlin W, Shattil SJ, Ginsberg MH (2009) RIAM activates integrins by linking talin to ras GTPase membrane-targeting sequences. J Biol Chem 284(8):5119–5127
48. Stritt S, Wolf K, Lorenz V et al (2015) Rap1-GTP-interacting adaptor molecule (RIAM) is dispensable for platelet integrin activation and function in mice. Blood 125(2):219–222
49. Vossler MR, Yao H, York RD et al (1997) cAMP activates MAP kinase and Elk-1 through a B-Raf- and Rap1-dependent pathway. Cell 89(1):73–82
50. York RD, Yao H, Dillon T et al (1998) Rap1 mediates sustained MAP kinase activation induced by nerve growth factor. Nature 392(6676):622–626

51. Shattil SJ, Newman PJ (2004) Integrins: dynamic scaffolds for adhesion and signaling in platelets. Blood 104(6):1606–1615
52. Andre P, Delaney SM, LaRocca T et al (2003) P2Y12 regulates platelet adhesion/activation, thrombus growth, and thrombus stability in injured arteries. J Clin Invest 112(3):398–406
53. Watson S, Daly M, Dawood B et al (2010) Phenotypic approaches to gene mapping in platelet function disorders – identification of new variant of P2Y12, TxA2 and GPVI receptors. Hamostaseologie 30(1):29–38
54. Boudreaux MK, Martin M (2011) P2Y12 receptor gene mutation associated with postoperative hemorrhage in a Greater Swiss Mountain dog. Vet Clin Pathol 40(2):202–206
55. Gloerich M, Bos JL (2011) Regulating Rap small G-proteins in time and space. Trends Cell Biol 21(10):615–623
56. Wang Z, Holly SP, Larson MK et al (2009) Rap1b is critical for glycoprotein VI-mediated but not ADP receptor-mediated α 2β 1activation. J Thromb Haemost 7(4):693–700
57. Fu Z, Lee SH, Simonetta A et al (2007) Differential roles of Rap1 and Rap2 small GTPases in neurite retraction and synapse elimination in hippocampal spiny neurons. J Neurochem 100(1):118–131
58. Stornetta RL, Zhu JJ (2011) Ras and Rap signaling in synaptic plasticity and mental disorders. Neuroscientist 17(1):54–78
59. Subramanian H, Zahedi RP, Sickmann A, Walter U, Gambaryan S (2013) Phosphorylation of CalDAG-GEFI by protein kinase A prevents Rap1b activation. J Thromb Haemost 11(8):1574–1582
60. Guidetti GF, Manganaro D, Consonni A et al (2013) Phosphorylation of the guanine-nucleotide-exchange factor CalDAG-GEFI by protein kinase A regulates Ca(2+)-dependent activation of platelet Rap1b GTPase. Biochem J 453(1):115–123
61. de Bruyn KMT, Zwartkruis FJT, de Rooij J, Akkerman JWN, Bos JL (2003) The small GTPase Rap1 is activated by turbulence and is involved in integrin IIb 3-mediated cell adhesion in human megakaryocytes. J Biol Chem 278(25):22412–22417
62. Watanabe N, Bodin L, Pandey M et al (2008) Mechanisms and consequences of agonist-induced talin recruitment to platelet integrin IIb 3. J Cell Biol 181(7):1211–1222
63. Balduini A, Pecci A, Lova P et al (2004) Expression, activation, and subcellular localization of the Rap1 GTPase in cord blood-derived human megakaryocytes. Exp Cell Res 300(1):84–93
64. Kaushansky K (2009) Determinants of platelet number and regulation of thrombopoiesis. Hematol Educ Prog Am Soc Hematol Am Soc Hemat Educ Prog 147–152
65. Kaushansky K (1997) Thrombopoietin the primary regulator of platelet production. Trends Endocrinol Metab TEM 8(2):45–50
66. Racke FK, Lewandowska K, Goueli S, Goldfarb AN (1997) Sustained activation of the extracellular signal-regulated kinase/mitogen-activated protein kinase pathway is required for megakaryocytic differentiation of K562 cells. J Biol Chem 272(37):23366–23370
67. Rouyez MC, Boucheron C, Gisselbrecht S, Dusanter-Fourt I, Porteu F (1997) Control of thrombopoietin-induced megakaryocytic differentiation by the mitogen-activated protein kinase pathway. Mol Cell Biol 17(9):4991–5000
68. Whalen AM, Galasinski SC, Shapiro PS, Nahreini TS, Ahn NG (1997) Megakaryocytic differentiation induced by constitutive activation of mitogen-activated protein kinase kinase. Mol Cell Biol 17(4):1947–1958
69. Melemed AS, Ryder JW, Vik TA (1997) Activation of the mitogen-activated protein kinase pathway is involved in and sufficient for megakaryocytic differentiation of CMK cells. Blood 90(9):3462–3470
70. Herrera R, Hubbell S, Decker S, Petruzzelli L (1998) A role for the MEK/MAPK pathway in PMA-induced cell cycle arrest: modulation of megakaryocytic differentiation of K562 cells. Exp Cell Res 238(2):407–414
71. Mazharian A, Watson SP, Séverin S (2009) Critical role for ERK1/2 in bone marrow and fetal liver-derived primary megakaryocyte differentiation, motility, and proplatelet formation. Exp Hematol 37(10):1238–1249.e5

72. Fichelson S, Freyssinier JM, Picard F et al (1999) Megakaryocyte growth and development factor-induced proliferation and differentiation are regulated by the mitogen-activated protein kinase pathway in primitive cord blood hematopoietic progenitors. Blood 94(5):1601–1613
73. Miyazaki R, Ogata H, Kobayashi Y (2001) Requirement of thrombopoietin-induced activation of ERK for megakaryocyte differentiation and of p38 for erythroid differentiation. Ann Hematol 80(5):284–291
74. Guerriero R, Parolini I, Testa U et al (2006) Inhibition of TPO-induced MEK or mTOR activity induces opposite effects on the ploidy of human differentiating megakaryocytes. J Cell Sci 119(Pt 4):744–752
75. Guerriero R, Mattia G, Testa U et al (2001) Stromal cell-derived factor 1alpha increases polyploidization of megakaryocytes generated by human hematopoietic progenitor cells. Blood 97(9):2587–2595
76. Minamiguchi H, Kimura T, Urata Y et al (2001) Simultaneous signalling through c-mpl, c-kit and CXCR4 enhances the proliferation and differentiation of human megakaryocyte progenitors: possible roles of the PI3-K, PKC and MAPK pathways. Br J Haematol 115(1):175–185
77. Severin S, Ghevaert C, Mazharian A (2010) The mitogen-activated protein kinase signaling pathways: role in megakaryocyte differentiation. J Thromb Haemost 8(1):17–26
78. Matsumura I, Nakajima K, Wakao H et al (1998) Involvement of prolonged ras activation in thrombopoietin-induced megakaryocytic differentiation of a human factor-dependent hematopoietic cell line. Mol Cell Biol 18(7):4282–4290
79. Garcia J, de Gunzburg J, Eychène A, Gisselbrecht S, Porteu F (2001) Thrombopoietin-mediated sustained activation of extracellular signal-regulated kinase in UT7-Mpl cells requires both Ras-Raf-1- and Rap1-B-Raf-dependent pathways. Mol Cell Biol 21(8):2659–2670
80. Porteu F, Rouyez MC, Cocault L et al (1996) Functional regions of the mouse thrombopoietin receptor cytoplasmic domain: evidence for a critical region which is involved in differentiation and can be complemented by erythropoietin. Mol Cell Biol 16(5):2473–2482
81. Goldfarb AN, Delehanty LL, Wang D, Racke FK, Hussaini IM (2001) Stromal inhibition of megakaryocytic differentiation correlates with blockade of signaling by protein kinase C- and ERK/MAPK. J Biol Chem 276(31):29526–29530
82. Delehanty LL, Mogass M, Gonias SL et al (2003) Stromal inhibition of megakaryocytic differentiation is associated with blockade of sustained Rap1 activation. Blood 101(5):1744–1751
83. Kamata T, Pritchard CA, Leavitt AD (2004) Raf-1 is not required for megakaryocytopoiesis or TPO-induced ERK phosphorylation. Blood 103(7):2568–2570
84. Kamata T, Kang J, Lee T-H et al (2005) A critical function for B-Raf at multiple stages of myelopoiesis. Blood 106(3):833–840
85. Pleines I, Dütting S, Cherpokova D et al (2013) Defective tubulin organization and proplatelet formation in murine megakaryocytes lacking Rac1 and Cdc42. Blood 122(18):3178–3187
86. Bluteau D, Lordier L, Di Stefano A et al (2009) Regulation of megakaryocyte maturation and platelet formation. J Thromb Haemost 7:227–234
87. Zhang L, Orban M, Lorenz M et al (2012) A novel role of sphingosine 1-phosphate receptor S1pr1 in mouse thrombopoiesis. J Exp Med 209(12):2165–2181
88. Gobbi G, Mirandola P, Carubbi C et al (2013) Proplatelet generation in the mouse requires PKC -dependent RhoA inhibition. Blood 122(7):1305–1311
89. Chang Y, Auradé F, Larbret F et al (2007) Proplatelet formation is regulated by the Rho/ROCK pathway. Blood 109(10):4229–4236
90. Suzuki A, Shin J-W, Wang Y et al (2013) RhoA is essential for maintaining normal megakaryocyte ploidy and platelet generation. PLoS ONE 8(7), e69315
91. Pleines I, Hagedorn I, Gupta S et al (2012) Megakaryocyte-specific RhoA deficiency causes macrothrombocytopenia and defective platelet activation in hemostasis and thrombosis. Blood 119(4):1054–1063

92. Gerard A, Mertens AEE, Van Der Kammen RA, Collard JG (2007) The Par polarity complex regulates Rap1- and chemokine-induced T cell polarization. J Cell Biol 176(6):863–875
93. Schwamborn JC, Püschel AW (2004) The sequential activity of the GTPases Rap1B and Cdc42 determines neuronal polarity. Nat Neurosci 7(9):923–929
94. Schwamborn JC, Müller M, Becker AH, Püschel AW (2007) Ubiquitination of the GTPase Rap1B by the ubiquitin ligase Smurf2 is required for the establishment of neuronal polarity. EMBO J 26(5):1410–1422
95. Takahashi M, Rikitake Y, Nagamatsu Y et al (2008) Sequential activation of Rap1 and Rac1 small G proteins by PDGF locally at leading edges of NIH3T3 cells. Genes Cells 13(6): 549–569
96. Choi W, Harris NJ, Sumigray KD, Peifer M (2013) Rap1 and canoe/afadin are essential for establishment of apical-basal polarity in the drosophila embryo. Mol Biol Cell 24(7):945–963
97. Li R, Pendergast AM (2011) Arg kinase regulates epithelial cell polarity by targeting β1-integrin and small GTPase pathways. Curr Biol 21(18):1534–1542
98. Iden S, Collard JG (2008) Crosstalk between small GTPases and polarity proteins in cell polarization. Nat Rev Mol Cell Biol 9(11):846–859
99. Molina-Ortiz P, Polizzi S, Ramery E et al (2014) Rasa3 controls megakaryocyte Rap1 activation, integrin signaling and differentiation into proplatelet. PLoS Genet 10(6), e1004420
100. Kupzig S, Deaconescu D, Bouyoucef D et al (2006) GAP1 family members constitute bifunctional Ras and Rap GTPase-activating proteins. J Biol Chem 281(15):9891–9900

Regulation of Megakaryocyte and Platelet Survival

9

Amanda E. Au, Marion Lebois, Irina Pleines, and Emma C. Josefsson

Abstract

Platelets play vital roles in hemostasis, wound healing, and a range of other processes [1]. Their number is tightly controlled within narrow physiological ranges. This occurs through a dynamic balance between platelet production and consumption/clearance rates, so as to ensure that the total platelet mass remains constant. Megakaryocytes in the bone marrow produce around 100 billion platelets per day. In a healthy individual, the majority of platelets are not consumed by hemostatic processes. It is therefore imperative that platelet lifespan is strictly regulated. Recent work has demonstrated that the survival of megakaryocytes and platelets is controlled by programmed cell death, apoptosis [2]. Both cell types possess a classical Bak- and Bax-mediated intrinsic, mitochondrial, apoptosis pathway that must be restrained in order for them to develop and survive. In addition, recent work has revealed that the glycosylation state of platelet surface proteins is an indicator of platelet age and that aged, desialylated platelets stimulate platelet production [3].

A.E. Au • E.C. Josefsson (✉)
Cancer & Haematology Division, The Walter and Eliza Hall Institute of Medical Research, 1G Royal Parade, Parkville, VIC 3052, Australia

Department of Medical Biology, The University of Melbourne, 1G Royal Parade, Melbourne, VIC 3052, Australia
e-mail: josefsson@wehi.edu.au

M. Lebois
Cancer & Haematology Division, The Walter and Eliza Hall Institute of Medical Research, 1G Royal Parade, Parkville, VIC 3052, Australia

I. Pleines
ACRF Chemical Biology Division and Department of Medical Biology, The Walter and Eliza Hall Institute of Medical Research, 1G Royal Parade, Parkville, VIC 3052, Australia

Department of Medical Biology, The University of Melbourne, 1G Royal Parade, Melbourne, VIC 3052, Australia

H. Schulze, J. Italiano (eds.), *Molecular and Cellular Biology of Platelet Formation*, DOI 10.1007/978-3-319-39562-3_9

9.1 Summary

Platelets play vital roles in hemostasis, wound healing, and a range of other processes [1]. Their number is tightly controlled within narrow physiological ranges. This occurs through a dynamic balance between platelet production and consumption/clearance rates, so as to ensure that the total platelet mass remains constant. Megakaryocytes in the bone marrow produce around 100 billion platelets per day. In a healthy individual, the majority of platelets are not consumed by hemostatic processes. It is therefore imperative that platelet lifespan is strictly regulated. Recent work has demonstrated that the survival of megakaryocytes and platelets is controlled by programmed cell death, apoptosis [2]. Both cell types possess a classical Bak- and Bax-mediated intrinsic, mitochondrial, apoptosis pathway that must be restrained in order for them to develop and survive. In addition, recent work has revealed that the glycosylation state of platelet surface proteins is an indicator of platelet age and that aged, desialylated platelets stimulate platelet production [3].

This chapter focuses on the molecular regulation of apoptosis in platelet and megakaryocyte survival, reviews acquired and inherited platelet disorders where this machinery could be of importance, examines implications of chemotherapy in targeting cell death pathways, as well as discusses alternative cell death paradigms that may exist in platelets and megakaryocytes.

9.2 Introduction

In humans, normal physiological platelet counts range from 150 to 450×10^9/L blood. Thrombocytopenia is diagnosed when platelet counts fall below 100×10^9/L, and emergency treatment, typically involving platelet transfusion, is necessary when platelet counts are diminished to levels below 30×10^9/L [4]. In recent years, thrombopoietin (TPO) mimetic drugs are increasingly being used in the clinic to drive platelet production and sustain platelet counts in patients with immune thrombocytopenia (ITP) [5]. The opposite extreme of thrombocytopenia is the production of excessive numbers of platelets which is referred to as thrombocythemia or thrombocytosis, being a primary disorder of the bone marrow or occurring secondary to other systemic conditions, respectively [6]. Abnormal circulating platelet counts can arise due to irregular platelet production by megakaryocytes, improper platelet destruction, and clearance by macrophages or a perturbation of the normal platelet lifespan in the circulation. These events may transpire independently, coincidentally, or consecutively.

9.2.1 Platelet Storage and Transfusion

To circumvent bleeding complications, platelet concentrates are transfused into patients at high risk of bleeding, after chemo- or radiotherapy treatment or in other settings of severe thrombocytopenia or dysfunctional platelets. To be clinically effective, transfused platelets must be able to circulate and to respond to an injury. Refrigerated platelets, however, are rapidly cleared upon transfusion [7]. Storage at 37 °C on the other hand leads to a prompt loss of platelet viability and function [8]. Significant effort has been made in attempt to optimize the storage and handling procedures of platelet concentrates. Hence, room temperature storage of platelet concentrates with agitation in bags (allowing for oxygen exchange) is the benchmark permitting platelet storage for 5 days or slightly longer if followed by pathogen inactivation [9]. However, even under optimized conditions, there is a gradual loss of quality known as the platelet storage lesion (PSL) that correlates with reductions in both platelet survival and hemostatic activity posttransfusion [10]. The PSL shares features of both platelet activation and induction of multiple cell death pathways and is characterized by P-selectin exposure, surface receptor shedding, microparticle formation, as well as signs of apoptosis, including loss of mitochondrial outer membrane potential (MOMP), externalization of phosphatidylserine (PS), and caspase activation [8, 10–12].

9.2.2 Assessment of Platelet Turnover

The lifespan of platelets in circulation is relatively short, averaging approximately 10 days in humans and 5 days in mice. Hence, a number of methods have been developed to assess platelet turnover. Measurements of platelet lifespan in humans were initially made with ex vivo radiolabeled platelets transfused into patients [13–15]. Today this method is still being employed in clinical trials [16]; however, since this is a costly method requiring specialized expertise, less invasive indirect measurements that allow an approximation of platelet turnover are now routinely used.

Newly produced platelets contain RNA that is quickly lost as they circulate. As such, analysis of this young RNA positive population can help to distinguish defects in clearance from those of production [17, 18]. Another method is based on the level of platelet destruction and involves measurement of the plasma level of glycocalicin, the shed extracellular domain of platelet glycoprotein Ibα [19, 20]. In experimental models, methods to determine platelet lifespan involve labeling with biotin, fluorescent dyes, or fluorescently conjugated antibodies [7, 21–25].

9.3 An Introduction to Apoptosis

It is now generally accepted that survival of megakaryocytes [26–30] and platelets [31] is regulated by intrinsic apoptosis, a form of programmed cell death [2]. This process plays a vital role in regulating cell numbers during hematopoiesis, allowing unwanted cells to be rapidly and furtively removed without triggering inflammation [32]. Kerr and colleagues initially characterized the apoptotic process by classical morphological features such as pyknosis (chromatin condensation), nuclear shrinkage, and membrane blebbing [33]. In light of advancements in biochemical and genetic exploration of cell death, today, apoptosis is defined on a molecular level. The apoptotic process can be broadly separated into the intrinsic (mitochondrial non-receptor-mediated) and extrinsic (receptor-mediated) pathway of apoptosis.

Intrinsic apoptosis can be initiated by a vast range of non-receptor-mediated stimuli such as nutrient deprivation, cellular stress, or exposure to chemotherapeutic agents, to name a few. For blood homeostasis to be effectively maintained, the apoptotic process is tightly controlled by BCL-2 family members, which balance the promotion or inhibition of apoptosis. The pro-death molecules Bak and Bax are the mediators of this pathway that ultimately dictate the fate of the cell. Bak and Bax are normally sequestered by five pro-survival BCL-2 proteins, namely, Bcl-2, Mcl-1, Bcl-x_L, Bcl-w, and A1 [32]. However, upon conditions of cellular stress, a third group of BCL-2 proteins, the BH3-only proteins, can inhibit and displace the pro-survival proteins. This in effect indirectly or directly [34–37] activates Bak and Bax to irreversibly cause MOMP, cytochrome *c* release, apoptosome assembly, caspase-9 activation, caspase-3/7 activation, and PS exposure (Fig. 9.1). The overexpression of pro-survival BCL-2 proteins has been identified as a mechanism for increased cell survival in certain malignancies and hence represents a potential target for cancer therapies.

The extrinsic arm of apoptosis involves the participation of death receptors. The execution of this pathway of apoptosis requires the association of ligands specific to death receptors on the cell surface, including the Fas receptor, and other death receptors of the tumor necrosis factor receptor (TNF) family. Activation of death receptors leads to downstream activation of caspase-8 and subsequent direct activation of effector caspase-3/7 in type 1 cells such as thymocytes [38] culminating in cell death. In type II cells, such as hepatocytes, Bid activation downstream of caspase-8 connects the extrinsic and intrinsic pathways, demonstrating cross talk between the two seemingly distinct pathways [39] (Fig. 9.1).

9.4 The Role of Apoptosis in Platelet Biology

The balance between pro-survival and pro-death proteins varies depending on cell type. Moreover, different components of the apoptotic cascade bear different importance in mediating apoptosis in different cell types. In the next section, we will explore mediators of apoptosis in platelets.

9.4.1 Regulation of Platelet Survival by Intrinsic Apoptosis

Apoptosis in platelets has been characterized in vitro by measuring cytochrome *c* release, caspase activation, and PS exposure [40]. Bcl-x_L is the critical pro-survival protein required to maintain platelet in vivo survival [31, 41]. The first indication for this finding originated from Wagner and colleagues who described thrombocytopenia in mice after *MMTV-Cre*-mediated deletion of *Bcl-x* in the hematopoietic system, skin, and various secretory tissues [42]. It has since been shown that megakaryocyte lineage-restricted *Pf4-Cre*-mediated deletion of *Bcl-x* in mice reduces platelet lifespan from ~5 days to ~5 h (Fig. 9.2a), with a resultant decrease in platelet counts to ~2 % of wild-type levels (Fig. 9.2b) [28, 43]. Accordingly, pharmacological blockade of Bcl-x_L with the BH3 mimetic drug ABT-737 [44] targeting Bcl-x_L, Bcl-2, and Bcl-w, or its oral version navitoclax (ABT-263) [45], prompts platelet death [31, 40, 46] and leads to thrombocytopenia [31, 47, 48]. Conceivably, deletion of *Bak* significantly extends platelet lifespan (Fig. 9.2a) and causes

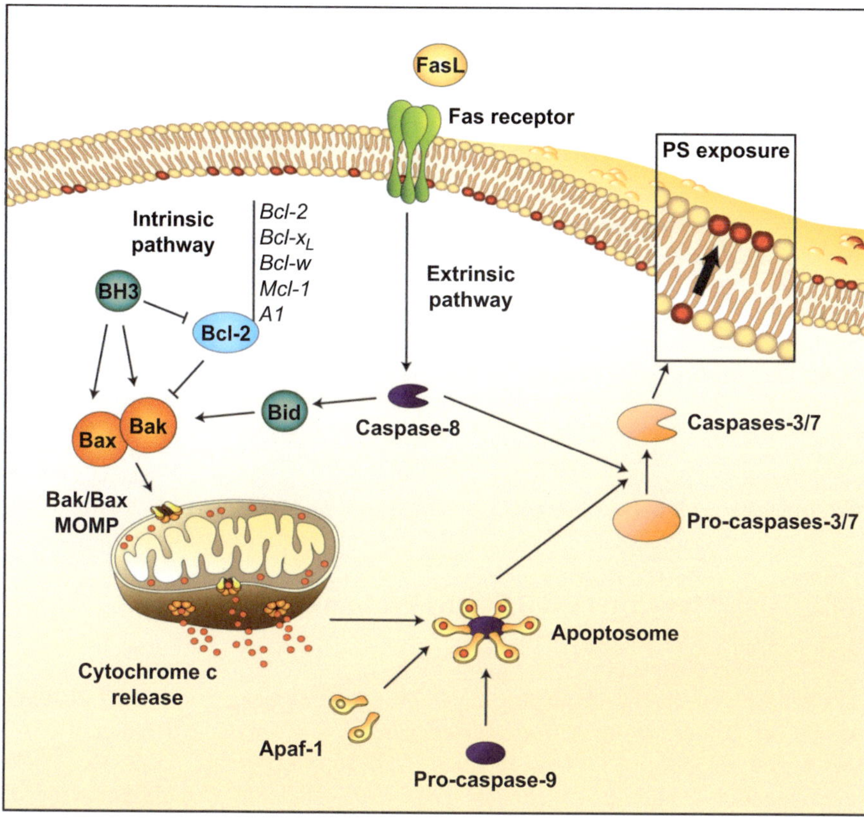

Fig. 9.1 Intrinsic and extrinsic apoptosis pathways. The intrinsic or mitochondrial pathway of apoptosis can be activated by cellular stresses including growth factor deprivation and irradiation or by chemotherapeutic agents. Five pro-survival members of the BCL-2 family, namely, Bcl-2, Mcl-1, Bcl-x_L, Bcl-w, and A1, operate by restraining Bak and Bax, the ultimate mediator of cell death. During apoptosis, BH3 proteins can inhibit and displace pro-survival proteins thereby directly or indirectly activating Bak and Bax to irreversibly cause mitochondrial outer membrane permeabilization (MOMP) leading to cytochrome c release, assembly of Apaf-1 and caspase-9 into the apoptosome and, subsequently, activation of caspase-9, caspase-3/7, as well as phosphatidylserine (PS) exposure. The extrinsic, death receptor-mediated, apoptosis pathway is activated by ligand binding to cell surface receptors, including Fas receptor and other death receptors of the TNF receptor family. It culminates in the activation of caspase-8 and subsequent direct effector caspase-3/7 activation. Alternatively, in certain cell types, Bid activation connects the extrinsic and intrinsic pathways downstream of caspase-8 activation

thrombocytosis in mice (Fig. 9.2b) [28, 31]. However, loss of *Bax* does not significantly influence platelet counts, indicating that Bak is the major player in regulating platelet lifespan at steady-state in vivo. Genetic deletion of *Bak* and *Bax* in mice doubles platelet lifespan [28] to the same extent as *Bak* deletion, rescues the thrombocytopenia caused by loss of *Bcl-x* [28, 30], and makes platelets resistant to the effects of ABT-737 [40]. In accordance with findings in mice, human genome-wide association studies (GWAS) have recently identified common variants in the *BAK1* gene associated with platelet count [49, 50].

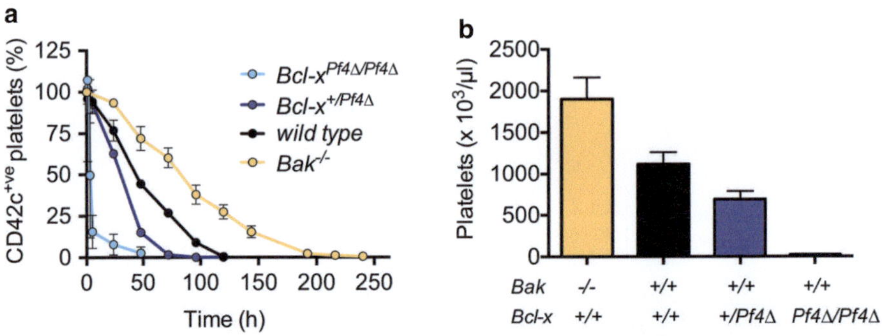

Fig. 9.2 Regulation of platelet lifespan in vivo. (**a**) Platelet survival curves in *Bcl-*$^{xPf4\Delta/Pf4\Delta}$, *Bcl-x*$^{+/Pf4\Delta}$, control, and *Bak*$^{-/-}$ mice. Platelets were labeled via *i.v.* injection of a DyLight 488-conjugated anti-CD42c (GPIbβ) antibody (Data are presented as mean ± s.d. ©Josefsson et al. 2011. Originally published in *Journal of Experimental Medicine*. doi:10.1084/jem.20110750 [28]). (**b**) Platelet counts in *Bak*$^{-/-}$, control, *Bcl-x*$^{+/Pf4\Delta}$, and *Bcl-*$^{xPf4\Delta/Pf4\Delta}$, mice at 7 weeks of age (©Josefsson et al. 2011. Originally published in *Journal of Experimental Medicine*. doi:10.1084/jem.20110750 [28])

9.4.2 Roles for Other Bcl-2 Family Proteins in Regulating Platelet Lifespan

Do other pro-survival proteins play a role in regulating platelet survival? Platelets express Bcl-2, but in contrast to *Bcl-x* deficiency, lineage-specific deletion of *Bcl-2* does not affect platelet lifespan in mice [51]. Supporting this result, the newly developed Bcl-2 inhibitor venetoclax (ABT-199) efficiently stimulates killing of leukemic cells without provoking thrombocytopenia [52]. Interestingly, overexpression of BCL-2 extends platelet survival [53] albeit not to the same extent as *Bak*/*Bax* deletion. Pro-survival BCL-w is expressed in platelets [30], but it is unlikely to play a significant role in regulating platelet lifespan since systemic *Bcl-w* knockout mice have normal platelet counts [54, 55]. A1 has been detected in young human platelets based on RNA expression [8], but little is known about the importance of A1 in platelets. Systemic *A1a* knockout mice were not reported to display a platelet phenotype [56]. Since there is a quadruplication of the *A1* gene locus in mice, recent studies employed constitutive as well as conditional expression of mi-shRNAs targeting all A1 isoforms expressed in the mouse. Even though roles of A1 were found in lymphocytes and granulocytes, there was no evidence of an impact on platelet counts [57]. Finally, neither conditional deletion nor hematopoietic overexpression of *Mcl-1* affected platelet lifespan [26, 29, 53]. The Mcl-1 protein is, however, known to have a brief half-life, in part due to rapid proteosomal degradation. Mcl-1 is poorly expressed in mouse and human platelets [26, 28–31, 41] but could be detected in platelets after injection of proteasome inhibitor into mice [29] indicating that some remnant protein could potentially exist in young platelets. Thus, despite being anuclear, platelets are capable of undergoing mitochondrially driven apoptosis; however, their failure in sustaining Mcl-1 levels makes them distinct to most other cell types and largely dependent on pro-survival Bcl-x_L (Fig. 9.3).

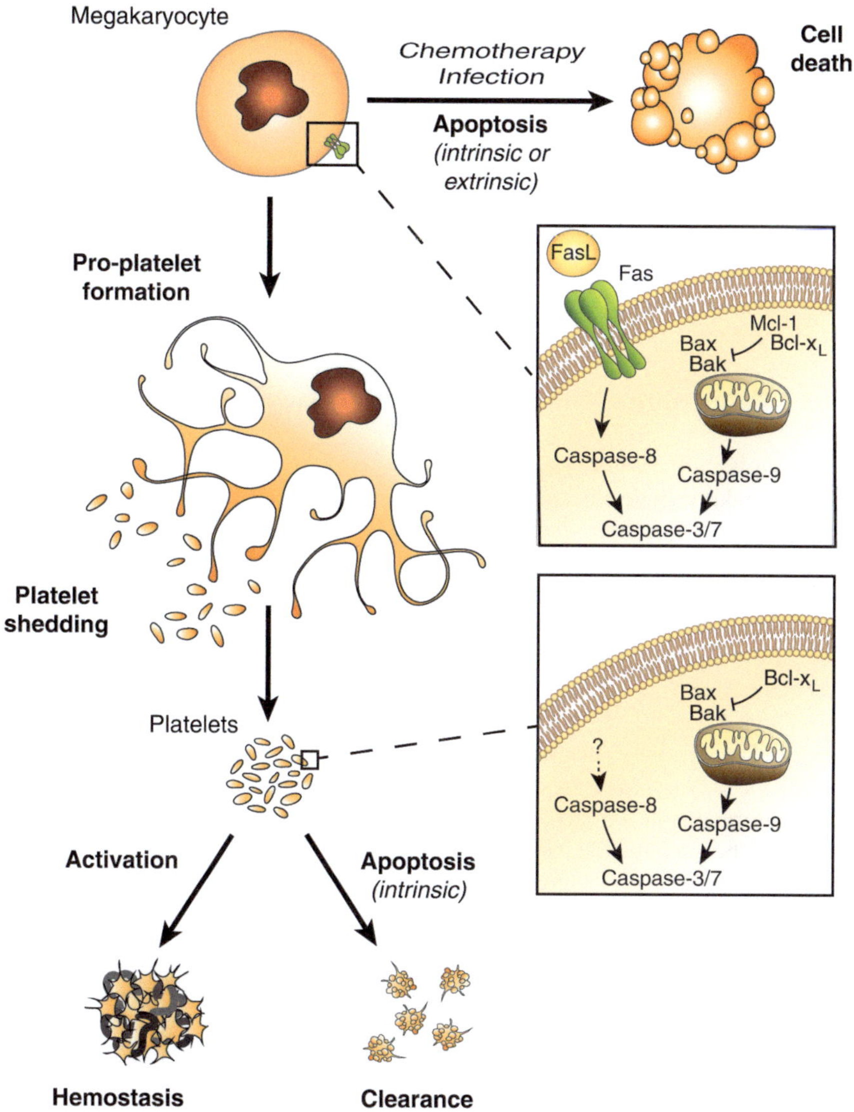

Fig. 9.3 Apoptosis in megakaryocyte and platelet biology. Megakaryocytes possess functional Bak-/Bax-mediated intrinsic and FasL-inducible extrinsic apoptosis pathways. Megakaryocyte apoptosis can be triggered in response to chemotherapy or infection but must be restrained during megakaryocyte development and maturation in order to allow efficient platelet production. In contrast to megakaryocytes, platelets lack the Fas receptor. While limited information is available on the extrinsic pathway in platelets, strong evidence suggests that the intrinsic apoptosis pathway regulates platelet lifespan in the circulation (©Josefsson et al. 2014. Originally published in *Nature Communications*. doi:10.1038/ncomms4455 [27])

The role of death-promoting BH3-only proteins in platelets is largely unexplored. Bim [30], Bid [27, 30], Bad [58], and Bik [59] have all been shown to be present in platelets. Puma [59] and Noxa [46] could not be detected, however *Puma* mRNA is present [60]. Kodama and colleagues took the approach of deleting both *Bim* and *Bid* in mice, but no effect on platelet count was observed [30]. So far, Bad is the only BH3-only protein that has been shown to play a role in influencing platelet lifespan: its deletion leading to a modest extension in platelet lifespan with concomitantly increased platelet counts [58]. Even though this effect is minor compared to *Bak* deletion, it is plausible that several BH3-only proteins are involved in triggering platelet apoptosis, and future research on this front is needed.

9.4.3 Regulation of the Extrinsic Apoptosis Pathway in Platelets

Caspase-8 is the critical regulator of extrinsic apoptosis and was recently deleted in the megakaryocytic lineage without affecting platelet count or lifespan [27]. This brings into question whether platelets have a functional extrinsic apoptosis pathway. In addition to caspase-8, they do express Fas-associated death domain (FADD) protein but failed to undergo classical extrinsic apoptosis after FasL stimulation in vitro [27]. This is somewhat unsurprising since several groups have reported the Fas receptor to be absent in platelets [8, 27, 61]. There are, however, other death ligands including TNF-related apoptosis-inducing ligand (TRAIL) and TNF-α that bind TRAIL receptors and the TNF receptor-1, respectively [62]. Plenchette et al. showed that human platelets do not express TRAIL receptor 1 and 2, while the expression of the decoy receptor DcR2 increased progressively during platelet storage [61]. Finally, two recent studies reported caspase-8 activation in platelets to be uncoupled from death receptor stimulation [63, 64]. Taken together, defined molecular insights into how caspase-8 is activated in platelets are currently lacking.

9.4.4 Apoptotic Caspases in Platelets

In addition to caspase-8, platelets contain caspase-9 and effector caspase-3 and. Even though activation of caspases is a hallmark of apoptosis, they are not needed for cell death and the apoptotic clearance of cells in vivo [65]. Deletion of *caspase-9* in hematopoietic cells did not affect platelet counts or lifespan at steady state; however, caspase-9-deficient platelets displayed significantly delayed PS-exposure in vitro after induction of intrinsic apoptosis [66]. Moreover, combined deletion of *caspase-3* and *caspase-7* or *Apaf-1* in hematopoietic cells did not affect platelet counts at steady state [27] indicating that all three proteins are superfluous for regulation of platelet lifespan. Nevertheless, it was recently shown that caspase activation is important to keep apoptosis immunologically silent. If Bak/Bax-mediated mitochondrial damage occurs in the absence of caspase activation, released mitochondrial DNA will trigger type-1 interferon production and cause a subsequent hematopoietic stem cell defect [67].

9.4.5 Platelet Procoagulant Function

It has long been recognized that the phospholipid membrane presents as one of the main contributors to the procoagulant function of platelets. Lipid asymmetry of the phospholipid membrane bilayer is maintained largely in an ATP-dependent directional lipid transport manner, and the exposure of PS is often used as an indicator of plasma membrane rearrangement. During intense platelet activation, lipid asymmetry disruption leads to the surface exposure of PS, which under normal conditions typically dominates the inner leaflet of the phospholipid bilayer [68]. These highly activated platelets are referred to as collagen- and thrombin-activated (COAT)/coated platelets [69] or sustained calcium-induced platelets (SCIP) [70]. Externalized PS assists in the assembly of highly active tenase and prothrombin complexes, which accelerates two successive reactions of the coagulation cascade comprising activation of factor X which in turn catalyzes the formation of thrombin, the master regulator of the coagulation cascade. The rapid loss of platelet asymmetry is regulated by proteins collectively referred to as protein scramblases [71, 72]. In platelets, two independent scramblase activation pathways exist, one of which is Ca^{2+}-dependent (non-apoptotic pathway) and the second being Ca^{2+}-independent and caspase-dependent (apoptotic pathway).

In the Ca^{2+}-dependent pathway of PS exposure, following platelet activation, the increase in cytosolic Ca^{2+} concentration leads to the reorganization of lipid asymmetry promoting increased scramblase activity. The lowering of Ca^{2+} levels conversely stops scrambling and restores lipid asymmetry. Therefore, a persistent elevation of Ca^{2+} levels is fundamental to initiate and complete restructuring of lipid asymmetry and subsequent stimulation of the coagulation cascade [73]. High levels of intracellular Ca^{2+}, can be induced through store- or receptor-mediated processes, respectively (reviewed in [74]). Two proteins, namely, Stim1 and Orai1 have emerged as key players in store-operated Ca^{2+} entry (SOCE) in platelets. Accordingly, SOCE was absent in platelets from *Stim1*$^{-/-}$ and *Orai1*$^{-/-}$ mice upon activation [75]. On the other hand, *Stim1Sax*$^{/+}$ mice that express mutant Stim1 displayed macrothrombocytopenia due to constantly elevated cytoplasmic Ca^{2+} levels, resulting in a pre-activated state and rapid clearance of the mutant platelets [76]. Moreover, it has been proposed that cyclophilin D prevents loss of $\Delta\Psi m$ in a process now recognized as mitochondrial permeability transition (MPT)-mediated regulated necrosis [77–79]. Accordingly, *cyclophilin D*-deficient mice have reduced ability to expose PS upon strong activation by dual agonist convulxin/thrombin [77]. Interestingly, a mutation in *TMEM16F* has been found in Scott syndrome patients that have impaired PS externalization on platelets leading to minor hemorrhagic complications [80–82]. TMEM16F has since been identified as a key player in mediating Ca^{2+}-dependent phospholipid scrambling [80]. This phenotype was recapitulated in *TMEM16F* knockout mice reported to have impaired scramblase activity and abnormal hemostasis [83].

The exposure of PS is often regarded as a hallmark of apoptosis and presents as an "eat-me signal" for recognition and engulfment by macrophages [84]. As described, platelets contain many of the main components of the intrinsic apoptotic machinery. When stimulated to undergo apoptosis via treatment with ABT-737 or navitoclax, platelets externalize PS in a dose- and temporal time-dependent fashion [41, 48]. These platelets are procoagulant and therefore support thrombin generation [40]. Accordingly, *Bak*$^{-/-}$*Bax*$^{-/-}$ platelets stimulated to undergo apoptosis by exposure to ABT-737 failed to induce surface exposure of PS and to promote thrombin generation [40]. Meanwhile, prominent PS exposure was detected in *Bak*$^{-/-}$*Bax*$^{-/-}$ platelets through intense activation using convulxin/thrombin thereby supporting the notion that two distinct pathways of PS exposure are evident in platelets [40]. Recently, the Xk-related protein 8 (Xkr8) has been identified as an enzyme pivotal for catalyzing caspase-dependent apoptosis-induced lipid scrambling [85].

9.5 Platelet Lifespan and Glycosylation

In addition to intrinsic apoptosis, platelet survival in the circulation also depends on the level of glycosylation of surface proteins [86]. In vertebrates, glycosylated surface proteins are normally capped with terminal sialic acid. Removal of sialic acid (desialylation) unmasks the underlying sugar, galactose. It has long been recognized that platelets lose sialic acid during storage [87, 88] and that cleavage of terminal sialic acid, by the bacterially derived enzyme neuraminidase (sialidase), causes swift platelet clearance in mice and rabbits [89–91]. Modifications in glycosylation after refrigerated storage of platelets [92] or in mice lacking the sialyltransferase ST3GalIV [93] lead to recognition and subsequent platelet clearance by the hepatic Ashwell-Morell receptor, a complex composed of asialoglycoprotein receptor (ASGPR) 1 and 2. Platelets also lose sialic acid as they age in the circulation [94] evidenced by microthrombocytosis in *Asgpr2*$^{-/-}$ mice despite increased levels of platelet surface terminal β-galactose [3]. Interestingly, it was recently shown that this removal system drives hepatic TPO mRNA expression via janus kinase 2 (JAK2) and signal transducer and activator of transcription 3 (STAT3), thereby regulating megakaryopoiesis [3]. Whether there is a direct link between the regulation of intrinsic apoptosis and altered platelet surface sialic acid content is unclear. Notably, apoptotic platelets are primarily cleared in the liver [41]; however, the exact mechanism remains to be established.

9.6 The Role of Apoptosis in Megakaryocyte Biology

9.6.1 Regulation of Megakaryocyte Survival by Intrinsic Apoptosis

Cultured mature bone marrow or fetal liver-derived megakaryocytes undergo apoptosis and fail to produce proplatelets after genetic loss of *Bcl-x*, or in response to ABT-737, and are fully apoptosis-resistant in the absence of Bak/Bax [28]. A similar effect on viability and proplatelet formation was seen after treatment of mature megakaryocytes with the topoisomerase II inhibitor etoposide [28]. In vivo, megakaryocyte-restricted deletion of *Bcl-x* led to severe macrothrombocytopenia, with release of atypical large vacuolar megakaryocytic fragments in the bone marrow sinusoids. Hence, mature megakaryocytes depend on the function of pro-survival Bcl-x_L, to proceed safely through the process of platelet shedding [28].

However, loss of Bcl-x_L did not affect the growth and maturation of megakaryocytes, suggesting redundancy with other pro-survival proteins. Single or dual megakaryocyte-specific deletion of pro-survival *Mcl-1* and *Bcl-2* did not influence megakaryopoiesis, platelet production, or platelet lifespan [26, 29, 51]. Thrombocytopenia in *Bclx*− deficient mice was not affected by additional genetic loss or pharmacological inhibition of Bcl-2 [51]. However, genetic deletion of both *Mcl-1* and *Bcl-x* in megakaryocytes resulted in pre-weaning lethality [26], rescued by ablation of *Bak* and *Bax* [29]. Megakaryopoiesis in *Bcl-x*/*Mcl-1*− deficient E12.5 embryos was severely compromised, where megakaryocytes were reduced in number and size and these animals presented with hemorrhage. A few *Bcl-x*− deficient mice with one functional *Mcl-1* allele reached adulthood. These mice revealed a subpopulation of pyknotic, caspase-3-positive megakaryocytes in bone marrow and spleen with cytoplasmic vacuolization. In order to induce full-blown megakaryocyte apoptosis in vivo, adult conditional *Mcl-1*-deficient mice were treated with ABT-737 (inhibiting Bcl-x_L, Bcl-2, and Bcl-w), resulting in ablation of megakaryocytes and platelets. 2.5 h postinjection apoptotic megakaryocytes could be detected with cytoplasmic vacuolization (Fig. 9.4a), pyknotic and dysmorphic nucleus, accumulation of short filamentous bundles, and fragmentation of the plasma membrane. Taken together, these studies indicate that the combination of pro-survival Bcl-x_L and Mcl-1 is essential for the viability of the megakaryocytic lineage, while Bcl-2 is dispensable (Fig. 9.3).

9.6.2 Regulation of Megakaryocyte Survival by Extrinsic Apoptosis

In contrast to platelets, megakaryocytes possess a functional FasL-inducible extrinsic apoptosis pathway [27, 97]. Activation of this pathway in mature megakaryocytes in vitro led to membrane blebbing (Fig. 9.4c), reduced viability, effector caspase-3/7 activation, and reduced proplatelet formation. This effect was blocked by genetic *caspase-8* deletion or nonfunctional Fas receptor [27]. Moreover, loss of *Bak*/*Bax* failed to protect FasL-treated megakaryocytes against caspase-3/7 activation or rescue viability and proplatelet formation indicating that megakaryocytes are type I cells, namely, the apoptotic signal mediated by caspase-8 does not require augmentation through the intrinsic apoptosis pathway to induce cell death (Fig. 9.3). In addition to Fas, TRAIL and TRAIL receptor 2 have also been shown to be expressed during human megakaryocyte development [98]. Blockage of the extrinsic apoptosis pathway in mice (megakaryocyte lineage-specific loss of *caspase-8*) did not affect platelet counts, platelet lifespan, or platelet production [27]. Since apoptosis had initially been suggested to promote platelet production [99, 100], investigations were taken to examine the effect of disabling the intrinsic and extrinsic apoptotic pathways in the megakaryocyte lineage by conditional loss of *Bak*/*Bax* and *caspase-8*. Strikingly, platelet production was normal in these triple knockout mice, and the extension of platelet lifespan observed in *Bak*/*Bax* mice was not further prolonged in the absence of *caspase-8* [27]. Thus, apoptosis is dispensable for platelet biogenesis in mice.

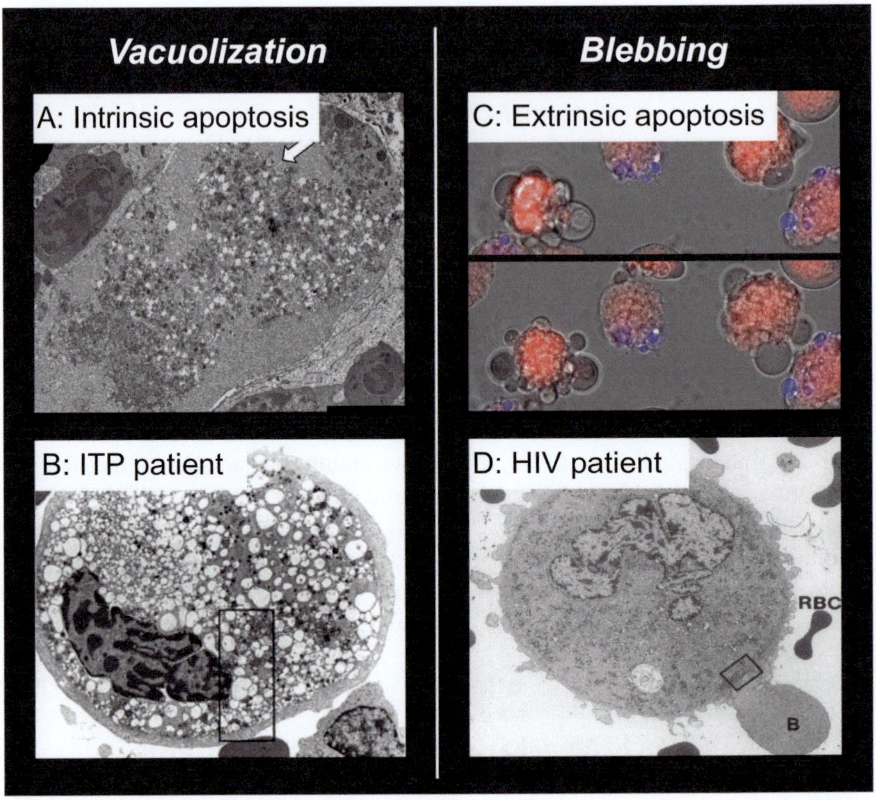

Fig. 9.4 Morphological features of megakaryocyte apoptosis. (**a**) Representative transmission electron microscopy image of a megakaryocyte undergoing intrinsic apoptosis in vivo induced by genetic loss of *Mcl-1* and pharmacological inhibition of Bcl-x_L (ABT-737). Cytoplasmic vacuolization is visible 2.5 h posttreatment (This research was originally published in *Blood*. Debrincat et al. [26]. © the American Society of Hematology). (**b**) Mature megakaryocyte from ITP patient showing ultrastructural features of para-apoptosis. Signs of cytoplasmic vacuolization are evident. The thick, enlarged peripheral margin in the cytoplasm lacks organelles (This research was originally published in *Blood*. Houwerzijl et al. [95]. © the American Society of Hematology). (**c**) Time-lapse microscopy of cultured mouse bone marrow megakaryocytes undergoing extrinsic apoptosis induced by cross-linked FasL. Cell permeable MitoTracker Red FM was used as a viability stain (*red*). Blebbing was followed by decreased red intensity as FasL-treated megakaryocytes died and became positive for the nuclear stain DAPI (*blue*) (© Josefsson et al. 2014. Originally published in *Nature Communications*. doi:10.1038/ncomms4455 [27]) (**d**) A blebbing megakaryocyte from an AIDS patient's marrow (Reprinted from Zucker-Franklin et al. [96]. Copyright 1989, with permission from Elsevier)

9.7 Chemotherapy Treatment Targeting Apoptosis

During chemotherapy treatment, the targeting of Bcl-x_L produces dose-limiting thrombocytopenia which is particularly deleterious in hematological malignancies where thrombocytopenia is a consequence of the disease at the late stage. This led to the development of agents that specifically target Bcl-2. Venetoclax [52] is one such agent that is currently used in clinical trials for treatment of a range of hematological malignancies with little impact on platelet counts [52, 101]. Despite dose-limiting thrombocytopenia, Bcl-x_L antagonism has shown encouraging results in certain solid tumors [102, 103], and Bcl-x_L-specific BH3 mimetics [104–106] were recently developed. As expected, they cause thrombocytopenia and can induce apoptosis in megakaryocytes in vitro [51]. Radiotherapy and myelosuppressive chemotherapy cause thrombocytopenia by directly targeting megakaryocytes and their progenitor cells to undergo apoptosis as evidenced by bone marrow chimeric mice deficient in *Bak*/*Bax* being resistant to carboplatin-induced thrombocytopenia. In light of the developments of new BH3 mimetics targeting different pro-survival proteins, it is important to carefully dissect out the molecular mechanism regulating apoptosis in the megakaryocytic lineage. Combination trials with Bcl-x_L inhibition and kinase inhibitors (MEK, RAF, and BRAF) are underway for advanced or metastatic solid tumors, and it will be imperative to closely monitor platelet counts, since both platelet production and platelet survival may be affected. As evidence indicates, the combined targeting of Mcl-1 and Bcl-x_L is likely to be detrimental to the megakaryocytic lineage [26, 29].

9.8 Implications of Extended Platelet Survival in Disease

In disease settings such as myeloproliferative neoplasms, can the increase in platelet counts due to increased progenitor proliferation be attributed to the overexpression of pro-survival proteins in platelets, which prolongs platelet lifespan? This is an interesting question since as previously described (Sect. 9.4), overexpression of BCL-2 in blood cells [53] or mice deficient in *Bak* and *Bax* [28, 31] have extended platelet lifespan. In megakaryoblastic cells, studies have demonstrated enhanced Bcl-x_L and Mcl-1 expression triggered by TPO through the JAK/STAT5 pathway [29, 107] and the same pro-survival proteins were shown to regulate survival in *JAK2V617F* mutant megakaryoblastic cell lines [108]. Additionally, bone marrow megakaryocytes from essential thrombocythemia patients showed increased Bcl-x_L expression [109]. Conversely, in ITP patients with low platelet counts, can platelet lifespan in circulation be extended therapeutically by administration of TPO mimetics? A study by Mitchell et al. recently invalidated this theory under chronic TPO-mimetic stimulation, but potentially it could be a contributing factor in the early phase of treatment [110]. Clearly, further investigations are needed to validate whether overexpression of pro-survival proteins contributes to increased megakaryocyte survival and/or extended platelet lifespan in disease settings with increased platelet count, and, importantly, whether platelet lifespan can be manipulated therapeutically to counter thrombocytopenia.

9.9 Acquired Thrombocytopenias Associated with Apoptosis

Thrombocytopenia induced by bacteria and viruses has been associated with megakaryocyte and platelet apoptosis. Platelets from dengue-virus-infected patients [111] and from *Escherichia coli* and *Staphylococcus* bacterial-infected sepsis patients [112] exhibit increased signs of apoptosis as do platelets exposed to *Helicobacter pylori* in vitro [113]. One of the mechanisms put forward for the fast clearance of platelets in sepsis is bacterial desialylation of platelet surface proteins and hepatic removal by the Ashwell-Morell receptor [114] (Sect. 9.5). Megakaryocytes have been shown to undergo apoptosis in dengue-virus- [115] and HIV-infected [116] patients, and ultrastructural features of megakaryocytes from AIDS patients have been shown to include membrane blebbing [96] (Fig. 9.4d), one of the hallmarks of apoptosis. Megakaryocytes have also been shown to die after infection with lymphocytic choriomeningitis virus (LCMV) [117], a murine model of HIV. In this model, combined genetic ablation of *caspase-8/Bak/Bax* could protect megakaryocytes, but not platelets [27]. Thus, megakaryocytes undergo apoptosis in LCMV-infected mice.

ITP is an autoimmune disorder characterized by low platelet counts due to accelerated platelet destruction and impaired platelet production. Platelets are targeted by antibodies and cleared by macrophages in the spleen and liver, or they are directly lysed by cytotoxic T cells [118]. Two case reports exist where platelet desialylation had occurred and sialidase inhibition had a positive outcome [119, 120]. Sialidase inhibition was recently put forward as a potential new treatment option for thrombocytopenia [121], and future studies will determine its clinical significance. Houwerzijl et al. classified megakaryocytes from ITP and myelodysplastic syndrome (MDS) patients into apoptotic, para-apoptotic, and necrosis-like based on platelet ultrastuctural characteristics [95, 122]. Figure 9.4b shows a megakaryocyte from an ITP patient with vacuolization of the cytoplasm (termed para-apoptosis) [95], a feature also seen in mouse megakaryocytes undergoing intrinsic apoptosis (Fig. 9.4a).

An increased number of denuded megakaryocyte nuclei have been detected in the bone marrow of AIDS [96] and MDS patients [123] and post-chemotherapy treatment [124]. It was suggested that denuded nuclei, surrounded by a thin ring of cytoplasm, are the remnant products leftover in the bone marrow after functional platelet release into the circulation. In AIDS patients, an accumulation of nuclei was speculated to be due to impaired macrophage clearance [96]. An increased number of bone marrow megakaryocytic denuded nuclei can be observed in mice with megakaryocytes unable to undergo intrinsic apoptosis in the absence of *Bak/Bax* [28] indicative of inefficient clearance and subsequent accumulation. Furthermore, in vitro, denuded nuclei have been shown to become annexin V-positive post-proplatelet formation [28]. Hence, it is conceivable that leftover denuded nuclei undergo apoptosis allowing swift recognition and clearance by macrophages.

9.10 Megakaryocyte and Platelet Survival in Inherited Thrombocytopenias

Congenital platelet disorders (CPDs) are a heterogenous group of conditions. They are rare and may only be described in single or a handful of families. The classification of CPDs is varied; however, more recently, CPDs have been classified based on the molecular classification of these syndromes [125]. Broadly, CPD can manifest as thrombocytopenia or thrombocytopathy (platelet functional defects/abnormalities) leading to impaired clot formation. Unlike thrombocytopenic disorders, platelet functional defects can often occur together with normal platelet counts. In the following section, we will present thrombocytopenic CPDs that have been allied with alteration of platelet lifespan and cell death events in megakaryocytes and/or platelets.

9.10.1 Direct Links to the Apoptotic Machinery

Thrombocytopenia Cargeeg (*Thrombocytopenia-4*) (THC4) is a rare autosomal dominant disease causing mild thrombocytopenia. It was first found to affect a single family in New Zealand for seven generations [126]. THC4 is caused by a G41S mutation of the cytochrome *c* (*CYCS*) gene. Interestingly, despite being a ubiquitously expressed protein, the only clinical phenotype apparent in THC4 patients was related to disrupted thrombopoiesis. In THC4, whereas patients have the same proportion of immature platelets as control patients, the absolute number of immature platelets is lower in affected individuals indicative of a platelet production defect. This was attributed to increased activity of cytochrome c leading to the premature ectopic release of platelets within the bone marrow space [126] with no indication on whether megakaryocyte survival was affected. More recently, a second naturally occurring mutation Y48H in *CYCS* was described in an Italian family by De Rocco and colleagues [127]. Irrespective of which somatic cytochrome *c* mutation, platelet size or morphology was not disrupted, and once again, the only disease phenotype observed was that related to impaired thrombopoiesis. In this study, it was additionally reported that mutations in *CYCS* also changed cellular bioenergetics [127] suggesting a potentially complex mechanism underlying cytochrome *c* mutation and its effects on megakaryocyte maturation and platelet production.

9.10.2 Indications of Increased Death in Megakaryocytes

MYH9-related disorders are autosomal dominantly inherited and encompass disorders that were previously classified as May-Hegglin anomaly, Fechtner syndrome, Sebastian syndrome, as well as Epstein syndrome. MYH9-related disorders are caused by mutations in the *MYH9* gene that encodes non-muscle myosin heavy chain IIa (NMMHC-IIA) [128]. NMMHC-IIA is an important component of the contractile cytoskeleton in platelets, megakaryocytes, as well as other tissues. Therefore, disruption of the *MYH9* leads to perturbation in the interactions of NMMHC-IIA to its binding partners or in the stability of protein itself [129]. While the mechanism behind poor platelet production in MYH9 disorders has not been fully elucidated, increased non-apoptotic cell death in megakaryocytes has been suggested to occur in *Myh9Δ* mice [130]. Meanwhile, platelet lifespan has been reported to be unaltered [131, 132].

Gray platelet syndrome (GPS) is a genetically heterogenous disorder caused by an autosomal recessive or dominant mutation in the *NBEAL2* gene. First described in 1971, platelets appeared large, misshaped, and deficient both of alpha granules and proteins synthesized within megakaryocytes. In the dominant form of GPS, dense granule deficiency has also been documented [133]. Defective alpha granule synthesis in GPS has been suggested to be due to the inability of NBEAL2-deficient megakaryocytes to efficiently package endogenously produced proteins into alpha granules [134]. In *Nbeal2*$^{-/-}$ mice, platelet lifespan was reported to be similar to that of control mice [135]. Interestingly, Kahr et al. has proposed a model in which survival of cultured *Nbeal2*$^{-/-}$ megakaryocytes may be affected thereby providing a potential explanation for the decreased platelet counts in GPS patients [136].

9.10.3 Indications of Increased Platelet Turnover

Filamin A deficiency caused by mutations in the *FLNA* gene can lead to a wide spectrum of diseases with the most prevalent two phenotypes being periventricular heterotopia and otopalatodigital syndromes [137]. Despite thrombocytopenia periodically occurring in filaminopathies, not much follow-up work on platelet survival has been done, and information on platelet size and counts are lacking [138]. In platelets, filamin A (FlnA) plays an important role in interactions between platelets and the vessel wall, particularly under conditions of high shear [139]. Accordingly, patients that have *FlnA* mutations display a variety of functional platelet disorders [140]. By generating *FlnAloxp Pf4-Cre* mice, Begonja et al. showed that FlnA-null megakaryocytes produced large fragile platelets that were rapidly cleared by macrophages leading to thrombocytopenia [141].

Bernard-Soulier syndrome (BSS) is an autosomal recessive inherited platelet disorder. It causes variable degrees of thrombocytopenia and is characterized by macrothrombocytopenia and a prolonged bleeding time [142]. BSS patients exhibit a decreased or complete absence of the GPIb/IX/V complex on the platelet surface resulting in impaired interactions of platelets with vWF and thrombin. Historically it was proposed that the diminished expression of this receptor complex led to the rapid removal of platelets [143, 144] despite platelet lifespan itself not being effected [145]. Indeed, studies using *GPIbβ*$^{-/-}$ mice as a model for BSS showed that platelet lifespan was no different from the control [146]. Collectively, this suggests that the major reason for the thrombocytopenia in BSS is a platelet production defect. However, other studies also showed that the existence of autologous labeled platelets was halved in BSS patients [147], proposing that there might be an additional component of increased platelet turnover. Moreover, newly emerging evidence suggests that apart from a defect in the GPIb/IX/V complex, intrinsic membrane abnormalities are present in BSS patients whereby total platelet phospholipid content was elevated, a notion that correlates with its enlarged size [148, 149]. It was further described that, in contrast to the control, a large proportion of activated platelets from BSS patients did not expose PS despite $\Delta\psi m$ disruption, suggestive of a PS-independent mode of cell death in BSS platelets that could underlie shortened platelet survival in BSS [149].

DiGeorge or velocardiofacial syndrome (DGS/VCFS) is caused by a hemizygous mutation of chromosome 22q11.2 passed down in an autosomal recessive pattern. Clinical features of DGS/VCFS include mild macrothrombocytopenia, cardiac abnormalities, impaired learning abilities, velopharyngeal insufficiency, immunodeficiency, facial dimorphisms, as well as thymic hypoplasia [150]. By and large, the features of this disorder are ascribable to haploinsufficiency of the *TBX1* gene, as evident in *Tbx*$^{-/-}$ mice [151], yet it is not fully clear how platelet counts are affected in this disorder. Given the location of the *TBX1* gene, which lies in relatively close proximity to the *GPIBB* gene, it has been proposed that the deletion of chromosome 22q11.2 may lead to a reduction in GPIβ expression. Indeed, validating this theory, the expression of GPIβ on platelets in thrombocytopenic patients with 22q11.2 mutations is less than that observed in controls [152]. As such, it has been postulated that a similar reduction of platelet lifespan seen in BSS patients may also occur in DGS/VCFS [153], although specific experimental evidence for this is lacking.

Type 2b von Willebrand disease (VWD2) is an autosomal dominant inherited congenital disorder arising from gain-of-function mutations in the *VWF* gene. This disease is characterized by an irregularly increased affinity of vWF to GPIb/IX/V on platelets. As a result, spontaneous platelet binding to vWF is enhanced leading to the rapid clearance of platelet aggregates. This mechanism is supported in mouse models of VWD2 in which macrophages were reported to show accelerated uptake of vWF/platelet complexes [154]. Additionally, studies performed in a small cohort of patients suggest that thrombocytopenia in VWD2 can also be partly due to impaired megakaryopoiesis, and they also reported that platelets from patients showed increased PARP cleavage compared to control [155].

Wiskott-Aldrich syndrome (WAS) is an X-linked disorder caused by hemizygous mutations in the *WAS* gene encoding the WAS protein (WASp). WASp plays a fundamental role in actin polymerization in hematopoietic cells and controls rearrangement of the actin cytoskeleton in response to cellular activation [156]. Functional disruption of WASp causes a myriad of effects leading to microthrombocytopenia, immunodeficiency, recurrent infections, eczema, and malignant lymphoma [157]. X-linked thrombocytopenia (XLT) which occasionally causes mild eczema and infections is a variant of WAS [158]. In WAS, the rapid clearance of platelets in human patients is proposed to be the main mechanism causing thrombocytopenia. This was corroborated in studies of *WASp*$^{-/-}$ mice whereby increased phagocytosis of isolated platelets was observed [159]. Moreover, high levels of antiplatelet antibodies were detected in *WASp*$^{-/-}$ mice; a separate line of evidence of rapid in vivo platelet clearance [160]. Clearly, evidence behind thrombocytopenia in WAS is associated with increased phagocytosis, yet studies also indicate that this may not be the sole contributing factor. For instance, shortened intrinsic platelet survival has been detected in WAS patients [161]. Moreover, mice deficient in the WASp-interacting protein Profilin1 (*Pln*$^{-/-}$) were reported to display a WAS-like phenotype. In *Pln*$^{-/-}$ mice, thrombocytopenia was attributed to hampered platelet production as well as to reduced platelet lifespan [162] demonstrating the complexity behind thrombocytopenia in this syndrome.

9.11 Controversies on the Role of Apoptosis in Thrombopoiesis

It has been described that in the absence of *Bak*/*Bax*, few mice survive into adulthood. These surviving mice have thrombocytopenia in addition to numerous developmental defects and accumulation of excess cells within the hematopoietic and central nervous systems [163]. Hematopoietic loss of *Bak*/*Bax* [28], deletion of *Bim* [164], or overexpression of *BCL-2* [165] in the hematopoietic compartment leads to thrombocytopenia and accumulation of lymphoid cells. The substantial drop in platelet counts was speculated in the late 1990s to be caused by impaired apoptosis thereby preventing efficient platelet shedding from megakaryocytes [164]. A series of publications followed in support of this notion [99, 100, 166], and localized caspase activation was identified during in vitro platelet production [100]. Nevertheless, more recent studies in genetically modified mice have demonstrated that deletion of *Bim*/*Bid* [30] or conditional loss of *Bak*/*Bax* [28, 30] in the megakaryocytic lineage is not associated with thrombocytopenia and splenectomy reverses the thrombocytopenia observed in mice lacking *Bak*/*Bax* in the hematopoietic system [28]. Furthermore, the reduced platelet counts in *vav-BCL-2* mice was lately shown to be largely due to expansion of the lymphoid compartment, since these mice had normal platelet counts on a B and T cell-deficient *Rag-1*$^{-/-}$ background [53]. Combined conditional blockade of intrinsic and extrinsic apoptosis pathways was recently achieved in mice, and platelet production was found to be normal both at steady state and under conditions of stress [27]. Recent advancement in knowledge provided by these new mouse models leaves little support to the previous theory that shedding from megakaryocytes depends on apoptosis. Recently, a IL-1α-dependent megakaryocyte rupture process was identified in response to acute platelet needs [97]. This inflammatory-associated response was shown to be morphologically and temporally different from the typical FasL-induced apoptosis, with no evidence of membrane blebbing, PS exposure, or TUNEL-positive staining, while caspase-3 was activated [97].

Still, there are unresolved questions such as the reason for the abnormal platelet production observed after *Pf4*-driven transgenic overexpression of Bcl-x_L in megakaryocytes and platelets [166]. These mice were described to have normal platelet counts, but a defect in response to stress-induced thrombocytopenia and decreased in vitro proplatelet formation. Interestingly, Bcl-x_L was recently shown to have additional death-unrelated metabolic functions [167]: and when overexpressed, stimulates autophagy [168]. Other examples are the two cytochrome *c* mutations [126, 127] causing a mild thrombocytopenia in patients with intramedullary platelet release [126] (Sect. 9.10). Enhanced intrinsic apoptosis activity was originally hypothesized to be the underlying reason for this abnormality and more recently effects on cellular bioenergetics have also been suggested [127, 169]. Thus, future studies are needed to gain mechanistic insights into how the described conditions influence platelet production.

9.12 Alternative Cell Death Pathways Implicated in Thrombopoiesis

In addition to intrinsic and extrinsic apoptosis, further, alternative cell death pathways might be of relevance in megakaryopoiesis and thrombopoiesis such as necrosis, necroptosis, pyroptosis, parthanatos, and autophagy [79].

Necrosis is not well defined on a molecular level and is mostly based on morphological characteristics including swelling of organelles, increased cell volume, and permeabilization of the plasma membrane. MPT-mediated regulated necrosis requires cyclophilin D, a component of the permeability transition pore complex. Mice deficient in *cyclophilin D* have normal platelet counts but exhibit a platelet functional defect with reduced procoagulant function (Sect. 9.4) [77, 78]. Necroptosis is a newly discovered pathway of programmed necrosis [79], with links to inflammation, that requires the protein receptor-interacting serine/threonine-protein kinase 3 (RIPK3) and mixed lineage kinase domain-like (MLKL) and is induced by death receptors, interferons, toll-like receptors, and intracellular RNA and DNA sensors. Since *Mlkl*-deficient mice displayed no hematopoietic anomalies [170], necroptosis is likely to be dispensable for thrombopoiesis at steady state. Pyroptosis is another inflammatory cell death pathway characterized by cellular swelling and plasma membrane permeabilization. Pyroptosis occurs after inflammasome stimulation, leading to caspase-1 and/or caspase-11 activation. Limited information is available on the role of pyroptosis in megakaryocytes, but dengue infection has been shown to trigger functional inflammasome assembly in platelets [171]. Continuous turnover of intracellular components by autophagy is necessary to preserve cellular homeostasis in all tissues, and there has been a recent spark in the interest surrounding the role of autophagy in megakaryocytes [172] and platelets [173–175]. Since genetic loss of *Atg7* in hematopoietic cells led to macrothrombocytopenia in mice [172], future molecular mechanistic insights into the role of autophagy and regulated necrosis [176] in megakaryopoiesis and platelet production will be of particular interest.

Acknowledgments This work was supported by Project Grants (1079250), Program Grants (1016647), and an Independent Research Institutes Infrastructure Support Scheme Grant (361646) from the Australian National Health and Medical Research Council, a fellowship from the German Research Foundation (I.P. DFG, PL707/1-1), and a Victorian State Government Operational Infrastructure Support Grant.

References

1. Weyrich AS (2014) Platelets: more than a sack of glue. Hematol Educ Prog Am Soc Hematol Am Soc Hematol Educ Prog 2014(1):400–403
2. Kile BT (2014) The role of apoptosis in megakaryocytes and platelets. Br J Haematol 165:217–226
3. Grozovsky R, Begonja AJ, Liu K, Visner G, Hartwig JH, Falet H et al (2015) The Ashwell-Morell receptor regulates hepatic thrombopoietin production via JAK2-STAT3 signaling. Nat Med 21(1):47–54
4. Michel M (2013) Immune thrombocytopenia nomenclature, consensus reports, and guidelines: what are the consequences for daily practice and clinical research? Semin Hematol 50(Suppl 1):S50–S54
5. Kuter DJ (2011) Romiplostim. Cancer Treat Res 157:267–288
6. Schafer AI (2001) Thrombocytosis and thrombocythemia. Blood Rev 15(4):159–166
7. Hoffmeister KM, Josefsson EC, Isaac NA, Clausen H, Hartwig JH, Stossel TP (2003) Glycosylation restores survival of chilled blood platelets. Science (New York, NY) 301(5639):1531–1534
8. Li J, Xia Y, Bertino AM, Coburn JP, Kuter DJ (2000) The mechanism of apoptosis in human platelets during storage. Transfusion 40(11):1320–1329

9. Cazenave JP, Isola H, Waller C, Mendel I, Kientz D, Laforet M et al (2011) Use of additive solutions and pathogen inactivation treatment of platelet components in a regional blood center: impact on patient outcomes and component utilization during a 3-year period. Transfusion 51(3):622–629
10. Ohto H, Nollet KE (2011) Overview on platelet preservation: better controls over storage lesion. Transfus Apher Sci Off J World Apher Assoc Off J Eur Soc Haemapheresis 44(3):321–325
11. Bertino AM, Qi XQ, Li J, Xia Y, Kuter DJ (2003) Apoptotic markers are increased in platelets stored at 37 degrees C. Transfusion 43(7):857–866
12. Hartley PS, Savill J, Brown SB (2006) The death of human platelets during incubation in citrated plasma involves shedding of CD42b and aggregation of dead platelets. Thromb Haemost 95(1):100–106
13. Cohen JA, Leeksma CH (1956) Determination of the life span of human blood platelets using labelled diisopropylfluorophosphonate. J Clin Invest 35(9):964–969
14. Hirsch EO, Gardner FH (1952) The transfusion of human blood platelets with a note on the transfusion of granulocytes. J Lab Clin Med 39(4):556–569
15. Leeksma CH, Cohen JA (1955) The life-span of thrombocytes determined with radioactive DFP. Ned Tijdschr Geneeskd 99(47):3563–3565
16. Slichter SJ, Bolgiano D, Corson J, Jones MK, Christoffel T (2010) Extended storage of platelet-rich plasma-prepared platelet concentrates in plasma or plasmalyte. Transfusion 50(10):2199–2209
17. Ault KA, Knowles C (1995) In vivo biotinylation demonstrates that reticulated platelets are the youngest platelets in circulation. Exp Hematol 23(9):996–1001
18. Dale GL, Friese P, Hynes LA, Burstein SA (1995) Demonstration that thiazole-orange-positive platelets in the dog are less than 24 hours old. Blood 85(7):1822–1825
19. Beer JH, Buchi L, Steiner B (1994) Glycocalicin: a new assay – the normal plasma levels and its potential usefulness in selected diseases. Blood 83(3):691–702
20. Kurata Y, Hayashi S, Kiyoi T, Kosugi S, Kashiwagi H, Honda S et al (2001) Diagnostic value of tests for reticulated platelets, plasma glycocalicin, and thrombopoietin levels for discriminating between hyperdestructive and hypoplastic thrombocytopenia. Am J Clin Pathol 115(5):656–664
21. Michelson AD, Barnard MR, Hechtman HB, MacGregor H, Connolly RJ, Loscalzo J et al (1996) In vivo tracking of platelets: circulating degranulated platelets rapidly lose surface P-selectin but continue to circulate and function. Proc Natl Acad Sci U S A 93(21):11877–11882
22. Baker GR, Sullam PM, Levin J (1997) A simple, fluorescent method to internally label platelets suitable for physiological measurements. Am J Hematol 56(1):17–25
23. Pleines I, Eckly A, Elvers M, Hagedorn I, Eliautou S, Bender M et al (2010) Multiple alterations of platelet functions dominated by increased secretion in mice lacking Cdc42 in platelets. Blood 115(16):3364–3373
24. Josefsson EC, White MJ, Dowling MR, Kile BT (2012) Platelet life span and apoptosis. Methods Mol Biol 788:59–71
25. Coupland LA, Cromer D, Davenport MP, Parish CR (2010) A novel fluorescent-based assay reveals that thrombopoietin signaling and Bcl-X(L) influence, respectively, platelet and erythrocyte lifespans. Exp Hematol 38(6):453–461 e451
26. Debrincat MA, Josefsson EC, James C, Henley KJ, Ellis S, Lebois M et al (2012) Mcl-1 and Bcl-x(L) coordinately regulate megakaryocyte survival. Blood 119(24):5850–5858
27. Josefsson EC, Burnett DL, Lebois M, Debrincat MA, White MJ, Henley KJ et al (2014) Platelet production proceeds independently of the intrinsic and extrinsic apoptosis pathways. Nat Commun 5:3455
28. Josefsson EC, James C, Henley KJ, Debrincat MA, Rogers KL, Dowling MR et al (2011) Megakaryocytes possess a functional intrinsic apoptosis pathway that must be restrained to survive and produce platelets. J Exp Med 208(10):2017–2031

29. Kodama T, Hikita H, Kawaguchi T, Shigekawa M, Shimizu S, Hayashi Y et al (2012) Mcl-1 and Bcl-xL regulate Bak/Bax-dependent apoptosis of the megakaryocytic lineage at multistages. Cell Death Differ 19(11):1856–1869
30. Kodama T, Takehara T, Hikita H, Shimizu S, Shigekawa M, Li W et al (2011) BH3-only activator proteins Bid and Bim are dispensable for Bak/Bax-dependent thrombocyte apoptosis induced by Bcl-xL deficiency: molecular requisites for the mitochondrial pathway to apoptosis in platelets. J Biol Chem 286(16):13905–13913
31. Mason KD, Carpinelli MR, Fletcher JI, Collinge JE, Hilton AA, Ellis S et al (2007) Programmed anuclear cell death delimits platelet life span. Cell 128(6):1173–1186
32. Youle RJ, Strasser A (2008) The BCL-2 protein family: opposing activities that mediate cell death. Nat Rev Mol Cell Biol 9(1):47–59
33. Kerr JF, Wyllie AH, Currie AR (1972) Apoptosis: a basic biological phenomenon with wide-ranging implications in tissue kinetics. Br J Cancer 26(4):239–257
34. Kuwana T, Bouchier-Hayes L, Chipuk JE, Bonzon C, Sullivan BA, Green DR et al (2005) BH3 domains of BH3-only proteins differentially regulate Bax-mediated mitochondrial membrane permeabilization both directly and indirectly. Mol Cell 17(4):525–535
35. Ren D, Tu HC, Kim H, Wang GX, Bean GR, Takeuchi O et al (2010) BID, BIM, and PUMA are essential for activation of the BAX- and BAK-dependent cell death program. Science (New York, NY) 330(6009):1390–1393
36. Villunger A, Labi V, Bouillet P, Adams J, Strasser A (2011) Can the analysis of BH3-only protein knockout mice clarify the issue of 'direct versus indirect' activation of Bax and Bak? Cell Death Differ 18(10):1545–1546
37. Willis SN, Fletcher JI, Kaufmann T, van Delft MF, Chen L, Czabotar PE et al (2007) Apoptosis initiated when BH3 ligands engage multiple Bcl-2 homologs, not Bax or Bak. Science (New York, NY) 315(5813):856–859
38. Scaffidi C, Fulda S, Srinivasan A, Friesen C, Li F, Tomaselli KJ et al (1998) Two CD95 (APO-1/Fas) signaling pathways. EMBO J 17(6):1675–1687
39. Jost PJ, Grabow S, Gray D, McKenzie MD, Nachbur U, Huang DC et al (2009) XIAP discriminates between type I and type II FAS-induced apoptosis. Nature 460(7258): 1035–1039
40. Schoenwaelder SM, Yuan Y, Josefsson EC, White MJ, Yao Y, Mason KD et al (2009) Two distinct pathways regulate platelet phosphatidylserine exposure and procoagulant function. Blood 114(3):663–666
41. Zhang H, Nimmer PM, Tahir SK, Chen J, Fryer RM, Hahn KR et al (2007) Bcl-2 family proteins are essential for platelet survival. Cell Death Differ 14(5):943–951
42. Wagner KU, Claudio E, Rucker EB 3rd, Riedlinger G, Broussard C, Schwartzberg PL et al (2000) Conditional deletion of the Bcl-x gene from erythroid cells results in hemolytic anemia and profound splenomegaly. Development (Camb Engl) 127(22):4949–4958
43. Kodama T, Takehara T, Hikita H, Shimizu S, Li W, Miyagi T et al (2010) Thrombocytopenia exacerbates cholestasis-induced liver fibrosis in mice. Gastroenterology 138(7):2487–2498, 2498 e2481–2487
44. Oltersdorf T, Elmore SW, Shoemaker AR, Armstrong RC, Augeri DJ, Belli BA et al (2005) An inhibitor of Bcl-2 family proteins induces regression of solid tumours. Nature 435(7042):677–681
45. Tse C, Shoemaker AR, Adickes J, Anderson MG, Chen J, Jin S et al (2008) ABT-263: a potent and orally bioavailable Bcl-2 family inhibitor. Cancer Res 68(9):3421–3428
46. Vogler M, Hamali HA, Sun XM, Bampton ET, Dinsdale D, Snowden RT et al (2011) BCL2/BCL-X(L) inhibition induces apoptosis, disrupts cellular calcium homeostasis, and prevents platelet activation. Blood 117(26):7145–7154
47. Roberts AW, Seymour JF, Brown JR, Wierda WG, Kipps TJ, Khaw SL et al (2012) Substantial susceptibility of chronic lymphocytic leukemia to BCL2 inhibition: results of a phase I study of navitoclax in patients with relapsed or refractory disease. J Clin Oncol Off J Am Soc Clin Oncol 30(5):488–496

48. Schoenwaelder SM, Jarman KE, Gardiner EE, Hua M, Qiao J, White MJ et al (2011) Bcl-xL-inhibitory BH3 mimetics can induce a transient thrombocytopathy that undermines the hemostatic function of platelets. Blood 118(6):1663–1674
49. Gieger C, Radhakrishnan A, Cvejic A, Tang W, Porcu E, Pistis G et al (2011) New gene functions in megakaryopoiesis and platelet formation. Nature 480(7376):201–208
50. Oh JH, Kim YK, Moon S, Kim YJ, Kim BJ (2014) Genome-wide association study identifies candidate Loci associated with platelet count in koreans. Genomics Inform 12(4):225–230
51. Debrincat MA, Pleines I, Lebois M, Lane RM, Holmes ML, Corbin J et al (2015) BCL-2 is dispensable for thrombopoiesis and platelet survival. Cell Death Dis 6:e1721
52. Souers AJ, Leverson JD, Boghaert ER, Ackler SL, Catron ND, Chen J et al (2013) ABT-199, a potent and selective BCL-2 inhibitor, achieves antitumor activity while sparing platelets. Nat Med 19(2):202–208
53. Vandenberg CJ, Josefsson EC, Campbell KJ, James C, Lawlor KE, Kile BT et al (2014) Loss of Bak enhances lymphocytosis but does not ameliorate thrombocytopaenia in BCL-2 transgenic mice. Cell Death Differ 21(5):676–684
54. Print CG, Loveland KL, Gibson L, Meehan T, Stylianou A, Wreford N et al (1998) Apoptosis regulator bcl-w is essential for spermatogenesis but appears otherwise redundant. Proc Natl Acad Sci U S A 95(21):12424–12431
55. Ross AJ, Waymire KG, Moss JE, Parlow AF, Skinner MK, Russell LD et al (1998) Testicular degeneration in Bclw-deficient mice. Nat Genet 18(3):251–256
56. Hamasaki A, Sendo F, Nakayama K, Ishida N, Negishi I, Nakayama K et al (1998) Accelerated neutrophil apoptosis in mice lacking A1-a, a subtype of the bcl-2-related A1 gene. J Exp Med 188(11):1985–1992
57. Ottina E, Grespi F, Tischner D, Soratroi C, Geley S, Ploner A et al (2012) Targeting antiapoptotic A1/Bfl-1 by in vivo RNAi reveals multiple roles in leukocyte development in mice. Blood 119(25):6032–6042
58. Kelly PN, White MJ, Goschnick MW, Fairfax KA, Tarlinton DM, Kinkel SA et al (2010) Individual and overlapping roles of BH3-only proteins Bim and Bad in apoptosis of lymphocytes and platelets and in suppression of thymic lymphoma development. Cell Death Differ 17(10):1655–1664
59. Liu ZJ, Hoffmeister KM, Hu Z, Mager DE, Ait-Oudhia S, Debrincat MA et al (2014) Expansion of the neonatal platelet mass is achieved via an extension of platelet lifespan. Blood 123(22):3381–3389
60. Delbridge AR, Chappaz S, Ritchie ME, Kile BT, Strasser A, Grabow S. (2016) Loss of PUMA (BBC3) does not prevent thrombocytopenia caused by the loss of BCL-XL (BCL2L1). Br J Haematol. doi: 10.1111/bjh.14155. [Epub ahead of print]
61. Plenchette S, Moutet M, Benguella M, N'Gondara JP, Guigner F, Coffe C et al (2001) Early increase in DcR2 expression and late activation of caspases in the platelet storage lesion. Leukemia 15(10):1572–1581
62. Vucic D, Dixit VM, Wertz IE (2011) Ubiquitylation in apoptosis: a post-translational modification at the edge of life and death. Nat Rev Mol Cell Biol 12(7):439–452
63. Lien LM, Su CC, Hsu WH, Lu WJ, Chung CL, Yen TL et al (2013) Mechanisms of andrographolide-induced platelet apoptosis in human platelets: regulatory roles of the extrinsic apoptotic pathway. Phytother Res PTR 27(11):1671–1677
64. Mutlu A, Gyulkhandanyan AV, Freedman J, Leytin V (2012) Activation of caspases-9, -3 and -8 in human platelets triggered by BH3-only mimetic ABT-737 and calcium ionophore A23187: caspase-8 is activated via bypass of the death receptors. Br J Haematol 159(5): 565–571
65. van Delft MF, Smith DP, Lahoud MH, Huang DC, Adams JM (2010) Apoptosis and non-inflammatory phagocytosis can be induced by mitochondrial damage without caspases. Cell Death Differ 17(5):821–832
66. White MJ, Schoenwaelder SM, Josefsson EC, Jarman KE, Henley KJ, James C et al (2012) Caspase-9 mediates the apoptotic death of megakaryocytes and platelets, but is dispensable for their generation and function. Blood 119(18):4283–4290

67. White MJ, McArthur K, Metcalf D, Lane RM, Cambier JC, Herold MJ et al (2014) Apoptotic caspases suppress mtDNA-induced STING-mediated type I IFN production. Cell 159(7): 1549–1562
68. Bevers EM, Comfurius P, Zwaal RF (1983) Changes in membrane phospholipid distribution during platelet activation. Biochim Biophys Acta 736(1):57–66
69. Dale GL (2005) Coated-platelets: an emerging component of the procoagulant response. J Thromb Haemost JTH 3(10):2185–2192
70. Kulkarni S, Jackson SP (2004) Platelet factor XIII and calpain negatively regulate integrin alphaIIbbeta3 adhesive function and thrombus growth. J Biol Chem 279(29): 30697–30706
71. Bevers EM, Williamson PL (2010) Phospholipid scramblase: an update. FEBS Lett 584(13):2724–2730
72. Sanyal S, Menon AK (2009) Flipping lipids: why an' what's the reason for? ACS Chem Biol 4(11):895–909
73. Keuren JF, Wielders SJ, Ulrichts H, Hackeng T, Heemskerk JW, Deckmyn H et al (2005) Synergistic effect of thrombin on collagen-induced platelet procoagulant activity is mediated through protease-activated receptor-1. Arterioscler Thromb Vasc Biol 25(7):1499–1505
74. Braun A, Vogtle T, Varga-Szabo D, Nieswandt B (2011) STIM and Orai in hemostasis and thrombosis. Front Biosci 16:2144–2160
75. Gilio K, van Kruchten R, Braun A, Berna-Erro A, Feijge MA, Stegner D et al (2010) Roles of platelet STIM1 and Orai1 in glycoprotein VI- and thrombin-dependent procoagulant activity and thrombus formation. J Biol Chem 285(31):23629–23638
76. Grosse J, Braun A, Varga-Szabo D, Beyersdorf N, Schneider B, Zeitlmann L et al (2007) An EF hand mutation in Stim1 causes premature platelet activation and bleeding in mice. J Clin Invest 117(11):3540–3550
77. Jobe SM, Wilson KM, Leo L, Raimondi A, Molkentin JD, Lentz SR et al (2008) Critical role for the mitochondrial permeability transition pore and cyclophilin D in platelet activation and thrombosis. Blood 111(3):1257–1265
78. Jackson SP, Schoenwaelder SM (2010) Procoagulant platelets: are they necrotic? Blood 116(12):2011–2018
79. Pasparakis M, Vandenabeele P (2015) Necroptosis and its role in inflammation. Nature 517(7534):311–320
80. Suzuki J, Umeda M, Sims PJ, Nagata S (2010) Calcium-dependent phospholipid scrambling by TMEM16F. Nature 468(7325):834–838
81. Toti F, Satta N, Fressinaud E, Meyer D, Freyssinet JM (1996) Scott syndrome, characterized by impaired transmembrane migration of procoagulant phosphatidylserine and hemorrhagic complications, is an inherited disorder. Blood 87(4):1409–1415
82. Castoldi E, Collins PW, Williamson PL, Bevers EM (2011) Compound heterozygosity for 2 novel TMEM16F mutations in a patient with Scott syndrome. Blood 117(16):4399–4400
83. Yang H, Kim A, David T, Palmer D, Jin T, Tien J et al (2012) TMEM16F forms a Ca2+-activated cation channel required for lipid scrambling in platelets during blood coagulation. Cell 151(1):111–122
84. Nagata S, Hanayama R, Kawane K (2010) Autoimmunity and the clearance of dead cells. Cell 140(5):619–630
85. Suzuki J, Denning DP, Imanishi E, Horvitz HR, Nagata S (2013) Xk-related protein 8 and CED-8 promote phosphatidylserine exposure in apoptotic cells. Science (New York, NY) 341(6144):403–406
86. Hoffmeister KM (2011) The role of lectins and glycans in platelet clearance. J Thromb Haemost JTH 9(Suppl 1):35–43
87. Soslau G, Giles J (1982) The loss of sialic acid and its prevention in stored human platelets. Thromb Res 26(6):443–455
88. Jansen AJ, Josefsson EC, Rumjantseva V, Liu QP, Falet H, Bergmeier W et al (2012) Desialylation accelerates platelet clearance after refrigeration and initiates GPIbalpha metalloproteinase-mediated cleavage in mice. Blood 119(5):1263–1273

89. Gasic GJ, Gasic TB, Stewart CC (1968) Antimetastatic effects associated with platelet reduction. Proc Natl Acad Sci U S A 61(1):46–52
90. Greenberg JP, Packham MA, Guccione MA, Rand ML, Reimers HJ, Mustard JF (1979) Survival of rabbit platelets treated in vitro with chymotrypsin, plasmin, trypsin, or neuraminidase. Blood 53(5):916–927
91. Stenberg PE, Levin J, Baker G, Mok Y, Corash L (1991) Neuraminidase-induced thrombocytopenia in mice: effects on thrombopoiesis. J Cell Physiol 147(1):7–16
92. Rumjantseva V, Grewal PK, Wandall HH, Josefsson EC, Sorensen AL, Larson G et al (2009) Dual roles for hepatic lectin receptors in the clearance of chilled platelets. Nat Med 15(11):1273–1280
93. Sorensen AL, Rumjantseva V, Nayeb-Hashemi S, Clausen H, Hartwig JH, Wandall HH et al (2009) Role of sialic acid for platelet life span: exposure of beta-galactose results in the rapid clearance of platelets from the circulation by asialoglycoprotein receptor-expressing liver macrophages and hepatocytes. Blood 114(8):1645–1654
94. Kotze HF, van Wyk V, Badenhorst PN, Heyns AD, Roodt JP, Lotter MG (1993) Influence of platelet membrane sialic acid and platelet-associated IgG on ageing and sequestration of blood platelets in baboons. Thromb Haemost 70(4):676–680
95. Houwerzijl EJ, Blom NR, van der Want JJ, Esselink MT, Koornstra JJ, Smit JW et al (2004) Ultrastructural study shows morphologic features of apoptosis and para-apoptosis in megakaryocytes from patients with idiopathic thrombocytopenic purpura. Blood 103(2): 500–506
96. Zucker-Franklin D, Termin CS, Cooper MC (1989) Structural changes in the megakaryocytes of patients infected with the human immune deficiency virus (HIV-1). Am J Pathol 134(6):1295–1303
97. Nishimura S, Nagasaki M, Kunishima S, Sawaguchi A, Sakata A, Sakaguchi H et al (2015) IL-1alpha induces thrombopoiesis through megakaryocyte rupture in response to acute platelet needs. J Cell Biol 209(3):453–466
98. Melloni E, Secchiero P, Celeghini C, Campioni D, Grill V, Guidotti L et al (2005) Functional expression of TRAIL and TRAIL-R2 during human megakaryocytic development. J Cell Physiol 204(3):975–982
99. Clarke MC, Savill J, Jones DB, Noble BS, Brown SB (2003) Compartmentalized megakaryocyte death generates functional platelets committed to caspase-independent death. J Cell Biol 160(4):577–587
100. De Botton S, Sabri S, Daugas E, Zermati Y, Guidotti JE, Hermine O et al (2002) Platelet formation is the consequence of caspase activation within megakaryocytes. Blood 100(4):1310–1317
101. Vandenberg CJ, Cory S (2013) ABT-199, a new Bcl-2-specific BH3 mimetic, has in vivo efficacy against aggressive Myc-driven mouse lymphomas without provoking thrombocytopenia. Blood 121:2285–2288
102. Gandhi L, Camidge DR, Ribeiro de Oliveira M, Bonomi P, Gandara D, Khaira D et al (2011) Phase I study of navitoclax (ABT-263), a novel Bcl-2 family inhibitor, in patients with small-cell lung cancer and other solid tumors. J Clin Oncol Off J Am Soc Clin Oncol 29(7):909–916
103. Rudin CM, Hann CL, Garon EB, Ribeiro de Oliveira M, Bonomi PD, Camidge DR et al (2012) Phase II study of single-agent navitoclax (ABT-263) and biomarker correlates in patients with relapsed small cell lung cancer. Clin Cancer Res Off J Am Assoc Cancer Res 18(11):3163–3169
104. Koehler MF, Bergeron P, Choo EF, Lau K, Ndubaku C, Dudley D et al (2014) Structure-guided rescaffolding of selective antagonists of BCL-XL. ACS Med Chem Lett 5(6): 662–667
105. Lessene G, Czabotar PE, Sleebs BE, Zobel K, Lowes KN, Adams JM et al (2013) Structure-guided design of a selective BCL-X(L) inhibitor. Nat Chem Biol 9(6):390–397
106. Tao ZF, Hasvold L, Wang L, Wang X, Petros AM, Park CH et al (2014) Discovery of a potent and selective BCL-XL inhibitor with in vivo activity. ACS Med Chem Lett 5(10):1088–1093

107. Kirito K, Watanabe T, Sawada K, Endo H, Ozawa K, Komatsu N (2002) Thrombopoietin regulates Bcl-xL gene expression through Stat5 and phosphatidylinositol 3-kinase activation pathways. J Biol Chem 277(10):8329–8337
108. Gozgit JM, Bebernitz G, Patil P, Ye M, Parmentier J, Wu J et al (2008) Effects of the JAK2 inhibitor, AZ960, on Pim/BAD/BCL-xL survival signaling in the human JAK2 V617F cell line SET-2. J Biol Chem 283(47):32334–32343
109. Florena AM, Tripodo C, Di Bernardo A, Iannitto E, Guarnotta C, Porcasi R et al (2009) Different immunophenotypical apoptotic profiles characterise megakaryocytes of essential thrombocythaemia and primary myelofibrosis. J Clin Pathol 62(4):331–338
110. Mitchell WB, Pinheiro MP, Boulad N, Kaplan D, Edison MN, Psaila B et al (2014) Effect of thrombopoietin receptor agonists on the apoptotic profile of platelets in patients with chronic immune thrombocytopenia. Am J Hematol 89(12):E228–E234
111. Alonzo MT, Lacuesta TL, Dimaano EM, Kurosu T, Suarez LA, Mapua CA et al (2012) Platelet apoptosis and apoptotic platelet clearance by macrophages in secondary dengue virus infections. J Infect Dis 205(8):1321–1329
112. Kraemer BF, Campbell RA, Schwertz H, Franks ZG, Vieira de Abreu A, Grundler K et al (2012) Bacteria differentially induce degradation of Bcl-xL, a survival protein, by human platelets. Blood 120(25):5014–5020
113. Yeh JJ, Tsai S, Wu DC, Wu JY, Liu TC, Chen A (2010) P-selectin-dependent platelet aggregation and apoptosis may explain the decrease in platelet count during Helicobacter pylori infection. Blood 115(21):4247–4253
114. Grewal PK, Uchiyama S, Ditto D, Varki N, Le DT, Nizet V et al (2008) The Ashwell receptor mitigates the lethal coagulopathy of sepsis. Nat Med 14(6):648–655
115. Clark KB, Noisakran S, Onlamoon N, Hsiao HM, Roback J, Villinger F et al (2012) Multiploid CD61+ cells are the pre-dominant cell lineage infected during acute dengue virus infection in bone marrow. PLoS ONE 7(12):e52902
116. Zauli G, Catani L, Gibellini D, Re MC, Vianelli N, Colangeli V et al (1996) Impaired survival of bone marrow GPIIb/IIa+ megakaryocytic cells as an additional pathogenetic mechanism of HIV-1-related thrombocytopenia. Br J Haematol 92(3):711–717
117. Binder D, Fehr J, Hengartner H, Zinkernagel RM (1997) Virus-induced transient bone marrow aplasia: major role of interferon-alpha/beta during acute infection with the noncytopathic lymphocytic choriomeningitis virus. J Exp Med 185(3):517–530
118. Olsson B, Andersson PO, Jernas M, Jacobsson S, Carlsson B, Carlsson LM et al (2003) T-cell-mediated cytotoxicity toward platelets in chronic idiopathic thrombocytopenic purpura. Nat Med 9(9):1123–1124
119. Shao L, Wu Y, Zhou H, Qin P, Ni H, Peng J et al (2014) Successful treatment with oseltamivir phosphate in a patient with chronic immune thrombocytopenia positive for anti-GPIb/IX autoantibody. Platelets 28:1–3
120. Alioglu B, Tasar A, Ozen C, Selver B, Dallar Y (2010) An experience of oseltamivir phosphate (tamiflu) in a pediatric patient with chronic idiopathic thrombocytopenic purpura: a case report. Pathophysiol Haemost Thromb 37(2–4):55–58
121. Jansen AJ, Peng J, Zhao HG, Hou M, Ni H (2015) Sialidase inhibition to increase platelet counts: a new treatment option for thrombocytopenia. Am J Hematol 90:E94–E95
122. Houwerzijl EJ, Blom NR, van der Want JJ, Vellenga E, de Wolf JT (2006) Megakaryocytic dysfunction in myelodysplastic syndromes and idiopathic thrombocytopenic purpura is in part due to different forms of cell death. Leukemia 20(11):1937–1942
123. Hatfill SJ, Fester ED, Steytler JG (1992) Apoptotic megakaryocyte dysplasia in the myelodysplastic syndromes. Hematol Pathol 6(2):87–93
124. Radley JM, Haller CJ (1983) Fate of senescent megakaryocytes in the bone marrow. Br J Haematol 53(2):277–287
125. Freson K, Wijgaerts A, van Geet C (2014) Update on the causes of platelet disorders and functional consequences. Int J Lab Hematol 36(3):313–325
126. Morison IM, Cramer Borde EM, Cheesman EJ, Cheong PL, Holyoake AJ, Fichelson S et al (2008) A mutation of human cytochrome c enhances the intrinsic apoptotic pathway but causes only thrombocytopenia. Nat Genet 40(4):387–389

127. De Rocco D, Cerqua C, Goffrini P, Russo G, Pastore A, Meloni F et al (2014) Mutations of cytochrome c identified in patients with thrombocytopenia THC4 affect both apoptosis and cellular bioenergetics. Biochim Biophys Acta 1842(2):269–274
128. Kunishima S, Kojima T, Tanaka T, Kamiya T, Ozawa K, Nakamura Y et al (1999) Mapping of a gene for May-Hegglin anomaly to chromosome 22q. Hum Genet 105(5):379–383
129. Seri M, Cusano R, Gangarossa S, Caridi G, Bordo D, Lo Nigro C et al (2000) Mutations in MYH9 result in the May-Hegglin anomaly, and Fechtner and Sebastian syndromes. The May-Heggllin/Fechtner syndrome consortium. Nat Genet 26(1):103–105
130. Eckly A, Rinckel JY, Laeuffer P, Cazenave JP, Lanza F, Gachet C et al (2010) Proplatelet formation deficit and megakaryocyte death contribute to thrombocytopenia in Myh9 knockout mice. J Thromb Haemost JTH 8(10):2243–2251
131. Hamilton RW, Shaikh BS, Ottie JN, Storch AE, Saleem A, White JG (1980) Platelet function, ultrastructure, and survival in the May-Hegglin anomaly. Am J Clin Pathol 74(5):663–668
132. Leon C, Eckly A, Hechler B, Aleil B, Freund M, Ravanat C et al (2007) Megakaryocyte-restricted MYH9 inactivation dramatically affects hemostasis while preserving platelet aggregation and secretion. Blood 110(9):3183–3191
133. Raccuglia G (1971) Gray platelet syndrome. A variety of qualitative platelet disorder. Am J Med 51(6):818–828
134. Breton-Gorius J, Vainchenker W, Nurden A, Levy-Toledano S, Caen J (1981) Defective alpha-granule production in megakaryocytes from gray platelet syndrome: ultrastructural studies of bone marrow cells and megakaryocytes growing in culture from blood precursors. Am J Pathol 102(1):10–19
135. Deppermann C, Cherpokova D, Nurden P, Schulz JN, Thielmann I, Kraft P et al (2013) Gray platelet syndrome and defective thrombo-inflammation in Nbeal2-deficient mice. J Clin Invest. pii: 69210. doi:10.1172/JCI69210.
136. Kahr WH, Lo RW, Li L, Pluthero FG, Christensen H, Ni R et al (2013) Abnormal megakaryocyte development and platelet function in Nbeal2(-/-) mice. Blood 122(19):3349–3358
137. Robertson SP, Twigg SR, Sutherland-Smith AJ, Biancalana V, Gorlin RJ, Horn D et al (2003) Localized mutations in the gene encoding the cytoskeletal protein filamin A cause diverse malformations in humans. Nat Genet 33(4):487–491
138. Nurden P, Debili N, Coupry I, Bryckaert M, Youlyouz-Marfak I, Sole G et al (2011) Thrombocytopenia resulting from mutations in filamin A can be expressed as an isolated syndrome. Blood 118(22):5928–5937
139. Cranmer SL, Ashworth KJ, Yao Y, Berndt MC, Ruggeri ZM, Andrews RK et al (2011) High shear-dependent loss of membrane integrity and defective platelet adhesion following disruption of the GPIbalpha-filamin interaction. Blood 117(9):2718–2727
140. Berrou E, Adam F, Lebret M, Fergelot P, Kauskot A, Coupry I et al (2013) Heterogeneity of platelet functional alterations in patients with filamin A mutations. Arterioscler Thromb Vasc Biol 33(1):e11–e18
141. Jurak Begonja A, Hoffmeister KM, Hartwig JH, Falet H (2011) FlnA-null megakaryocytes prematurely release large and fragile platelets that circulate poorly. Blood 118(8):2285–2295
142. Bernard J, Soulier JP [Not Available] La semaine des hopitaux: organe fonde par l'Association d'enseignement medical des hopitaux de Paris 1948;24(Spec No):3217–3223
143. Grottum KA, Solum NO (1969) Congenital thrombocytopenia with giant platelets: a defect in the platelet membrane. Br J Haematol 16(3):277–290
144. Cullum C, Cooney DP, Schrier SL (1967) Familial thrombocytopenic thrombocytopathy. Br J Haematol 13(2):147–159
145. Heyns Adu P, Badenhorst PN, Wessels P, Pieters H, Lotter MG (1985) Kinetics, in vivo redistribution and sites of sequestration of indium-111-labelled platelets in giant platelet syndromes. Br J Haematol 60(2):323–330
146. Strassel C, Eckly A, Leon C, Petitjean C, Freund M, Cazenave JP et al (2009) Intrinsic impaired proplatelet formation and microtubule coil assembly of megakaryocytes in a mouse model of Bernard-Soulier syndrome. Haematologica 94(6):800–810

147. Tomer A, Scharf RE, McMillan R, Ruggeri ZM, Harker LA (1994) Bernard-Soulier syndrome: quantitative characterization of megakaryocytes and platelets by flow cytometric and platelet kinetic measurements. Eur J Haematol 52(4):193–200
148. Perret BP, Plantavid M, Chap H, Douste-Blazy L (1983) Are polyphosphoinositides involved in platelet activation? Biochem Biophys Res Commun 110(2):660–667
149. Rand ML, Wang H, Bang KW, Teitel JM, Blanchette VS, Freedman J et al (2010) Phosphatidylserine exposure and other apoptotic-like events in Bernard-Soulier syndrome platelets. Am J Hematol 85(8):584–592
150. Ryan AK, Goodship JA, Wilson DI, Philip N, Levy A, Seidel H et al (1997) Spectrum of clinical features associated with interstitial chromosome 22q11 deletions: a European collaborative study. J Med Genet 34(10):798–804
151. Jerome LA, Papaioannou VE (2001) DiGeorge syndrome phenotype in mice mutant for the T-box gene, Tbx1. Nat Genet 27(3):286–291
152. Kato T, Kosaka K, Kimura M, Imamura S, Yamada O, Iwai K et al (2003) Thrombocytopenia in patients with 22q11.2 deletion syndrome and its association with glycoprotein Ib-beta. Genetics Med Off J Am Coll Med Genetics 5(2):113–119
153. Van Geet C, Devriendt K, Eyskens B, Vermylen J, Hoylaerts MF (1998) Velocardiofacial syndrome patients with a heterozygous chromosome 22q11 deletion have giant platelets. Pediatr Res 44(4):607–611
154. Casari C, Du V, Wu YP, Kauskot A, de Groot PG, Christophe OD et al (2013) Accelerated uptake of VWF/platelet complexes in macrophages contributes to VWD type 2B-associated thrombocytopenia. Blood 122(16):2893–2902
155. Nurden P, Debili N, Vainchenker W, Bobe R, Bredoux R, Corvazier E et al (2006) Impaired megakaryocytopoiesis in type 2B von Willebrand disease with severe thrombocytopenia. Blood 108(8):2587–2595
156. Thrasher AJ (2009) New insights into the biology of Wiskott-Aldrich syndrome (WAS). Hematol Educ Prog Am Soc Hematol Am Soc Hematol Educ Prog 1:132–138
157. Aldrich RA, Steinberg AG, Campbell DC (1954) Pedigree demonstrating a sex-linked recessive condition characterized by draining ears, eczematoid dermatitis and bloody diarrhea. Pediatrics 13(2):133–139
158. Zhu Q, Zhang M, Blaese RM, Derry JM, Junker A, Francke U et al (1995) The Wiskott-Aldrich syndrome and X-linked congenital thrombocytopenia are caused by mutations of the same gene. Blood 86(10):3797–3804
159. Prislovsky A, Marathe B, Hosni A, Bolen AL, Nimmerjahn F, Jackson CW et al (2008) Rapid platelet turnover in WASP(-) mice correlates with increased ex vivo phagocytosis of opsonized WASP(-) platelets. Exp Hematol 36(5):609–623
160. Marathe BM, Prislovsky A, Astrakhan A, Rawlings DJ, Wan JY, Strom TS (2009) Antiplatelet antibodies in WASP(-) mice correlate with evidence of increased in vivo platelet consumption. Exp Hematol 37(11):1353–1363
161. Murphy S (1972) Intrinsic platelet defects in hereditary thrombocytopenia. Ann N Y Acad Sci 201:421–428
162. Bender M, Stritt S, Nurden P, van Eeuwijk JM, Zieger B, Kentouche K et al (2014) Megakaryocyte-specific profilin1-deficiency alters microtubule stability and causes a Wiskott-Aldrich syndrome-like platelet defect. Nat Commun 5:4746
163. Lindsten T, Ross AJ, King A, Zong WX, Rathmell JC, Shiels HA et al (2000) The combined functions of proapoptotic Bcl-2 family members bak and bax are essential for normal development of multiple tissues. Mol Cell 6(6):1389–1399
164. Bouillet P, Metcalf D, Huang DC, Tarlinton DM, Kay TW, Kontgen F et al (1999) Proapoptotic Bcl-2 relative Bim required for certain apoptotic responses, leukocyte homeostasis, and to preclude autoimmunity. Science (New York, NY) 286(5445):1735–1738
165. Ogilvy S, Metcalf D, Print CG, Bath ML, Harris AW, Adams JM (1999) Constitutive Bcl-2 expression throughout the hematopoietic compartment affects multiple lineages and enhances progenitor cell survival. Proc Natl Acad Sci U S A 96(26):14943–14948

166. Kaluzhny Y, Yu G, Sun S, Toselli PA, Nieswandt B, Jackson CW et al (2002) BclxL overexpression in megakaryocytes leads to impaired platelet fragmentation. Blood 100(5):1670–1678
167. Michels J, Kepp O, Senovilla L, Lissa D, Castedo M, Kroemer G et al (2013) Functions of BCL-X L at the interface between cell death and metabolism. Int J Cell Biol 2013:705294
168. Priault M, Hue E, Marhuenda F, Pilet P, Oliver L, Vallette FM (2010) Differential dependence on Beclin 1 for the regulation of pro-survival autophagy by Bcl-2 and Bcl-xL in HCT116 colorectal cancer cells. PLoS ONE 5(1):e8755
169. Liptak MD, Fagerlund RD, Ledgerwood EC, Wilbanks SM, Bren KL (2011) The proapoptotic G41S mutation to human cytochrome c alters the heme electronic structure and increases the electron self-exchange rate. J Am Chem Soc 133(5):1153–1155
170. Murphy JM, Czabotar PE, Hildebrand JM, Lucet IS, Zhang JG, Alvarez-Diaz S et al (2013) The pseudokinase MLKL mediates necroptosis via a molecular switch mechanism. Immunity 39(3):443–453
171. Hottz ED, Lopes JF, Freitas C, Valls-de-Souza R, Oliveira MF, Bozza MT et al (2013) Platelets mediate increased endothelium permeability in dengue through NLRP3-inflammasome activation. Blood 122(20):3405–3414
172. Cao Y, Cai J, Zhang S, Yuan N, Li X, Fang Y et al (2015) Loss of autophagy leads to failure in megakaryopoiesis, megakaryocyte differentiation, and thrombopoiesis in mice. Exp Hematol 43:488–494
173. Feng W, Chang C, Luo D, Su H, Yu S, Hua W et al (2014) Dissection of autophagy in human platelets. Autophagy 10(4):642–651
174. Maugeri N, Campana L, Gavina M, Covino C, De Metrio M, Panciroli C et al (2014) Activated platelets present high mobility group box 1 to neutrophils, inducing autophagy and promoting the extrusion of neutrophil extracellular traps. J Thromb Haemost JTH 12(12):2074–2088
175. Pieczarka EM, Yamaguchi M, Wellman ML, Radin MJ (2014) Platelet vacuoles in a dog with severe nonregenerative anemia: evidence of platelet autophagy. Vet Clin Pathol Am Soc Vet Clin Pathol 43(3):326–329
176. Vanden Berghe T, Linkermann A, Jouan-Lanhouet S, Walczak H, Vandenabeele P (2014) Regulated necrosis: the expanding network of non-apoptotic cell death pathways. Nat Rev Mol Cell Biol 15(2):135–147

Platelet Functions Beyond Hemostasis 10

Rick Kapur and John W. Semple

Abstract

Platelets are small cellular fragments lacking a nucleus, derived from megakaryocytes, and are well known to have a major role in maintaining hemostasis. Apart from this well-established role, it is now becoming evident that platelets also have other important functions, besides hemostasis, during infection and inflammation. This chapter will focus on these nonhemostatic functions of platelets, in general, outlined as "platelets versus pathogens" and "platelet-target cell communication." Platelets actively contribute to protection against invading pathogens and are capable of regulating immune functions in various target cells, all through an array of sophisticated mechanisms. These relatively novel features will be discussed, demonstrating an important multifunctional role of platelets in an inflammatory setting.

R. Kapur
Toronto Platelet Immunobiology Group,
Keenan Research Centre for Biomedical Science, Toronto, ON, Canada

Canadian Blood Services, St. Michael's Hospital,
30 Bond Street, Toronto, ON M5B 1W8, Canada

J.W. Semple (✉)
Toronto Platelet Immunobiology Group,
Keenan Research Centre for Biomedical Science, Toronto, ON, Canada

Canadian Blood Services, St. Michael's Hospital,
30 Bond Street, Toronto, ON M5B 1W8, Canada

Departments of Pharmacology, University of Toronto, Toronto, ON, Canada

Departments of Medicine, University of Toronto, Toronto, ON, Canada

Departments of Laboratory Medicine and Pathobiology,
University of Toronto, Toronto, ON, Canada
e-mail: semplej@smh.ca

H. Schulze, J. Italiano (eds.), *Molecular and Cellular Biology of Platelet Formation*,
DOI 10.1007/978-3-319-39562-3_10

10.1 Introduction

Platelets are small (~2–4 μm in diameter) anucleate cellular fragments derived from bone marrow-resident megakaryocytes (~50–100 μm in diameter) and are classically known to be indispensable for hemostasis [1]. Platelet generation is a complex and highly regulated process [2], especially considering the multifunctional aspects of platelets. The importance of this tight regulation is demonstrated in vivo, where it is essential to prevent bleedings under conditions when platelet counts are low, but it is also imperative to prevent serious organ damage and vascular occlusion due to elevated platelet counts. It is becoming increasingly clear that platelets are involved in several other functional processes, apart from their well-known role in hemostasis, in both health and disease [1, 3–9]. These nonhemostatic functions, generally outlined as "platelets versus pathogens" and "platelet-target cell communication," will be highlighted in this chapter and will shed light on the versatile role of platelets in an inflammatory setting.

10.2 Platelets Versus Pathogens

10.2.1 Pathogen Recognition by Platelets

Platelets have the ability to "sense danger," such as pathogens or damage in case of sterile inflammation, as they express functional immune receptors called pattern recognition receptors. These include Ig- or complement receptors and Toll-like receptors (TLRs) [1]. Via these receptors, platelets are able to bind invading pathogens and microbes, including their derived materials. Pathogens first encounter TLRs on professional phagocytes, such as neutrophils, dendritic cells (DCs), or macrophages [1, 10, 11]. TLRs are germ line-encoded proteins, capable of binding various infectious molecular structures and potently stimulate innate immune mechanisms [1, 10, 11]. Many groups have reported the expression of TLRs 1–9 on human as well as on murine platelets, and some of these TLRs have been shown to be functional [1]. For instance, TLR4, which has been shown to mediate lipopolysaccharide (LPS, a gram-negative endotoxin)-induced thrombocytopenia and TNF-α production in vivo [12–18]. Engagement of TLR2 in human platelets by Pam_3CSK_4, a synthetic ligand mimicking bacterial lipopeptide, was shown to induce a thromboinflammatory response through activation of phosphoinositide 3-kinase [19]. Platelet TLR2 and TLR4 were also shown to be of importance in a study regarding periodontitis, which is associated with an increased risk for cardiovascular diseases [20]. The periodontopathogens involved (*A. actinomycetemcomitans* and *P. gingivalis*) were shown to induce expression of CD40L, known to mediate thrombotic and inflammatory processes, on human platelets via TLR2 and TLR4 [20]. Recently, new insights have emerged regarding platelet TLR3 and TRLR7, as human platelets expressing TLR3 were shown to respond to poly I:C, indicating an effect on innate immune responses when detecting viral dsRNA [21]. Platelet TLR7 was shown to mediate host survival and platelet counts during infection with encephalomyocarditis virus

(EMCV) in mice, independently of thrombosis [22]. In contrast, platelet TLR9 seems to be important as sensor of internal danger signals, rather than external signals. It was shown that platelet TLR9 was functionally associated with oxidative stress, innate immunity, and thrombosis, as carboxy(alkylpyrrole) protein adducts, altered-self ligands produced during oxidative stress, through engagement of platelet TLR9, could enhance platelet activation, granule secretion, aggregation in vitro, and thrombosis in vivo [23]. Not many studies have been conducted investigating the role of TLRs on megakaryocytes. It was shown that endotoxemia could increase thrombopoietin (TPO) levels in vivo, accompanied by an increase in circulating young reticulated platelets with enhanced platelet-neutrophil aggregates [24]. In addition, the bone marrow treated with LPS demonstrated increased TPO levels, implicating a key role of infection and inflammation in thrombopoiesis [25]. Also, TLR4-knockout mice displayed decreased circulating platelet counts and reticulated platelets, suggesting TLR4 to be of importance in platelet production [12, 26]. A lot of evidence has accumulated over the years suggesting an important role for platelets as pathogen sensors within the blood, due to their expression of several receptors without any clear relation to hemostatic functions.

10.2.2 Pathogen Retainment by Platelets

Platelets are capable of harboring pathogens on their plasma membrane as well as internally [4, 27], as has been demonstrated for viruses [27, 28], bacteria [29–31], and parasites [5]. Platelets were also shown to be involved in acute and chronic hepatic disease due to hepatitis B virus, via upregulation of virus-specific CD8+ T cells and nonspecific inflammatory cells into the liver [32]. Interestingly, activated platelets were shown to surround or encapsulate *Staphylococcus aureus*, driving the pathogens into clusters resulting in reduced bacterial growth [33]. This occurred through secretion of the antimicrobial peptide β-defensin and signaling of neutrophil extracellular trap (NET) formation [33], which has now shown to be involved in many pathologies including thrombosis, transfusion-related acute lung injury, sickle cell disease, storage of red blood cells, and very recently diabetes [34–40]. Bacteria (methicillin-resistant *Staphylococcus aureus* and *Bacillus cereus*) were also trapped on the hepatic Kupffer cells, which were dependent upon interactions with platelet-adhesion receptor GP1b [6]. In that study, infected GP1bα-deficient mice suffered more endothelial cell and Kupffer cell damage and displayed more vascular leakage and rapid mortality [6]. Activation of platelets during sepsis can contribute to disseminated intravascular coagulation, which can lead to blood vessel occlusion, increased ischemia, and multiple organ failure, and it can also contribute to stimulation of pro- and anti-inflammatory cytokine production [41]. This platelet activation is evident from increased surface P-selectin expression [42, 43] and increased levels of triggering receptor expressed on myeloid cells-like transcript-1 [44] or PF-4 in mice [45]. During sepsis, neutrophils were also shown to be activated by platelet TLR4, causing the release of NETs, which subsequently trapped bacteria in blood vessels primarily of liver sinusoids and lung capillaries [7]. It was proposed that platelets act as circulating sentinels, sensing infectious agents and presenting them to neutrophils and/or the reticuloendothelial system [14–18]. Although platelet-dependent NET formation is an effective strategy in trapping bacteria, it may be detrimental as it may occur at the expense of injury to the host. When LPS-activated neutrophils come into contact with endothelium, there is little damage; however, if bound neutrophils encounter LPS-bearing platelets, neutrophil activation is enhanced resulting in NET formation together with reactive oxygen species release, which damages the underlying endothelium [7]. Furthermore, it was demonstrated that neutrophils are able to scan platelets for activation in the bloodstream via P-selectin ligand signaling, resulting in inflammation [8]. Also, platelet P-selectin, soluble or cellular, was found to trigger NET formation in mice through binding to neutrophil P-selectin glycoprotein ligand-1 (PSGL-1) [46]. Recently, in diabetes, in which neutrophils are more susceptible to NET formation, NETs were found to impair wound healing. Therefore, it was suggested that cleaving NETs or inhibiting NET formation may improve wound healing and reduce inflammation in diabetes [40].

10.2.3 Pathogen Elimination by Platelets

Platelets have been implicated to be involved in the clearance of bacterial infections. In infective endocarditis, for instance, thrombin-stimulated platelets were shown to facilitate clearance of streptococci [47]. Additionally, in mice infected with *P. gingivalis*, platelet TLR2 was implicated in the formation of platelet-neutrophil aggregates [19], and later it was demonstrated that phagocytosis of periodontopathogens mediated by neutrophils was dependent upon platelets, plasma factors, and TLR2 [48]. It was elegantly shown by McMorran and colleagues that activated platelets can kill the malarial parasite *Plasmodium falciparum* inside the red blood cell [5]. In a follow-up study, it was further elucidated that the mechanism of this platelet-mediated parasite killing was dependent upon platelet factor 4 (PF4 or CXCL4) and the erythrocyte Duffy-antigen receptor (Fy) [49]. This implies that in Duffy-negative individuals, thus lacking Fy, platelets would be incapable of eliminating this intraerythrocytic malarial parasite.

10.2.4 Pathogen Escape from Platelets

On the other hand, viruses and bacteria appear to have developed countermeasures to evade these immune responses elicited by platelets. This can be supported by the fact that acute viral or bacterial infections often lead to low platelet counts or thrombocytopenia. This has frequently been observed in immune thrombocytopenia (ITP), an autoimmune bleeding disease in which platelets are destroyed [50]. The pathogenesis of infection-related platelet targeting is incompletely understood, but several mechanisms have been described. These include molecular mimicry between viral/bacterial antigens and platelet antigens, resulting in cross-reactive autoantibody generation [51–55]. In addition, ITP patients infected with the gram-negative bacteria *Helicobacter pylori* demonstrated increased platelet counts following *Helicobacter pylori*-eradication therapy [56]. Similarly, the gram-negative bacterial endotoxin LPS also enhanced antiplatelet antibody-mediated platelet phagocytosis in vitro [15], as well as an increased platelet clearance in vivo, when antiplatelet antibodies and LPS were coinjected in mice [57]. Furthermore, C-reactive protein (CRP) was found to be a novel serum factor which potentiated antibody-mediated platelet destruction both in vitro as well as in vivo in mice [58]. CRP is an acute-phase protein, also present in healthy individuals, but known to vastly increase during acute bacterial or viral infections. CRP was found to be increased in children suffering from ITP and treatment with IVIg was correlated with increased platelet counts, with decreased levels of CRP, and with reduced clinical bleeding severity [58]. Interestingly, an elevated CRP value at diagnosis appeared to be predictive for slower platelet count recovery after 3 months [58]. From a functional perspective, the mechanism appeared to be independent of the platelet FcγRIIA but occurred via platelet oxidation triggered by antiplatelet antibodies and the phagocyte NADPH oxidase system, resulting in platelet-membrane phosphorylcholine exposure, to which CRP could bind and then subsequently enhance antibody-mediated platelet phagocytosis through interaction with FcγR [58].

10.3 Platelet-Target Cell Communication

10.3.1 Platelet Release of Mediators

CD40L (CD40L/CD154) and CD40, besides their role in costimulation and perhaps thrombotic diseases [59], are also relevant in platelet immune reactions. When platelets are activated, most of their expressed CD40L is released, generating the soluble form (sCD40L), which is in fact the vast majority of all sCD40L in circulation [60]. Platelet CD40L can engage with endothelial cell-CD40 (in the membrane), which results in a cascade of inflammatory reactions leading to the release of several adhesion molecules including VCAM1, ICAM1, and CCL2 [61]. Platelet-secreted sCD40L, when interacting with CD40 positive vascular cells (including endothelial cells), can enhance the expression of adhesion molecules like P-selectin and E-selectin and stimulate the release of tissue factor and IL-6 [62, 63]. Therefore, a central role for platelet CD40L-CD40 interactions between endothelium/coagulation and inflammation is becoming apparent. Additionally, platelet-derived CD40L was shown to enhance CD8+ T cell responses and to stimulate T cell responses following infection with *Listeria monocytogenes* [64, 65], demonstrating a clear link between innate and adaptive immunity. Platelet CD40L was also shown to bind to dendritic cells (DCs) and thereby impair their differentiation, suppress the pro-inflammatory cytokines IL-12p70 and TNF by DCs, and increase IL-10 production by DCs [66]. Furthermore, activated platelets were shown to enhance lymphocyte adhesion to endothelial cells [67] and facilitate homing of lymphocytes in high endothelial venules [68] and migration toward inflammatory environments. Also, platelets can enable B cell differentiation and Ab class switching via their CD40L [69, 70]. Several other signaling pathways have been linked to platelet activation via the CD40L-CD40 axis, including NF-kB [71–76], illustrating that platelets have in fact several strategies for modulating adaptive immune responses through their CD40L and/or their derived sCD40L.

Platelets have many different cytokines and chemokines in their system, all differently impacting hemostasis and wound repair [77], and also pro-inflammatory and anti-inflammatory reactions, for instance, the immunosuppressant TGF-β [78]. Platelets appear to control the levels of TGF-β as is evident from patients suffering from immune thrombocytopenia (ITP), where low levels of TGF-β were observed during active disease, on one hand, but those levels normalized again upon treatment which increased platelet counts, on the other hand [79, 80]. Most of the platelet chemokines and cytokines are located within the different platelet granules. The α granules contain several immunomodulatory soluble factors, like chemokines, which included PF (CXCL4), RANTES (CCL5), β-thromboglobulin (β-TG, an isoform of CXCL7), and MIP-1α (CCL3) [81]. Platelet activation triggers release of these chemokines causing a diverse response of cellular interactions and responses. PF-4, for instance, renders monocytes resistant to apoptosis and stimulates their differentiation into macrophages [82]. Besides that, PF-4 is capable of enhancing neutrophil adhesion to unstimulated endothelial cells and of release of granule content [83]. In contrast, platelet-derived β-TGs, which are proteolytic products of inactive precursors, can either stimulate or inhibit neutrophil activity [84]. Also, platelet-derived MIP-1α can enhance histamine release from basophils [85] and is chemotactic for T cells [86].

10.3.2 Platelet Microparticle Shedding

Platelet microparticles (also referred to as microvesicles) are small extracellular vesicles produced by cell cytoplasmic blebbing and fission. Originally, they were described as "dust" released from activated platelets which supported thrombin generation, even without the presence of intact platelets [87]. Generally, the size of microparticles ranges from ~100 to 1,000 nm in diameter, although the majority are ~200 nm, and they are distinct from exosomes, which are ~50–100 nm in diameter and thus smaller in size and originating from multivesicular bodies via exocytosis [88]. The minimal experimental requirements for the definitions of extracellular vesicles and their functions are described by the International Society for Extracellular Vesicles [89]. Platelets appear to be particularly effective in the formation of microparticles, as compared to other cell types, as was demonstrated by the high abundance of platelet microparticles in circulation using cryotransmission electron microscopy and gold nanospheres conjugated to antibodies against the platelet CD41 [90]. Microparticle formation is associated with elevated intracellular calcium levels, cytoskeletal rearrangement, and membrane phosphatidylserine (PS) exposure [91], which supports coagulation considering its anionic properties, but platelet microparticles express modest levels of tissue factor (TF) and seem to be less procoagulant than monocyte-derived microparticles, which express PS as well as TF [92]. Examination of platelet activation under physiological flow conditions revealed elongated membrane strands (up to 250 μM) emerging from platelets, so-called flow-induced protrusions (FLIPRs) [93]. FLIPRs also expose PS, recruit monocytes and neutrophils, and appear to shed off PS+ microparticles [93]. Interestingly, PS-microparticles have also been described in body fluids [90, 94–96], demonstrating the complexity of platelet microparticle production. It has been challenging to properly assess microparticles in biological fluids, due to their small dimensions. Platelet microparticles have been described in various inflammatory conditions, in which platelets become activated [97, 98], and their clinically their levels were often associated with disease progression. For example, in blood and synovial fluid of patients suffering from rheumatoid arthritis (RA), platelet microparticles were found to be elevated [9, 99–102]. Using a murine model of RA, it was demonstrated that depletion of platelets attenuates inflammation [9, 103]. However, microparticles are observed in sterile, as well as in inflammatory diseases, making it unclear what triggers the platelets to produce microparticles. Several activation pathways may be driving the production of microparticles during inflammation, such as apoptosis, high shear forces, or platelet receptor signaling. The disease setting, at least partly, determines the route leading to microparticle formation, as in RA activation of the collagen receptor glycoprotein VI (GPVI) is a trigger, while in sepsis microparticles are produced via TLR-4 signaling through LPS [9, 104]. Both these signals, however, are accompanied by an increase in IL-1, indicating their role in enhancing inflammation. Additionally, signaling of immune complexes, consisting of bacterial components and well-conserved epitopes expressed by influenza viruses, through the platelet FcγRIIA [105, 106], was shown to lead to the production of microparticles.

From a functional perspective, platelet microparticles are thought to facilitate cell-cell communication. The platelet microparticle cargo is substantial and can consist of various cytokines and chemokines (e.g., IL-1, RANTES), potent lipid mediators (e.g., thromboxane A_2), enzymes (e.g., inducible NO synthase), surface receptors (e.g., CD40L), autoantigens (e.g., citrullinated fibrinogen), nucleic acids (e.g., microRNA), transcription factors (e.g., PPARγ, RuvB-like2, STAT3, STAT5a), and interestingly even respiratory competent mitochondria, all of them potentially targeting and impacting a cell [96–98, 107–110]. As the microparticles can express PS and surface receptors, they interact with other cells through integrin and via the PS-binding proteins lactadherin [111] and developmental endothelial locus-1 (Del-1) [112]. These proteins appear to be involved in microparticle clearance and microparticle interaction with other cells, as Del-1 -/- and lactadherin -/- mice express elevated levels of plasma microparticles [111, 112]. Transcription factors transported within platelet microparticles can enable transcellular effects, such as PPARγ, which was demonstrated to be transported inside platelet microparticles and transferred to monocytes where it elicited transcellular effects [109]. Currently, however, more research is required to establish if specific internalization signals are required beyond the initial contact between microparticles and the cellular recipient. Microparticles appear to be important biomarkers in inflammatory disorders, but, further, delineation of their function, mechanisms of their generation, and technical improvements in their assessment are warranted, in order to better understand their role in health and disease.

10.3.3 Platelet RNA Transfer

Platelets are known to express and secrete many different molecules during platelet activation, and they do so via different signaling mechanisms [81, 113–115]. These molecules have different origins, such as those inherited from megakaryocytes, adsorbed from plasma or synthesized de novo. Despite being anucleate, platelets have been shown to express significant amounts of RNA, including mRNAs (e.g., (pre)mature RNA), structural and catalytic RNAs (e.g., ribosomal and tRNA), regulatory RNAs (e.g., microRNA), and noncoding RNA (e.g., antisense RNA) [107, 116–131]. Moreover, it was described that platelets also possess the molecular machinery for mRNA translation into proteins and that they are able to transfer RNA to recipient cells, in order to regulate cellular functions, such as platelet microRNA-223 transfer to human umbilical vein endothelial cells [107, 127–129, 131]. As mentioned before, intercellular transfer of platelet RNA to target cells can occur via platelet microparticles. However, the content of platelet RNA transcript does not fully match to the platelet proteome content [132]. These molecular tools have opened up a new area of investigation of platelet mRNA and its impact on platelet function in both health and disease [133].

10.3.4 Platelet MHC Class I Signaling

Major histocompatibility complex (MHC) class I molecules are present on both the platelet plasma membrane and intracellularly [134]. The MHC class I molecules on the platelet plasma membrane are mainly adsorbed from plasma and basically consist of denatured H chains. The platelet-membrane MHC class I molecules appear to be somewhat instable, as they can passively dissociate from the platelet during storage or can be eluted from the membrane due to chloroquine diphosphate or acid treatment, without affecting the platelet membrane integrity [135–141]. Interestingly, denatured MHC class I can elicit faulty interactions with CD8+ T cells, anergizing Cytotoxic T-lymphocytes (CTLs), following transfusions. For instance, allogeneic platelet MHC class I molecules are incapable of stimulating CTL-mediated cytotoxicity on their own [141] but can facilitate the so-called transfusion effect, an immunosuppressive-like reaction to transfused blood cells. CBA mice transfused with allogeneic BALB/c platelets accepted donor-specific skin grafts, in contrast to nontransfused recipients [142]. This implies that allogeneic platelets may inhibit T cell-mediated cytotoxicity reactions, like skin graft rejections. On the other hand, intracellular platelet MHC class I molecules are associated with α granules and generally consist of intact integral membrane proteins associated with β2-microglobulin [143]. Furthermore, it was also demonstrated that platelets contain the entire proteasome system, including TAP molecules, but the endoplasmic reticulum is absent. In syngeneic settings, platelet activation can lead to expression of nascent MHC class I molecules, which are capable of presenting antigens to CD8+ T cells. Activated platelets were shown to present malarial peptides to malaria-specific T cells, resulting in enhanced immunity against the parasite [144]. Therefore, the type of platelet MHC class I (platelet plasma membrane-bound or intracellularly) will determine the effect on T cells (suppression or activation).

A summary of the key nonhemostatic functions of platelets is summarized in Fig. 10.1.

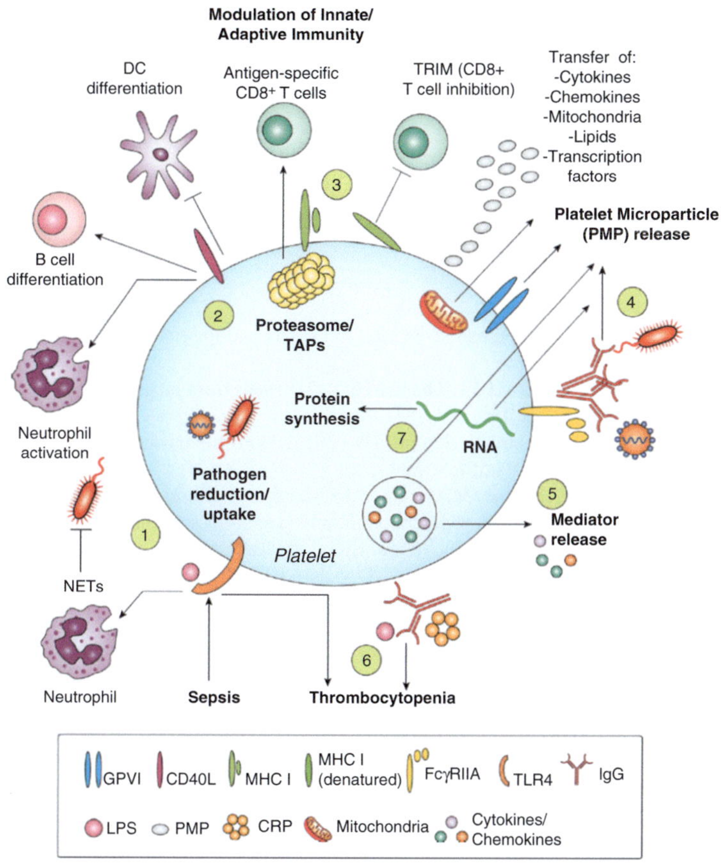

Fig. 10.1 The key roles of platelets in modulating inflammatory processes. **1** Platelets can uptake infectious agents, and, via the expression of TLR, they can activate neutrophils to, for example, secrete NETs. **2** Platelet CD40L expression allows them to interact with different cells of the immune system and either activate (*arrows*) and/or suppress (T bar) them. **3** Intact platelet MHC class I molecules are located intracellularly but upon activation are expressed and can activate antigen (e.g., malaria)-specific CD8+ T cells. On the other hand, the MHC class I molecules on the surface of resting platelet are denatured and lead to CD8+ T cell inhibition. **4** Platelets release Platelet Microparticles (PMPs) under a variety of stress conditions and these PMPs can carry multiple cargos to other cells and sites of inflammation. **5** Platelets contain many different pro-inflammatory and anti-inflammatory cytokines and chemokines and upon activation can release them to the extracellular space. **6** Immune interactions with platelets can lead to severe thrombocytopenic states such as in the case of sepsis, where infections can bind to platelets and cause their sequestration and/or destruction or immune thrombocytopenia (ITP) where the combination of antibody and infectious particles or CRP leads to increased platelet destruction. **7** Platelets contain several different species of RNA and these can be exported via PMPs of mRNAs can be translated into nascent protein synthesis. The culmination of these events makes platelets a formidable immunomodulatory host (Reproduced from reference [3])

Conclusions

Platelets have traditionally been viewed as primary regulators of hemostasis and thrombin generation. However, it has become increasingly clear that platelets have multiple functions in inflammation and immunity. Platelets can not only enforce sophisticated protection mechanisms against invading pathogens but are also capable of impacting immune functions in a large variety of recipient cells. They do so by utilizing numerous mechanisms, via diverse surface molecules, through secretion of several pro- and inflammatory mediators and shedding of platelet microparticles carrying a heterogeneous cargo. These relatively novel aspects have shed new light on platelet functions beyond hemostasis.

References

1. Semple JW, Italiano JE Jr, Freedman J (2011) Platelets and the immune continuum. Nat Rev Immunol 11:264–274
2. Machlus KR, Italiano JE Jr (2013) The incredible journey: from megakaryocyte development to platelet formation. J Cell Biol 201:785–796
3. Kapur R, Zufferey A, Boilard E, Semple JW (2015) Nouvelle cuisine: platelets served with inflammation. J Immunol 194:5579–5587
4. Youssefian T, Drouin A, Masse JM, Guichard J, Cramer EM (2002) Host defense role of platelets: engulfment of HIV and Staphylococcus aureus occurs in a specific subcellular compartment and is enhanced by platelet activation. Blood 99:4021–4029
5. McMorran BJ, Marshall VM, de Graaf C, Drysdale KE, Shabbar M, Smyth GK, Corbin JE, Alexander WS, Foote SJ (2009) Platelets kill intraerythrocytic malarial parasites and mediate survival to infection. Science 323:797–800
6. Wong CH, Jenne CN, Petri B, Chrobok NL, Kubes P (2013) Nucleation of platelets with blood-borne pathogens on Kupffer cells precedes other innate immunity and contributes to bacterial clearance. Nat Immunol 14:785–792
7. Clark SR, Ma AC, Tavener SA, McDonald B, Goodarzi Z, Kelly MM, Patel KD, Chakrabarti S, McAvoy E, Sinclair GD, Keys EM, Allen-Vercoe E, Devinney R, Doig CJ, Green FH, Kubes P (2007) Platelet TLR4 activates neutrophil extracellular traps to ensnare bacteria in septic blood. Nat Med 13:463–469
8. Sreeramkumar V, Adrover JM, Ballesteros I, Cuartero MI, Rossaint J, Bilbao I, Nacher M, Pitaval C, Radovanovic I, Fukui Y, McEver RP, Filippi MD, Lizasoain I, Ruiz-Cabello J, Zarbock A, Moro MA, Hidalgo A (2014) Neutrophils scan for activated platelets to initiate inflammation. Science 346:1234–1238
9. Boilard E, Nigrovic PA, Larabee K, Watts GF, Coblyn JS, Weinblatt ME, Massarotti EM, Remold-O'Donnell E, Farndale RW, Ware J, Lee DM (2010) Platelets amplify inflammation in arthritis via collagen-dependent microparticle production. Science 327:580–583
10. Janeway CA Jr (1992) The immune system evolved to discriminate infectious nonself from noninfectious self. Immunol Today 13:11–16
11. Janeway CA Jr, Medzhitov R (2002) Innate immune recognition. Annu Rev Immunol 20:197–216
12. Andonegui G, Kerfoot SM, McNagny K, Ebbert KV, Patel KD, Kubes P (2005) Platelets express functional Toll-like receptor-4. Blood 106:2417–2423
13. Cognasse F, Hamzeh H, Chavarin P, Acquart S, Genin C, Garraud O (2005) Evidence of Toll-like receptor molecules on human platelets. Immunol Cell Biol 83:196–198
14. Aslam R, Speck ER, Kim M, Crow AR, Bang KW, Nestel FP, Ni H, Lazarus AH, Freedman J, Semple JW (2006) Platelet Toll-like receptor expression modulates lipopolysaccharide-induced thrombocytopenia and tumor necrosis factor-alpha production in vivo. Blood 107:637–641
15. Semple JW, Aslam R, Kim M, Speck ER, Freedman J (2007) Platelet-bound lipopolysaccharide enhances Fc receptor-mediated phagocytosis of IgG-opsonized platelets. Blood 109:4803–4805
16. Patrignani P, Di FC, Tacconelli S, Moretta V, Baccante G, Sciulli MG, Ricciotti E, Capone ML, Antonucci I, Guglielmi MD, Stuppia L, Porreca E (2006) Reduced thromboxane biosynthesis in carriers of toll-like receptor 4 polymorphisms in vivo. Blood 107:3572–3574

17. Stahl AL, Svensson M, Morgelin M, Svanborg C, Tarr PI, Mooney JC, Watkins SL, Johnson R, Karpman D (2006) Lipopolysaccharide from enterohemorrhagic Escherichia coli binds to platelets through TLR4 and CD62 and is detected on circulating platelets in patients with hemolytic uremic syndrome. Blood 108:167–176
18. Zhang G, Han J, Welch EJ, Ye RD, Voyno-Yasenetskaya TA, Malik AB, Du X, Li Z (2009) Lipopolysaccharide stimulates platelet secretion and potentiates platelet aggregation via TLR4/MyD88 and the cGMP-dependent protein kinase pathway. J Immunol 182:7997–8004
19. Blair P, Rex S, Vitseva O, Beaulieu L, Tanriverdi K, Chakrabarti S, Hayashi C, Genco CA, Iafrati M, Freedman JE (2009) Stimulation of Toll-like receptor 2 in human platelets induces a thromboinflammatory response through activation of phosphoinositide 3-kinase. Circ Res 104:346–354
20. Assinger A, Laky M, Badrnya S, Esfandeyari A, Volf I (2012) Periodontopathogens induce expression of CD40L on human platelets via TLR2 and TLR4. Thromb Res 130:e73–e78
21. Anabel AS, Eduardo PC, Pedro Antonio HC, Carlos SM, Juana NM, Honorio TA, Nicolas VS, Sergio Roberto AR (2014) Human platelets express toll-like receptor 3 and respond to poly I:C. Hum Immunol 75:1244–1251
22. Koupenova M, Vitseva O, MacKay CR, Beaulieu LM, Benjamin EJ, Mick E, Kurt-Jones EA, Ravid K, Freedman JE (2014) Platelet-TLR7 mediates host survival and platelet count during viral infection in the absence of platelet-dependent thrombosis. Blood 124:791–802
23. Panigrahi S, Ma Y, Hong L, Gao D, West XZ, Salomon RG, Byzova TV, Podrez EA (2013) Engagement of platelet toll-like receptor 9 by novel endogenous ligands promotes platelet hyperreactivity and thrombosis. Circ Res 112:103–112
24. Stohlawetz P, Folman CC, von dem Borne AE, Pernerstorfer T, Eichler HG, Panzer S, Jilma B (1999) Effects of endotoxemia on thrombopoiesis in men. Thromb Haemost 81:613–617
25. Pick M, Perry C, Lapidot T, Guimaraes-Sternberg C, Naparstek E, Deutsch V, Soreq H (2006) Stress-induced cholinergic signaling promotes inflammation-associated thrombopoiesis. Blood 107:3397–3406
26. Jayachandran M, Brunn GJ, Karnicki K, Miller RS, Owen WG, Miller VM (2007) In vivo effects of lipopolysaccharide and TLR4 on platelet production and activity: implications for thrombotic risk. J Appl Physiol (1985) 102:429–433
27. Flaujac C, Boukour S, Cramer-Borde E (2010) Platelets and viruses: an ambivalent relationship. Cell Mol Life Sci 67:545–556
28. Assinger A (2014) Platelets and infection – an emerging role of platelets in viral infection. Front Immunol 5:649
29. Yeaman MR (2010) Bacterial-platelet interactions: virulence meets host defense. Future Microbiol 5:471–506
30. Yeaman MR (2010) Platelets in defense against bacterial pathogens. Cell Mol Life Sci 67:525–544
31. Kerrigan SW, Cox D (2010) Platelet-bacterial interactions. Cell Mol Life Sci 67:513–523
32. Aiolfi R, Sitia G (2015) Chronic hepatitis B: role of anti-platelet therapy in inflammation control. Cell Mol Immunol 12:264–268
33. Kraemer BF, Campbell RA, Schwertz H, Cody MJ, Franks Z, Tolley ND, Kahr WH, Lindemann S, Seizer P, Yost CC, Zimmerman GA, Weyrich AS (2011) Novel anti-bacterial activities of beta-defensin 1 in human platelets: suppression of pathogen growth and signaling of neutrophil extracellular trap formation. PLoS Pathog 7, e1002355
34. Fuchs TA, Brill A, Duerschmied D, Schatzberg D, Monestier M, Myers DD Jr, Wrobleski SK, Wakefield TW, Hartwig JH, Wagner DD (2010) Extracellular DNA traps promote thrombosis. Proc Natl Acad Sci U S A 107:15880–15885
35. Thomas GM, Carbo C, Curtis BR, Martinod K, Mazo IB, Schatzberg D, Cifuni SM, Fuchs TA, von Andrian UH, Hartwig JH, Aster RH, Wagner DD (2012) Extracellular DNA traps are associated with the pathogenesis of TRALI in humans and mice. Blood 119:6335–6343
36. Demers M, Krause DS, Schatzberg D, Martinod K, Voorhees JR, Fuchs TA, Scadden DT, Wagner DD (2012) Cancers predispose neutrophils to release extracellular DNA traps that contribute to cancer-associated thrombosis. Proc Natl Acad Sci U S A 109:13076–13081

37. Fuchs TA, Alvarez JJ, Martinod K, Bhandari AA, Kaufman RM, Wagner DD (2013) Neutrophils release extracellular DNA traps during storage of red blood cell units. Transfusion 53:3210–3216
38. Chen G, Zhang D, Fuchs TA, Manwani D, Wagner DD, Frenette PS (2014) Heme-induced neutrophil extracellular traps contribute to the pathogenesis of sickle cell disease. Blood 123:3818–3827
39. Caudrillier A, Kessenbrock K, Gilliss BM, Nguyen JX, Marques MB, Monestier M, Toy P, Werb Z, Looney MR (2012) Platelets induce neutrophil extracellular traps in transfusion-related acute lung injury. J Clin Invest 122:2661–2671
40. Wong SL, Demers M, Martinod K, Gallant M, Wang Y, Goldfine AB, Kahn CR, Wagner DD (2015) Diabetes primes neutrophils to undergo NETosis, which impairs wound healing. Nat Med 21:815–819
41. de Stoppelaar SF, Van V, van der Poll T (2014) The role of platelets in sepsis. Thromb Haemost 112:666–677
42. Gawaz M, Dickfeld T, Bogner C, Fateh-Moghadam S, Neumann FJ (1997) Platelet function in septic multiple organ dysfunction syndrome. Intensive Care Med 23:379–385
43. Russwurm S, Vickers J, Meier-Hellmann A, Spangenberg P, Bredle D, Reinhart K, Losche W (2002) Platelet and leukocyte activation correlate with the severity of septic organ dysfunction. Shock 17:263–268
44. Washington AV, Gibot S, Acevedo I, Gattis J, Quigley L, Feltz R, De La Mota A, Schubert RL, Gomez-Rodriguez J, Cheng J, Dutra A, Pak E, Chertov O, Rivera L, Morales J, Lubkowski J, Hunter R, Schwartzberg PL, McVicar DW (2009) TREM-like transcript-1 protects against inflammation-associated hemorrhage by facilitating platelet aggregation in mice and humans. J Clin Invest 119:1489–1501
45. de Stoppelaar SF, Van't Veer C, van den Boogaard FE, Nieuwland R, Hoogendijk AJ, de Boer OJ, Roelofs JJ, van der Poll T (2013) Protease activated receptor 4 limits bacterial growth and lung pathology during late stage Streptococcus pneumoniae induced pneumonia in mice. Thromb Haemost 110:582–592
46. Etulain J, Martinod K, Wong SL, Cifuni SM, Schattner M, Wagner DD (2015) P-selectin promotes neutrophil extracellular trap formation in mice. Blood 126:242–246
47. Dankert J, van der Werff J, Zaat SA, Joldersma W, Klein D, Hess J (1995) Involvement of bactericidal factors from thrombin-stimulated platelets in clearance of adherent viridans streptococci in experimental infective endocarditis. Infect Immun 63:663–671
48. Assinger A, Laky M, Schabbauer G, Hirschl AM, Buchberger E, Binder BR, Volf I (2011) Efficient phagocytosis of periodontopathogens by neutrophils requires plasma factors, platelets and TLR2. J Thromb Haemost 9:799–809
49. McMorran BJ, Wieczorski L, Drysdale KE, Chan JA, Huang HM, Smith C, Mitiku C, Beeson JG, Burgio G, Foote SJ (2012) Platelet factor 4 and Duffy antigen required for platelet killing of Plasmodium falciparum. Science 338:1348–1351
50. Cines DB, Cuker A, Semple JW (2014) Pathogenesis of immune thrombocytopenia. Presse Med 43:e49–e59
51. Zhang W, Nardi MA, Borkowsky W, Li Z, Karpatkin S (2009) Role of molecular mimicry of hepatitis C virus protein with platelet GPIIIa in hepatitis C-related immunologic thrombocytopenia. Blood 113:4086–4093
52. Wright JF, Blanchette VS, Wang H, Arya N, Petric M, Semple JW, Chia WK, Freedman J (1996) Characterization of platelet-reactive antibodies in children with varicella-associated acute immune thrombocytopenic purpura (ITP). Br J Haematol 95:145–152
53. Takahashi T, Yujiri T, Shinohara K, Inoue Y, Sato Y, Fujii Y, Okubo M, Zaitsu Y, Ariyoshi K, Nakamura Y, Nawata R, Oka Y, Shirai M, Tanizawa Y (2004) Molecular mimicry by Helicobacter pylori CagA protein may be involved in the pathogenesis of H. pylori-associated chronic idiopathic thrombocytopenic purpura. Br J Haematol 124:91–96
54. Li Z, Nardi MA, Karpatkin S (2005) Role of molecular mimicry to HIV-1 peptides in HIV-1-related immunologic thrombocytopenia. Blood 106:572–576
55. Chia WK, Blanchette V, Mody M, Wright JF, Freedman J (1998) Characterization of HIV-1-specific antibodies and HIV-1-crossreactive antibodies to platelets in HIV-1-infected haemophiliac patients. Br J Haematol 103:1014–1022

56. Asahi A, Nishimoto T, Okazaki Y, Suzuki H, Masaoka T, Kawakami Y, Ikeda Y, Kuwana M (2008) Helicobacter pylori eradication shifts monocyte Fcgamma receptor balance toward inhibitory FcgammaRIIB in immune thrombocytopenic purpura patients. J Clin Invest 118:2939–2949
57. Tremblay T, Aubin E, Lemieux R, Bazin R (2007) Picogram doses of lipopolysaccharide exacerbate antibody-mediated thrombocytopenia and reduce the therapeutic efficacy of intravenous immunoglobulin in mice. Br J Haematol 139:297–302
58. Kapur R, Heitink-Polle KM, Porcelijn L, Bentlage AE, Bruin MC, Visser R, Roos D, Schasfoort RB, de Hass M, van der Schoot CE, Vidarsson G (2015) C-reactive protein enhances IgG-mediated phagocyte responses and thrombocytopenia. Blood 125:1793–1802
59. Grewal IS, Flavell RA (1998) CD40 and CD154 in cell-mediated immunity. Annu Rev Immunol 16:111–135
60. Henn V, Steinbach S, Buchner K, Presek P, Kroczek RA (2001) The inflammatory action of CD40 ligand (CD154) expressed on activated human platelets is temporally limited by coexpressed CD40. Blood 98:1047–1054
61. Andre P, Nannizzi-Alaimo L, Prasad SK, Phillips DR (2002) Platelet-derived CD40L: the switch-hitting player of cardiovascular disease. Circulation 106:896–899
62. Hammwohner M, Ittenson A, Dierkes J, Bukowska A, Klein HU, Lendeckel U, Goette A (2007) Platelet expression of CD40/CD40 ligand and its relation to inflammatory markers and adhesion molecules in patients with atrial fibrillation. Exp Biol Med (Maywood) 232:581–589
63. Anand SX, Viles-Gonzalez JF, Badimon JJ, Cavusoglu E, Marmur JD (2003) Membrane-associated CD40L and sCD40L in atherothrombotic disease. Thromb Haemost 90:377–384
64. Elzey BD, Schmidt NW, Crist SA, Kresowik TP, Harty JT, Nieswandt B, Ratliff TL (2008) Platelet-derived CD154 enables T-cell priming and protection against Listeria monocytogenes challenge. Blood 111:3684–3691
65. Iannacone M, Sitia G, Isogawa M, Marchese P, Castro MG, Lowenstein PR, Chisari FV, Ruggeri ZM, Guidotti LG (2005) Platelets mediate cytotoxic T lymphocyte-induced liver damage. Nat Med 11:1167–1169
66. Kissel K, Berber S, Nockher A, Santoso S, Bein G, Hackstein H (2006) Human platelets target dendritic cell differentiation and production of proinflammatory cytokines. Transfusion 46:818–827
67. Diacovo TG, Catalina MD, Siegelman MH, von Andrian UH (1998) Circulating activated platelets reconstitute lymphocyte homing and immunity in L-selectin-deficient mice. J Exp Med 187:197–204
68. Diacovo TG, Puri KD, Warnock RA, Springer TA, von Andrian UH (1996) Platelet-mediated lymphocyte delivery to high endothelial venules. Science 273:252–255
69. von Hundelshausen P, Weber C (2007) Platelets as immune cells: bridging inflammation and cardiovascular disease. Circ Res 100:27–40
70. Elzey BD, Tian J, Jensen RJ, Swanson AK, Lees JR, Lentz SR, Stein CS, Nieswandt B, Wang Y, Davidson BL, Ratliff TL (2003) Platelet-mediated modulation of adaptive immunity. A communication link between innate and adaptive immune compartments. Immunity 19:9–19
71. Hachem A, Yacoub D, Zaid Y, Mourad W, Merhi Y (2012) Involvement of nuclear factor kappaB in platelet CD40 signaling. Biochem Biophys Res Commun 425:58–63
72. Malaver E, Romaniuk MA, D'Atri LP, Pozner RG, Negrotto S, Benzadon R, Schattner M (2009) NF-kappaB inhibitors impair platelet activation responses. J Thromb Haemost 7:1333–1343
73. Spinelli SL, Casey AE, Pollock SJ, Gertz JM, McMillan DH, Narasipura SD, Mody NA, King MR, Maggirwar SB, Francis CW, Taubman MB, Blumberg N, Phipps RP (2010) Platelets and megakaryocytes contain functional nuclear factor-kappaB. Arterioscler Thromb Vasc Biol 30:591–598
74. Gambaryan S, Kobsar A, Rukoyatkina N, Herterich S, Geiger J, Smolenski A, Lohmann SM, Walter U (2010) Thrombin and collagen induce a feedback inhibitory signaling pathway in platelets involving dissociation of the catalytic subunit of protein kinase A from an NFkappaB-IkappaB complex. J Biol Chem 285:18352–18363

75. Karim ZA, Zhang J, Banerjee M, Chicka MC, Al HR, Hamilton TR, Roche PA, Whiteheart SW (2013) IkappaB kinase phosphorylation of SNAP-23 controls platelet secretion. Blood 121:4567–4574
76. Liu F, Morris S, Epps J, Carroll R (2002) Demonstration of an activation regulated NF-kappaB/I-kappaBalpha complex in human platelets. Thromb Res 106:199–203
77. Mazzucco L, Borzini P, Gope R (2010) Platelet-derived factors involved in tissue repair-from signal to function. Transfus Med Rev 24:218–234
78. Assoian RK, Komoriya A, Meyers CA, Miller DM, Sporn MB (1983) Transforming growth factor-beta in human platelets. Identification of a major storage site, purification, and characterization. J Biol Chem 258:7155–7160
79. Andersson PO, Stockelberg D, Jacobsson S, Wadenvik H (2000) A transforming growth factor-beta1-mediated bystander immune suppression could be associated with remission of chronic idiopathic thrombocytopenic purpura. Ann Hematol 79:507–513
80. Andersson PO, Olsson A, Wadenvik H (2002) Reduced transforming growth factor-beta1 production by mononuclear cells from patients with active chronic idiopathic thrombocytopenic purpura. Br J Haematol 116:862–867
81. Blair P, Flaumenhaft R (2009) Platelet alpha-granules: basic biology and clinical correlates. Blood Rev 23:177–189
82. Gleissner CA (2012) Macrophage phenotype modulation by CXCL4 in Atherosclerosis. Front Physiol 3:1
83. Petersen F, Bock L, Flad HD, Brandt E (1999) Platelet factor 4-induced neutrophil-endothelial cell interaction: involvement of mechanisms and functional consequences different from those elicited by interleukin-8. Blood 94:4020–4028
84. Brandt E, Petersen F, Ludwig A, Ehlert JE, Bock L, Flad HD (2000) The beta-thromboglobulins and platelet factor 4: blood platelet-derived CXC chemokines with divergent roles in early neutrophil regulation. J Leukoc Biol 67:471–478
85. Alam R, Forsythe PA, Stafford S, Lett-Brown MA, Grant JA (1992) Macrophage inflammatory protein-1 alpha activates basophils and mast cells. J Exp Med 176:781–786
86. Schall TJ, Bacon K, Camp RD, Kaspari JW, Goeddel DV (1993) Human macrophage inflammatory protein alpha (MIP-1 alpha) and MIP-1 beta chemokines attract distinct populations of lymphocytes. J Exp Med 177:1821–1826
87. Wolf P (1967) The nature and significance of platelet products in human plasma. Br J Haematol 13:269–288
88. Buzas EI, Gyorgy B, Nagy G, Falus A, Gay S (2014) Emerging role of extracellular vesicles in inflammatory diseases. Nat Rev Rheumatol 10:356–364
89. Lotvall J, Hill AF, Hochberg F, Buzas EI, Di VD, Gardiner C, Gho YS, Kurochkin IV, Mathivanan S, Quesenberry P, Sahoo S, Tahara H, Wauben MH, Witwer KW, Thery C (2014) Minimal experimental requirements for definition of extracellular vesicles and their functions: a position statement from the International Society for Extracellular Vesicles. J Extracell Vesicles 3:26913
90. Arraud N, Linares R, Tan S, Gounou C, Pasquet JM, Mornet S, Brisson AR (2014) Extracellular vesicles from blood plasma: determination of their morphology, size, phenotype and concentration. J Thromb Haemost 12:614–627
91. Morel O, Jesel L, Freyssinet JM, Toti F (2011) Cellular mechanisms underlying the formation of circulating microparticles. Arterioscler Thromb Vasc Biol 31:15–26
92. Owens AP III, Mackman N (2011) Microparticles in hemostasis and thrombosis. Circ Res 108:1284–1297
93. Tersteeg C, Heijnen HF, Eckly A, Pasterkamp G, Urbanus RT, Maas C, Hoefer IE, Nieuwland R, Farndale RW, Gachet C, de Groot PG, Roest M (2014) FLow-induced PRotrusions (FLIPRs): a platelet-derived platform for the retrieval of microparticles by monocytes and neutrophils. Circ Res 114:780–791
94. Connor DE, Exner T, Ma DD, Joseph JE (2010) The majority of circulating platelet-derived microparticles fail to bind annexin V, lack phospholipid-dependent procoagulant activity and demonstrate greater expression of glycoprotein Ib. Thromb Haemost 103:1044–1052

95. Perez-Pujol S, Marker PH, Key NS (2007) Platelet microparticles are heterogeneous and highly dependent on the activation mechanism: studies using a new digital flow cytometer. Cytometry A 71:38–45
96. Cloutier N, Tan S, Boudreau LH, Cramb C, Subbaiah R, Lahey L, Albert A, Shnayder R, Gobezie R, Nigrovic PA, Farndale RW, Robinson WH, Brisson A, Lee DM, Boilard E (2013) The exposure of autoantigens by microparticles underlies the formation of potent inflammatory components: the microparticle-associated immune complexes. EMBO Mol Med 5:235–249
97. Nurden AT (2011) Platelets, inflammation and tissue regeneration. Thromb Haemost 105(Suppl 1):S13–S33
98. Reid VL, Webster NR (2012) Role of microparticles in sepsis. Br J Anaesth 109:503–513
99. Gyorgy B, Szabo TG, Turiak L, Wright M, Herczeg P, Ledeczi Z, Kittel A, Polgar A, Toth K, Derfalvi B, Zelenak G, Borocz I, Carr B, Nagy G, Vekey K, Gay S, Falus A, Buzas EI (2012) Improved flow cytometric assessment reveals distinct microvesicle (cell-derived microparticle) signatures in joint diseases. PLoS One 7, e49726
100. Rousseau M, Belleannee C, Duchez AC, Cloutier N, Levesque T, Jacques F, Perron J, Nigrovic PA, Dieude M, Hebert MJ, Gelb MH, Boilard E (2015) Detection and quantification of microparticles from different cellular lineages using flow cytometry. Evaluation of the impact of secreted phospholipase A2 on microparticle assessment. PLoS One 10:e0116812
101. Gitz E, Pollitt AY, Gitz-Francois JJ, Alshehri O, Mori J, Montague S, Nash GB, Douglas MR, Gardiner EE, Andrews RK, Buckley CD, Harrison P, Watson SP (2014) CLEC-2 expression is maintained on activated platelets and on platelet microparticles. Blood 124:2262–2270
102. Boilard E, Blanco P, Nigrovic PA (2012) Platelets: active players in the pathogenesis of arthritis and SLE. Nat Rev Rheumatol 8:534–542
103. Mott PJ, Lazarus AH (2013) CD44 antibodies and immune thrombocytopenia in the amelioration of murine inflammatory arthritis. PLoS One 8, e65805
104. Brown GT, McIntyre TM (2011) Lipopolysaccharide signaling without a nucleus: kinase cascades stimulate platelet shedding of proinflammatory IL-1beta-rich microparticles. J Immunol 186:5489–5496
105. Boilard E, Pare G, Rousseau M, Cloutier N, Dubuc I, Levesque T, Borgeat P, Flamand L (2014) Influenza virus H1N1 activates platelets through FcgammaRIIA signaling and thrombin generation. Blood 123:2854–2863
106. Sun D, Popescu NI, Raisley B, Keshari RS, Dale GL, Lupu F, Coggeshall KM (2013) Bacillus anthracis peptidoglycan activates human platelets through FcgammaRII and complement. Blood 122:571–579
107. Laffont B, Corduan A, Ple H, Duchez AC, Cloutier N, Boilard E, Provost P (2013) Activated platelets can deliver mRNA regulatory Ago2*microRNA complexes to endothelial cells via microparticles. Blood 122:253–261
108. Boudreau LH, Duchez AC, Cloutier N, Soulet D, Martin N, Bollinger J, Pare A, Rousseau M, Naika GS, Levesque T, Laflamme C, Marcoux G, Lambeau G, Farndale RW, Pouliot M, Hamzeh-Cognasse H, Cognasse F, Garraud O, Nigrovic PA, Guderley H, Lacroix S, Thibault L, Semple JW, Gelb MH, Boilard E (2014) Platelets release mitochondria serving as substrate for bactericidal group IIA-secreted phospholipase A2 to promote inflammation. Blood 124:2173–2183
109. Ray DM, Spinelli SL, Pollock SJ, Murant TI, O'Brien JJ, Blumberg N, Francis CW, Taubman MB, Phipps RP (2008) Peroxisome proliferator-activated receptor gamma and retinoid X receptor transcription factors are released from activated human platelets and shed in microparticles. Thromb Haemost 99:86–95
110. Garcia BA, Smalley DM, Cho H, Shabanowitz J, Ley K, Hunt DF (2005) The platelet microparticle proteome. J Proteome Res 4:1516–1521
111. Dasgupta SK, Abdel-Monem H, Niravath P, Le A, Bellera RV, Langlois K, Nagata S, Rumbaut RE, Thiagarajan P (2009) Lactadherin and clearance of platelet-derived microvesicles. Blood 113:1332–1339

112. Dasgupta SK, Le A, Chavakis T, Rumbaut RE, Thiagarajan P (2012) Developmental endothelial locus-1 (Del-1) mediates clearance of platelet microparticles by the endothelium. Circulation 125:1664–1672
113. Italiano JE Jr, Richardson JL, Patel-Hett S, Battinelli E, Zaslavsky A, Short S, Ryeom S, Folkman J, Klement GL (2008) Angiogenesis is regulated by a novel mechanism: pro- and antiangiogenic proteins are organized into separate platelet alpha granules and differentially released. Blood 111:1227–1233
114. Sehgal S, Storrie B (2007) Evidence that differential packaging of the major platelet granule proteins von Willebrand factor and fibrinogen can support their differential release. J Thromb Haemost 5:2009–2016
115. White GC, Rompietti R (2007) Platelet secretion: indiscriminately spewed forth or highly orchestrated? J Thromb Haemost 5:2006–2008
116. Rowley JW, Oler AJ, Tolley ND, Hunter BN, Low EN, Nix DA, Yost CC, Zimmerman GA, Weyrich AS (2011) Genome-wide RNA-seq analysis of human and mouse platelet transcriptomes. Blood 118:e101–e111
117. Rowley JW, Schwertz H, Weyrich AS (2012) Platelet mRNA: the meaning behind the message. Curr Opin Hematol 19:385–391
118. Lood C, Amisten S, Gullstrand B, Jonsen A, Allhorn M, Truedsson L, Sturfelt G, Erlinge D, Bengtsson AA (2010) Platelet transcriptional profile and protein expression in patients with systemic lupus erythematosus: up-regulation of the type I interferon system is strongly associated with vascular disease. Blood 116:1951–1957
119. Healy AM, Pickard MD, Pradhan AD, Wang Y, Chen Z, Croce K, Sakuma M, Shi C, Zago AC, Garasic J, Damokosh AI, Dowie TL, Poisson L, Lillie J, Libby P, Ridker PM, Simon DI (2006) Platelet expression profiling and clinical validation of myeloid-related protein-14 as a novel determinant of cardiovascular events. Circulation 113:2278–2284
120. Goodall AH, Burns P, Salles I, Macaulay IC, Jones CI, Ardissino D, de Bono B, Bray SL, Deckmyn H, Dudbridge F, Fitzgerald DJ, Garner SF, Gusnanto A, Koch K, Langford C, O'Connor MN, Rice CM, Stemple D, Stephens J, Trip MD, Zwaginga JJ, Samani NJ, Watkins NA, Maguire PB, Ouwehand WH (2010) Transcription profiling in human platelets reveals LRRFIP1 as a novel protein regulating platelet function. Blood 116:4646–4656
121. Simon LM, Edelstein LC, Nagalla S, Woodley AB, Chen ES, Kong X, Ma L, Fortina P, Kunapuli S, Holinstat M, McKenzie SE, Dong JF, Shaw CA, Bray PF (2014) Human platelet microRNA-mRNA networks associated with age and gender revealed by integrated plateletomics. Blood 123:e37–e45
122. Edelstein LC, Simon LM, Montoya RT, Holinstat M, Chen ES, Bergeron A, Kong X, Nagalla S, Mohandas N, Cohen DE, Dong JF, Shaw C, Bray PF (2013) Racial differences in human platelet PAR4 reactivity reflect expression of PCTP and miR-376c. Nat Med 19:1609–1616
123. Ple H, Maltais M, Corduan A, Rousseau G, Madore F, Provost P (2012) Alteration of the platelet transcriptome in chronic kidney disease. Thromb Haemost 108:605–615
124. McManus DD, Beaulieu LM, Mick E, Tanriverdi K, Larson MG, Keaney JF Jr, Benjamin EJ, Freedman JE (2013) Relationship among circulating inflammatory proteins, platelet gene expression, and cardiovascular risk. Arterioscler Thromb Vasc Biol 33:2666–2673
125. Freedman JE, Larson MG, Tanriverdi K, O'Donnell CJ, Morin K, Hakanson AS, Vasan RS, Johnson AD, Iafrati MD, Benjamin EJ (2010) Relation of platelet and leukocyte inflammatory transcripts to body mass index in the Framingham heart study. Circulation 122:119–129
126. Raghavachari N, Xu X, Harris A, Villagra J, Logun C, Barb J, Solomon MA, Suffredini AF, Danner RL, Kato G, Munson PJ, Morris SM Jr, Gladwin MT (2007) Amplified expression profiling of platelet transcriptome reveals changes in arginine metabolic pathways in patients with sickle cell disease. Circulation 115:1551–1562
127. Risitano A, Beaulieu LM, Vitseva O, Freedman JE (2012) Platelets and platelet-like particles mediate intercellular RNA transfer. Blood 119:6288–6295
128. Clancy L, Freedman JE (2014) New paradigms in thrombosis: novel mediators and biomarkers platelet RNA transfer. J Thromb Thrombolysis 37:12–16

129. Gidlof O, van der Brug M, Ohman J, Gilje P, Olde B, Wahlestedt C, Erlinge D (2013) Platelets activated during myocardial infarction release functional miRNA, which can be taken up by endothelial cells and regulate ICAM1 expression. Blood 121:3908–3926
130. Landry P, Plante I, Ouellet DL, Perron MP, Rousseau G, Provost P (2009) Existence of a microRNA pathway in anucleate platelets. Nat Struct Mol Biol 16:961–966
131. Rondina MT, Weyrich AS (2015) Regulation of the genetic code in megakaryocytes and platelets. J Thromb Haemost 13(Suppl 1):S26–S32
132. Burkhart JM, Vaudel M, Gambaryan S, Radau S, Walter U, Martens L, Geiger J, Sickmann A, Zahedi RP (2012) The first comprehensive and quantitative analysis of human platelet protein composition allows the comparative analysis of structural and functional pathways. Blood 120:e73–e82
133. Schubert S, Weyrich AS, Rowley JW (2014) A tour through the transcriptional landscape of platelets. Blood 124:493–502
134. Shulman NR, Aster RH, Pearson HA, Hiller MC (1962) Immunoreactions involving platelet. VI. Reactions of maternal isoantibodies responsible for neonatal purpura. Differentiation of a second platelet antigen system. J Clin Invest 41:1059–1069
135. Blumberg N, Masel D, Mayer T, Horan P, Heal J (1984) Removal of HLA-A, B antigens from platelets. Blood 63:448–450
136. Kao KJ, Cook DJ, Scornik JC (1986) Quantitative analysis of platelet surface HLA by W6/32 anti-HLA monoclonal antibody. Blood 68:627–632
137. Kao KJ (1987) Plasma and platelet HLA in normal individuals: quantitation by competitive enzyme-linked immunoassay. Blood 70:282–286
138. Kao KJ (1988) Selective elution of HLA antigens and beta 2-microglobulin from human platelets by chloroquine diphosphate. Transfusion 28:14–17
139. Neumuller J, Tohidast-Akrad M, Fischer M, Mayr WR (1993) Influence of chloroquine or acid treatment of human platelets on the antigenicity of HLA and the 'thrombocyte-specific' glycoproteins Ia/IIa, IIb, and IIb/IIIa. Vox Sang 65:223–231
140. Ghio M, Contini P, Mazzei C, Brenci S, Barberis G, Filaci G, Indiveri F, Puppo F (1999) Soluble HLA class I, HLA class II, and Fas ligand in blood components: a possible key to explain the immunomodulatory effects of allogeneic blood transfusions. Blood 93:1770–1777
141. Gouttefangeas C, Diehl M, Keilholz W, Hornlein RF, Stevanovic S, Rammensee HG (2000) Thrombocyte HLA molecules retain nonrenewable endogenous peptides of megakaryocyte lineage and do not stimulate direct allocytotoxicity in vitro. Blood 95:3168–3175
142. Aslam R, Speck ER, Kim M, Freedman J, Semple JW (2008) Transfusion-related immunomodulation by platelets is dependent on their expression of MHC class I molecules and is independent of white cells. Transfusion 48:1778–1786
143. Zufferey A, Schvartz D, Nolli S, Reny JL, Sanchez JC, Fontana P (2014) Characterization of the platelet granule proteome: evidence of the presence of MHC1 in alpha-granules. J Proteomics 101:130–140
144. Chapman LM, Aggrey AA, Field DJ, Srivastava K, Ture S, Yui K, Topham DJ, Baldwin WM III, Morrell CN (2012) Platelets present antigen in the context of MHC class I. J Immunol 189:916–923

Mouse Models for Platelet Production and Function

11

Markus Bender and Bernhard Nieswandt

Abstract

Highly polyploid, mature bone marrow megakaryocytes convert their cytoplasm into long protrusions (proplatelets) which extend into the lumen of a vessel in bone marrow sinusoids where so-called pre-platelets are released and final platelet shaping and sizing occurs in the bloodstream. Platelets never undergo firm adhesion in the circulation, and only at sites of vascular injury, the adhesion potential becomes evident. This is critical for the formation of a platelet plug which seals the vessel and limits excessive blood loss, but is also a key step in the pathogenesis of ischemic cardio- and cerebrovascular diseases which represent the leading causes of death and severe disability worldwide. Static as well as shear stress-dependent in vitro assays helped us to identify and better understand the role of proteins and pathways in platelet biogenesis and function. However, in vitro experiments only provide limited mechanistic information as they cannot completely mimic these complex processes. The generation of genetically modified mice and the availability of large-scale mouse knockout programs producing a continuous resource of targeted mutations in all protein-encoding genes has been a major step forward in order to investigate these processes under in vivo conditions. Furthermore, the establishment of multiple assays to analyze platelet production and function, in combination with improved imaging techniques, such as intravital two-photon microscopy, helped us to better understand the underlying mechanisms. This chapter summarizes the most important mouse models that contributed significantly to our current knowledge.

M. Bender (✉) • B. Nieswandt (✉)
Department of Experimental Biomedicine, University of Würzburg,
University Hospital and Rudolf Virchow Center, Würzburg, Germany
e-mail: bender_m1@ukw.de; bernhard.nieswandt@virchow.uni-wuerzburg.de

H. Schulze, J. Italiano (eds.), *Molecular and Cellular Biology of Platelet Formation*,
DOI 10.1007/978-3-319-39562-3_11

11.1 Introduction

The formation of anucleated blood platelets from their bone marrow resident precursors, the megakaryocytes (MKs), is a unique process and only found in mammalian physiology. Terminally differentiated, polyploid MKs are the largest cells evolving from hematopoietic stem cells. MKs contain a highly organized cytoplasm with a membranous network, the demarcation membrane system (DMS), which converts into cytoplasmic protrusions during proplatelet formation [1]. The tips of these proplatelet protrusions extend into the lumen of bone marrow sinusoids and are shed by shear forces generated by the blood flow.

Platelets are essential players in thrombosis and hemostasis, as they "survey" the integrity of the vascular system. After vessel wall injury, they rapidly adhere to exposed components of the extracellular matrix and form a hemostatic plug. However, in disease states, uncontrolled thrombus formation may occur and cause irreversible occlusion of the vessel resulting in, e.g., myocardial infarction or stroke [2, 3], which are the leading causes of disability and mortality in the Western world [4]. Despite recent advances, there is still a strong demand for the development and production of selective, powerful, yet safe antithrombotic drugs targeting platelet receptor or signaling proteins to prevent and treat these thrombotic events [5, 6].

Under conditions of high shear, as found in small arteries, arterioles, and moderately stenosed vessels, the initial tethering of platelets to the extracellular matrix (ECM) is mediated by the interaction between the platelet receptor glycoprotein (GP) Ib and collagen-bound von Willebrand factor (vWF) (Fig. 11.1). The binding of GPIb to vWF has a fast off-rate and does not allow stable platelet adhesion but retains platelets in close contact to the vessel wall, allowing continuous platelet rolling in the direction of blood flow. While translocating along the vessel wall, platelets establish contacts with exposed subendothelial extracellular matrix components, such as collagen. GPVI binds collagen and triggers intracellular signals that shift platelet integrins to a high-affinity state and induce granule release. Released adenosine diphosphate (ADP) and synthesized thromboxane A2 (TxA2) contribute, together with locally produced thrombin, to the reinforcement of platelet activation by stimulating G-protein-coupled receptors [7]. All these signaling pathways synergize to induce complex cellular responses, such as activation of integrins, release of granule contents, and coagulant activity [8, 9]. Finally, firm platelet adhesion to the extracellular matrix is mediated by inside-out upregulation of β1- and β3-integrin affinity resulting in the recruitment, activation, and incorporation of additional platelets into the growing thrombus [9].

In vitro studies have been very important to understand the contribution of multiple factors and detailed mechanisms to MK maturation, thrombopoiesis, platelet function, and thrombus formation. The advantages of in vitro systems are the relatively low costs, quick performance, and the well-controlled experimental setup. Moreover, new devices allow, e.g., to apply shear stress resembling the blood flow or to perform high-throughput screenings. However, in vitro studies have limitations and their findings are difficult to directly correlate with the in vivo situation. They cannot mimic the myriad of hemodynamic and localized cellular and molecular interactions that occur during the generation and propagation of thrombi in vivo. In the past decade, one major

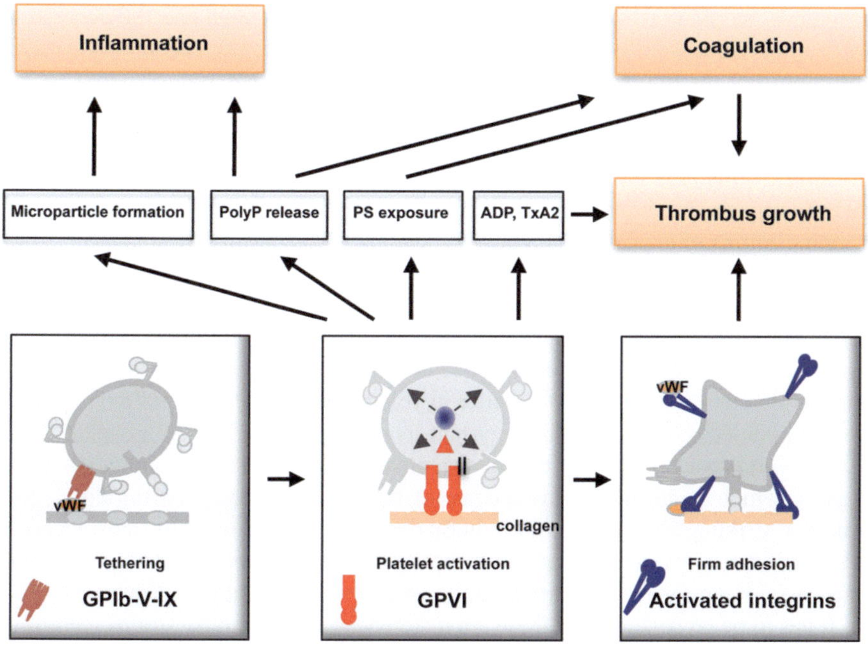

Fig. 11.1 Simplified model of thrombus formation. At sites of vessel wall injury, platelet tethering is mediated by the reversible GPIb–von Willebrand factor (vWF) interaction which enables GPVI dimerization and subsequently binding to collagen. GPVI is the major platelet-activating receptor and mediates inside-out integrin activation, which is essential for firm platelet adhesion and aggregation. Released adenosine diphosphate (ADP) and thromboxane A2 (TxA2) reinforce platelet activation and drive thrombus growth. In addition, GPVI-stimulated platelets facilitate coagulation and inflammation by releasing inorganic polyphosphates (PolyP). Furthermore, GPVI-activated platelets release microparticles and expose negatively charged phosphatidylserine (PS) on their surface providing high-affinity binding sites for key coagulation factors (not shown), thus enabling the generation of thrombin (not shown) which in turn supports coagulation and thrombus growth

breakthrough in murine platelet studies was the real-time observation on how platelets are produced into the circulation by intravital two-photon microscopy [10–12]. Similarly, much effort has been spent on the development of animal models specific for platelet-related hemostatic and thrombotic processes. Moreover, it is now possible to observe in vivo single cell–cell interaction or the contribution of different factors in a growing thrombus by advanced imaging techniques [13].

Mice are the most frequently used species in thrombosis research because of their small size, high fertility, exceptional reproductive capacity, and the similarity to humans in anatomy, physiology, and genetics [14]. Moreover, different genetic methods allow targeted manipulations in the mouse genome and have been exceptionally successful in unraveling protein function in platelets. The complete sequence and analysis of the mouse genome facilitated the ability to manipulate the mouse genome and establish a large number of mouse mutants from public resources. In addition, a major step forward in platelet research was the generation of platelet factor 4 (Pf4)-Cre transgenic mice, carrying the Cre recombinase under the control of the Pf4 promoter, which replaced the inducible, non-platelet-specific Mx-Cre/loxP system [15] and allows the generation of lineage-restricted gene knockouts for studying MK and platelet function in vivo [16]. The latest major advance in this area is the CRISPR (clustered, regularly interspaced, short palindromic repeats)/Cas9 system which represents a highly efficient technology for genome editing of mouse zygotes that will greatly accelerate in vivo studies [17, 18].

Although 95 % of the mouse genome is similar to the human genome, making mouse genetic research particularly applicable to human disease, one has to be cautious with directly extrapolating data obtained from studies in mice to the human system [14, 19]. Mice have roughly three to four times more platelets than humans. Mouse platelets are about 50 % smaller in volume and the life span of a circulating mouse platelet is also approximately half as compared to a human platelet (5 vs. 10 days). It was also reported that mouse platelets display a higher granule heterogeneity which might be a consequence of different membrane partitioning [20]. Another striking difference is the alternative use of protease-activated receptors (PARs) and the absence of FcγRIIa receptor on mouse platelets. However, in general, the similarities between mouse and human exceed by far the differences.

It has to be considered that there are differences among mouse strains regarding platelet counts and their response to vascular injury. Genetic variation has been observed in the integrin α2-subunit for the FVB mouse strain [21]. Therefore, it is generally recommended to establish congenic colonies, although backcross breeding is cost intensive and time consuming.

Besides mice, zebrafish, rabbits, pigs, dogs, and nonhuman primates have been used to analyze platelet function and to perform thrombosis models. Particularly, the latter have been used to test promising antithrombotic agents. Nevertheless, basic platelet research will continue to primarily use in vivo mouse models.

In the following, we will summarize results obtained by analysis of genetically modified mice with altered platelet biogenesis or function.

11.2 Mouse Models for Platelet Production

11.2.1 Role of the TPO/c-mpl Axis on Megakaryocyte and Platelet Development

The discovery of the growth factor thrombopoietin (TPO) in 1994 was a major achievement in the understanding of the regulation of platelet formation [22]. TPO is crucial for platelet biogenesis by inducing MK production. It was shown that TPO has an effect on MK maturation and injection into mice results in an increase of platelet and MK numbers [23]. TPO acts via the receptor c-Mpl on hematopoietic stem cells and primitive MK progenitors. Mice deficient for either TPO or the TPO receptor, c-Mpl, display a drop of platelet and MK numbers down to 10–15 % as compared to controls [24, 25]. Interestingly, these mice do not suffer from any major bleedings, although bleeding time is threefold to fourfold prolonged. The ultrastructure of MKs and platelets as well as platelet function is normal in these animals [26]. However, the number of MK progenitor cells is reduced in *Tpo*$^{-/-}$- and *c-Mpl*$^{-/-}$ mice [25, 27]. These data suggest that the TPO/c-Mpl axis is required to stimulate the production of MKs and platelets in normal numbers [28]. Recent studies on mice with megakaryocyte-/platelet-specific (Pf4-Cre/loxP system) deficiency in Mpl or Jak2 (essential mediator of TPO signaling) further revealed that the key regulatory step controlled by TPO/Mpl is to stimulate MK-biased stem and progenitor cells and thereby to determine the number of committed megakaryocytic progenitors and undifferentiated megakaryocytic precursors [29–31].

The mechanisms regulating steady-state TPO production have been controversially discussed for years. The prevailing model that the circulating TPO concentration is inversely proportional to the "Mpl mass" contributed by the total number of MKs and platelets was challenged by an elegant use of mouse models lacking asialoglycoprotein receptor 2 (ASGPR2) and sialyltransferase ST3GalIV. This study revealed that the hepatic Ashwell–Morell receptor recognizes circulating desialylated, senile platelets under steady-state conditions and functions as a direct feedback pathway between senile platelet removal and hepatic TPO mRNA expression and secretion. Together, this study identified desialylated, senile platelets and the Ashwell–Morell receptor as the physiological ligand-receptor pair regulating steady-state hepatic TPO production, thereby regulating platelet production [32].

11.2.2 Role of Apoptosis in Megakaryocytes and Platelets

The intrinsic apoptosis pathway is known to be another mechanism to regulate platelet life span [33]. Considerable evidence has accumulated suggesting that localized caspase activation in MKs is causal to proplatelet formation [34, 35]. However, with the help of genetically modified mouse lines, more recent data revealed that apoptosis is not required for platelet production and rather has to be restrained by MKs [36]. MK-specific deletions of the essential mediators of the extrinsic (Caspase-8) and intrinsic (BAK/BAX) apoptosis pathways do not impair thrombopoiesis [37]. Interestingly, mice lacking Bak and Bax in the hematopoietic system are protected against thrombocytopenia induced by the chemotherapeutic agent carboplatin. These data indicate that MKs do not activate the intrinsic pathway for platelet formation but rather progress safely through proplatelet formation and platelet shedding. In contrast, mice with a MK-specific deletion of the anti-apoptotic protein Bcl-xL, exhibit severe thrombocytopenia owing to a dramatically reduced platelet life span. Megakaryopoiesis is dramatically increased in these mice [36].

Taken together, it has been shown by using multiple animal knockout models that the intrinsic and extrinsic apoptosis pathways are dispensable for thrombopoiesis.

11.2.3 Role of Transcription Factors in Megakaryopoiesis

Over the years, numerous transcription factors have been identified that control megakaryocytic differentiation. These factors include GATA-1, GATA-2, friend of

GATA-1 (FOG-1), nuclear factor-erythroid 2 (NF-E2), RUNX-1, SCL/Tal1, FLI-1, and others. How these transcription factors act together to coordinate MK maturation is complex and remains incompletely understood. In the following, only some animal models for transcription factors will be mentioned. For a comprehensive overview, please see references Schulze et al. and Tijssen et al. [38, 39] or Chaps. 2 and 3 in this book.

The X chromosome-linked transcription factor GATA-1 is expressed specifically in erythroid, mast, MK, and eosinophil lineages, as well as in hematopoietic progenitors, and is a key transcription factor regulating MK maturation. GATA-1-deficient mice have been shown to die during embryonic development (E10.5–11.5) and the lack of the transcription factor leads to MK hyperproliferation and the cells fail to undergo terminal differentiation or produce platelets [40–42]. Subsequently, sequences upstream of the GATA-1 locus were modified in embryonic stem cells resulting in generation of mice with a selective loss of MK GATA-1 expression, but sufficient GATA-1 levels in erythroid cells to avoid lethal anemia. The mutant mice have markedly reduced platelet counts, associated with deregulated MK proliferation and severely impaired cytoplasmic maturation. These results revealed a critical role for GATA-1 in MK development and platelet biogenesis [43]. In humans, mutations in the transcription factor GATA-1 gene were found in a series of families resulting in GATA1-related thrombocytopenia with dyserythropoiesis [44].

FOG-1 is an essential cofactor that acts via the formation of a heterodimer with transcription factors of the GATA family and thereby regulates transcriptional activity. Loss of FOG-1 leads to an ablation of the megakaryocytic lineage at a very early stage pointing to an absolute requirement for FOG during early megakaryopoiesis [45].

The heterodimer formation of FOG-1 with GATA proteins is essential to activate expression of genes such as the transcription factor NF-E2. *p45 NF-E2*$^{-/-}$ mice lack circulating platelets and die of hemorrhage where less than 10 % of the live-born homozygotes survive to adulthood. The highly abundant MKs of these mice show a disorganized DMS and a late-stage maturation arrest [46, 47]. This suggests that NF-E2 acts downstream of GATA-1.

RUNX1$^{-/-}$ mice die between embryonic day E11.5–12.5 and it was shown that Runx1 is essential for definitive hematopoiesis of all lineages [48, 49]. Inducible RUNX1 loss (Mx-Cre/loxP) results in impaired MK differentiation and an 80 % reduced platelet count [50, 51]. In humans, a propensity to develop myeloid leukemia is linked to mutations or intragenic deletion of one allele of the RUNX1 gene [52, 53].

Together, animal models provided us more insight into how transcription factors regulate hematopoiesis, and disturbances of transcriptional regulation may lead to human diseases, such as platelet disorder or myeloproliferative neoplasms.

11.2.4 Role of the Cytoskeleton and Cytoskeletal Regulatory Proteins in Platelet Production

Cytoskeletal rearrangement in MKs is primarily crucial in the terminal phase of platelet production [54, 55]. The mature MK with the highly organized DMS converts the mass into cytoplasmic protrusions, the so-called proplatelets. The extension of cytoplasmic protrusions from MKs is mechanically driven by microtubule sliding (--> Chap. 1) [56, 57].

In the past decade, the analysis of various mouse lines lacking cytoskeleton regulatory proteins deepened our understanding of the contribution of cytoskeletal components in proplatelet formation, platelet release, and size determination.

Mice lacking β1-tubulin produce approximately 60 % less platelets than wild-type control animals, and the cells display functional defects, and are spherical with fewer microtubule coilings around the platelet periphery [58]. Similarly, mutations in human β1-tubulin result in an autosomal dominant macrothrombocytopenia [59]. Profilin 1 (Pfn1), historically well known as an actin-regulating protein, was recently reported to play a role in the regulation of microtubule dynamics. Mice with a MK/platelet-specific profilin 1 deficiency display a microthrombocytopenia due to accelerated turnover of platelets and premature platelet release into the bone marrow. The small platelets in these mice contain abnormally organized microtubules that are refractory to depolymerization. Interestingly, a similar observation was made in platelets of Wiskott–Aldrich syndrome (WAS) patients but not in WAS protein-deficient mouse platelets indicating that the platelet phenotype in WAS patients might be based on aberrant profilin 1 activity in these cells [12]. Other proteins were also described to play a role in microtubule dynamics and platelet formation. Mice deficient in the tubulin-binding protein RanBP10 have normal platelet numbers, but the cells show altered microtubule filament numbers and localization that translates into defective agonist-induced granule release and CD62P exposure [60]. A more dramatic phenotype was observed in MKs specifically lacking the small GTPases Rac1 and Cdc42. This combined deletion results in a macrothrombocytopenia with abnormal platelet morphology and impaired platelet function. Double-deficient bone marrow MKs mature normally in vivo but display highly abnormal morphology and uncontrolled fragmentation due to severely defective tubulin organization [61].

However, not only tubulin reorganization but also actin filament turnover was shown to be important for proper proplatelet formation. Actin-depolymerizing factor (ADF) and n-cofilin are regulators of actin depolymerization and severing. MK-/platelet-specific ADF/n-cofilin-deficient mice suffer from a massive MK hyperplasia and thrombocytopenia. The analysis of the ultrastructure of bone marrow MKs revealed actin accumulation and that fetal-liver cell-derived MKs fail to produce normal proplatelets [62].

Filamin A is a crucial protein that cross-links actin filaments and links membrane receptors such as the adhesion receptor GPIb to the actin network. Filamin A constitutive deficiency results in embryonic lethality. Mice conditionally lacking filamin A via the GATA1-Cre/loxP system display a macrothrombocytopenia, altered surface distribution of GPIbα, multiple platelet function defects, and increased tail bleeding times [63]. Subsequent studies on mice in which filamin A was deleted by the PF4-Cre/loxP system also show a severe macrothrombocytopenia due to removal of platelets by macrophages. Further, megakaryopoiesis is increased and mutant proplatelets release premature platelets [64]. Similarly, platelet counts are decreased, and platelet shape and function altered in patients with filaminopathy A caused by dominant mutations of the X-linked filamin A gene [65, 66]. Together, these studies showed that filamin A has diverse and versatile functions in MKs and platelets and that mutations in this gene can lead to human diseases. Another human disorder related to an actin-regulatory protein is the Wiskott–Aldrich syndrome (WAS) which is a rare disorder caused by mutations in the WAS protein (WASp) resulting in protein loss. WAS is characterized by the triad, microthrombocytopenia, eczema, and immunodeficiency. WASp interacts with many intracellular proteins and regulates actin polymerization. Interestingly, *Wasp*$^{-/-}$ mice do not completely reproduce the clinical symptoms described for WAS patients, particularly with respect to the severe microthrombocytopenia. *Wasp*$^{-/-}$- mice exhibit only a mildly reduced platelet count and platelets are normal in size [67]. As mentioned before, however, *Pfn1*$^{-/-}$ mice display a microthrombocytopenia, thereby strongly resembling the platelet phenotype of WAS patients [12]. May–Hegglin anomaly, Fechtner syndrome, Sebastian platelet syndrome, and Epstein syndrome constitute a group of disorders with autosomal dominant inheritance and mutations in the *MYH9* gene encoding the non-muscle myosin heavy-chain II protein. Several mutations in this gene lead to premature release of platelets from the bone marrow, macrothrombocytopenia, and cytoplasmic inclusion bodies in leukocytes. Besides macrothrombocytopenia, some patients develop additional non-hematologic manifestations such as renal failure, hearing loss, and presenile cataracts [44, 68]. Mouse models of MYH9-related diseases have been generated to better understand the molecular basis of these syndromes. Mice carrying mutations in the *Myh9* gene reproduce the human phenotype, for instance, with respect to the platelet defects [69]. Thus, these mouse models can serve as important in vivo models for studying the pathomechanisms underlying MYH9-related disorders [69, 70].

Taken together, research on human megakaryopoiesis has been limited by technical problems in obtaining sufficient quantities of megakaryocytic cells for experimental purposes. Animal models help us to better analyze and understand the complex process of megakaryocyte maturation and platelet production. In addition, new imaging techniques, such as two-photon microscopy of the bone marrow, light-sheet microscopy of intact bone, and immunofluorescence staining on whole femora cryosections, allow us a detailed temporal and spatial analysis of the mouse bone marrow. In summary, these technical advances will enable the knowledge transfer between the mouse and the human system to finally better diagnose and treat patients with defects in megakaryo- and thrombopoiesis.

11.3 Mouse Models to Study Platelet Function

In the following sections of this chapter, a short overview will be given of the technical aspects of assays to determine the hemostatic function and in vivo thrombus formation. After this, we will summarize mouse models that significantly contributed to a better understanding of platelet biology. We will mainly focus on the most important membrane receptors and partially on their direct signaling pathways and provide cross-references when certain topics were only briefly addressed.

11.3.1 Animal Models to Assess Hemostasis

The major challenge in the development of antithrombotic agents has been the demonstration of strong efficacy, but at the same time providing a high safety level meaning only minor or ideally no increase in bleeding tendency. The tail-tip bleeding assay is a widely used model to determine the overall hemostatic function in mice. Typical parameters that are measured include time to cessation of bleeding, blood volume loss, rebleeding, and animal survival after challenge. Three models are mainly used after tail-cutting: immersion into warm saline, collecting blood with a Whatman paper, or tail observation (Fig. 11.2). However, the bleeding time model is well known for its considerable variation, even between different groups with the same mouse genotype. Therefore, multiple factors have to be considered: (I) The genetic background can influence the inter-animal variability. 129S1/SvmJ, BALB/c mice C57BL/6JBomTac, and, to a certain extent, C57BL/6NCrl mice showed almost no blood loss or inter-animal variability in the tail-tip bleeding model, whereas C57BL/6JCrl and C57BL/6JOlaHsd mice had a clearly higher total blood loss with significantly higher inter-animal variability [71]. (II) Furthermore, it was reported that gender plays an important role in this model since male mice tend to bleed longer than their female counterparts. (III) Finally, age, weight, the anesthetic drug, and the way of tail-cutting are factors that have to be considered [72]. In general, it is highly recommended to perform those experiments with an operator blinded to the mouse genotype. Interestingly, bleeding times tend to be normal in a variety of mice deficient for coagulation factors and are rather prolonged in mice deficient for platelet proteins [72, 73]. It is tempting to speculate that vessel spasm and a platelet plug are able to control bleeding from a mouse tail vein with its low blood pressure [72]. Taken together, the murine tail-tip assay is currently the most frequently used assay to determine the hemostatic function in mice.

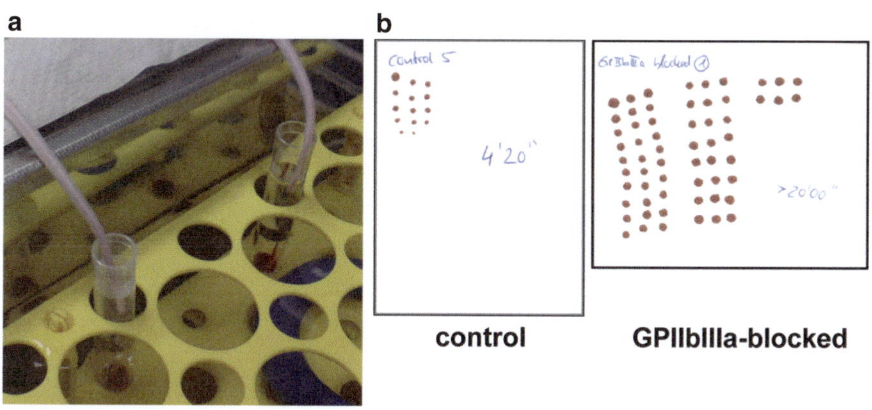

Fig. 11.2 Models to assess hemostasis in mice. (**a**) Immersion of clipped tail tip into warm saline; (**b**) collection of blood drops with a Whatman paper. *Left*, control mouse; *right*, mouse was injected with 100 µg blocking αIIbβ3 (GPIIbIIIa) antibody Fab fragments prior to experiment. Bleeding was manually stopped after 20 min

11.3.2 Experimental In Vivo Thrombosis Models

In parallel, many experimental models have been established to allow in vivo observation and kinetic analysis of thrombus formation. Arterial lesions can be induced in different parts of the vascular bed and by different methods [74] (Fig. 11.3):

(I) One widely used assay is the laser-injury model that injures the endothelium of the microcirculation and triggers thrombus formation in the range of minutes; however, these thrombi do typically not completely occlude the vessel [75]. This

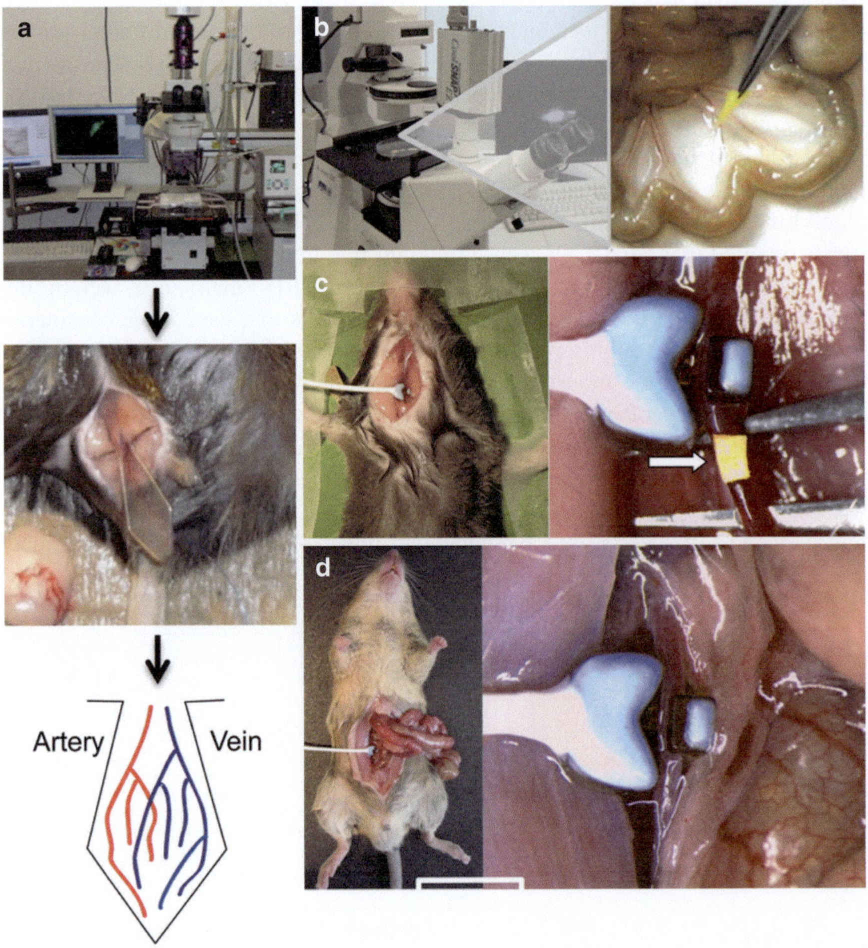

Fig. 11.3 Arterial thrombosis models. (**a**) *Top*, overview microscopic setup image of laser-induced thrombosis model; *middle*, preparation of the cremaster; *bottom*, scheme of the isolated cremaster. (**b**) $FeCl_3$-induced mesenteric arteriole thrombosis model. (**c**) $FeCl_3$-induced carotid artery thrombosis model. Doppler flow probe measures blood flow. *Arrow* indicates $FeCl_3$-soaked filter paper. (**d**) Mechanically injured (e.g., compression with forceps) abdominal aorta. Doppler flow probe measures blood flow

model allows simultaneous observation, e.g., of platelet deposition, tissue factor accumulation, and fibrin generation after laser-induced endothelial injury in a single developing thrombus [75]. Moreover, the laser power and exposure time can be adjusted which allows different types and levels of injury. Here, *Plcγ2*$^{-/-}$ mice display defective thrombus formation in superficial lesions but productive thrombosis after a more severe laser injury. In contrast, resistance to thrombus formation was observed in *Gαq*$^{-/-}$ mice after superficial and severe injuries in the laser-induced mesenteric artery model of thrombosis [76]. These data demonstrate that in this thrombosis model the severity of laser-induced injury determines the pathway of platelet activation at the site of injury. (II) Endothelial rupture can be mechanically induced in large arteries (e.g., aorta, carotid artery, femoral artery), either by direct damage of the luminal side of the vessel using a guide wire [77, 78] or indirectly by compression or ligation of the vessel using forceps or a filament [79, 80] or by a combination of these two methods [81]. It was shown that GPVI-deficient or GPVI-depleted mice are completely protected from occlusive thrombus formation after mechanical injury of large arteries by measuring the blood flow with a Doppler flow probe [82]. (III) Another common procedure to induce endothelial injury is direct, topical application of ferric chloride ($FeCl_3$), dropwise or with mounting a soaked filter paper onto the adventitial surface of an artery. Here, thrombus formation can be analyzed by direct mesenteric blood vessel visualization using intravital microscopy or a Doppler flow probe to monitor carotid artery occlusion. Ultrastructural analysis revealed that $FeCl_3$ diffused through the vessel wall, resulting in endothelial cell denudation and exposure of basement membrane components. In addition, numerous ferric ion-filled spherical bodies with large amounts of tissue factor appeared on the endothelial cells enabling platelet adhesion and aggregation [83]. It has been suggested that in the $FeCl_3$-induced injury model after severe injury, thrombus formation is mainly triggered by tissue factor exposure and thrombin generation [84]. Thus, possibly high concentrations of tissue factor and thrombin can bypass the collagen-dependent activation pathway in platelets. However, a role of the collagen-GPVI-induced pathway can be observed in this model by using moderate concentrations and exposure times of $FeCl_3$ [82]. Another chemical method to trigger arterial thrombosis is the intravenous injection of the photoreactive substance Rose Bengal, which rapidly accumulates in the membranes of endothelial and other cells. Subsequently, an arterial segment is exposed to green light (540 nm) that triggers a photochemical reaction, resulting in the formation of reactive oxygen species and damage of the endothelium [85, 86].

It has to be mentioned that each arterial thrombosis model has its specific advantages and limitations (Table 11.1). All models are normally performed in healthy mice which are not prone to atherothrombosis, whereas arterial thrombosis in humans typically occurs in diseased vessels. Nonetheless, studies of in vivo thrombus formation offer much more information than in vitro studies and help to better understand the molecular mechanisms involved in fatal occlusive thrombus formation in humans.

In general, it is highly recommended to use established protocols and more than one model to study the significance of certain platelet receptors and signaling

Table 11.1 Summary of widely used in vivo thrombus formation models

Type of injury	Determination of thrombus formation	Vascular bed	Pathway of platelet activation	Information
Laser injury	Intravital microscopy	Microcirculation	Laser power and exposure time (superficial vs. severe injury) dependent	No complete vessel occlusion
Mechanical injury:	Doppler flow probe, intravital microscopy	Aorta, carotid artery, femoral artery	GPVI dependent	
Compression by forceps				
Ligation with filament				
Guide wire				
Chemical injury with ferric chloride	Intravital microscopy, Doppler flow probe	Carotid artery, mesenteric arterioles, femoral artery	Severe injury: tissue factor and thrombin dependent	
			Moderate injury: GPVI dependent	

Note: Photochemical injury with Rose Bengal was not included into the table
For further information, see also Day et al. [74] and Westrick et al. [87]

pathways in arterial thrombus formation since high variability in the experimental protocols has caused controversial results and discussions on the relevance of individual molecules and pathways for intravascular thrombus formation.

11.3.3 Platelet Adhesion and Activation

11.3.3.1 GPIb: vWF

Platelet adhesion and rolling at sites of vascular injury is mediated by GPIb-vWF reversible interactions. GPIb is part of the GPIb/V/IX complex, and besides interacting with vWF to initiate thrombus formation, it also binds other ligands, including P-selectin, macrophage antigen 1 (Mac-1), and coagulation factor XII and thrombin. In humans, the lack or dysfunction of GPIb/V/IX is associated with a rare congenital bleeding disorder, the Bernard–Soulier syndrome (BSS), which is characterized by a bleeding phenotype, thrombocytopenia, and giant platelets (--> Chap. 14) [88]. Mice deficient in GPIbα [89] or GPIbβ [90] lack the entire receptor complex and reflect the human BSS accompanied with severe bleeding. Notably, no mutations within the *GP5* gene have been described in BSS patients and GPV deficiency in mice do not cause a BSS-like phenotype [91].

The essential role of GPIb in arterial thrombus formation was revealed by in vivo studies in mice. Transgenic mice in which the extracellular domain of GPIbα was replaced by the human interleukin-4 (IL-4) receptor ectodomain show severely prolonged tail bleeding [92] and are profoundly protected from vessel occlusion following chemical injury of mesenteric arterioles [93]. Absence of the GPIb ligand vWF leads to prolonged bleeding times, defective thrombus formation in chemically injured arterioles, and sometimes even to spontaneous bleeding in newborns [94]. The antithrombotic effect in GPIb mutant mice is more profound than that seen in *Vwf*$^{-/-}$ mice suggesting that also other ligands of GPIb may be involved in thrombus formation [93]. Similarly, inhibition of the vWF-binding site on GPIb with Fab fragments of the antibody p0p/B in wild-type mice exhibits abolished platelet adhesion to the injured carotid artery and these mice are protected from occlusive thrombus formation in vivo, again confirming a mandatory role of the receptor in this process [78].

The antithrombotic potential of targeting the vWF-GPIb axis has also been demonstrated in other animal models. For example, treatment of baboons with an anti-vWF antibody [95] or the application of Fab fragments of a blocking anti-GPIb antibody [96] exerts strong antithrombotic effects, whereas bleeding times are not or only moderately prolonged.

GPIb-vWF interaction induces intracellular signaling events that eventually lead to weak integrin activation, and phospholipase D1 (PLD1) seems to play a central function in this process [97]. Platelets from *Pld1*$^{-/-}$ mice display impaired activation in response to major agonists and defective GPIb-dependent aggregate formation under high shear conditions. These defects result in protection from arterial thrombosis without affecting tail bleeding times [97]. In line with these findings, inhibition of PLD isoforms with the small molecule PLD inhibitor, 5-fluoro-2-indolyl des-chlorohalopemide (FIPI), reproduces this protection in wild-type mice [98].

Ultra-large vWF (>20 million Da) is the most thrombogenic form of vWF and is cleaved to smaller, less thrombogenic forms by the multidomain structured 185 kDa protein *a disintegrin-like and metalloprotease with thrombospondin type I repeats-13* (ADAMTS13) in the plasma. Analysis of genetically modified mice revealed that ADAMTS13 downregulates platelet adhesion and aggregation in vivo, and ADAMTS13 deficiency can induce enhanced thrombus formation at the site of vascular lesions. This shows that ADAMTS13 regulates vWF size and is a major modulator of thrombus growth in vivo [99]. Another strategy to passivate GPIb function could be the controlled downregulation of the receptor from the platelet surface. *A disintegrin and metalloproteinase 17* (ADAM17) has been identified as the major sheddase to mediate GPIbα proteolysis in vitro and in vivo [100, 101]. Based on these results, one may speculate that the direct selective activation of ADAM17 in platelets might have strong antithrombotic potential.

11.3.3.2 GPVI

The major activating collagen receptor on the platelet surface is GPVI. GPVI is a MK-/platelet-specific transmembrane type I receptor that non-covalently associates with the FcR γ-chain. Upon ligand-induced GPVI clustering, the immunoreceptor tyrosine-based activation motif (ITAM) becomes tyrosine phosphorylated and initiates a series of phosphorylation events finally resulting in cellular activation [102].

A few patients with GPVI-related defects caused by autoantibody-induced receptor loss [103], compound heterozygous [104, 105], or homozygous mutations [106] have been reported suffering only from a mild bleeding tendency, but their platelets are unresponsive to collagen. Interestingly, this phenotype could be reproduced in different mouse models of GPVI deficiency or with blocking anti-GPVI antibody Fab fragments (reviewed in [102]), revealing that GPVI is largely dispensable for normal hemostasis but has a predominant role in the formation of experimental arterial thrombi in vivo. Although some controversies exist concerning the functional role of GPVI in certain thrombosis models [82, 83], the vast majority of analyses emphasized a prominent role of the receptor in thrombus formation in injured arteries or on ruptured atherosclerotic lesions [102]. It can be summarized that thrombus formation in the mechanically injured aorta and the $FeCl_3$-injured carotid artery are strongly dependent on functional GPVI signaling. However, longer exposure time or high concentration of $FeCl_3$ generates large amounts of thrombin which overcomes the importance of GPVI signaling [81, 82]. Moreover, mice deficient for GPVI were reported to display normal thrombus formation after laser-induced injury in small vessels. These data suggest that the initiating event can determine the relative importance of GPVI-mediated versus thrombin-mediated platelet activation in thrombus formation [84].

Over the last years, several other experimental strategies have been developed to interfere with GPVI function. One promising approach turned out to be the direct blockade of the ligand-binding site on platelet GPVI by a monovalent agent, such as an antibody Fab fragment, without altering its surface expression. This approach was efficient to inhibit collagen-induced thrombus formation ex vivo in rats and cynomolgus monkeys [107–109] and occlusive thrombus formation in injured arteries in vivo in mice and rats [78, 107] without causing a significant bleeding defect. Another targeting strategy may be based on the observation that GPVI can be specifically and irreversibly removed from circulating platelets in vivo by antibody treatment. Monoclonal anti-GPVI antibody injection in mice results in an acquired GPVI deficiency [110, 111] similar to that seen in patients with anti-GPVI autoantibodies and lasted for several days. However, this effect is accompanied by a short transient thrombocytopenia and reduced *protease-activated receptor* (PAR) 4 activity at early time points after treatment [112]. These mice display only moderately increased bleeding times, but a long-term antithrombotic protection in different experimental arterial thrombosis models. Interestingly, antibody-induced removal of GPVI could also be reproduced in human platelets circulating in NOD/SCID mice, demonstrating that this mechanism could be a powerful antithrombotic strategy in humans [113]. GPVI can principally be downregulated by internalization followed by degradation, or ectodomain shedding, both of which require signaling through the FcR γ-chain ITAM [114]. In vitro studies revealed ADAM10 and ADAM17 as the GPVI sheddases in platelets depending on the shedding-inducing stimulus, whereby ADAM10 seems to play the major role [101, 115]. However, the targeted ("therapeutic") downregulation of GPVI in vivo appears to occur through a more complex mechanism, since the antibody-induced GPVI downregulation is not prevented in mice with ADAM10/ADAM17 double-deficient platelets [101]. It has to be noted that an anti-GPVI therapy principally has to be carefully evaluated since anti-GPVI treatment severely compromises hemostasis in mice lacking CLEC-2 [116], integrin α2β1, or wild-type mice concomitantly treated with aspirin [80]. This indicates that other receptors can compensate for defective GPVI function in hemostasis and might explain the observed mild bleeding tendencies in humans and mice lacking GPVI. These findings could be very important with regard to potential anti-GPVI treatment of patients with other inherited or acquired platelet defects. Another possible strategy relies on the competitive inhibition of GPVI binding sites on its substrate collagen by a soluble dimeric GPVI-Fc fusion protein [117]. The antithrombotic potential of GPVI-Fc fusion proteins, however, seems to be less efficient in different arteriole injury models as compared to direct targeting of GPVI [118].

11.3.3.3 Integrin α2β1

Another major collagen receptor on the platelet surface is integrin α2β1 which plays an important but not essential role for the platelet adhesion on the matrix protein. Integrin α2-deficient mice display normal tail bleeding times and are able to form occlusive thrombi in injured arteries, although this was in one case reported to be delayed compared to the wild type [119–122]. This rather mild defect can be explained by multiple integrin–ligand interactions contributing to platelet adhesion at sites of injury since also mice lacking the β1 integrin (β1(fl/fl)/Mx-cre+) show largely unaltered platelet adhesion and aggregation in injured vessels [119]. The dominant integrin on the platelet surface, αIIbβ3, mediates platelet aggregation through binding of plasma fibrinogen and serves as the principal receptor for platelet adhesion to the ECM in vivo. The mandatory role of αIIbβ3 for hemostasis and thrombosis is well documented in humans and in experimental animals (reviewed in [9]). The shift of integrin αIIbβ3 from a low- to a high-affinity state is considered the final common pathway of platelet activation.

11.3.3.4 G-Protein-Coupled Receptors: PAR-3/PAR-4, P2Y1/P2Y12, and TP

Following initial adhesion and activation of platelets on the subendothelial ECM via GPIb and GPVI, respectively, thrombus growth requires locally produced and released soluble agonists, including thrombin, ADP, thromboxane A2 (TxA2), ATP, and epinephrine, which amplify the signaling response and recruit circulating platelets to the thrombus. Platelet activation by thrombin is largely mediated by members of the PAR family, with PAR-1 and PAR-4 being expressed in human platelets and PAR-3 and PAR-4 in mouse platelets [123]. Absence of PAR-3 in mice causes a markedly delayed and reduced, but not absent, response to thrombin and a protection against $FeCl_3$-induced thrombosis of mesenteric arterioles [124, 125]. PAR-4-deficient platelets show a similar degree of protection in this model, but these cells are completely resistant to thrombin [125, 126].

ADP is released from dense granules of activated cells and potentiates many platelet responses. Platelets express two different ADP receptors, P2Y1 and P2Y12.

In vivo, P2Y1-null mice have moderately increased bleeding times and are resistant against collagen/epinephrine- or ADP-induced thromboembolism [127]. P2Y12-deficient animals revealed a prolonged bleeding time and are protected from arterial thrombosis as measured in the $FeCl_3$ model [128]. These findings identified P2Y12 as the major ADP receptor amplifying platelet activation. TxA2 is produced by platelets from arachidonic acid via the cyclooxygenase pathway. Thromboxane-prostanoid (TP)-null mice show prolonged bleeding times and display delayed aggregation in response to collagen [129]. Platelet activation via G-protein-coupled receptors, such as P2Y1/P2Y12, TP, and PAR-1/PAR-3/PAR-4, involves three major G-protein-mediated signaling pathways that are initiated by activation of the G proteins: Gq, G12/G13, and Gi (see also review [130]). Mouse platelets that lack Gαq fail to aggregate and secrete their granule contents in response to thrombin, ADP, and TxA2. Accordingly, these mice display massively prolonged bleeding times and are protected from thrombus formation [131]. Mice deficient in Gα13 show a markedly increased bleeding time and fail to form stable occlusive thrombi in a model of carotid artery injury [132]. These data demonstrate that Gα13 is essentially required for thrombus formation, whereas Gα12 is dispensable for this process [132].

11.3.3.5 Munc13-4 and NBEAL2

Members of the Munc family control the fusion potential of granules with the membrane. Munc13-4 (Unc13d) was shown to be an important regulator of platelet granule secretion with in vivo implications [133, 134]. Mice lacking this protein show impaired hemostatic function and reduced platelet aggregation and thrombus formation in vitro and ex vivo. All of these deficits can be rescued by addition of ADP, demonstrating that Munc13-4 is crucial for ADP secretion from dense granules [134, 135]. On the other hand, mutations in *NBEAL2* have been linked to the gray platelet syndrome (GPS) [136–138], a rare bleeding disorder characterized by macrothrombocytopenia, with platelets lacking α-granules. Nbeal2-deficient murine platelets reproduce all hallmarks of GPS platelets in that they lack α-granules and display impaired adhesion, aggregation, and coagulant activity ex vivo which translates into defective hemostasis and arterial thrombus formation [139, 140].

11.3.3.6 Passivators of Platelet Signaling

Nitric oxide and prostacyclin are well-known exogenous inhibitors of platelet activation. In the past few years, analysis of genetically modified mice identified several other molecules that dampen platelet signaling. Constitutive signaling through GPVI is thought to be prevented by immunoreceptor tyrosine-based inhibition motif (ITIM) containing receptors, such as PECAM-1, CEACAM1/2, or G6b-B and the action of protein-tyrosine phosphatases (--> Chap. 7) [141, 142]. Moreover, intracellular proteins, such as SLAP/SLAP2 [143] or CLP-36 [144], have been shown to act as negative regulators of GPVI/FcR γ-chain-mediated signaling in platelets. Mice lacking the endothelial cell-specific adhesion molecule (ESAM) [145] develop larger thrombi in vivo as compared to control mice. These results show that platelet proteins can function as signaling passivators to avoid marked platelet hyperreactivity.

11.3.4 Thrombus Formation and Coagulation

11.3.4.1 αIIbβ3, Fibrinogen, and αIIbβ3 Regulators

Platelet aggregation is essential for the formation of the hemostatic plug at sites of vascular injury. This process is primarily mediated by αIIbβ3. Inherited deficiency or dysfunction of integrin αIIbβ3 in humans results in a disorder termed Glanzmann thrombasthenia, characterized by defective platelet aggregation and a severe bleeding diathesis [146]. Mice lacking either the αIIb (GPIIb) or β3 (GPIIIa) integrin subunit display a similar phenotype, namely, defective platelet aggregation and clot retraction in vitro, as well as a severe bleeding phenotype and defective experimental thrombus formation in vivo [147, 148]. Fibrinogen is a ligand of αIIbβ3 and bridges adjacent, activated platelets. Fibrinogen-null mice develop overt bleeding after birth and show neither platelet aggregation nor blood clotting in vitro [149]. Thus, fibrinogen is necessary for a permanent linkage of activated neighboring platelets, whereas vWF is necessary for thrombus growth once local shear rates become high. However, vWF alone is not sufficient to achieve stable platelet aggregation, supporting the hypothesis that concurrent binding of vWF to αIIbβ3 and GPIbα allows initial interplatelet contacts [150].

Many intracellular proteins have been identified to be involved in integrin regulation, among which talin-1 and kindlin-3 are crucial as shown by the analysis of the respective knockout mouse platelets. Both proteins bind to integrin β tails and are indispensable for integrin inside-out activation in platelets [151–153]. The defective integrin activation in those platelets translates into abolished thrombus formation in $FeCl_3$-injured vessels and infinite bleeding times in vivo. Several mutations in *KINDLIN-3* have been identified in humans resulting in a rare disorder called leukocyte adhesion deficiency type III (LAD-III) [154–156] with defective inside-out activation of hematopoietic cells causing among other defects severe bleeding.

Besides talin-1 and kindlin-3, the small GTPase Rap1 [157] and the balance between CalDAG-GEFI [158] and RASA3 [159] play essential roles in platelet aggregation and stable thrombus formation. However, the Rap1-GTP-interacting adaptor molecule (RIAM) seems to be dispensable for this process in platelets as shown by a RIAM knockout approach [160].

In addition to integrins, a wide range of other receptors on the platelet surface have been reported to further stabilize the thrombus, such as ephrins/Eph kinases, junctional adhesion molecules (JAMs), SEM4D, CD150, or CD84 [161]. A role for the latter one could, however, not be confirmed by studies on CD84-deficient mice [162]. A comprehensive discussion of the proteins involved in thrombus stability is beyond the scope of this book chapter and is discussed in detail in the review by Brass LF et al. [161].

11.3.4.2 Coagulation Factors

At sites of vascular injury, exposed tissue factor (TF) triggers local thrombin generation, which is further supported by the exposure of negatively charged phosphatidylserine on the platelet surface, providing a platform for the prothrombinase complex [163] and the release of inorganic polyphosphates which trigger the intrinsic pathway of coagulation through the activation of coagulation factor XII (FXII) [164]. It has been demonstrated that mice lacking FXII are protected against collagen- and epinephrine-induced thromboembolism and show defective thrombus formation and stabilization. However, they do not exhibit prolonged bleeding times or spontaneous bleedings [165]. Similarly, mice deficient in the FXII substrate FXI were similarly protected from vessel-occluding fibrin-rich clot formation, suggesting that FXII indeed contributes to pathological clotting through the "intrinsic pathway". In accordance with this finding, absence of FXI was previously demonstrated to protect mice from occlusion of the carotid artery in a $FeCl_3$ model [166, 167]. Taken together, coagulation factor-deficient mice, such as FXI [167] and FXII [165], subjected to the bleeding time model show normal hemostatic function. In contrast, blood coagulation FX deficiency in mice causes partial embryonic lethality and fatal neonatal bleeding [168], FVII deficiency causes fatal postnatal bleeding [169], and also FVIII deficiency causes increased bleeding [170].

Conclusion

The mouse is the premier animal model for biomedical research since the complete sequence of the mouse genome is available and made it possible to easily manipulate the mouse genome. Moreover, investigators have access to a large number of mouse mutants from public resources that are readily available for analysis. The generation of the Pf4-Cre transgenic mouse line was a major breakthrough to produce MK-/platelet-lineage-restricted knockouts. These tools allow us to get new insights into how platelets are produced and how they contribute to hemostasis and arterial thrombosis, but also provide tools to study their (patho-) physiological roles beyond these classical functions, i.e., in the orchestration of inflammatory processes.

Further studies will very likely include, besides classic knockout mice, mutant mice with functionally altered proteins to better mimic the situation in patients. Mouse models can help to take data obtained from genome-wide association study (GWAS) to the next level by investigating the function of associated genes in vivo [171]. In addition, the introduction of the CRISPR/Cas9 technique will allow very efficient genome editing of mouse zygotes that will greatly accelerate in vivo studies in the future.

Acknowledgments Prof. Dr. Bernhard Nieswandt holds a Chair of Experimental Biomedicine at the University Clinic and Rudolf Virchow Center, Würzburg Germany. This work was supported by the Deutsche Forschungsgemeinschaft (SFB688) and the Rudolf Virchow Centre.

Dr. Markus Bender is supported by the Deutsche Forschungsgemeinschaft (DFG) as a group leader in the Emmy Noether program of the DFG (BE 5084/3-1).

References

1. Hartwig J, Italiano J (2003) The birth of the platelet. J Thromb Haemost: JTH 1:1580–1586
2. Jackson S (2011) Arterial thrombosis – insidious, unpredictable and deadly. Nat Med 17: 1423–1436

3. Nieswandt B, Pleines I, Bender M (2011) Platelet adhesion and activation mechanisms in arterial thrombosis and ischaemic stroke. J Thromb Haemost: JTH 9(Suppl 1):92–104
4. Lopez A, Mathers C, Ezzati M, Jamison D, Murray C (2006) Global burden of disease and risk factors. World Bank, Washington, DC
5. Hagedorn I, Vögtle T, Nieswandt B (2010) Arterial thrombus formation. Novel mechanisms and targets. Hamostaseologie 30:127–135
6. Stoll G, Kleinschnitz C, Nieswandt B (2008) Molecular mechanisms of thrombus formation in ischemic stroke: novel insights and targets for treatment. Blood 112:3555–3562
7. Varga-Szabo D, Pleines I, Nieswandt B (2008) Cell adhesion mechanisms in platelets. Arterioscler Thromb Vasc Biol 28:403–412
8. Varga-Szabo D, Braun A, Nieswandt B (2009) Calcium signaling in platelets. J Thromb Haemost: JTH 7:1057–1066
9. Nieswandt B, Varga-Szabo D, Elvers M (2009) Integrins in platelet activation. J Thromb Haemost: JTH 7(Suppl 1):206–209
10. Junt T et al (2007) Dynamic visualization of thrombopoiesis within bone marrow. Science (New York, NY) 317:1767–1770
11. Zhang L et al (2012) A novel role of sphingosine 1-phosphate receptor S1pr1 in mouse thrombopoiesis. J Exp Med 209:2165–2181
12. Bender M et al (2014) Megakaryocyte-specific Profilin1-deficiency alters microtubule stability and causes a Wiskott-Aldrich syndrome-like platelet defect. Nat Commun 5:4746
13. Stalker T et al (2013) Hierarchical organization in the hemostatic response and its relationship to the platelet-signaling network. Blood 121:1875–1885
14. Sachs U, Nieswandt B (2007) In vivo thrombus formation in murine models. Circ Res 100: 979–991
15. Kühn R, Schwenk F, Aguet M, Rajewsky K (1995) Inducible gene targeting in mice. Science (New York, NY) 269:1427–1429
16. Tiedt R, Schomber T, Hao-Shen H, Skoda R (2007) Pf4-Cre transgenic mice allow the generation of lineage-restricted gene knockouts for studying megakaryocyte and platelet function in vivo. Blood 109:1503–1506
17. Sander J, Joung J (2014) CRISPR-Cas systems for editing, regulating and targeting genomes. Nat Biotechnol 32:347–355
18. Wang H et al (2013) One-step generation of mice carrying mutations in multiple genes by CRISPR/Cas-mediated genome engineering. Cell 153:910–918
19. Kim I, Shin J, Seong J (2010) Mouse phenogenomics, toolbox for functional annotation of human genome. BMB Rep 43:79–90
20. Schmitt A, Guichard J, Massé J, Debili N, Cramer E (2001) Of mice and men: comparison of the ultrastructure of megakaryocytes and platelets. Exp Hematol 29:1295–1302
21. Li T-T et al (2004) Genetic variation responsible for mouse strain differences in integrin alpha 2 expression is associated with altered platelet responses to collagen. Blood 103: 3396–3402
22. Kuter D (2014) Milestones in understanding platelet production: a historical overview. Br J Haematol 165:248–258
23. Hitchcock I, Kaushansky K (2014) Thrombopoietin from beginning to end. Br J Haematol 165:259–268
24. Gurney A, Carver-Moore K, de Sauvage F, Moore M (1994) Thrombocytopenia in c-mpl-deficient mice. Science (New York, NY) 265:1445–1447
25. de Sauvage F et al (1996) Physiological regulation of early and late stages of megakaryocytopoiesis by thrombopoietin. J Exp Med 183:651–656
26. Bunting S et al (1997) Normal platelets and megakaryocytes are produced in vivo in the absence of thrombopoietin. Blood 90:3423–3429
27. Carver-Moore K et al (1996) Low levels of erythroid and myeloid progenitors in thrombopoietin-and c-mpl-deficient mice. Blood 88:803–808
28. de Sauvage F, Villeval J, Shivdasani R (1998) Regulation of megakaryocytopoiesis and platelet production: lessons from animal models. J Lab Clin Med 131:496–501

29. Ng A et al (2014) Mpl expression on megakaryocytes and platelets is dispensable for thrombopoiesis but essential to prevent myeloproliferation. Proc Natl Acad Sci U S A 111: 5884–5889
30. Meyer S et al (2014) Genetic studies reveal an unexpected negative regulatory role for Jak2 in thrombopoiesis. Blood 124:2280–2284
31. Skoda R (2014) Less Jak2 makes more platelets. Blood 124:2168–2169
32. Grozovsky R et al (2015) The Ashwell-Morell receptor regulates hepatic thrombopoietin production via JAK2-STAT3 signaling. Nat Med 21:47–54
33. Kile B (2014) The role of apoptosis in megakaryocytes and platelets. Br J Haematol 165:217–226
34. De Botton S et al (2002) Platelet formation is the consequence of caspase activation within megakaryocytes. Blood 100:1310–1317
35. Clarke M, Savill J, Jones D, Noble B, Brown S (2003) Compartmentalized megakaryocyte death generates functional platelets committed to caspase-independent death. J Cell Biol 160:577–587
36. Josefsson E et al (2011) Megakaryocytes possess a functional intrinsic apoptosis pathway that must be restrained to survive and produce platelets. J Exp Med 208:2017–2031
37. Josefsson E et al (2014) Platelet production proceeds independently of the intrinsic and extrinsic apoptosis pathways. Nat Commun 5:3455
38. Schulze H, Shivdasani R (2004) Molecular mechanisms of megakaryocyte differentiation. Semin Thromb Hemost 30:389–398
39. Tijssen M, Ghevaert C (2013) Transcription factors in late megakaryopoiesis and related platelet disorders. J Thromb Haemost: JTH 11:593–604
40. Fujiwara Y, Browne C, Cunniff K, Goff S, Orkin S (1996) Arrested development of embryonic red cell precursors in mouse embryos lacking transcription factor GATA-1. Proc Natl Acad Sci U S A 93:12355–12358
41. Pevny L et al (1991) Erythroid differentiation in chimaeric mice blocked by a targeted mutation in the gene for transcription factor GATA-1. Nature 349:257–260
42. Pevny L et al (1995) Development of hematopoietic cells lacking transcription factor GATA-1. Development 121:163–172
43. Shivdasani R, Fujiwara Y, McDevitt M, Orkin S (1997) A lineage-selective knockout establishes the critical role of transcription factor GATA-1 in megakaryocyte growth and platelet development. EMBO J 16:3965–3973
44. Nurden A (2005) Qualitative disorders of platelets and megakaryocytes. J Thromb Haemost: JTH 3:1773–1782
45. Tsang A, Fujiwara Y, Hom D, Orkin S (1998) Failure of megakaryopoiesis and arrested erythropoiesis in mice lacking the GATA-1 transcriptional cofactor FOG. Genes Dev 12:1176–1188
46. Shivdasani R et al (1995) Transcription factor NF-E2 is required for platelet formation independent of the actions of thrombopoietin/MGDF in megakaryocyte development. Cell 81:695–704
47. Shivdasani R, Orkin S (1995) Erythropoiesis and globin gene expression in mice lacking the transcription factor NF-E2. Proc Natl Acad Sci U S A 92:8690–8694
48. Okuda T, van Deursen J, Hiebert S, Grosveld G, Downing J (1996) AML1, the target of multiple chromosomal translocations in human leukemia, is essential for normal fetal liver hematopoiesis. Cell 84:321–330
49. Wang Q et al (1996) Disruption of the Cbfa2 gene causes necrosis and hemorrhaging in the central nervous system and blocks definitive hematopoiesis. Proc Natl Acad Sci U S A 93:3444–3449
50. Ichikawa M et al (2004) AML-1 is required for megakaryocytic maturation and lymphocytic differentiation, but not for maintenance of hematopoietic stem cells in adult hematopoiesis. Nat Med 10:299–304
51. Growney J et al (2005) Loss of Runx1 perturbs adult hematopoiesis and is associated with a myeloproliferative phenotype. Blood 106:494–504

52. Michaud J et al (2002) In vitro analyses of known and novel RUNX1/AML1 mutations in dominant familial platelet disorder with predisposition to acute myelogenous leukemia: implications for mechanisms of pathogenesis. Blood 99:1364–1372
53. Song W et al (1999) Haploinsufficiency of CBFA2 causes familial thrombocytopenia with propensity to develop acute myelogenous leukaemia. Nat Genet 23:166–175
54. Hartwig J, Italiano J (2006) Cytoskeletal mechanisms for platelet production. Blood Cells Mol Dis 36:99–103
55. Machlus K, Italiano J (2013) The incredible journey: from megakaryocyte development to platelet formation. J Cell Biol 201:785–796
56. Bender M et al (2014) Microtubule sliding drives proplatelet elongation and is dependent on cytoplasmic dynein. Blood 125:860–868
57. Patel S et al (2005) Differential roles of microtubule assembly and sliding in proplatelet formation by megakaryocytes. Blood 106:4076–4085
58. Schwer H et al (2001) A lineage-restricted and divergent beta-tubulin isoform is essential for the biogenesis, structure and function of blood platelets. Curr Biol: CB 11:579–586
59. Kunishima S, Kobayashi R, Itoh T, Hamaguchi M, Saito H (2009) Mutation of the beta1-tubulin gene associated with congenital macrothrombocytopenia affecting microtubule assembly. Blood 113:458–461
60. Kunert S et al (2009) The microtubule modulator RanBP10 plays a critical role in regulation of platelet discoid shape and degranulation. Blood 114:5532–5540
61. Pleines I et al (2013) Defective tubulin organization and proplatelet formation in murine megakaryocytes lacking Rac1 and Cdc42. Blood 122:3178–3187
62. Bender M et al (2010) ADF/n-cofilin-dependent actin turnover determines platelet formation and sizing. Blood 116:1767–1775
63. Falet H et al (2010) A novel interaction between FlnA and Syk regulates platelet ITAM-mediated receptor signaling and function. J Exp Med 207:1967–1979
64. Jurak Begonja A, Hoffmeister K, Hartwig J, Falet H (2011) FlnA-null megakaryocytes prematurely release large and fragile platelets that circulate poorly. Blood 118:2285–2295
65. Nurden P et al (2011) Thrombocytopenia resulting from mutations in filamin A can be expressed as an isolated syndrome. Blood 118:5928–5937
66. Berrou E et al (2013) Heterogeneity of platelet functional alterations in patients with filamin A mutations. Arterioscler Thromb Vasc Biol 33:8
67. Snapper S et al (1998) Wiskott-Aldrich syndrome protein-deficient mice reveal a role for WASP in T but not B cell activation. Immunity 9:81–91
68. Althaus K, Greinacher A (2009) MYH9-related platelet disorders. Semin Thromb Hemost 35:189–203
69. Zhang Y et al (2012) Mouse models of MYH9-related disease: mutations in nonmuscle myosin II-A. Blood 119:238–250
70. Chen Z et al (2007) The May-Hegglin anomaly gene MYH9 is a negative regulator of platelet biogenesis modulated by the Rho-ROCK pathway. Blood 110:171–179
71. Schiviz A et al (2014) Influence of genetic background on bleeding phenotype in the tail-tip bleeding model and recommendations for standardization: communication from the SSC of the ISTH. J Thromb Haemost: JTH 12:1940–1942
72. Greene T et al (2010) Towards a standardization of the murine tail bleeding model. J Thromb Haemost: JTH 8:2820–2822
73. Broze G, Yin Z, Lasky N (2001) A tail vein bleeding time model and delayed bleeding in hemophiliac mice. Thromb Haemost 85:747–748
74. Day S, Reeve J, Myers D, Fay W (2004) Murine thrombosis models. Thromb Haemost 92:486–494
75. Falati S, Gross P, Merrill-Skoloff G, Furie B, Furie B (2002) Real-time in vivo imaging of platelets, tissue factor and fibrin during arterial thrombus formation in the mouse. Nat Med 8:1175–1181
76. Nonne C et al (2005) Importance of platelet phospholipase Cgamma2 signaling in arterial thrombosis as a function of lesion severity. Arterioscler Thromb Vasc Biol 25:1293–1298

77. Konishi H et al (2002) Platelets activated by collagen through immunoreceptor tyrosine-based activation motif play pivotal role in initiation and generation of neointimal hyperplasia after vascular injury. Circulation 105:912–916
78. Massberg S et al (2003) A crucial role of glycoprotein VI for platelet recruitment to the injured arterial wall in vivo. J Exp Med 197:41–49
79. Massberg S et al (2002) A critical role of platelet adhesion in the initiation of atherosclerotic lesion formation. J Exp Med 196:887–896
80. Grüner S et al (2004) Anti-glycoprotein VI treatment severely compromises hemostasis in mice with reduced alpha2beta1 levels or concomitant aspirin therapy. Circulation 110:2946–2951
81. Mangin P et al (2006) Thrombin overcomes the thrombosis defect associated with platelet GPVI/FcRgamma deficiency. Blood 107:4346–4353
82. Bender M, Hagedorn I, Nieswandt B (2011) Genetic and antibody-induced glycoprotein VI deficiency equally protects mice from mechanically and FeCl(3) -induced thrombosis. J Thromb Haemost: JTH 9:1423–1426
83. Eckly A et al (2011) Mechanisms underlying FeCl3-induced arterial thrombosis. J Thromb Haemost: JTH 9:779–789
84. Dubois C, Panicot-Dubois L, Merrill-Skoloff G, Furie B, Furie B (2006) Glycoprotein VI-dependent and -independent pathways of thrombus formation in vivo. Blood 107:3902–3906
85. Matsuno H, Uematsu T, Nagashima S, Nakashima M (1991) Photochemically induced thrombosis model in rat femoral artery and evaluation of effects of heparin and tissue-type plasminogen activator with use of this model. J Pharmacol Methods 25:303–317
86. Kikuchi S, Umemura K, Kondo K, Saniabadi A, Nakashima M (1998) Photochemically induced endothelial injury in the mouse as a screening model for inhibitors of vascular intimal thickening. Arterioscler Thromb Vasc Biol 18:1069–1078
87. Westrick R, Winn M, Eitzman D (2007) Murine models of vascular thrombosis (Eitzman series). Arterioscler Thromb Vasc Biol 27:2079–2093
88. Canobbio I, Balduini C, Torti M (2004) Signalling through the platelet glycoprotein Ib-V-IX complex. Cell Signal 16:1329–1344
89. Ware J, Russell S, Ruggeri Z (2000) Generation and rescue of a murine model of platelet dysfunction: the Bernard-Soulier syndrome. Proc Natl Acad Sci U S A 97:2803–2808
90. Kato K et al (2004) Genetic deletion of mouse platelet glycoprotein Ibbeta produces a Bernard-Soulier phenotype with increased alpha-granule size. Blood 104:2339–2344
91. Kahn M et al (1999) Glycoprotein V-deficient platelets have undiminished thrombin responsiveness and do not exhibit a Bernard-Soulier phenotype. Blood 94:4112–4121
92. Kanaji T, Russell S, Ware J (2002) Amelioration of the macrothrombocytopenia associated with the murine Bernard-Soulier syndrome. Blood 100:2102–2107
93. Bergmeier W et al (2006) The role of platelet adhesion receptor GPIbalpha far exceeds that of its main ligand, von Willebrand factor, in arterial thrombosis. Proc Natl Acad Sci U S A 103:16900–16905
94. Denis C et al (1998) A mouse model of severe von Willebrand disease: defects in hemostasis and thrombosis. Proc Natl Acad Sci U S A 95:9524–9529
95. Wu D et al (2002) Inhibition of the von Willebrand (VWF)-collagen interaction by an antihuman VWF monoclonal antibody results in abolition of in vivo arterial platelet thrombus formation in baboons. Blood 99:3623–3628
96. Cauwenberghs N et al (2000) Antithrombotic effect of platelet glycoprotein Ib-blocking monoclonal antibody Fab fragments in nonhuman primates. Arterioscler Thromb Vasc Biol 20:1347–1353
97. Elvers M et al (2010) Impaired alpha(IIb)beta(3) integrin activation and shear-dependent thrombus formation in mice lacking phospholipase D1. Sci Signal 3:ra1
98. Stegner D et al (2013) Pharmacological inhibition of phospholipase D protects mice from occlusive thrombus formation and ischemic stroke – brief report. Arterioscler Thromb Vasc Biol 33:2212–2217
99. Banno F et al (2009) The distal carboxyl-terminal domains of ADAMTS13 are required for regulation of in vivo thrombus formation. Blood 113:5323–5329

100. Bergmeier W et al (2004) Tumor necrosis factor-alpha-converting enzyme (ADAM17) mediates GPIbalpha shedding from platelets in vitro and in vivo. Circ Res 95:677–683
101. Bender M et al (2010) Differentially regulated GPVI ectodomain shedding by multiple platelet-expressed proteinases. Blood 116:3347–3355
102. Dütting S, Bender M, Nieswandt B (2012) Platelet GPVI: a target for antithrombotic therapy?! Trends Pharmacol Sci 33:583–590
103. Arthur J, Dunkley S, Andrews R (2007) Platelet glycoprotein VI-related clinical defects. Br J Haematol 139:363–372
104. Dumont B et al (2009) Absence of collagen-induced platelet activation caused by compound heterozygous GPVI mutations. Blood 114:1900–1903
105. Hermans C et al (2009) A compound heterozygous mutation in glycoprotein VI in a patient with a bleeding disorder. J Thromb Haemost: JTH 7:1356–1363
106. Matus V et al (2013) An adenine insertion in exon 6 of human GP6 generates a truncated protein associated with a bleeding disorder in four Chilean families. J Thromb Haemost: JTH 11:1751–1759
107. Li H et al (2007) The Fab fragment of a novel anti-GPVI monoclonal antibody, OM4, reduces in vivo thrombosis without bleeding risk in rats. Arterioscler Thromb Vasc Biol 27: 1199–1205
108. Matsumoto Y et al (2006) Ex vivo evaluation of anti-GPVI antibody in cynomolgus monkeys: dissociation between anti-platelet aggregatory effect and bleeding time. Thromb Haemost 96:167–175
109. Ohlmann P et al (2008) Ex vivo inhibition of thrombus formation by an anti-glycoprotein VI Fab fragment in non-human primates without modification of glycoprotein VI expression. J Thromb Haemost: JTH 6:1003–1011
110. Nieswandt B et al (2001) Long-term antithrombotic protection by in vivo depletion of platelet glycoprotein VI in mice. J Exp Med 193:459–469
111. Schulte V et al (2003) Targeting of the collagen-binding site on glycoprotein VI is not essential for in vivo depletion of the receptor. Blood 101:3948–3952
112. Schulte V et al (2006) Two-phase antithrombotic protection after anti-glycoprotein VI treatment in mice. Arterioscler Thromb Vasc Biol 26:1640–1647
113. Boylan B, Berndt M, Kahn M, Newman P (2006) Activation-independent, antibody-mediated removal of GPVI from circulating human platelets: development of a novel NOD/SCID mouse model to evaluate the in vivo effectiveness of anti-human platelet agents. Blood 108:908–914
114. Rabie T et al (2007) Diverging signaling events control the pathway of GPVI down-regulation in vivo. Blood 110:529–535
115. Gardiner E et al (2007) Controlled shedding of platelet glycoprotein (GP)VI and GPIb-IX-V by ADAM family metalloproteinases. J Thromb Haemost: JTH 5:1530–1537
116. Bender M et al (2013) Combined in vivo depletion of glycoprotein VI and C-type lectin-like receptor 2 severely compromises hemostasis and abrogates arterial thrombosis in mice. Arterioscler Thromb Vasc Biol 33:926–934
117. Massberg S et al (2004) Soluble glycoprotein VI dimer inhibits platelet adhesion and aggregation to the injured vessel wall in vivo. FASEB J: Off Publ Fed Am Soc Exp Biol 18:397–399
118. Grüner S et al (2005) Relative antithrombotic effect of soluble GPVI dimer compared with anti-GPVI antibodies in mice. Blood 105:1492–1499
119. Grüner S et al (2003) Multiple integrin-ligand interactions synergize in shear-resistant platelet adhesion at sites of arterial injury in vivo. Blood 102:4021–4027
120. Holtkötter O et al (2002) Integrin alpha 2-deficient mice develop normally, are fertile, but display partially defective platelet interaction with collagen. J Biol Chem 277: 10789–10794
121. Chen J, Diacovo T, Grenache D, Santoro S, Zutter M (2002) The alpha(2) integrin subunit-deficient mouse: a multifaceted phenotype including defects of branching morphogenesis and hemostasis. Am J Pathol 161:337–344

122. He L et al (2003) The contributions of the alpha 2 beta 1 integrin to vascular thrombosis in vivo. Blood 102:3652–3657
123. Coughlin S (2001) Protease-activated receptors in vascular biology. Thromb Haemost 86:298–307
124. Kahn M et al (1998) A dual thrombin receptor system for platelet activation. Nature 394:690–694
125. Weiss E, Hamilton J, Lease K, Coughlin S (2002) Protection against thrombosis in mice lacking PAR3. Blood 100:3240–3244
126. Sambrano G, Weiss E, Zheng Y, Huang W, Coughlin S (2001) Role of thrombin signalling in platelets in haemostasis and thrombosis. Nature 413:74–78
127. Léon C et al (1999) Defective platelet aggregation and increased resistance to thrombosis in purinergic P2Y(1) receptor-null mice. J Clin Invest 104:1731–1737
128. Andre P et al (2003) P2Y12 regulates platelet adhesion/activation, thrombus growth, and thrombus stability in injured arteries. J Clin Invest 112:398–406
129. Thomas D et al (1998) Coagulation defects and altered hemodynamic responses in mice lacking receptors for thromboxane A2. J Clin Invest 102:1994–2001
130. Offermanns S (2006) Activation of platelet function through G protein-coupled receptors. Circ Res 99:1293–1304
131. Offermanns S, Toombs C, Hu Y, Simon M (1997) Defective platelet activation in G alpha(q)-deficient mice. Nature 389:183–186
132. Moers A et al (2003) G13 is an essential mediator of platelet activation in hemostasis and thrombosis. Nat Med 9:1418–1422
133. Shirakawa R et al (2004) Munc13-4 is a GTP-Rab27-binding protein regulating dense core granule secretion in platelets. J Biol Chem 279:10730–10737
134. Ren Q et al (2010) Munc13-4 is a limiting factor in the pathway required for platelet granule release and hemostasis. Blood 116:869–877
135. Savage J et al (2013) Munc13-4 is critical for thrombosis through regulating release of ADP from platelets. J Thromb Haemost: JTH 11:771–775
136. Albers C et al (2011) Exome sequencing identifies NBEAL2 as the causative gene for gray platelet syndrome. Nat Genet 43:735–737
137. Kahr W et al (2011) Mutations in NBEAL2, encoding a BEACH protein, cause gray platelet syndrome. Nat Genet 43:738–740
138. Gunay-Aygun M et al (2011) NBEAL2 is mutated in gray platelet syndrome and is required for biogenesis of platelet α-granules. Nat Genet 43:732–734
139. Deppermann C et al (2013) Gray platelet syndrome and defective thrombo-inflammation in Nbeal2-deficient mice. J Clin Invest 123:3331–3342
140. Kahr W et al (2013) Abnormal megakaryocyte development and platelet function in Nbeal2 (-/-) mice. Blood 122:3349–3358
141. Gibbins J (2002) The negative regulation of platelet function: extending the role of the ITIM. Trends Cardiovasc Med 12:213–219
142. Senis Y (2013) Protein-tyrosine phosphatases: a new frontier in platelet signal transduction. J Thromb Haemost: JTH 11:1800–1813
143. Cherpokova D et al (2015) SLAP/SLAP2 prevent excessive platelet (hem)ITAM signaling in thrombosis and ischemic stroke in mice. Blood 125:185–194
144. Gupta S et al (2012) CLP36 is a negative regulator of glycoprotein VI signaling in platelets. Circ Res 111:1410–1420
145. Stalker T et al (2009) Endothelial cell specific adhesion molecule (ESAM) localizes to platelet-platelet contacts and regulates thrombus formation in vivo. J Thromb Haemost: JTH 7:1886–1896
146. Nurden A (2006) Glanzmann thrombasthenia. Orphanet J Rare Dis 1:10
147. Hodivala-Dilke K et al (1999) Beta3-integrin-deficient mice are a model for Glanzmann thrombasthenia showing placental defects and reduced survival. J Clin Invest 103:229–238
148. Tronik-Le Roux D et al (2000) Thrombasthenic mice generated by replacement of the integrin alpha(IIb) gene: demonstration that transcriptional activation of this megakaryocytic locus precedes lineage commitment. Blood 96:1399–1408

149. Suh T et al (1995) Resolution of spontaneous bleeding events but failure of pregnancy in fibrinogen-deficient mice. Genes Dev 9:2020–2033
150. Ni H et al (2000) Persistence of platelet thrombus formation in arterioles of mice lacking both von Willebrand factor and fibrinogen. J Clin Invest 106:385–392
151. Moser M, Nieswandt B, Ussar S, Pozgajova M, Fässler R (2008) Kindlin-3 is essential for integrin activation and platelet aggregation. Nat Med 14:325–330
152. Petrich B et al (2007) Talin is required for integrin-mediated platelet function in hemostasis and thrombosis. J Exp Med 204:3103–3111
153. Tadokoro S et al (2003) Talin binding to integrin beta tails: a final common step in integrin activation. Science (New York, NY) 302:103–106
154. Malinin N et al (2009) A point mutation in KINDLIN3 ablates activation of three integrin subfamilies in humans. Nat Med 15:313–318
155. Mory A et al (2008) Kindlin-3: a new gene involved in the pathogenesis of LAD-III. Blood 112:2591
156. Svensson L et al (2009) Leukocyte adhesion deficiency-III is caused by mutations in KINDLIN3 affecting integrin activation. Nat Med 15:306–312
157. Chrzanowska-Wodnicka M, Smyth S, Schoenwaelder S, Fischer T, White G (2005) Rap1b is required for normal platelet function and hemostasis in mice. J Clin Invest 115:680–687
158. Crittenden J et al (2004) CalDAG-GEFI integrates signaling for platelet aggregation and thrombus formation. Nat Med 10:982–986
159. Stefanini L et al (2015) RASA3 is a critical inhibitor of RAP1-dependent platelet activation. J Clin Invest 125:1419–1432
160. Stritt S et al (2015) Rap1-GTP-interacting adaptor molecule (RIAM) is dispensable for platelet integrin activation and function in mice. Blood 125:219–222
161. Brass L, Zhu L, Stalker T (2005) Minding the gaps to promote thrombus growth and stability. J Clin Invest 115:3385–3392
162. Hofmann S et al (2014) Mice lacking the SLAM family member CD84 display unaltered platelet function in hemostasis and thrombosis. PLoS One 9:e115306
163. Heemskerk J, Mattheij N, Cosemans JME (2013) Platelet-based coagulation: different populations, different functions. J Thromb Haemost: JTH 11:2–16
164. Müller F et al (2009) Platelet polyphosphates are proinflammatory and procoagulant mediators in vivo. Cell 139:1143–1156
165. Renné T et al (2005) Defective thrombus formation in mice lacking coagulation factor XII. J Exp Med 202:271–281
166. Müller F, Gailani D, Renné T (2011) Factor XI and XII as antithrombotic targets. Curr Opin Hematol 18:349–355
167. Wang X et al (2005) Effects of factor IX or factor XI deficiency on ferric chloride-induced carotid artery occlusion in mice. J Thromb Haemost: JTH 3:695–702
168. Dewerchin M et al (2000) Blood coagulation factor X deficiency causes partial embryonic lethality and fatal neonatal bleeding in mice. Thromb Haemost 83:185–190
169. Rosen E et al (1997) Mice lacking factor VII develop normally but suffer fatal perinatal bleeding. Nature 390:290–294
170. Bi L et al (1996) Further characterization of factor VIII-deficient mice created by gene targeting: RNA and protein studies. Blood 88:3446–3450
171. Cox R, Church C (2011) Mouse models and the interpretation of human GWAS in type 2 diabetes and obesity. Dis Model Mech 4:155–164

Part III

Platelets in Health and Disease

Fetal and Neonatal Megakaryopoiesis and Platelet Biology

12

Christof Dame, Viola Lorenz, and Martha Sola-Visner

Abstract

Megakaryopoiesis and platelet biology undergo significant changes during development. Major differences in fetal/neonatal compared to adult megakaryopoiesis include the smaller size of neonatal megakaryocytes, their cytoplasmic maturity despite lower ploidy levels, and the tenfold higher proliferative rate of neonatal megakaryocyte progenitors. Such differences are accompanied by relatively high thrombopoietin (Thpo) plasma levels in fetuses and neonates and a higher sensitivity of neonatal megakaryocyte progenitors to Thpo in vitro. Particularly during the first week after birth, neonatal platelets are hyporesponsive to most agonists, a finding that is most pronounced in very preterm infants. Interestingly, this platelet hyporeactivity is compensated by a number of factors in neonatal blood that enhance primary hemostasis, such as higher hematocrit values, higher von Willebrand factor (vWF) concentrations, and predominance of ultralong vWF polymers in the blood of neonates. This delicate balance at each step of fetal development is required for optimal hemostasis and also for the proper development of the vascular and hematopoietic system, particularly in the bone marrow. The delicate neonatal hemostatic balance is easily disturbed, especially by inflammatory conditions that can cause significant bleeding, often with high mortality or severe long-term sequelae. Research on the developmental biology of megakaryopoiesis has led to the discovery of the molecular mechanisms of various inherited thrombocytopenias, such as congenital amegakaryocytic thrombocytopenia, thrombocytopenia-absent radii syndrome, and Down syndrome-associated transient myeloproliferative disorder.

C. Dame, MD (✉)
Department of Neonatology, Charité – Universitätsmedizin Berlin,
Augustenburger Platz 1, 13553 Berlin, Germany
e-mail: christof.dame@charite.de

V. Lorenz • M. Sola-Visner
Division of Newborn Medicine, Boston Children's Hospital, Harvard Medical School,
Boston, MA, USA

H. Schulze, J. Italiano (eds.), *Molecular and Cellular Biology of Platelet Formation*,
DOI 10.1007/978-3-319-39562-3_12

12.1 Introduction

Over the last two decades, a mounting body of evidence has supported the existence of substantial differences between fetal/neonatal and adult megakaryocytes and platelets, both regarding their morphology and biology. Developmental stage-specific differences are ontogenetically important, because they allow the fetus to generate rising platelet counts while the blood volume is rapidly expanding, in a time period characterized by extraordinary growth. The developmental stage-specific functional characteristics of fetal and neonatal platelets are part of a delicate balance between platelets and coagulation factors that provides optimal hemostasis. There is early evidence in support of the fact that transfusion of adult donor platelets might result in a prothrombotic phenotype in neonates. As developmental processes are easily to disturb, sick neonates, in particular very-low-birth-weight infants (birth weight <1500 g), are at high risk of thrombocytopenia. The pathogenesis of some inherited disorders of megakaryopoiesis, including those that selectively affect fetuses and neonates (but not adults), could be elucidated in recent years by identifying the underlying molecular mechanisms. Beyond newborn medicine, a better understanding of the biology of the megakaryocytes that originate from cord blood-derived stem cells may be important to optimize engraftment in adult recipients of cord blood stem cell transplants. These considerations are the rationale underlying the focus of this chapter exclusively on fetal and neonatal megakaryopoiesis and platelet biology.

12.2 Cellular Development of Megakaryopoiesis in the Human Embryo, Fetus, and Neonate

12.2.1 Sites of Embryonic, Fetal, and Neonatal Megakaryopoiesis

In the murine embryo, hematopoiesis starts shortly after the onset of gastrulation in the blood islands of the yolk sac (YS) [74]. Here, primitive hematopoietic and endothelial precursors originate from a common progenitor cell, the hemangioblast [51], from which the megakaryocyte lineage also initiates [102]. In the murine embryo, megakaryocyte (MK) progenitors are first detected at low numbers by the early- to mid-neural plate stages and then increase exclusively within the YS (Fig. 12.1a). After the onset of circulation, low numbers of MK progenitors are found in the embryo proper and in the blood stream. By day E10.5 of gestation, the number of megakaryocyte colony-forming cells (Meg-CFCs) starts to decrease in the YS and to increase in the fetal liver (Fig. 12.1b). At that stage, colonies containing both primitive erythroid (βH1-globin-positive) and megakaryocyte (glycoprotein GPIbβ-positive) cells are identified, indicating the existence of bipotential primitive megakaryocyte-erythroid progenitors (MEP), in addition to pure erythroid or MK colonies. MK maturation within these colonies is evident by the production of proplatelets (Fig. 12.1c) [102].

Two recent murine studies have shown that the first platelet-forming cells of the YS are not polyploid MKs, but diploid platelet-forming cells (DPFC) [88]. DPFC appear in the yolk sac at E8.5 and produce the first platelets at E9.5, whereas the terminal differentiation of MKs is accomplished in the fetal liver by E16.5, representing

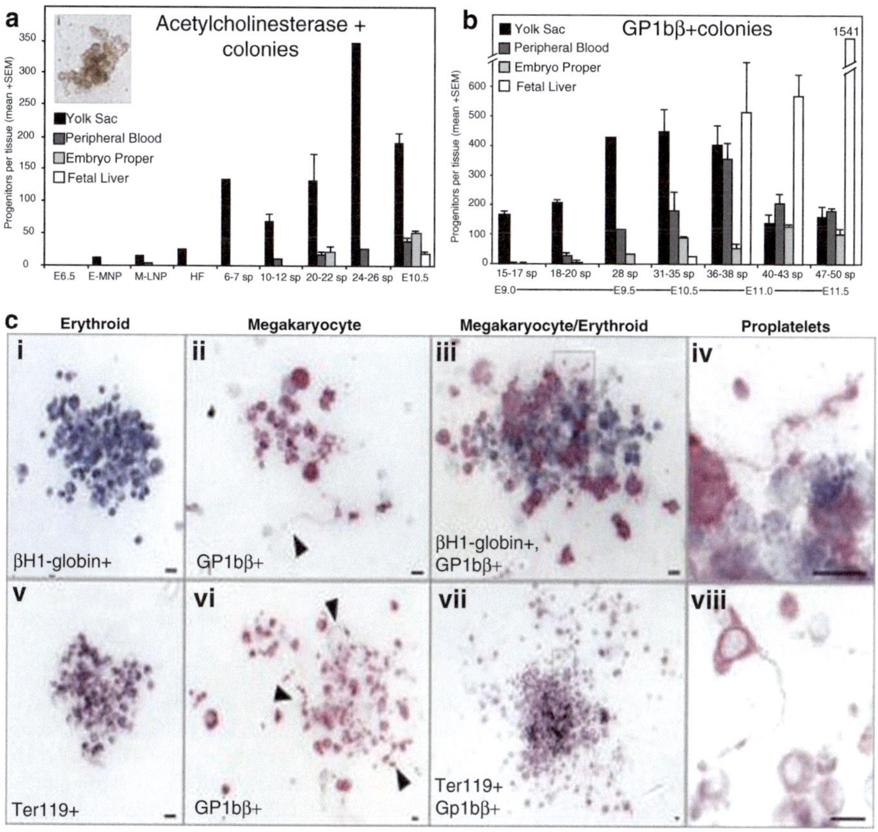

Fig. 12.1 Meg-CFCs and bipotential MEPs emerge during murine embryonic hematopoiesis. (**a**) Formation of acetylcholinesterase-positive MK colonies from individual embryonic tissues by stage. The inset (*i*) shows an acetylcholinesterase-positive colony; notably, this marker cannot be used in human MK colonies because they do not express the receptor. (**b**) Formation of primitive erythroid, megakaryocyte, and bipotential primitive megakaryocyte/erythroid progenitor (MEP) colonies from plated whole embryos (PS-LNP) and yolk sac (HF-46 sp) as determined by immunohistochemistry. (**c**) Erythroid and megakaryocyte progenitor-derived colonies by immunohistochemical staining. *Top*: primitive erythroid (*i*), megakaryocyte (*ii*), and bipotential primitive MEP (*iii*) colonies from neural plate-stage cultures stained with primitive erythroid-specific βH1-globin (*blue*) and megakaryocyte-specific GPIbβ (*pink*) antibodies. *Bottom*: erythroid (*v*), megakaryocyte (*vi*), and bipotential definitive MEP (*vii*) colonies from E9.5 to E10.5 yolk sac stained with pan-erythroid Ter119 (*blue*) and with GPIbβ (*pink*) antibodies. Boxed areas in panels iii and vii highlight proplatelet formation at 100× shown in panels iv and viii, respectively. Arrowheads indicate proplatelet formation in panels ii and vi. Scale bars represent 10 μm [102] (Reproduced with permission of Blood)

a replacement of platelet production from DPFC to MK derived. It has also been shown that DPFC formation and thrombopoiesis are thrombopoietin (Thpo) independent in vitro and in vivo, as Thpo-receptor-deficient animals (*c-mpl*$^{-/-}$) have normal DPFC and peripheral blood (PB) platelet numbers at E10.5 (Fig. 12.2) [87].

In humans, the yolk sac also appears to be the first hematopoietic site [74]. MKs have been observed in the YS by 5 weeks of gestation [42] and are found in the

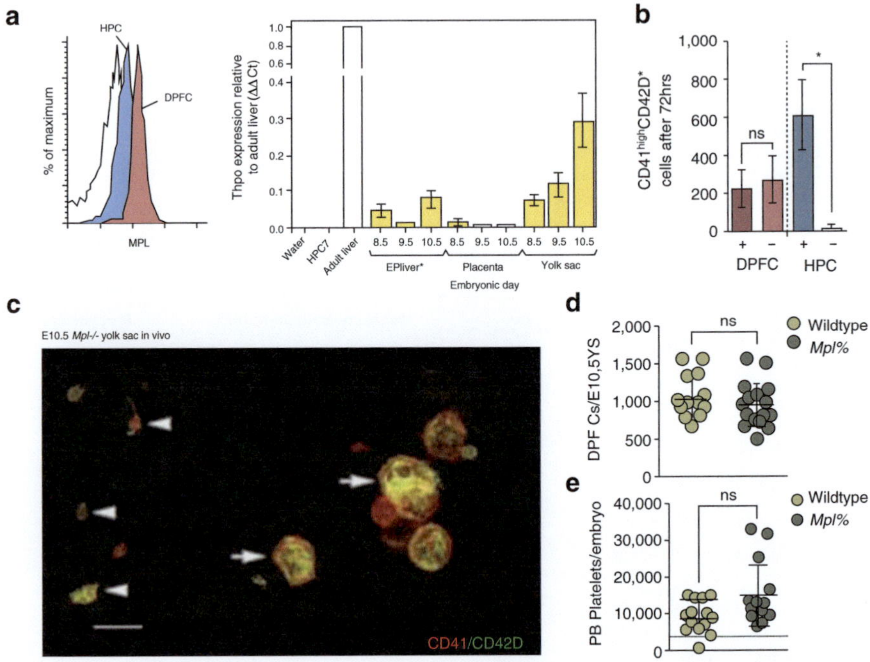

Fig. 12.2 Thrombopoietin (Thpo) is dispensable for DPFC formation and thrombopoiesis. (**a**) (*i*) Flow cytometric analysis of diploid platelet-forming cells (DPFC, *red*) and yolk sac (YS) hematopoietic progenitor cells (HPC, *blue*) showed that both cells express c-Mpl on their surface. (*ii*) The thrombopoietin (Thpo) expression analysis at embryonic day E10.5 by qRT-PCR showed that it was expressed in the YS at 30 % of the adult level. (**b**) E10.5 YS DPFC and HPC were cultured for 72 h in the presence or absence of Thpo. DPCF numbers were not increased, but Thpo stimulation is critical for MK formation from HPCs. (**c**) In whole mount preparations of YS E10.5 of *c-mpl*$^{-/-}$ CD41^{high}CD42D^{+} DPFC (*arrow*) and platelets (*arrow head*) are visible. (**d**, **e**) Quantifications reveal that in *c-mpl*$^{-/-}$ embryos (E10.5) DPFCs and circulating peripheral blood (PB) platelet are present in wild-type numbers [87] (Reproduced with permission of Blood)

connective tissue and in the vessels of almost every organ at 8 weeks of gestation. The transition to hepatic hematopoiesis is thought to involve migration of stem cells from the YS to the liver [73]. MKs are present in the liver by 8 weeks of gestation (Fig. 12.3a) and in the spleen by 10 weeks [9, 62]. Every stage of MK development is found in the liver phase, but the cells are smaller and of lower ploidy than adult MKs [27, 37, 58]. Fetal bone marrow (BM) hematopoiesis has been documented as early as 10 weeks of gestation, when MKs constitute 0.58 % of the nucleated cells (this percentage increases to 1.2 % by week 22) [16]. By 19 weeks of gestation, hematopoiesis is well established in the fetal BM (Fig. 12.3b) [1, 30, 57, 58].

12.2.2 Fetal Megakaryocyte Progenitors

In humans, studies culturing Meg-CFCs derived from term and preterm cord blood (CB), fetal blood (18–22 weeks of gestation), or fetal BM have documented a significantly higher proliferative potential of fetal/neonatal MK progenitors, compared

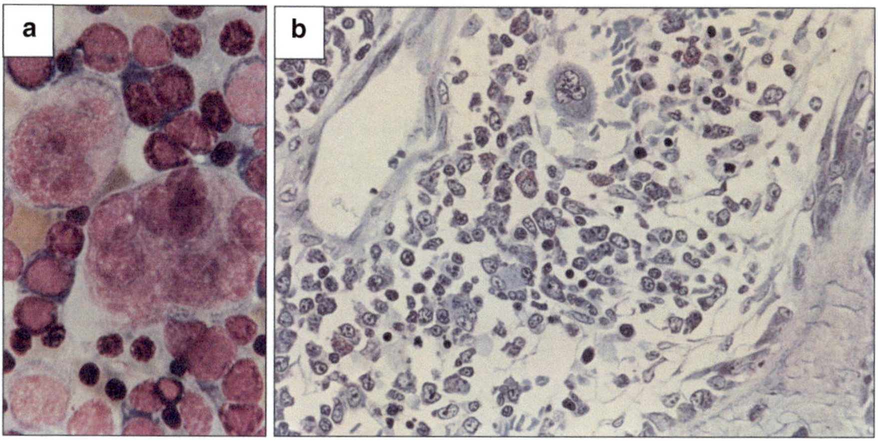

Fig. 12.3 Megakaryocytes in the developing human liver and bone marrow. (**a**) Liver smear from an 8-weeks-old human embryo (14 mm CR length). An early megakaryocyte with basophilic cytoplasm is seen in the upper left part and a mature megakaryocyte with eight nuclei is located in the middle of the picture (H&E, × 1000). (**b**) Femoral bone marrow from a 22-weeks-old human fetus. At the right, a bone trabecula is surrounded by osteoblasts. At the left is an arteriole. At the upper center a sinusoid is seen, containing erythrocytes, and a megakaryocyte with a large lobulated nucleus and well-demarcated granular cytoplasm (Methacylate Giemsa, × 1250) [62] (Reproduced with permission of Springer, Heidelberg)

to their adult counterparts. In Meg-CFCs assays, fetal Meg-CFCs generate significantly larger colonies, containing more MKs than adult Meg-CFCs [82, 111]. Furthermore, a high proliferative potential cell-megakaryocyte (HPPC-MK) that gives rise to large, unifocal colonies containing >300 cells has been found in fetal BM [13, 81]. This cell, not observed in adult BM cultures, may represent a more primitive MK progenitor. When cultured in liquid systems, neonatal progenitors (CD34+ cells) derived from cord blood (CB) give rise to tenfold more MKs than adult progenitors cultured under identical conditions (Fig. 12.4a) [69, 85].

The concentration of circulating MK progenitors, similar to other hematopoietic progenitors, decreases with advancing gestational age [19, 78, 91]. This has been thought to reflect the progressive transfer of progenitors from the liver to the BM.

12.2.3 Fetal and Neonatal Megakaryocytes

Regardless of the source of MKs (blood, fetal liver, bone marrow) and the techniques used (cultured MKs or histologic studies of tissue samples), fetal MKs are substantially smaller and have lower ploidy levels than adult MKs (Fig. 12.4b, c). During human gestation, a shift toward larger megakaryocytes occurs. Despite this shift, however, 78 % of MKs in the bone marrow of fetuses at 7–8 months' gestation have a ploidy of 8N or less, compared with only 33 % of the MKs in adult BM [70]. In the marrow of preterm infants, MKs are also smaller than in adult bone marrow samples [96].

The mechanisms underlying the small size of neonatal MKs are not clearly understood, but likely involve a combination of cell-intrinsic factors and factors in the fetal/neonatal microenvironment. In studies transplanting neonatal liver cells or adult

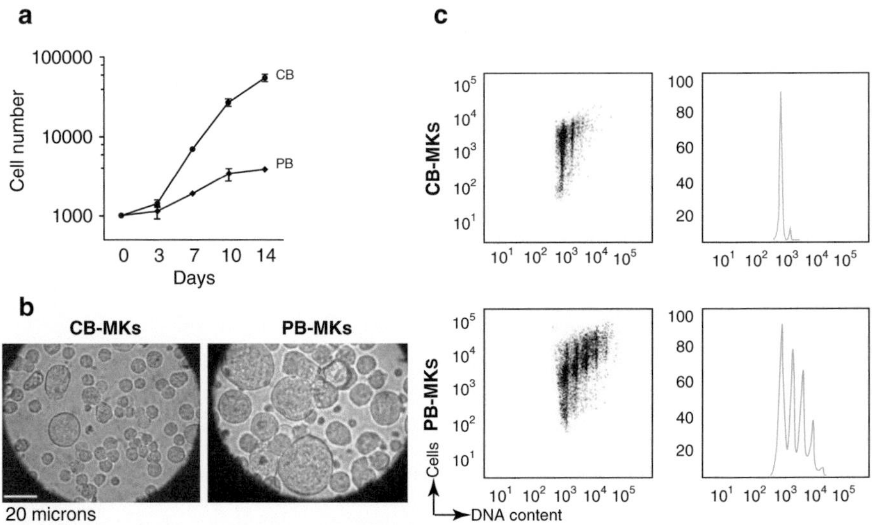

Fig. 12.4 Differences between fetal/neonatal and adult megakaryocyte proliferation, size, and ploidy. (**a**) Megakaryocytes progenitors isolated from umbilical cord blood (CB) are highly proliferative and generate tenfold more megakaryocytes than adult peripheral blood (PB) progenitors in an in vitro liquid culture system. However, CB-derived megakaryocytes are much smaller in size (**b**) and have lower ploidy levels measured by flow cytometry (**c**) than PB-derived megakaryocytes [69] (Reproduced with permission of Current Opinion in Hematology)

bone marrow cells from green fluorescent protein (GFP) transgenic mice into wild-type adult recipients, MKs derived from neonatal liver were significantly larger and of higher ploidy than neonatal MKs in their original environment [95]. However, they were still smaller than posttransplant MKs derived from adult BM cells. In vitro studies culturing CB hematopoietic progenitor (CD34+) cells in adult BM stromal conditioned media also yielded MKs with higher ploidy levels than those cultured in fetal BM stromal or in unconditioned serum-free media [85]. Taken together, these findings suggest that environmental factors influence the size and ploidy of MKs, but cell-intrinsic factors limit the ultimate size and ploidy that neonatal MKs achieve.

Given the small size and low-ploidy or fetal/neonatal MKs, these were traditionally considered to be immature compared to adult MKs [11]. However, it was shown that CB-derived MKs have significantly higher levels of CD42b expression (GPIbα, a marker of mature MKs) than peripheral blood (PB)-derived MKs [32, 66]. Further analysis of CD42b expression at different ploidy levels revealed that these differences were particularly pronounced among low-ploidy MKs (Fig. 12.5a, b). Neonatal MKs also contained α-granules, consistent with a mature cytoplasm (Fig. 12.5c), and electron microscopic examination of flow-sorted 2N/4N megakaryocytes confirmed that most CB-derived 2N/4N megakaryocytes were cytoplasmically mature (58 %), compared with only 23 % of PB-derived MKs with the same ploidy (Fig. 12.5d), thus demonstrating that neonatal MKs undergo full cytoplasmic maturation without polyploidization. The net result of this process is the production of large numbers of low-ploidy but highly mature MKs, with which fetuses and

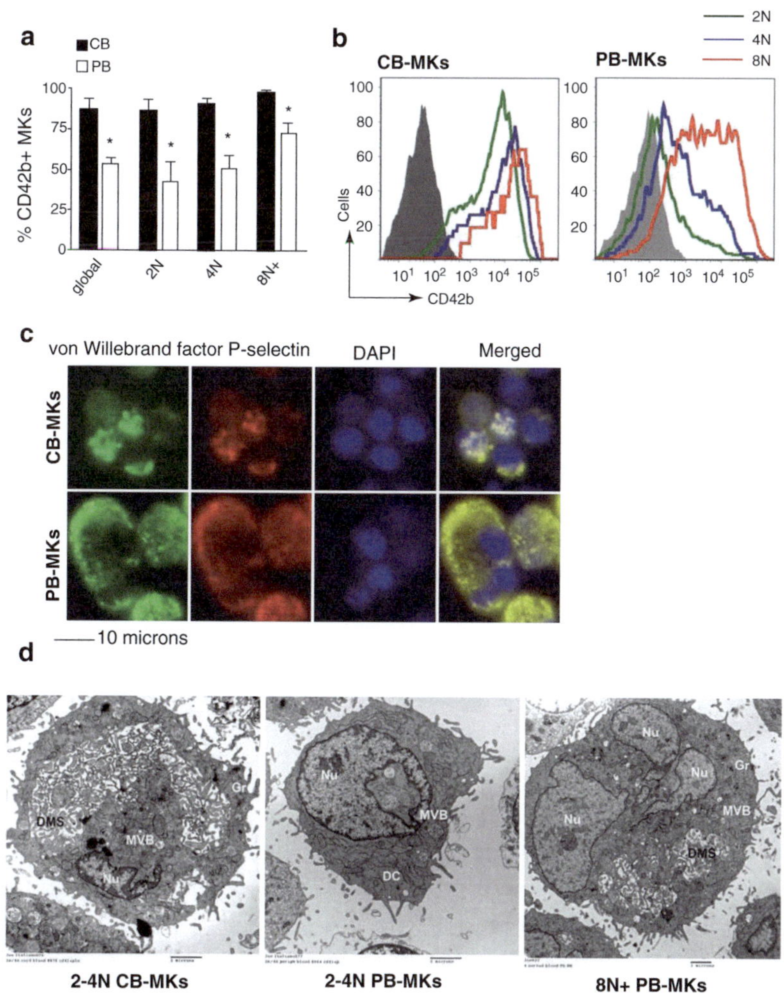

Fig. 12.5 Cytoplasmic maturation of CB-megakaryocytes (CB-MKs). Megakaryocyte polyploidization and cytoplasmic maturation of different sources were analyzed by flow cytometry and immunofluorescence at day 14 of culture. (**a**) Compared to PB-MKs, CB-MKs exhibited significantly higher percentages of CD42b+ (mature) cells at each ploidy level. The *bars* represent the means ±SEM of three independent experiments. $*P<0.05$. (**b**) When the intensity of CD42b staining was measured by fluorescence, CB-MKs exhibited almost equally intense expression at all ploidy levels. In contrast, among PB-MKs, the intensity of the staining increased gradually in parallel to the ploidy level. (**c**) Like mature PB-MKs, low-ploidy CB-MKs formed α-granules co-expressing P-selectin and vWF, as indicated by immunofluorescence staining. (**d**) Human MKs of different ploidy levels were flow sorted after antihuman CD41-FITC and Hoechst 33342 staining and examined by transmission electron microscopy. *Left*: representative flow-sorted 2N/4N CB-MK. The majority of MKs were mature and exhibited abundant platelet granules and a well-developed double membrane system. *Middle*: in contrast, 77 % of flow-sorted 2N/4N PB-MKs were immature, as evidenced by the absence of a DMS and the paucity of granules (Gr) seen in this representative electron ultramicrograph. These immature forms also exhibited multivesicular bodies (MVBs) and dense compartments (DCs), representing the precursors of platelet granules and of the DMS, respectively. *Right*: 54 % of all flow-sorted 8N PB-MKs were ultrastructurally similar to the mature neonatal 2N/4N MKs, although they were larger [69] (Author's own material, reproduced with permission of Blood)

neonates populate their rapidly expanding BM space and blood volume while maintaining normal platelet counts.

More recently, the timing of the transition from a neonatal to an adult phenotype has been studied [41]. This study showed that neonates have MKs of uniform and small size, which diverge into separate clusters of smaller and larger cells beginning at 2 years, and finally transit to larger (adult-like) MKs by 4 years.

12.2.4 Molecular Mechanisms for the Cellular Characteristics of Fetal and Neonatal Megakaryocytes

At the molecular level, these developmental differences were associated with higher expression levels of the Thpo receptor (c-Mpl) and upregulated Thpo signaling through Janus kinase 2 (Jak2) and mTOR (mammalian target of rapamycin) in neonatal MKs. The protein level of the GATA1 transcription factor was also threefold higher in CB- vs. PB-MKs (Fig. 12.6). The high GATA1 level was consistent with the high cytoplasmic maturation of neonatal MKs, since overexpression of

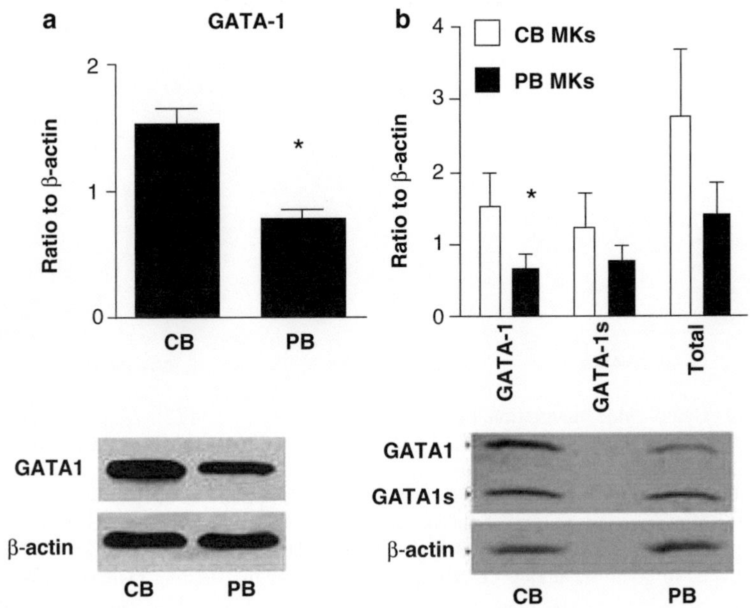

Fig. 12.6 Differences in GATA1 protein expression levels in cord blood (CB) vs. peripheral blood (PB) megakaryocytes (MKs). (**a**) GATA1 protein in CB- and PB-derived MKs was measured by Western blot analysis using an antibody targeting the N-terminal motif, which only recognizes full-length GATA1. *$P<0.05$ (**b**) An antibody recognizing the C-terminal motif was then used to differentiate between the full-length GATA1 and GATA1s. Only the levels of full-length GATA1 were significantly higher in CB-MKs compared with PB-MKs. *$P<0.05$. The bars represent the means ± SEM of 4 independent experiments, and the immunoblots show representative results of each set of experiments [69] (Author's won material, reproduced with permission of Blood)

GATA1 in murine adult MKs accelerates and enhances its maturation [50]. However, GATA1 overexpression also promoted polyploidization in murine adult MKs, an effect not seen in human neonatal MKs. Transcriptome analyses of neonatal- and fetal-derived vs. adult PB-derived MKs found no changes in *GATA1* mRNA expression levels during development, but an enrichment in downstream targets of GATA1 in fetal MKs [10]. Further studies evaluating downstream targets of GATA1 revealed that CB-MKs had elevated protein levels of cyclin D1, but low levels of its partner Cdk4 (cyclin-dependent kinase 4). Interestingly, in a prior study, the overexpression of cyclin D1 in the absence of Cdk4 overexpression resulted in a paradoxical *decrease* in MK ploidy and size [76]. Thus, the combination of elevated GATA1 and cyclin D1 levels with low Cdk4 could be one of the mechanisms mediating the high cytoplasmic maturation *without polyploidization* of neonatal MKs, although this remains to be tested.

Clinically, a better understanding of the molecular mechanisms underlying the developmental differences in megakaryopoiesis is essential to elucidate the pathogenesis of a number of MK and platelet disorders that exclusively, or more severely, affect neonates. These disorders include the trisomy 21 (Down syndrome)-associated transient myeloproliferative disorder (DS-TMD), which exclusively affects fetuses and neonates and resolves spontaneously by 4 months of age, and the thrombocytopenia-absent radius (TAR, MIM #274000) syndrome, in which the thrombocytopenia improves significantly by 1 year of age [39, 46]. Genetically, DS-TMD results from the combination of trisomy 21 and a GATA1 mutation leading to the exclusive expression of a short GATA1 isoform, termed GATA1s [15], which lacks the N-terminal transactivation domain. The reason why this combination leads to a myeloproliferative disorder only during fetal and neonatal (but not adult) life has been elucidated. In 2010, Klusmann and colleagues found an upregulation of insulin-like growth factor (IGF) signaling in human DS-AMKL (Down syndrome – acute megakaryoblastic leukemia) cells, as well as in a mouse model of DS-AMKL [63]. Following this observation, it was found that fetal (but not adult) MK progenitors are dependent on IGF signaling, which activates the E2F transcriptional network through mTOR. In normal fetal MK progenitors, full-length GATA1 restricts the IGF-mediated activation of the E2F transcription network to coordinate proliferation and differentiation. However, in the absence of full-length GATA1 (i.e., in mutated GATA1s), the overactive IGF signaling is "uncontrolled" and leads to unrestricted proliferation of fetal MK progenitors and to the transformation into DS-TMD. More recently, a separate study showed a significant upregulation of type 1 interferon (IFN) – responsive genes in both murine and human BM – compared to fetal liver-MK progenitors [110]. Importantly, exogenous IFNα markedly reduced the hyperproliferation of fetal liver MKs obtained from GATA1s mice. Conversely, genetic or pharmacologic neutralization of IFN signaling increased the proliferation of MK progenitors in the BM of adult GATA1s mice. These observations suggested that increased type 1 IFN signaling in the adult bone marrow abolishes the developmental stage-specific effects of GATA1s and possibly contributes to the spontaneous resolution of DS-TMD. Taken together, these studies have shed light on the mechanisms underlying the developmental stage-specific effects of GATA1s mutations on fetal/neonatal megakaryopoiesis.

Finally, micro-RNAs (miRNAs) have been increasingly recognized as key regulators of multiple steps in the process of megakaryopoiesis, and significant differences have been reported between neonatal and adult megakaryocytes in their miRNA expression profiles, particularly involving let-7 family members and miR-9 [10, 38]. The exact contribution of these miRNA profiles to the developmental differences between fetal/neonatal and adult megakaryocytes is the subject of ongoing research.

12.3 Regulation of Fetal and Neonatal Megakaryopoiesis and Platelet Production

Thpo is the primary humoral regulator of megakaryopoiesis from fetal life onward and induces lineage-restricted MK progenitor proliferation, differentiation of those committed progenitors to megakaryoblasts, and subsequent differentiation of these cells into mature MKs [31, 61]. However, Thpo is not required for proplatelet formation and platelet release, since mutant mice lacking c-Mpl on mature MKs and platelets exhibit even elevated platelet counts [80].

Particularly during the hepatic period of human fetal hematopoiesis, THPO is produced at the site of megakaryopoiesis (Fig. 12.7a). As early as by 8 weeks of gestation, *THPO* mRNA expression is detected in the human liver [98]. Between 17 and 36 weeks of gestation, the liver contributes more than 90 % to the combined *THPO* mRNA expression in the liver, kidneys, and spleen as calculated by semi-quantitative RT-PCR analysis [109]. The liver remains the primary site of THPO production in parenchymal organs throughout the entire life [7], but THPO is subsequently also locally produced in BM stroma cells [100]. In some fetal BM specimens, *THPO* mRNA expression levels were as high as in the corresponding liver or kidneys, suggesting that THPO may play a role in the ontogeny of medullary

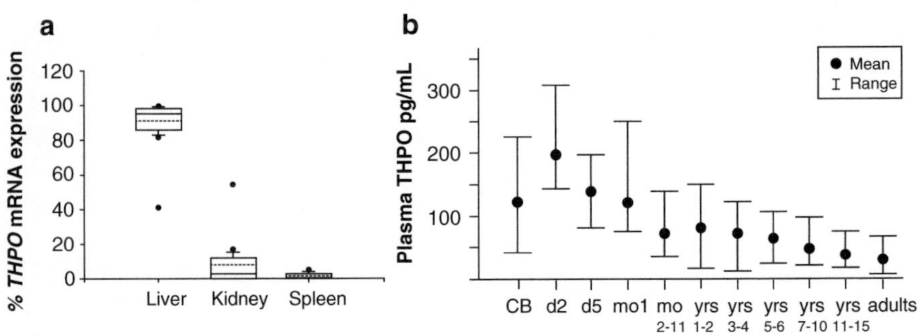

Fig. 12.7 (**a**) Organ distribution of *THPO* mRNA expression in human fetuses and preterm neonates. The amount of *THPO* mRNA (per microgram of total RNA) was equal in the liver, spleen, and bone marrow, slightly lower in the kidney, and significantly lower in the lung (not shown). When adjusted for grams of tissue, *THPO* mRNA levels were highest in the liver. Considering the total amount of *THPO* mRNA produced in the liver, kidney, and spleen (in total defined as 100 %), the liver accounted for 95.3 % [109]. (Author's won material, reproduced with permission from by Blood). (**b**) Age-related changes in THPO serum concentrations in healthy children (Data obtained from Ishiguro et al. [53])

megakaryopoiesis [109]. This hypothesis deserves further attention, because in another context Thpo contributes to the accumulation and persistence of plasma cells in murine BM niches [108]. Indeed, *THPO* gene expression in BM stromal cells increases in response to thrombocytopenia [79]. At the molecular level, various platelet granular proteins are involved in the regulation of this process. Primary cultures of adult BM indicate that platelet-derived growth factor (PDGF), fibroblast growth factor 2 (FGF2), and interleukin-11 (IL11) increase *THPO* expression, while transforming growth factor β (TGFβ), platelet factor 4 (PF4), and thrombospondin (TSP) lead to a diminished *THPO* expression in stroma cells [100].

Besides the putative developmental role of THPO in regulating the seeding of hematopoietic stem cells (HSC) in the developing BM niches, patients with congenital amegakaryocytic thrombocytopenia (CAMT; MIM #604498) provide evidence of the pivotal role of THPO in maintaining HSCs in the BM [6, 52]. Patients with homozygous or compound heterozygous mutations in the *MPL* gene exhibit pancytopenia early in life, with subsequent need for BM transplantation if the mutations result in total deficiency of MPL signaling (CAMT Type 1) [6].

Under steady-state conditions, hepatic THPO production is thought to be mainly constitutive [40]. This concept of end-cell-mediated regulation of circulating THPO concentrations results from the observation that hepatic *THPO* mRNA expression does normally not change in thrombocytopenia and that THPO protein concentrations depend on the mass of c-Mpl-bearing cells, in particular platelets and MK progenitor cells. Most recently, new evidence indicated that the hepatic Ashwell-Morell receptor (AMR) can bind aged, desialylated platelets and thereby induce hepatic *THPO* transcription and translation [45]. While this mechanism has specific implications in certain platelet disorders (e.g., infections by neuraminidase-producing organisms and certain immune thrombocytopenias), its role in developmental megakaryopoiesis is still unknown.

Considering the putative role of THPO in establishing medullary megakaryopoiesis, it is of interest that healthy, nonthrombocytopenic neonates exhibit higher THPO plasma concentrations than adults. Normal CB THPO concentrations are <300 pg/ml in plasma samples and <320 pg/ml in serum samples (2 SD above the mean) [28]. Some studies reported higher THPO concentrations in preterm compared to term neonates, but this finding has not been confirmed in the majority of the studies [28]. Relevant developmental changes in circulating THPO concentrations during gestation were excluded by the longitudinal analysis of fetal blood samples [22, 59, 86, 89]. During the neonatal period, however, THPO plasma concentrations exhibit a characteristic course (Fig. 12.7b). Starting from a median concentration of 124 pg/mL (range 48–225 pg/mL) at birth, THPO concentrations increase and peak by day 2–5, before they decrease by the end of the neonatal period to levels that are only slightly higher than those found in infants [53].

Notably, the responsiveness of human BM-derived Meg-CFCs to recombinant human THPO (rTHPO) is also age-dependent. A higher in vitro and in vivo sensitivity of neonatal Meg-CFCs has been demonstrated in comparison to adult Meg-CFCs (Fig. 12.8) [96, 97]. A possible mechanism might be the higher expression of c-mpl in CB-derived MKs compared to those derived from adult BM [69]. As mentioned above, the higher rTHPO concentrations and responsiveness did not support the

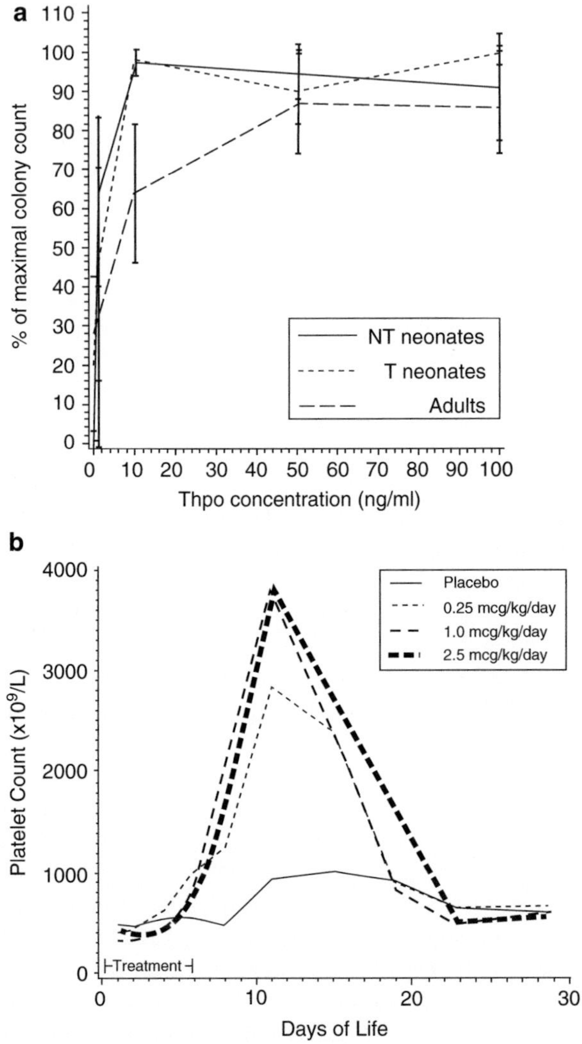

Fig. 12.8 Response of neonatal progenitors to rTHPO in vitro and in vivo. (**a**) In vitro responsiveness of human bone marrow-derived MK progenitors to increasing concentrations of recombinant human THPO. The MK progenitors originated either from nonthrombocytopenic neonates (NT), thrombocytopenic neonates (T), or nonthrombocytopenic adults were cultured in a collagen-based semisolid system with different concentrations of human THPO [97]. (Author's won work, reproduced with permission by the British Journal of hematology). (**b**) Dose-response of Pegylated-recombinant human megakaryocyte growth and development factor (PEG-rHuMGDF, a THPO derivative) on the blood platelet concentration in newborn rhesus monkeys treated with different daily subcutaneous injections of placebo or PEG-rHuMGDF. The platelet count increased on day 6 of treatment, peaked on day 11, and returned to baseline by day 23 (Sola MC et al., Pediatr Res 2000; 47(2):208–214; author's own work, reproduced with permission by Pediatric Research)

polyploidization of neonatal MKs [85], although rTHPO increased the ploidy of adult MKs. Thus, the developmental differences in the response to THPO may play a critical role in thrombocytopenia. While adults upregulate thrombopoiesis by increasing the number and size of their MKs in response to elevated THPO concentrations, the neonatal response to thrombocytopenia is limited to increasing the number and proliferative rate of MK progenitors. This may explain the limited ability of neonates to correct thrombocytopenia in situations of drastically increased demand, i.e., bacterial sepsis. This conclusion is supported by current data on the immature platelet fraction (IPF), a novel measure of newly released platelets that are still reticulated and contain high amounts of RNA. As expected based on the THPO concentrations, IPF values in nonthrombocytopenic neonates are higher than in children and adults [23]. But in very-low-birth-weight infants with severe thrombocytopenia due to sepsis or necrotizing enterocolitis, the IPF numbers are decreased by approximately 50 % [26].

12.4 Platelet Function in Neonates

The first evidence of developmental differences in platelet function between neonates and adults came from platelet aggregation studies in full-term CB-derived platelet-rich plasma. These studies demonstrated that platelets from neonatal CB are less responsive than adult platelets to agonists such as adenosine diphosphate (ADP), epinephrine, collagen, thrombin, and thromboxane analogues [56]. Similar results were obtained in flow cytometric platelet activation studies, which showed decreased expression of surface activation markers in neonatal platelets stimulated with thrombin, ADP, and epinephrine [56, 94]. Different mechanisms account for the hyporeactivity of neonatal platelets to various agents [2]: The hyporesponsiveness to epinephrine is due to fewer α2-adrenergic receptors, the binding sites for epinephrine [20]; the reduced response to collagen likely results from impaired calcium mobilization [43]; and the decreased response to thromboxane results from differences in signaling downstream from the receptor in neonatal platelets [55].

Surprisingly, while the hypofunctional platelet phenotype in vitro would predict a bleeding tendency, healthy full-term neonates display *enhanced* primary hemostasis, compared to older children or adults. Bleeding times performed on healthy term neonates are shorter than bleeding times in adults [4]. Similarly, studies using the platelet function analyzer (PFA-100®, an in vitro test of primary hemostasis) found that CB samples from term neonates exhibited shorter closure times than samples from older children or adults [12, 54]. Overall, these studies point toward an enhanced platelet/vessel wall interaction in neonates, likely related to the higher hematocrit levels, higher mean corpuscular volumes, higher von Willebrand Factor (vWF) concentrations [5], and predominance of longer vWF polymers in the blood of neonates, all of which enhance primary hemostasis and compensate for the hyporeactivity of neonatal platelets. Using a Cone and Platelet Analyzer, which examines whole blood platelet adhesion and aggregation on an extracellular matrix-coated plate under physiological arterial flow conditions, healthy full-term neonatal platelets also demonstrated more extensive adhesion properties than adult platelets, a finding mediated by the higher amounts and enhanced adhesive activity of vWF in neonatal plasma [92]. Taken together, the available evidence strongly suggests that the in vitro platelet hyporeactivity of healthy full-term infants is an integral part of a carefully balanced and well-functioning neonatal hemostatic system, rather than a developmental deficiency.

Since the neonates with the highest bleeding risk are those born preterm, particularly at less than 30 weeks of gestation, studies of platelet function and hemostatic balance in this subset of infants are of particular clinical importance. At the time of birth, the in vitro platelet hyporeactivity is more pronounced in preterm infants compared to term infants. These differences are most evident among infants born at the lowest gestational ages (i.e., <30 weeks), suggesting a correlation between platelet reactivity and gestational age [94, 103]. Two studies using the Cone and Platelet Analyzer also found that platelets from healthy preterm infants exhibited decreased platelet adhesion compared with platelets from term infants (although it was still better than in healthy adults) [67, 68] and that adherence correlated with gestational age in the first 48 h after birth. Importantly, these differences in platelet adhesion were not related to lower levels of vWF antigen or ristocetin cofactor activity in preterm compared to term infants, suggesting that they are due to intrinsic platelet function developmental differences [68]. Consistent with these findings, bleeding times performed on the first day of life were longer in preterm compared to term infants, with neonates born at <33 weeks of gestation exhibiting bleeding times twice as long as those from term neonates [33].

In regard to the duration of the neonatal hyporeactive phenotype, it is clear that the platelet hyporeactivity is still present 3–4 days after birth in term as well as preterm infants [8]. After that, the results of available studies have been somewhat inconsistent, with some reporting prolonged platelet hyporeactivity [48, 94], but most observing significant improvement to full normalization in the platelet function of term as well as preterm infants by day 10–14 after birth [8, 94, 103]. The latter are consistent with the finding that, by day 10, all infants have shorter bleeding times than at birth, and gestational age-dependent differences present on the first day have disappeared. Little or no further shortening of the bleeding time occurs between days 10 and 30 [33].

12.5 Thrombocytopenia in Preterm and Term Neonates

Thrombocytopenia is a frequent problem among sick term and preterm neonates. In most cases, thrombocytopenia results from acquired processes, and most resolve with time and/or treatment of the underlying illness. In neonates, thrombocytopenia has traditionally been defined as a platelet count $<150 \times 10^9$/L and has been classified as mild ($100–150 \times 10^9$/L), moderate ($50–99 \times 10^9$/L), and severe ($<50 \times 10^9$/L). Based on those definitions, large studies in unselected populations established an overall incidence of neonatal thrombocytopenia of 0.7–0.9 % [35, 104]. However, when focusing on neonates admitted to the neonatal intensive care unit (NICU), the incidence of thrombocytopenia is much higher, ranging from 18 to 35 % [17, 72, 83]. The incidence of thrombocytopenia is also inversely correlated to the gestational age, so that the most immature neonates are the most frequently affected. In extremely low-birth-weight infants (<1000 g), the incidence of thrombocytopenia reaches up to 73 % [18]. However, there is debate on the lower reference range of platelet counts after a very large cohort that included more than 47,000 neonates between 23 and 42 weeks of gestation has been examined. Using automated blood cell counters, the 5th percentile for platelet counts in infants born ≤32 weeks of gestation was 104×10^9/L and 123×10^9/L for late preterm and term neonates [107].

The reasons underlying the predisposition of preterm neonates to develop thrombocytopenia are not entirely clear, but may be associated with the above reported developmental stage-specific differences in megakaryopoiesis and platelet biology compared to children and adults. The clinically most important issue remains that the critical platelet count for spontaneous bleeding is not known, especially for very preterm infants. Currently, it is thought that the risk for bleeding increases if the platelet count falls below 20×10^9/L [3].

When clinically approaching a neonate with thrombocytopenia, five key points may be helpful diagnostic clues (Table 12.1): (a) time of onset of thrombocytopenia, delineating "early onset" (within 72 h after birth) and "late onset"; (b) the clinical aspect ("sick" vs. "otherwise healthy"); (c) the severity and the kinetics of thrombocytopenia ($<50\times10^9$/L, fast drop in platelet numbers of more than 50×10^9/L per d); (d) the likelihood of maternal, placental, or fetal/neonatal cause [24]; and (e) dysmorphic stigmata or malformations suggestive of multisystem disorders.

The most common etiology of transient mild early-onset thrombocytopenia in non-septic appearing neonates (particularly those born preterm) is placental insufficiency. Thrombocytopenia associated with chronic (or acute) intrauterine hypoxia (preeclampsia, HELLP syndrome, maternal diabetes, or intrauterine growth restriction) typically resolves within 3–10 days and is associated with decreased numbers of circulating MK progenitors and IPF, suggesting decreased platelet production [25, 78]. Notably, the platelet reactivity in response to in vitro agonists is reduced at birth in infants with intrauterine growth restriction (IUGR) due to maternal hypertension or preeclampsia if compared to age-matched infants with nonhypertensive IUGR [75].

However, in an ill-appearing term or preterm neonate or in a neonate with abnormal liver enzymes and/or coagulation tests, early-onset thrombocytopenia of any severity should prompt an evaluation for bacterial (i.e., Group B streptococci) or viral sepsis (i.e., HSV, cytomegalovirus), parasitic, or fungal infections (i.e., toxoplasma gondii, candida). The most common cause of congenital viral infection is CMV. This virus directly affects megakaryopoiesis by infecting BM stromal cells [99] as well as MKs [21].

There are also other specific disorders with reduced megakaryopoiesis (i.e., severe fetal/neonatal hemolytic disease due to anti-Kell, anti-D, or anti-c antibodies, nonimmune hydrops, maternal medications) or increased platelet consumption (i.e., Kasabach-Merritt syndrome) to consider, if the neonate exhibits thrombocytopenia. After perinatal asphyxia, disseminated intravascular coagulation (DIC) may cause severe early-onset thrombocytopenia. The pathopyhsiology of these and other acquired thrombocytopenias (i.e., associated with thrombosis, hypersplenism), which can affect both neonates and adults, is explained in Chaps. 15 and 16.

Table 12.1 Classification of neonatal thrombocytopenias based on the time of onset and disease severity

Onset	Clinical aspect	Severity of thrombocytopenia	Differential diagnosis
Early onset (<72 h)	"Not sick"	Mild to moderate	Chronic intrauterine hypoxia (preeclampsia, diabetes, intrauterine growth restriction)
			Immune (allo- or autoimmune)
			Genetic disorders[a]
			Thrombosis (renal vein, sagittal sinus)
		Severe	Alloimmune
			Genetic disorders[a]
	"Sick"	Variable	Infections (bacterial, viral)
			Congenital infections (TORCH)
			Perinatal asphyxia
Late onset (>72 h)	"Not sick"	Variable	Drug-induced thrombocytopenia
			Thrombosis (renal vein, sagittal sinus)
	"Sick"	Moderate to severe	Sepsis (bacterial, viral, fungal) and necrotizing enterocolitis
			Metabolic disorders[a]

Adapted from Saxonhouse and Sola-Visner. NeoReviews.org, 2009
TORCH: Toxoplasma, rubella, cytomegalovirus, herpes simplex virus
[a]Congenital thrombocytopenia

Severe early-onset thrombocytopenia in an otherwise healthy or mildly symptomatic infant should always trigger suspicion for neonatal immune-mediated thrombocytopenia. There are two forms of immune thrombocytopenia: (a) *allo*- and (b) *auto*-/isoimmune.

Fetal and neonatal alloimmune thrombocytopenia (FAIT/NAIT) results from an incompatibility between parental platelet antigens. After the mother has been sensitized to paternal antigens expressed on the fetal platelets, the resulting IgG antibodies cross the placenta, enter the fetal circulation, and coat the fetal platelets, so that their removal from the circulation is triggered. Almost all human platelet antigens (HPA) are due to a single nucleotide polymorphism in genes that encode for a glycoprotein on the platelet membrane. NAIT occurs with a frequency of 1:1000 in Caucasians, and most frequently alloantibodies are directed against HPA-1a (approx. 80 %) or against HPA-5b (approx. 20 %). In contrast, HPA-4 is most frequently implicated among Asians [44]. Less frequently, cases of NAIT are mediated by antibodies against HPA-3, HPA-2, and HPA-6 [60, 64, 105, 106]. FAIT can manifest as early as 16–20 weeks of gestation and – in contrast to fetal hemolytic disease – already in the first pregnancy. Intracranial hemorrhage, which often manifests itself already during gestation with subsequent severe brain damage, posthemorrhagic hydrocephalus, and/or periventricular leukomalacia resulting in severe lifelong disabilities, is the most perilous complication of FAIT/NAIT and occurs in 10–30 % of cases [14]. Interestingly, the likelihood of clinical manifestation of FAIT in a given risk constellation of parental platelet incompatibility is associated with the presence of certain human leukocyte antigens (HLA type). For example, HPA-1a alloimmunization is 140 times more frequent in HLA DRB3*01:01- and HLA-DQB1*02:01-positive women than in women who are negative for these HLA types [65]. HPA-5b-associated fetal/neonatal thrombocytopenia is more frequent in HLA-DRw6 women [60, 84]. These associations deserve further attention in research on developmental platelet biology.

Autoimmune thrombocytopenia in the fetus and neonate results from the transplacental transfer of maternal autoantibodies in chronic immune thrombocytopenia or other maternal autoimmune disorders (i.e., systemic lupus erythematodes or Sjögren syndrome). Neonatal thrombocytopenia is present in about 10–50 % of cases at risk, with an overall incidence of 1:10,000 to 1: 15,000 live births. While neonatal platelet counts are often normal at birth, the nadir occurs between day 2 and 5. Platelet counts $<20 \times 10^9$/L were found in <2 % of affected infants. Significant bleeding occurs in less than 1 % of the entire group [90].

In sick infants, early- and late-onset sepsis or necrotizing enterocolitis (NEC) are the most frequent causes of thrombocytopenia. Infection-induced thrombocytopenia usually lasts an average of 6 days (range 1–10 days) and must be treated appropriately. Sepsis-induced thrombocytopenia is mainly due to increased platelet consumption, secondary to either endothelial damage with subsequent platelet adhesion and aggregation, bacterial-induced degradation of the platelet anti-apoptotic protein Bcl-xL, platelet removal by the reticulo-endothelial system or by the hepatic Ashwell-Morrell receptor, and/or immune-mediated platelet destruction [47, 101]. However, megakaryopoiesis can be also significantly inhibited [26], and a small number of neonates with late-onset sepsis or NEC develop severe thrombocytopenia that persists longer than 2 weeks [77]. Sepsis accounts for approximately one-third of NICU patients that qualify as very-high platelet users (i.e., >20 transfusions) [34].

Inherited thrombocytopenias are a clinically and genetically very heterogeneous group of rare diseases that are often diagnosed beyond the neonatal period. Currently, mutations in more than 30 genes have been found to be implicated, affecting megakaryopoiesis, platelet formation and turnover, and platelet function [93]. Inherited thrombocytopenia can result from an isolated blood disorder, but also be a part of a complex multisystem disease. Over the past two decades, the knowledge of the underlying molecular mechanisms has rapidly increased, and a separate chapter of this textbook is dedicated to inherited thrombocytopenias (see Chap. 14). While a classification of inherited thrombocytopenias based on pathophysiological mechanisms is most desirable, current diagnostic approaches still primarily consider (a) the size of platelets and (b) the presence of specific clinical signs associated with thrombocytopenia. Inherited thrombocytopenias, such as CAMT, TAR syndrome, and DS-TMD/AMKL (all with normal platelet size/volume), have been already integrated in our chapter because their mechanisms are directly linked to developmental stage-specific characteristics of fetal and neonatal megakaryopoiesis and platelet biology. Table 12.2 summarizes inherited thrombocytopenias that are typically diagnosed during the neonatal period or during infancy. The reference number in the

Table 12.2 Inherited thrombocytopenia, typically diagnosed in the neonatal period or early infancy

	OMIM #	Gene	Inheritance	Clinical and laboratory signs
Normal platelet size				
T21	190685	*GATA1*	IC	Down syndrome with TMD/AMKL
CAMT	604498	*MPL*	AR	Development of pancytopenia
TAR	274000	*RBM8A*	IC/Digenic	Bilateral absent radii
CTRUS	605432	*HOXA11*	AD	Radioulnar synostosis, development of pancytopenia
HHS	305000	*DKC1*	X	Microcephaly, cerebellar hypoplasia, intrauterine growth restriction, immunodeficiency
		TERT	AR	
		RTEL1	AR	
		TINF2	AD	
Large platelet size				
T13, T18		?/PACAP	IC	Pätau syndrome, Edwards syndrome
MYH9-RD	155100	*MYH9*	AD	Neutrophil inclusions of myosin-9; deafness, cataract, glomerulonephritis, and others
	153650			
	153640			
	605249			
GPS	139090	*NBEAL2*	AR	Pale platelets, α-granule deficiency
		GFI1B	AD	
BBS	231200	*GP1BA*	AR	Defective GPIb/IX/V complex
		GP1BB		
		GP9		
XLTDA/XLTT		*GATA1*	X	Dyserythropoiesis, anemia, thalassemia
DGS/VCFS	188400	*GP1BB*	AD	Cardiac malformation, thymus hypoplasia, immunodeficiency, facial dysmorphia, hypocalcemia
	192430			
JBS, TCPT	147791	*FLI1*	AD	Dysmorphia, cardiac malformation, mental retardation
	188025			
FLNA-RT	300017	*FLNA*	X	Periventricular nodular heterotopia
Small platelet size				
WAS	301000	*WAS*	X	Immunodeficiency, eczema, autoimmune disorders, predisposition for malignancies
XLT	313900	*THC1*	X	Autoimmune disorders

Disorders: T13, T18, T21 (trisomy 13, 18, 21). *CAMT* congenital amegakaryocytic thrombocytopenia, *TAR* thrombocytopenia-absent radii syndrome, *CTRUS* congenital thrombocytopenia with radioulnar synostosis, *HHS* Hoyeraal-Hreidarsson syndrome, *MYH9-RD* MYH9-related diseases, formerly May-Hegglin anomaly, Epstein-, Fechtner-, Sebastian-platelet syndrome, *GPS* Gray platelet syndrome, *BBS* Bernard-Soulier syndrome, *XLTDA* X-linked thrombocytopenia and dyserythropoietic anemia, *XLTT* X-linked thrombocytopenia and thalassemia, *DGS* DiGeorge syndrome, *VCFS* velo-cardio-facial syndrome, *JBS* Jacobson syndrome, *TCPT* Paris-Trousseau syndrome, *FLNA-RT* Filamin-A-related thrombocytopenia, *WAS* Wiskott-Aldrich syndrome, *XLT* X-linked thrombocytopenia

Inheritance: *IC* incidental occurrence, *AD* autosomal dominant, *AR* autosomal recessive, *X* X-chromosomal

online library of Mendelian Inheritance in Man (OMIM) is provided for further reading, particularly on clinical details that are far from the scope of this textbook. Notably, developmental disorders of the bone and the central nervous system are frequently associated with inherited thrombocytopenias. This implicates the interaction of cells and chemokines involved in megakaryopoiesis and bone development [71], but also of normal THPO/Mpl signaling in the developing brain [29, 36, 49].

In summary, developmental differences in megakaryopoiesis and platelet function are important in clinical decision making regarding the diagnostic approach to thrombocytopenia and its management, particularly in sick preterm and term neonates. The potential developmental mismatch associated with transfusing adult platelets into the neonatal environment, particularly when the neonatal thrombocytopenia is associated with inflammatory conditions, is a topic that deserves future clinical and experimental research.

Acknowledgment C. D. and M. S-V. acknowledge the support of the National Institutes of Health (NIH) for collaborative research in PO1 HL 046925 "Neonatal Anemia and Thrombocytopenia" and R01 HL069990 (M. S-V.).

References

1. Allen Graeve JL, de Alarcon PA (1989) Megakaryocytopoiesis in the human fetus. Arch Dis Child 64(4):481–484
2. Andres O, Schulze H, Speer CP (2015) Platelets in neonates: central mediators in haemostasis, antimicrobial defence and inflammation. Thromb Haemost 113(1):3–12
3. Andrew M, Castle V, Saigal S, Carter C, Kelton JG (1987) Clinical impact of neonatal thrombocytopenia. J Pediatr 110(3):457–464
4. Andrew M, Paes B, Bowker J, Vegh P (1990) Evaluation of an automated bleeding time device in the newborn. Am J Hematol 35(4):275–277
5. Andrew M, Paes B, Johnston M (1990) Development of the hemostatic system in the neonate and young infant. Am J Pediatr Hematol Oncol 12(1):95–104
6. Ballmaier M, Germeshausen M, Schulze H, Cherkaoui K, Lang S, Gaudig A, Krukemeier S, Eilers M, Strauss G, Welte K (2001) c-mpl mutations are the cause of congenital amegakaryocytic thrombocytopenia. Blood 97(1):139–146
7. Bartley TD, Bogenberger J, Hunt P, Li YS, Lu HS, Martin F, Chang MS, Samal B, Nichol JL, Swift S et al (1994) Identification and cloning of a megakaryocyte growth and development factor that is a ligand for the cytokine receptor Mpl. Cell 77(7):1117–1124
8. Bednarek FJ, Bean S, Barnard MR, Frelinger AL, Michelson AD (2009) The platelet hyporeactivity of extremely low birth weight neonates is age-dependent. Thromb Res 124(1):42–45
9. Bleyer WA, Hakami N, Shepard TH (1971) The development of hemostasis in the human fetus and newborn infant. J Pediatr 79(5):838–853
10. Bluteau O, Langlois T, Rivera-Munoz P, Favale F, Rameau P, Meurice G, Dessen P, Solary E, Raslova H, Mercher T, Debili N, Vainchenker W (2013) Developmental changes in human megakaryopoiesis. J Thromb Haemost 11(9):1730–1741
11. Bornstein R, Garcia-Vela J, Gilsanz F, Auray C, Cales C (2001) Cord blood megakaryocytes do not complete maturation, as indicated by impaired establishment of endomitosis and low expression of G1/S cyclins upon thrombopoietin-induced differentiation. Br J Haematol 114(2):458–465
12. Boudewijns M, Raes M, Peeters V, Mewis A, Cartuyvels R, Magerman K, Rummens JL (2003) Evaluation of platelet function on cord blood in 80 healthy term neonates using the

Platelet Function Analyser (PFA-100); shorter in vitro bleeding times in neonates than adults. Eur J Pediatr 162(3):212–213
13. Bruno E, Murray LJ, DiGiusto R, Mandich D, Tsukamoto A, Hoffman R (1996) Detection of a primitive megakaryocyte progenitor cell in human fetal bone marrow. Exp Hematol 24(4):552–558
14. Bussel JB, Zacharoulis S, Kramer K, McFarland JG, Pauliny J, Kaplan C (2005) Clinical and diagnostic comparison of neonatal alloimmune thrombocytopenia to non-immune cases of thrombocytopenia. Pediatr Blood Cancer 45(2):176–183
15. Calligaris R, Bottardi S, Cogoi S, Apezteguia I, Santoro C (1995) Alternative translation initiation site usage results in two functionally distinct forms of the GATA-1 transcription factor. Proc Natl Acad Sci U S A 92(25):11598–11602
16. Carbonell F, Calvo W, Fliedner TM (1982) Cellular composition of human fetal bone marrow. Histologic study in methacrylate sections. Acta Anat (Basel) 113(4):371–375
17. Castle V, Andrew M, Kelton J, Giron D, Johnston M, Carter C (1986) Frequency and mechanism of neonatal thrombocytopenia. J Pediatr 108(5 Pt 1):749–755
18. Christensen RD, Henry E, Wiedmeier SE, Stoddard RA, Sola-Visner MC, Lambert DK, Kiehn TI, Ainsworth S (2006) Thrombocytopenia among extremely low birth weight neonates: data from a multihospital healthcare system. J Perinatol 26(6):348–353
19. Clapp DW, Baley JE, Gerson SL (1989) Gestational age-dependent changes in circulating hematopoietic stem cells in newborn infants. J Lab Clin Med 113(4):422–427
20. Corby DG, O'Barr TP (1981) Decreased alpha-adrenergic receptors in newborn platelets: cause of abnormal response to epinephrine. Dev Pharmacol Ther 2(4):215–225
21. Crapnell K, Zanjani ED, Chaudhuri A, Ascensao JL, Jeor SS, Maciejewski JP (2000) In vitro infection of megakaryocytes and their precursors by human cytomegalovirus. Blood 95(2):487–493
22. Cremer M, Dame C, Schaeffer HJ, Giers G, Bartmann P, Bald R (2003) Longitudinal thrombopoietin plasma concentrations in fetuses with alloimmune thrombocytopenia treated with intrauterine PLT transfusions. Transfusion 43(9):1216–1222
23. Cremer M, Paetzold J, Schmalisch G, Hammer H, Loui A, Dame C, Weimann A (2009) Immature platelet fraction as novel laboratory parameter predicting the course of neonatal thrombocytopenia. Br J Haematol 144(4):619–621
24. Cremer M, Sallmon H, Kling PJ, Buhrer C, Dame C (2016) Thrombocytopenia and platelet transfusion in the neonate. Semin Fetal Neonatal Med 21(1):10–18
25. Cremer M, Weimann A, Schmalisch G, Hammer H, Buhrer C, Dame C (2010) Immature platelet values indicate impaired megakaryopoietic activity in neonatal early-onset thrombocytopenia. Thromb Haemost 103(5):1016–1021
26. Cremer M, Weimann A, Szekessy D, Hammer H, Buhrer C, Dame C (2013) Low immature platelet fraction suggests decreased megakaryopoiesis in neonates with sepsis or necrotizing enterocolitis. J Perinatol 33(8):622–626
27. Daimon T, David H (1982) An automatic image analysis of megakaryocytes in fetal liver and adult bone marrow. Z Mikrosk Anat Forsch 96(3):454–460
28. Dame C (2002) Developmental biology of thrombopoietin in the human fetus and neonate. Acta Paediatr Suppl 91(438):54–65
29. Dame C, Wolber EM, Freitag P, Hofmann D, Bartmann P, Fandrey J (2003) Thrombopoietin gene expression in the developing human central nervous system. Brain Res Dev Brain Res 143(2):217–223
30. de Alarcon PA, Graeve JL (1996) Analysis of megakaryocyte ploidy in fetal bone marrow biopsies using a new adaptation of the feulgen technique to measure DNA content and estimate megakaryocyte ploidy from biopsy specimens. Pediatr Res 39(1):166–170
31. de Sauvage FJ, Carver-Moore K, Luoh SM, Ryan A, Dowd M, Eaton DL, Moore MW (1996) Physiological regulation of early and late stages of megakaryocytopoiesis by thrombopoietin. J Exp Med 183(2):651–656
32. Debili N, Issaad C, Masse J, Guichard J, Katz A, Breton-Gorius J, Vainchenker W (1992) Expression of CD34 and platelet glycoproteins during human megakaryocytic differentiation. Blood 80:3022–3035

33. Del Vecchio A, Latini G, Henry E, Christensen RD (2008) Template bleeding times of 240 neonates born at 24 to 41 weeks gestation. J Perinatol 28(6):427–431
34. Dohner ML, Wiedmeier SE, Stoddard RA, Null D Jr, Lambert DK, Burnett J, Baer VL, Hunt JC, Henry E, Christensen RD (2009) Very high users of platelet transfusions in the neonatal intensive care unit. Transfusion 49(5):869–872
35. Dreyfus M, Kaplan C, Verdy E, Schlegel N, Durand-Zaleski I, Tchernia G (1997) Frequency of immune thrombocytopenia in newborns: a prospective study. Immune Thrombocytopenia Working Group. Blood 89(12):4402–4406
36. Ehrenreich H, Hasselblatt M, Knerlich F, von Ahsen N, Jacob S, Sperling S, Woldt H, Vehmeyer K, Nave KA, Siren AL (2005) A hematopoietic growth factor, thrombopoietin, has a proapoptotic role in the brain. Proc Natl Acad Sci U S A 102(3):862–867
37. Enzan H, Takahashi H, Kawakami M, Yamashita S, Ohkita T, Yamamoto M (1980) Light and electron microscopic observations of hepatic hematopoiesis of human fetuses. II. Megakaryocytopoiesis. Acta Pathol Jpn 30(6):937–954
38. Ferrer-Marin F, Gutti R, Liu ZJ, Sola-Visner M (2014) MiR-9 contributes to the developmental differences in CXCR-4 expression in human megakaryocytes. J Thromb Haemost 12(2):282–285
39. Fiedler J, Strauss G, Wannack M, Schwiebert S, Seidel K, Henning K, Klopocki E, Schmugge M, Gaedicke G, Schulze H (2012) Two patterns of thrombopoietin signaling suggest no coupling between platelet production and thrombopoietin reactivity in thrombocytopenia-absent radii syndrome. Haematologica 97(1):73–81
40. Fielder PJ, Gurney AL, Stefanich E, Marian M, Moore MW, Carver-Moore K, de Sauvage FJ (1996) Regulation of thrombopoietin levels by c-mpl-mediated binding to platelets. Blood 87(6):2154–2161
41. Fuchs DA, McGinn SG, Cantu CL, Klein RR, Sola-Visner MC, Rimsza LM (2012) Developmental differences in megakaryocyte size in infants and children. Am J Clin Pathol 138(1):140–145
42. Fukuda T (1974) Fetal hemopoiesis. II. Electron microscopic studies on human hepatic hemopoiesis. Virchows Arch B Cell Pathol 16(3):249–270
43. Gelman B, Setty BN, Chen D, Amin-Hanjani S, Stuart MJ (1996) Impaired mobilization of intracellular calcium in neonatal platelets. Pediatr Res 39(4 Pt 1):692–696
44. George D, Bussel JB (1995) Neonatal thrombocytopenia. Semin Thromb Hemost 21(3):276–293
45. Grozovsky R, Begonja AJ, Liu K, Visner G, Hartwig JH, Falet H, Hoffmeister KM (2015) The Ashwell-Morell receptor regulates hepatic thrombopoietin production via JAK2-STAT3 signaling. Nat Med 21(1):47–54
46. Hall JG, Levin J, Kuhn JP, Ottenheimer EJ, van Berkum KA, McKusick VA (1969) Thrombocytopenia with absent radius (TAR). Medicine (Baltimore) 48(6):411–439
47. Harker LA, Slichter SJ (1972) Platelet and fibrinogen consumption in man. N Engl J Med 287(20):999–1005
48. Hezard N, Potron G, Schlegel N, Amory C, Leroux B, Nguyen P (2003) Unexpected persistence of platelet hyporeactivity beyond the neonatal period: a flow cytometric study in neonates, infants and older children. Thromb Haemost 90(1):116–123
49. Hoffmann O, Rung O, Im AR, Freyer D, Zhang J, Held J, Stenzel W, Dame C (2011) Thrombopoietin contributes to neuronal damage in experimental bacterial meningitis. Infect Immun 79(2):928–936
50. Huang Z, Richmond TD, Muntean AG, Barber DL, Weiss MJ, Crispino JD (2007) STAT1 promotes megakaryopoiesis downstream of GATA-1 in mice. J Clin Invest 117(12):3890–3899
51. Huber K, Wolf H, Van Lindern M, Worofka B, Rosen A, Hafner E, Beug H, Philipp K, Bauer K (1996) Development of a rapid means of estimating the haemoglobin F content of candidate fetal cells isolated from maternal blood using HPLC. Prenat Diagn 16(11):1011–1019
52. Ihara K, Ishii E, Eguchi M, Takada H, Suminoe A, Good RA, Hara T (1999) Identification of mutations in the c-mpl gene in congenital amegakaryocytic thrombocytopenia. Proc Natl Acad Sci U S A 96(6):3132–3136

53. Ishiguro A, Nakahata T, Matsubara K, Hayashi Y, Kato T, Suzuki Y, Shimbo T (1999) Age-related changes in thrombopoietin in children: reference interval for serum thrombopoietin levels. Br J Haematol 106(4):884–888
54. Israels SJ, Cheang T, McMillan-Ward EM, Cheang M (2001) Evaluation of primary hemostasis in neonates with a new in vitro platelet function analyzer. J Pediatr 138(1):116–119
55. Israels SJ, Odaibo FS, Robertson C, McMillan EM, McNicol A (1997) Deficient thromboxane synthesis and response in platelets from premature infants. Pediatr Res 41(2):218–223
56. Israels SJ, Rand ML, Michelson AD (2003) Neonatal platelet function. Semin Thromb Hemost 29(4):363–372
57. Izumi T (1987) Morphometric studies of megakaryocytes in human and rat fetal, infantile and adult hematopoiesis. II. Observations on rat fetuses and adult rats. Hiroshima J Med Sci 36(1):31–37
58. Izumi T, Kawakami M, Enzan H, Ohkita T (1983) The size of megakaryocytes in human fetal, infantile and adult hematopoiesis. Hiroshima J Med Sci 32(3):257–260
59. Jilma-Stohlawetz P, Homoncik M, Jilma B, Folman CC, von dem Borne AE, Bernaschek G, Deutinger J, Ulm B, Eppel W, Panzer S (2001) High levels of reticulated platelets and thrombopoietin characterize fetal thrombopoiesis. Br J Haematol 112(2):466–468
60. Kaplan C, Morel-Kopp MC, Kroll H, Kiefel V, Schlegel N, Chesnel N, Mueller-Eckhardt C (1991) HPA-5b (Br(a)) neonatal alloimmune thrombocytopenia: clinical and immunological analysis of 39 cases. Br J Haematol 78(3):425–429
61. Kaushansky K, Lok S, Holly RD, Broudy VC, Lin N, Bailey MC, Forstrom JW, Buddle MM, Oort PJ, Hagen FS et al (1994) Promotion of megakaryocyte progenitor expansion and differentiation by the c-Mpl ligand thrombopoietin. Nature 369(6481):568–571
62. Kelemen E, Calvo W, Fliedner TM (1979) Atlas of human hemopoietic development. Springer, Berlin
63. Klusmann JH, Godinho FJ, Heitmann K, Maroz A, Koch ML, Reinhardt D, Orkin SH, Li Z (2010) Developmental stage-specific interplay of GATA1 and IGF signaling in fetal megakaryopoiesis and leukemogenesis. Genes Dev 24(15):1659–1672
64. Kroll H, Muntean W, Kiefel V, Giptner A, Schluter C, Santoso S, Mueller-Eckhardt C (1994) Anti Ko(a) as a cause of neonatal alloimmune thrombocytopenia. Beitr Infusionsther Transfusionsmed 32:244–246
65. L'Abbe D, Tremblay L, Filion M, Busque L, Goldman M, Decary F, Chartrand P (1992) Alloimmunization to platelet antigen HPA-1a (PIA1) is strongly associated with both HLA-DRB3*0101 and HLA-DQB1*0201. Hum Immunol 34(2):107–114
66. Lepage A, Leboeuf M, Cazenave JP, de la Salle C, Lanza F, Uzan G (2000) The alpha(IIb) beta(3) integrin and GPIb-V-IX complex identify distinct stages in the maturation of CD34(+) cord blood cells to megakaryocytes. Blood 96(13):4169–4177
67. Levy-Shraga Y, Maayan-Metzger A, Lubetsky A, Shenkman B, Kuint J, Martinowitz U, Kenet G (2006) Platelet function of newborns as tested by cone and plate(let) analyzer correlates with gestational Age. Acta Haematol 115(3–4):152–156
68. Linder N, Shenkman B, Levin E, Sirota L, Vishne TH, Tamarin I, Dardik R, Lubin D, Savion N, Varon D (2002) Deposition of whole blood platelets on extracellular matrix under flow conditions in preterm infants. Arch Dis Child Fetal Neonatal Ed 86(2):F127–F130
69. Liu ZJ, Italiano J Jr, Ferrer-Marin F, Gutti R, Bailey M, Poterjoy B, Rimsza L, Sola-Visner M (2011) Developmental differences in megakaryocytopoiesis are associated with up-regulated TPO signaling through mTOR and elevated GATA-1 levels in neonatal megakaryocytes. Blood 117(15):4106–4117
70. Ma DC, Sun YH, Chang KZ, Zuo W (1996) Developmental change of megakaryocyte maturation and DNA ploidy in human fetus. Eur J Haematol 57(2):121–127
71. Mansour A, Abou-Ezzi G, Sitnicka E, Jacobsen SE, Wakkach A, Blin-Wakkach C (2012) Osteoclasts promote the formation of hematopoietic stem cell niches in the bone marrow. J Exp Med 209(3):537–549
72. Mehta P, Vasa R, Neumann L, Karpatkin M (1980) Thrombocytopenia in the high-risk infant. J Pediatr 97(5):791–794

73. Migliaccio G, Migliaccio AR, Petti S, Mavilio F, Russo G, Lazzaro D, Testa U, Marinucci M, Peschle C (1986) Human embryonic hemopoiesis. Kinetics of progenitors and precursors underlying the yolk sac – liver transition. J Clin Invest 78(1):51–60
74. Moore MA, Metcalf D (1970) Ontogeny of the haemopoietic system: yolk sac origin of in vivo and in vitro colony forming cells in the developing mouse embryo. Br J Haematol 18(3):279–296
75. Mullers SM, Burke N, Flood K, Cowman J, O'Connor H, Cotter B, Kearney M, Dempsey M, Dicker P, Tully E, Geary MP, Kenny D, Malone FD (2016) Altered platelet function in intrauterine growth restriction: a cause or a consequence of uteroplacental disease? Am J Perinatol 33(8):791–799
76. Muntean AG, Pang L, Poncz M, Dowdy SF, Blobel GA, Crispino JD (2007) Cyclin D-Cdk4 is regulated by GATA-1 and required for megakaryocyte growth and polyploidization. Blood 109(12):5199–5207
77. Murray NA, Howarth LJ, McCloy MP, Letsky EA, Roberts IA (2002) Platelet transfusion in the management of severe thrombocytopenia in neonatal intensive care unit patients. Transfus Med 12(1):35–41
78. Murray NA, Roberts IA (1995) Circulating megakaryocytes and their progenitors (BFU-MK and CFU-MK) in term and pre-term neonates. Br J Haematol 89(1):41–46
79. Nagata Y, Shozaki Y, Nagahisa H, Nagasawa T, Abe T, Todokoro K (1997) Serum thrombopoietin level is not regulated by transcription but by the total counts of both megakaryocytes and platelets during thrombocytopenia and thrombocytosis. Thromb Haemost 77(5):808–814
80. Ng AP, Kauppi M, Metcalf D, Hyland CD, Josefsson EC, Lebois M, Zhang JG, Baldwin TM, Di Rago L, Hilton DJ, Alexander WS (2014) Mpl expression on megakaryocytes and platelets is dispensable for thrombopoiesis but essential to prevent myeloproliferation. Proc Natl Acad Sci U S A 111(16):5884–5889
81. Nishihira H, Toyoda Y, Miyazaki H, Kigasawa H, Ohsaki E (1996) Growth of macroscopic human megakaryocyte colonies from cord blood in culture with recombinant human thrombopoietin (c-mpl ligand) and the effects of gestational age on frequency of colonies. Br J Haematol 92(1):23–28
82. Olson TA, Levine RF, Mazur EM, Wright DG, Salvado AJ (1992) Megakaryocytes and megakaryocyte progenitors in human cord blood. Am J Pediatr Hematol Oncol 14(3):241–247
83. Oren H, Irken G, Oren B, Olgun N, Ozkan H (1994) Assessment of clinical impact and predisposing factors for neonatal thrombocytopenia. Indian J Pediatr 61(5):551–558
84. Panzer S, Auerbach L, Cechova E, Fischer G, Holensteiner A, Kitl EM, Mayr WR, Putz M, Wagenbichler P, Walchshofer S (1995) Maternal alloimmunization against fetal platelet antigens: a prospective study. Br J Haematol 90(3):655–660
85. Pastos KM, Slayton WB, Rimsza LM, Young L, Sola-Visner MC (2006) Differential effects of recombinant thrombopoietin and bone marrow stromal-conditioned media on neonatal versus adult megakaryocytes. Blood 108(10):3360–3362
86. Porcelijn L, Folman CC, de Haas M, Kanhai HH, Murphy MF, von dem Borne AE, Bussel JB (2002) Fetal and neonatal thrombopoietin levels in alloimmune thrombocytopenia. Pediatr Res 52(1):105–108
87. Potts KS, Sargeant TJ, Dawson CA, Josefsson EC, Hilton DJ, Alexander WS, Taoudi S (2015) Mouse prenatal platelet-forming lineages share a core transcriptional program but divergent dependence on MPL. Blood 126(6):807–816
88. Potts KS, Sargeant TJ, Markham JF, Shi W, Biben C, Josefsson EC, Whitehead LW, Rogers KL, Liakhovitskaia A, Smyth GK, Kile BT, Medvinsky A, Alexander WS, Hilton DJ, Taoudi S (2014) A lineage of diploid platelet-forming cells precedes polyploid megakaryocyte formation in the mouse embryo. Blood 124(17):2725–2729
89. Sainio S, Javela K, Kekomaki R, Teramo K (2000) Thrombopoietin levels in cord blood plasma and amniotic fluid in fetuses with alloimmune thrombocytopenia and healthy controls. Br J Haematol 109(2):330–335

90. Sainio S, Kekomaki R, Riikonen S, Teramo K (2000) Maternal thrombocytopenia at term: a population-based study. Acta Obstet Gynecol Scand 79(9):744–749
91. Saxonhouse MA, Christensen RD, Walker DM, Hutson AD, Sola MC (2004) The concentration of circulating megakaryocyte progenitors in preterm neonates is a function of post-conceptional age. Early Hum Dev 78(2):119–124
92. Shenkman B, Linder N, Savion N, Tamarin I, Dardik R, Kennet G, German B, Varon D (1999) Increased neonatal platelet deposition on subendothelium under flow conditions: the role of plasma von Willebrand factor. Pediatr Res 45(2):270–275
93. Simeoni I, Stephens JC, Hu F, Deevi SV, Megy K, Bariana TK, Lentaigne C, Schulman S, Sivapalaratnam S, Vries MJ, Westbury SK, Greene D, Papadia S, Alessi MC, Attwood AP, Ballmaier M, Baynam G, Bermejo E, Bertoli M, Bray PF, Bury L, Cattaneo M, Collins P, Daugherty LC, Favier R, French DL, Furie B, Gattens M, Germeshausen M, Ghevaert C, Goodeve A, Guerrero J, Hampshire DJ, Hart DP, Heemskerk JW, Henskens YM, Hill M, Hogg N, Jolley JD, Kahr WH, Kelly AM, Kerr R, Kostadima M, Kunishima S, Lambert MP, Liesner R, Lopez J, Mapeta RP, Mathias M, Millar CM, Nathwani A, Neerman-Arbez M, Nurden AT, Nurden P, Othman M, Peerlinck K, Perry DJ, Poudel P, Reitsma P, Rondina M, Smethurst PA, Stevenson W, Szkotak A, Tuna S, van Geet C, Whitehorn D, Wilcox DA, Zhang B, Revel-Vilk S, Gresele P, Bellissimo D, Penkett CJ, Laffan MA, Mumford AD, Rendon A, Gomez K, Freson K, Ouwehand WH, Turro E (2016) A comprehensive high-throughput sequencing test for the diagnosis of inherited bleeding, thrombotic and platelet disorders. Blood 127(23):2791–2803
94. Sitaru AG, Holzhauer S, Speer CP, Singer D, Obergfell A, Walter U, Grossmann R (2005) Neonatal platelets from cord blood and peripheral blood. Platelets 16(3–4):203–210
95. Slayton WB, Wainman DA, Li XM, Hu Z, Jotwani A, Cogle CR, Walker D, Fisher RC, Wingard JR, Scott EW, Sola MC (2005) Developmental differences in megakaryocyte maturation are determined by the microenvironment. Stem Cells 23(9):1400–1408
96. Sola-Visner MC, Christensen RD, Hutson AD, Rimsza LM (2007) Megakaryocyte size and concentration in the bone marrow of thrombocytopenic and nonthrombocytopenic neonates. Pediatr Res 61(4):479–484
97. Sola MC, Du Y, Hutson AD, Christensen RD (2000) Dose-response relationship of megakaryocyte progenitors from the bone marrow of thrombocytopenic and non-thrombocytopenic neonates to recombinant thrombopoietin. Br J Haematol 110(2):449–453
98. Sola MC, Juul SE, Meng YG, Garg S, Sims P, Calhoun DA, Dame JB, Christensen D (1999) Thrombopoietin (Tpo) in the fetus and neonate: Tpo concentrations in preterm and term neonates, and organ distribution of Tpo and its receptor (c-mpl) during human fetal development. Early Hum Dev 53(3):239–250
99. Steinberg HN, Anderson J Jr, Lim B, Chatis PA (1993) Cytomegalovirus infection of the BS-1 human stroma cell line: effect on murine hemopoiesis. Virology 196(2):427–432
100. Sungaran R, Markovic B, Chong BH (1997) Localization and regulation of thrombopoietin mRNa expression in human kidney, liver, bone marrow, and spleen using in situ hybridization. Blood 89(1):101–107
101. Tate DY, Carlton GT, Johnson D, Sorenson RL, Nesbit M, White J, Thompson T, Krivit W (1981) Immune thrombocytopenia in severe neonatal infections. J Pediatr 98(3):449–453
102. Tober J, Koniski A, McGrath KE, Vemishetti R, Emerson R, de Mesy-Bentley KKL, Waugh R, Palis J (2007) The megakaryocyte lineage originates from hemangioblast precursors and is an integral component both of primitive and of definitive hematopoiesis. Blood 109:1433–1441
103. Ucar T, Gurman C, Arsan S, Kemahli S (2005) Platelet aggregation in term and preterm newborns. Pediatr Hematol Oncol 22(2):139–145
104. Uhrynowska M, Maslanka K, Zupanska B (1997) Neonatal thrombocytopenia: incidence, serological and clinical observations. Am J Perinatol 14(7):415–418
105. von dem Borne AE, von Riesz E, Verheugt FW, ten Cate JW, Koppe JG, Engelfriet CP, Nijenhuis LE (1980) Baka, a new platelet-specific antigen involved in neonatal allo-immune thrombocytopenia. Vox Sang 39(2):113–120

106. Westman P, Hashemi-Tavoularis S, Blanchette V, Kekomaki S, Laes M, Porcelijn L, Kekomaki R (1997) Maternal DRB1*1501, DQA1*0102, DQB1*0602 haplotype in fetomaternal alloimmunization against human platelet alloantigen HPA-6b (GPIIIa-Gln489). Tissue Antigens 50(2):113–118
107. Wiedmeier SE, Henry E, Sola-Visner MC, Christensen RD (2009) Platelet reference ranges for neonates, defined using data from over 47,000 patients in a multihospital healthcare system. J Perinatol 29(2):130–136
108. Winter O, Moser K, Mohr E, Zotos D, Kaminski H, Szyska M, Roth K, Wong DM, Dame C, Tarlinton DM, Schulze H, MacLennan IC, Manz RA (2010) Megakaryocytes constitute a functional component of a plasma cell niche in the bone marrow. Blood 116(11):1867–1875
109. Wolber EM, Dame C, Fahnenstich H, Hofmann D, Bartmann P, Jelkmann W, Fandrey J (1999) Expression of the thrombopoietin gene in human fetal and neonatal tissues. Blood 94(1):97–105
110. Woo AJ, Wieland K, Huang H, Akie TE, Piers T, Kim J, Cantor AB (2013) Developmental differences in IFN signaling affect GATA1s-induced megakaryocyte hyperproliferation. J Clin Invest 123(8):3292–3304
111. Zauli G, Vitale L, Brunelli MA, Bagnara GP (1992) Prevalence of the primitive megakaryocyte progenitors (BFU-meg) in adult human peripheral blood. Exp Hematol 20(7):850–854

Platelets in Ischemic Stroke

13

Deya Cherpokova and Bernhard Nieswandt

Abstract

Ischemic stroke is among the leading causes of mortality and permanent disability worldwide. Treatment approaches in acute stroke patients aim at rapid recanalization of occluded extra- and intracranial vessels. However, progressive stroke can still develop in patients despite successful recanalization, a process termed "reperfusion injury". It has long been recognized that pathological platelet activity essentially contributes to initiation and progression of ischemic stroke. However, the exact mechanisms and signaling pathways governing the detrimental effects of platelets in reperfusion injury have remained elusive. This chapter summarizes recent experimental data which have revealed that the pathophysiology of ischemic stroke critically depends on the activity of the von Willebrand factor receptor glycoprotein (GP)Ib and the major collagen receptor GPVI which are both involved in early platelet adhesion and activation. By contrast, interference with GPIIb/IIIa-dependent platelet aggregation and thrombus formation does not improve the outcome of acute brain ischemia and dramatically increases the susceptibility to intracranial hemorrhage. Furthermore, this chapter focuses on recent experimental data documenting a complex cascade of pathological events, involving interactions between platelets, immune cells, endothelial cells and the contact activation system in ischemic stroke progression. The potential translational impact of several key findings is briefly discussed.

D. Cherpokova (✉) • B. Nieswandt
Department of Experimental Biomedicine, University of Würzburg, University Hospital and Rudolf Virchow Center, Wurzburg, Germany
e-mail: Deya.Cherpokova@childrens.harvard.edu; bernhard.nieswandt@virchow.uni-wuerzburg.de

H. Schulze, J. Italiano (eds.), *Molecular and Cellular Biology of Platelet Formation*,
DOI 10.1007/978-3-319-39562-3_13

13.1 Introduction

Stroke is the second leading cause of death and represents the primary reason for permanent disability worldwide [27]. Intracranial hemorrhage accounts for approximately 20 % of all strokes, whereas about 80 % result from cerebral ischemia induced by thromboembolism to the brain, mostly derived from extracranial artery stenosis or atrial fibrillation [36]. Focal occlusion of a major or multiple smaller intracranial vessels leads to severe impairment or cessation of the blood flow in the downstream cerebral vasculature which initiates a cascade of pathophysiological processes, eventually resulting in infarct growth. Therefore, the primary therapeutic goal is the rapid recanalization of occluded intracerebral vessels. Current therapeutic options for acute ischemic stroke are limited to immediate thrombolysis within 4.5 h of stroke onset by recombinant tissue plasminogen activator (tPA) [26] and more recently, mechanical thrombectomy [32]. Despite successful vessel recanalization, a significant number of patients suffer from progressive infarct growth, a process referred to as "reperfusion injury" [18, 64].

In recent years, rodent models of experimental stroke have considerably advanced our understanding of the complex pathophysiology underlying ischemic stroke. Several experimental stroke models have been established that cover different aspects of the complex and heterogeneous nature of human stroke [11]. In experimental stroke research, transient middle cerebral artery occlusion (tMCAO) by an intraluminal thread and subsequent withdrawing of the filament is considered to be the model which most closely reflects the pathology of reperfusion injury [64]. This model has therefore been instrumental in elucidating principal pathomechanisms that govern secondary infarct growth in the downstream microvasculature, while translation of therapeutic approaches from the animal model to patient care has proved challenging due to different variables [11]. In the tMCAO model, a silicon-coated thread is advanced through the carotid artery up to the origin of the middle cerebral artery (MCA), thereby reducing cerebral flow by >90 % and causing an MCA infarction. Removal of the filament after an occlusion time of 30–60 min allows reperfusion of the MCA territory, however, progressive strokes develops within the following 24 h which strongly resembles the reperfusion injury observed in human patients. All experimental studies summarized in this chapter refer to results obtained using the tMCAO model.

Although the exact mechanisms by which platelets contribute to the development of brain infarction are still poorly understood, compelling experimental evidence suggests that acute ischemic stroke does not solely result from local thrombotic vessel occlusion. Rather, ischemic stroke has been increasingly recognized as a complex disease in which platelets orchestrate multiple molecular and cellular interactions, involving immune cells, endothelial cells and the contact activation system, finally leading to disruption of the blood-brain barrier and neurological damage. The observations that both thrombotic and inflammatory processes govern stroke progression in the ischemic brain led to the novel concept of "thrombo-inflammation" [57, 65].

This chapter will summarize recent progress that has been made in understanding the pathophysiological role of platelet receptors, downstream signaling pathways as well as the intrinsic coagulation system and T cells in experimental cerebral ischemia.

13.2 Platelet Adhesion and Activation Receptors

The initial capture of circulating platelets to exposed components of the extracellular matrix (ECM) at sites of vascular injury is mediated by the interaction of the glycoprotein (GP) Ib-V-IX receptor complex with von Willebrand factor (VWF) immobilized on collagen [59]. GPIbα-VWF interactions are essential for initial platelet tethering under conditions of high shear flow (>500 s^{-1}), found e.g. in stenosed arteries, but cannot mediate firm platelet adhesion to the ECM and lead instead to rapid deceleration of the cells [59]. This in turn enables the interaction of collagen with GPVI, the central activating platelet collagen receptor [56], which initiates an intracellular signaling cascade, eventually culminating in powerful cellular activation characterized by a rise in cytosolic Ca^{2+} concentration, cytoskeletal rearrangements, mobilization of alpha and dense granules and subsequent release of secondary platelet agonists, including adenosine diphosphate (ADP) and thromboxane A_2 (TxA_2). These agonists together with locally generated thrombin act on G protein-coupled receptors (GPCRs) and thus reinforce cellular activation and recruit additional circulating platelets. Finally, the extra- and intracellular signaling events induce the conformational change of integrin adhesion receptors, most notably αIIbβ3 (GPIIb/IIIa), from a low to a high affinity state, thereby mediating firm platelet adhesion, aggregation and thrombus growth [69].

Remarkably, however, recent studies on the involvement of platelet receptor-ligand interactions in acute experimental stroke in mice have demonstrated that interference with early steps of platelet adhesion and activation, but notably not aggregation, significantly reduces infarct growth without adverse effects on intracranial bleeding [42]. These findings emphasize that the mechanisms by which platelets contribute to stroke development may at least partially differ from those involved in classical thrombus formation.

13.2.1 GPIb

GPIb is part of the GPIb-V-IX complex and binds to several ligands, including VWF, P-selectin, thrombospondin, macrophage antigen 1 (Mac-1), the coagulation factors XI, XII, thrombin and high-molecular-weight kininogen [3]. In humans, lack or dysfunction of GPIb-V-IX is associated with a rare autosomal recessive disorder, the Bernard-Soulier syndrome, which is characterized by a bleeding phenotype, thrombocytopenia and giant platelets [3].

A critical role of GPIb in arterial thrombus formation and the antithrombotic potential of the GPIb-VWF axis have been well documented in different experimental arterial thrombosis models in mice [7, 23, 50] and baboons [15, 70]. Notably, GPIb has only recently emerged as a promising potential target for powerful and safe prevention and treatment of acute stroke. In the mouse tMCAO model, both the prophylactic (1 h before tMCAO) and the therapeutic administration of an anti-GPIbα Fab fragment that blocks the VWF binding site on GPIbα (p0p/B) resulted in a marked protection from stroke development which was accompanied by a significantly better functional outcome [42]. Importantly, GPIbα-blockade was not associated with an increased incidence of intracranial bleeding despite prolonged tail bleeding times in Fab-treated mice [42]. Significantly, treatment of aged and comorbid animals (e.g. atherosclerotic *Ldlr*$^{-/-}$, diabetic or hypertensive mice) with p0p/B Fab fragments provided a comparable protection from reperfusion injury, thus substantiating the notion that GPIbα-blockade may have important therapeutic implications in the future for patients with accompanying common cardiovascular diseases [47].

Notably, transgenic mice in which the extracellular domain of GPIbα was replaced by the α-subunit of the human IL-4 receptor [38] displayed a significant decrease in infarct size without any signs of intracranial hemorrhage [20]. Likewise, VWF-deficient mice developed significantly smaller infarcts than wild-type control animals and accordingly, neurological scores assessing motor function and coordination were considerably better [43]. Of note, VWF-deficiency did not increase the risk of intracranial bleeding after focal cerebral ischemia, emphasizing that interference with the GPIb-VWF axis during stroke may indeed provide a safe and effective antithrombotic approach. Reconstitution of plasma VWF by hydrodynamic gene transfer restored the susceptibility of *Vwf*$^{-/-}$ mice to ischemic stroke [19, 43]. Reconstitution experiments further revealed that the binding sites on VWF for GPIb (A1-domain) and collagen, but not that for GPIIb/IIIa, are essential for stroke development [19]. The importance of the GPIbα binding site on VWF for stroke progression has been supported by a recent study in which both the prophylactic and therapeutic pharmacological blockade of the A1-domain on VWF by a divalent humanized nanobodies in guinea pigs prevented acute thrombotic occlusion without inducing intracranial hemorrhage in two different models of experimental stroke [51]. Indeed, significant effort has been made in recent years in developing agents which block VWF-GPIbα or VWF-collagen interactions. The therapeutic potential of some of these inhibitors has been demonstrated in the prevention of VWF-mediated thrombosis in different experimental disease models [21].

Importantly, VWF levels have been identified as a predictive factor for long-term mortality after acute stroke [14]. However, not only VWF levels, but also its activity essentially contributes to the risk of stroke. VWF is a large, multimeric protein (up to 20,000 kDa) synthesized in endothelial cells and megakaryocytes that is highly thrombogenic and gets rapidly cleaved in the plasma to less active VWF by the enzyme A disintegrin and metalloprotease with thrombospondin type 1 repeats 13 (ADAMTS13) which thereby critically controls VWF activity. ADAMTS13-deficiency in mice severely worsened ischemic brain damage characterized by increased accumulation of thrombi and inflammatory cells [28, 72]. Conversely, intravenous administration of recombinant ADAMTS13 into mice 10 min prior to reperfusion reduced infarct size and improved functional outcome after focal cerebral ischemia [72]. Mutations in the ADAMTS13 gene as found in patients with hereditary thrombotic thrombocytopenic purpura (TTP) or autoantibodies against ADAMTS13 as detected in most patients with acquired TTP are associated with an increased risk of thrombotic occlusion of microvessels in different organs, including the brain [73].

In further support of a crucial role of the GPIb-VWF axis in stroke development, mice lacking phospholipase D1 (PLD1) which transduces signaling downstream of VWF-occupied GPIb and thus modulates GPIb-dependent GPIIb/IIIa activation and platelet adhesion under high shear conditions were markedly protected from ischemic stroke without any signs of increased intracranial bleeding [25]. Similarly, pharmacological inhibition of PLD isoforms by the small molecule PLD inhibitor, 5-fluoro-2-indolyl des-chlorohalopemide (FIPI), reproduced this profound protection, thus confirming the notion that interference with the GPIb-VWF axis and modulators of the downstream signaling cascade may indeed become a potential approach for future treatment strategies in ischemic stroke [62].

13.2.2 GPVI

The central platelet collagen receptor GPVI is a megakaryocyte-/platelet-specific transmembrane type I receptor that is non-covalently associated with the FcRγ-chain which contains an immunoreceptor tyrosine-based activation motif (ITAM) [56]. GPVI can be specifically and irreversibly removed from the surface of circulating platelets by antibodies in both humans [10] and mice [55]. In the latter, GPVI immunodepletion leads to a prolonged GPVI-knockout-like phenotype accompanied by a long-term antithrombotic protection in models of experimental arterial thrombosis [50]. Likewise, GPVI-depleted mice developed significantly smaller infarcts in the tMCAO model and GPVI depletion did not increase the risk of intracranial hemorrhage [42]. Contrariwise, elevated GPVI expression and a pronounced hyperreactivity in the GPVI signaling pathway in platelets translated into a dramatic increase in infarct volume and severely worsened neurological outcome in SLAP/SLAP2-deficient mice [17], thus further emphasizing the role of GPVI as a critical modulator of stroke severity. Similarly, increased GPVI expression levels have been linked to an increased risk of stroke occurrence in humans [8] and indirect evidence for increased GPVI activity has been provided by detection of elevated levels of soluble GPVI in patients suffering from acute ischemic stroke [2].

Current GPVI targeting strategies under development focus on three different approaches, including blockade of the GPVI-collagen interaction, immunodepletion of GPVI via monoclonal antibodies and interference with key signaling molecules downstream of GPVI, such as the tyrosine kinase Syk [63]. For instance, mice with megakaryocyte-/platelet-specific deficiency in Syk displayed a marked protection from ischemic stroke and likewise, pharmacolgocial inhibition of Syk improved clinical outcome [67]. However, recently demonstrated functional redundancies between GPVI and the second ITAM-bearing receptor in mouse platelets C-type lectin-like receptor 2 (CLEC-2) in hemostasis [6] and maintenance of the vascular integrity during inflammation [9] emphasize the importance of evaluating the functional status of CLEC-2 prior to and during GPVI targeting. This requirement obviously also applies to currently tested therapeutic agents which block common downstream signaling proteins, including Syk inhibitors.

13.2.3 GPIIb/IIIa

In sharp contrast to the promising results obtained in experimental stroke models after blockade of GPIb and depletion of GPVI, treatment of mice 1 h before tMCAO with $F(ab')_2$ fragments of an anti-GPIIb/IIIa antibody, which prevent platelet aggregation, was associated with increased mortality due to intracranial hemorrhages [42]. Furthermore, infarct volumes in the surviving animals were similar to those in control animals [42], thereby clearly excluding interference with final steps of platelet aggregation via GPIIb/IIIa as a safe and effective approach in stroke therapy. Importantly, treatment of stroke patients with the GPIIb/IIIa inhibitors abciximab [1] or tirofiban [39] did not improve stroke outcome and additionally resulted in increased incidence of fatal intracranial bleedings, which eventually led to the premature termination of a large international phase III trial assessing the effects of abciximab in acute stroke patients [1].

Strikingly, defective platelet dense granule secretion [61] or absence of alpha granules [24] were associated with a marked protection from acute ischemic stroke in mice, but clearly not with increased intracranial hemorrhage, thus demonstrating a critical role of the platelet secretome in infarct progression, but likely not in the maintenance of vascular integrity in thrombo-inflammatory conditions in the ischemic brain. Likewise, mice deficient in key components of the machinery mediating store-operated Ca^{2+} entry were less susceptible to secondary infarct growth [12, 68]. Further studies evaluating the role of platelets in experimental models of focal cerebral ischemia identified that reduction of platelet counts is not protective in stroke, since only approximately 10 % of normal platelet counts were still sufficient to mediate full brain infarction and were sufficient to prevent bleeding [52]. By contrast, intracranial hemorrhage occurred in severely thrombocytopenic mice (peripheral platelet counts of less than 2.5 % of control) [31], revealing that platelets fulfill essential functions in the maintenance of the integrity of the blood-brain barrier during acute brain ischemia.

13.3 The Intrinsic Coagulation System

Platelet activation leads to a marked increase in cytosolic Ca^{2+} concentration and subsequent surface exposure of negatively charged procoagulant phosphatidylserine (PS) which provides high-affinity binding sites for coagulation factors, serves as a membrane substrate for the tenase and prothrombinase complexes and thus facilitates rapid thrombin generation [34]. On the other hand, activated platelets release negatively charged inorganic polyphosphates (PolyP) that potently activate coagulation factor XII (FXII, Hageman factor) – the starting point of the intrinsic coagulation pathway [54]. FXII has been identified as a crucial regulator of thrombotic-inflammatory processes. *F12*$^{-/-}$ mice displayed, similarly to FXII-deficient humans, unaltered hemostatic functions and were protected in different models of occlusive arterial thrombosis [58]. Likewise, both genetic FXII-deficiency [41] and pharmacological inhibition of activated FXII (FXIIa) by the inhibitor rHA-infestin-4 before tMCAO [33] mediated a profound reduction in infarct size without increasing the susceptibility to intracranial hemorrhage, thus providing first experimental evidence that FXII may become a novel target for stroke treatment and prevention. Meanwhile, several FXII and FXIIa inhibitors have been developed and their mode of action, in particular in different animal models of thrombosis and ischemia-reperfusion injury, is currently a subject of intense research [40].

Besides its central function in initiating the intrinsic coagulation (contact activation) systems, FXIIa plays a pivotal role as a starting point for the kallikrein-kinin system which culminates in the generation of the pro-inflammatory peptide hormone bradykinin [53]. FXIIa converts plasma prekallikrein into the active form of the protease plasma kallikrein (PK). Both genetic PK-deficiency and the therapeutic treatment of mice with an anti-PK antibody up to 3 h after induction of tMCAO resulted in smaller infarcts, reduced tissue permeability and edema formation and less immune cell infiltration without increasing the risk of intracranial bleeding [30]. Similarly, deficiency in plasma kininogen, which is cleaved by PK upon activation of the contact system, induced a marked protection from microvascular thrombosis, blood-brain-barrier disruption and inflammation without affecting hemostatic function in mice [48]. Finally, deletion of the bradykinin receptor B1 (B1R) or its blockade by a selective B1R-inhibitor administered 1 h after tMCAO ameliorated intracerebral thrombosis and edema formation in mice [4]. Importantly, therapeutic administration of a C1-esterase inhibitor (C1Inh), which is a major endogenous regulator of the contact system by blocking FXII and PK, equally reduced microvascular thrombosis and blood-brain-barrier damage and inflammation in mice and rats following focal cerebral ischemia [22, 29, 35, 66]. Of note, C1Inh has been approved for many years now for treatment of C1Inh-deficiency in the setting of hereditary or acquired angioedema which from a translational perspective may facilitate approval for different indications [5]. Collectively, these findings in experimental stroke research support the notion that interference with key components of the contact activation systems could become an effective and safe approach for the treatment of thrombo-inflammatory disease conditions.

13.4 Platelet-Immune Cell Interactions in Early Stroke Development

Ischemic stroke leads to recruitment of immune cells which participate in the pathophysiological processes that occur both in the acute phase after disease onset and at later stages. The contribution of the immune system to stroke severity, progression and functional outcome as well as the complex interactions of immune cells with other cell types in the ischemic brain have been reviewed elsewhere [16, 37]. Here, we specifically focus on the emerging role of T cells and on their platelet-dependent functions in reperfusion injury of the ischemic brain. Studies in *Rag1*$^{-/-}$ [44, 71] and *Rag2*$^{-/-}$ [49, 60] mice, which both lack mature T and B cells, demonstrated that these animals are protected from focal cerebral ischemia. Adoptive transfer experiments revealed that susceptibility to stroke progression in the tMCAO model was fully restored after infusion of wild-type T cells, but not after reconstitution of the B cell population [44, 71]. Remarkably, the detrimental function of T cells in acute ischemic stroke has been demonstrated to occur independently of antigen recognition or co-stimulation [44]. In line with these findings, administration of FTY720 (fingolimod) to mice prior to reperfusion in the tMCAO model significantly reduced infarct size and improved neurological outcome [46]. FTY720 is a modulator of sphingosine-1-phosphate receptors that prevents egress of lymphocytes from lymphoid organs and it is approved for treatment of relapsing multiple sclerosis [13]. Of note, a recent pilot study with acute stroke patients demonstrated that thrombolysis in combination with FTY720 treatment was superior to thrombolysis alone in the prevention of reperfusion injury progression [74]. These experimental and clinical data reveal that immune modulation may have an important (supportive) function in stroke therapy, but clearly, further investigations on the exact mode of action are required.

Importantly, platelet adhesion on collagen under flow conditions and occlusion times in the ferric chloride-injured mesenteric arterioles were unaltered in *Rag1*$^{-/-}$ mice [44], demonstrating that the protection from stroke progression in this mouse model cannot be ascribed to altered thrombus formation. However, evidence for a functional cross-talk between platelets and T cells in ischemic brain damage came from an experimental study which revealed that adoptive transfer of T cells into platelet-depleted *Rag1*$^{-/-}$ mice is not detrimental [45]. Collectively, these experimental data led to the introduction of the concept of "thrombo-inflammation" (Fig. 13.1) in the progression of ischemic stroke [57, 65]. The mechanisms underlying the deleterious effects of platelet-T cell interactions in reperfusion injury of the brain are only beginning to be understood and highlight the importance of further investigations to elucidate the role of both cell types in thrombo-inflammatory brain infarction.

13.5 Concluding Remarks

In recent years, experimental stroke research has established an important role for receptors involved in platelet adhesion and activation in the progression of reperfusion injury. These findings in experimental animal models strongly suggest that interference with early steps of platelet activation could become a novel therapeutic approach in acute ischemic stroke. However, further studies are required to prove if the promising experimental data are likely to have important implications in patient care in the future. Likewise, a better understanding of the concerted action of platelets and immune cells in acute brain infarction may as well provide the basis for the development of safe and effective anti-thrombo-inflammatory therapies.

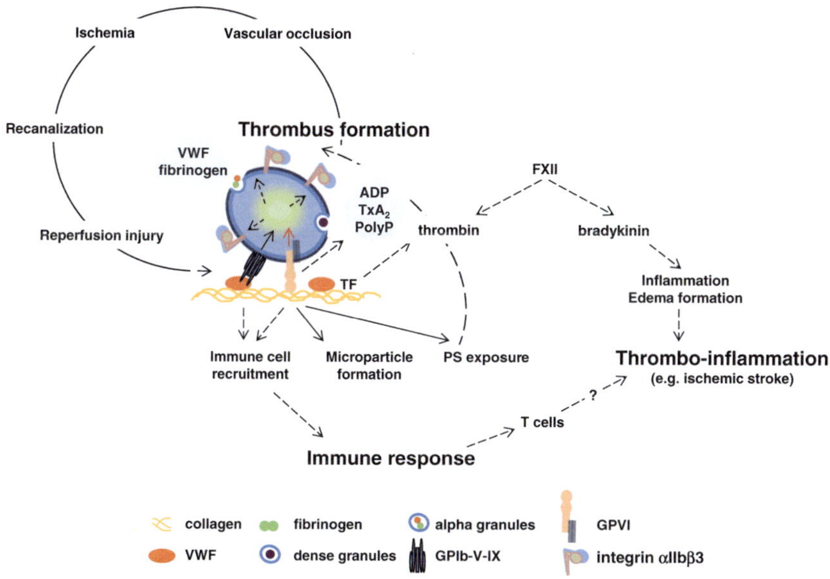

Fig. 13.1 Mechanisms governing thrombus formation and thrombo-inflammatory responses. VWF-GPIb interactions are essential for platelet tethering at the site of vascular injury. Additionally, GPIb has been implicated in immune cell recruitment during inflammatory responses. The central activating collagen receptor on the platelet surface GPVI critically contributes to platelet activation and supports granule release which mediates the release of ADP and TxA_2 that together with locally generated thrombin potentiate cellular activation. Furthermore, strong platelet activation via engagement of GPVI eventually leads to the exposure of procoagulant phosphatidylserine (PS) and the secretion of inorganic polyphosphates (PolyP) which triggers coagulation and activates FXII, respectively. GPVI signaling may also induce immune cell recruitment and fuel inflammation via microparticle formation. Integrin αIIbβ3 is indispensable for stable thrombus formation which may lead to vascular occlusion in diseased vessels and subsequent ischemia. Reperfusion of a previously occluded cerebral vessel can paradoxically induce reperfusion injury characterized by damage to the surrounding tissue and platelet recruitment to the site of injury. Importantly, downstream signaling events may differ from those culminating in thrombus formation. For example, activation of FXII can also promote the synthesis of the pro-inflammatory mediator bradykinin which triggers a profound thrombo-inflammatory response, such as found in ischemic stroke. The mechanisms underlying T cell recruitment and function as well as the downstream effector signaling cascade in this thrombo-inflammatory setting are still poorly understood. *TF* tissue factor

Acknowledgments Prof. Dr. Bernhard Nieswandt holds the Chair of Experimental Biomedicine at the University Hospital and the Rudolf Virchow Center, University of Würzburg, Germany. This work was supported by the Deutsche Forschungsgemeinschaft (SFB 688) and the Rudolf Virchow Center. Dr. Deya Cherpokova was supported by a grant of the German Excellence Initiative to the Graduate School of Life Sciences, University of Würzburg. The authors thank Prof. Dr. Guido Stoll and Prof. Dr. Christoph Kleinschnitz for continuous support and longstanding collaboration.

References

1. Adams HP Jr, Effron MB, Torner J, Davalos A, Frayne J, Teal P, Leclerc J, Oemar B, Padgett L, Barnathan ES, Hacke W, Ab E-III (2008) Emergency administration of abciximab for treatment of patients with acute ischemic stroke: results of an international phase III trial: abciximab in emergency treatment of stroke trial (AbESTT-II). Stroke J Cereb Circ 39(1):87–99. doi:10.1161/STROKEAHA.106.476648
2. Al-Tamimi M, Gardiner EE, Thom JY, Shen Y, Cooper MN, Hankey GJ, Berndt MC, Baker RI, Andrews RK (2011) Soluble glycoprotein VI is raised in the plasma of patients with acute ischemic stroke. Stroke J Cereb Circ 42(2):498–500. doi:10.1161/STROKEAHA.110.602532
3. Andrews RK, Berndt MC (2013) Bernard-Soulier syndrome: an update. Semin Thromb Hemost 39(6):656–662. doi:10.1055/s-0033-1353390
4. Austinat M, Braeuninger S, Pesquero JB, Brede M, Bader M, Stoll G, Renne T, Kleinschnitz C (2009) Blockade of bradykinin receptor B1 but not bradykinin receptor B2 provides protection from cerebral infarction and brain edema. Stroke J Cereb Circ 40(1):285–293. doi:10.1161/STROKEAHA.108.526673
5. Banerji A (2010) Current treatment of hereditary angioedema: an update on clinical studies. Allergy Asthma Proc: Off J Reg State Allergy Soc 31(5):398–406. doi:10.2500/aap.2010.31.3387
6. Bender M, May F, Lorenz V, Thielmann I, Hagedorn I, Finney BA, Vogtle T, Remer K, Braun A, Bosl M, Watson SP, Nieswandt B (2013) Combined in vivo depletion of glycoprotein VI and C-type lectin-like receptor 2 severely compromises hemostasis and abrogates arterial thrombosis in mice. Arterioscler Thromb Vasc Biol 33(5):926–934. doi:10.1161/ATVBAHA.112.300672
7. Bergmeier W, Piffath CL, Goerge T, Cifuni SM, Ruggeri ZM, Ware J, Wagner DD (2006) The role of platelet adhesion receptor GPIbalpha far exceeds that of its main ligand, von Willebrand factor, in arterial thrombosis. Proc Natl Acad Sci U S A 103(45):16900–16905. doi:10.1073/pnas.0608207103
8. Bigalke B, Stellos K, Weig HJ, Geisler T, Seizer P, Kremmer E, Potz O, Joos T, May AE, Lindemann S, Gawaz M (2009) Regulation of platelet glycoprotein VI (GPVI) surface expression and of soluble GPVI in patients with atrial fibrillation (AF) and acute coronary syndrome (ACS). Basic Res Cardiol 104(3):352–357. doi:10.1007/s00395-009-0779-7
9. Boulaftali Y, Hess PR, Getz TM, Cholka A, Stolla M, Mackman N, Owens AP 3rd, Ware J, Kahn ML, Bergmeier W (2013) Platelet ITAM signaling is critical for vascular integrity in inflammation. J Clin Invest 123(2):908–916. doi:10.1172/JCI65154
10. Boylan B, Chen H, Rathore V, Paddock C, Salacz M, Friedman KD, Curtis BR, Stapleton M, Newman DK, Kahn ML, Newman PJ (2004) Anti-GPVI-associated ITP: an acquired platelet disorder caused by autoantibody-mediated clearance of the GPVI/FcRgamma-chain complex from the human platelet surface. Blood 104(5):1350–1355. doi:10.1182/blood-2004-03-0896
11. Braeuninger S, Kleinschnitz C (2009) Rodent models of focal cerebral ischemia: procedural pitfalls and translational problems. Exp Transl Stroke Med 1:8. doi:10.1186/2040-7378-1-8
12. Braun A, Varga-Szabo D, Kleinschnitz C, Pleines I, Bender M, Austinat M, Bosl M, Stoll G, Nieswandt B (2009) Orai1 (CRACM1) is the platelet SOC channel and essential for pathological thrombus formation. Blood 113(9):2056–2063. doi:10.1182/blood-2008-07-171611
13. Brinkmann V, Billich A, Baumruker T, Heining P, Schmouder R, Francis G, Aradhye S, Burtin P (2010) Fingolimod (FTY720): discovery and development of an oral drug to treat multiple sclerosis. Nat Rev Drug Discov 9(11):883–897. doi:10.1038/nrd3248
14. Carter AM, Catto AJ, Mansfield MW, Bamford JM, Grant PJ (2007) Predictive variables for mortality after acute ischemic stroke. Stroke J Cereb Circ 38(6):1873–1880. doi:10.1161/STROKEAHA.106.474569
15. Cauwenberghs N, Meiring M, Vauterin S, van Wyk V, Lamprecht S, Roodt JP, Novak L, Harsfalvi J, Deckmyn H, Kotze HF (2000) Antithrombotic effect of platelet glycoprotein

Ib-blocking monoclonal antibody Fab fragments in nonhuman primates. Arterioscler Thromb Vasc Biol 20(5):1347–1353

16. Chamorro A, Meisel A, Planas AM, Urra X, van de Beek D, Veltkamp R (2012) The immunology of acute stroke. Nat Rev Neurol 8(7):401–410. doi:10.1038/nrneurol.2012.98
17. Cherpokova D, Bender M, Morowski M, Kraft P, Schuhmann MK, Akbar SM, Sultan CS, Hughes CE, Kleinschnitz C, Stoll G, Dragone LL, Watson SP, Tomlinson MG, Nieswandt B (2015) SLAP/SLAP2 prevent excessive platelet (hem)ITAM signaling in thrombosis and ischemic stroke in mice. Blood 125(1):185–194. doi:10.1182/blood-2014-06-580597
18. Coutts SB, Goyal M (2009) When recanalization does not improve clinical outcomes. Stroke J Cereb Circ 40(8):2661. doi:10.1161/STROKEAHA.109.557447
19. De Meyer SF, Schwarz T, Deckmyn H, Denis CV, Nieswandt B, Stoll G, Vanhoorelbeke K, Kleinschnitz C (2010) Binding of von Willebrand factor to collagen and glycoprotein Ibalpha, but not to glycoprotein IIb/IIIa, contributes to ischemic stroke in mice – brief report. Arterioscler Thromb Vasc Biol 30(10):1949–1951. doi:10.1161/ATVBAHA.110.208918
20. De Meyer SF, Schwarz T, Schatzberg D, Wagner DD (2011) Platelet glycoprotein Ibalpha is an important mediator of ischemic stroke in mice. Exp Transl Stroke Med 3:9. doi:10.1186/2040-7378-3-9
21. De Meyer SF, Stoll G, Wagner DD, Kleinschnitz C (2012) von Willebrand factor: an emerging target in stroke therapy. Stroke J Cereb Circ 43(2):599–606. doi:10.1161/STROKEAHA.111.628867
22. De Simoni MG, Storini C, Barba M, Catapano L, Arabia AM, Rossi E, Bergamaschini L (2003) Neuroprotection by complement (C1) inhibitor in mouse transient brain ischemia. J Cereb Blood Flow Metab: Off J Int Soc Cereb Blood Flow Metab 23(2):232–239
23. Denis C, Methia N, Frenette PS, Rayburn H, Ullman-Cullere M, Hynes RO, Wagner DD (1998) A mouse model of severe von Willebrand disease: defects in hemostasis and thrombosis. Proc Natl Acad Sci U S A 95(16):9524–9529
24. Deppermann C, Cherpokova D, Nurden P, Schulz JN, Thielmann I, Kraft P, Vogtle T, Kleinschnitz C, Dutting S, Krohne G, Eming SA, Nurden AT, Eckes B, Stoll G, Stegner D et al (2013) Gray platelet syndrome and defective thrombo-inflammation in Nbeal2-deficient mice. J Clin Invest 123(8):3331–3342. doi:10.1172/JCI69210
25. Elvers M, Stegner D, Hagedorn I, Kleinschnitz C, Braun A, Kuijpers ME, Boesl M, Chen Q, Heemskerk JW, Stoll G, Frohman MA, Nieswandt B (2010) Impaired alpha(IIb)beta(3) integrin activation and shear-dependent thrombus formation in mice lacking phospholipase D1. Sci Signal 3(103):ra1. doi:10.1126/scisignal.2000551
26. Emberson J, Lees KR, Lyden P, Blackwell L, Albers G, Bluhmki E, Brott T, Cohen G, Davis S, Donnan G, Grotta J, Howard G, Kaste M, Koga M, von Kummer R et al (2014) Effect of treatment delay, age, and stroke severity on the effects of intravenous thrombolysis with alteplase for acute ischaemic stroke: a meta-analysis of individual patient data from randomised trials. Lancet 384(9958):1929–1935. doi:10.1016/S0140-6736(14)60584-5
27. Feigin VL, Forouzanfar MH, Krishnamurthi R, Mensah GA, Connor M, Bennett DA, Moran AE, Sacco RL, Anderson L, Truelsen T, O'Donnell M, Venketasubramanian N, Barker-Collo S, Lawes CM, Wang W et al (2014) Global and regional burden of stroke during 1990–2010: findings from the Global Burden of Disease Study 2010. Lancet 383(9913):245–254
28. Fujioka M, Hayakawa K, Mishima K, Kunizawa A, Irie K, Higuchi S, Nakano T, Muroi C, Fukushima H, Sugimoto M, Banno F, Kokame K, Miyata T, Fujiwara M, Okuchi K et al (2010) ADAMTS13 gene deletion aggravates ischemic brain damage: a possible neuroprotective role of ADAMTS13 by ameliorating postischemic hypoperfusion. Blood 115(8):1650–1653. doi:10.1182/blood-2009-06-230110
29. Gesuete R, Storini C, Fantin A, Stravalaci M, Zanier ER, Orsini F, Vietsch H, Mannesse ML, Ziere B, Gobbi M, De Simoni MG (2009) Recombinant C1 inhibitor in brain ischemic injury. Ann Neurol 66(3):332–342. doi:10.1002/ana.21740
30. Gob E, Reymann S, Langhauser F, Schuhmann MK, Kraft P, Thielmann I, Gobel K, Brede M, Homola G, Solymosi L, Stoll G, Geis C, Meuth SG, Nieswandt B, Kleinschnitz C (2015)

Blocking of plasma kallikrein ameliorates stroke by reducing thromboinflammation. Ann Neurol 77(5):784–803. doi:10.1002/ana.24380
31. Goerge T, Ho-Tin-Noe B, Carbo C, Benarafa C, Remold-O'Donnell E, Zhao BQ, Cifuni SM, Wagner DD (2008) Inflammation induces hemorrhage in thrombocytopenia. Blood 111(10):4958–4964. doi:10.1182/blood-2007-11-123620
32. Grotta JC, Hacke W (2015) Stroke neurologist's perspective on the new endovascular trials. Stroke J Cereb Circ 46(6):1447–1452. doi:10.1161/STROKEAHA.115.008384
33. Hagedorn I, Schmidbauer S, Pleines I, Kleinschnitz C, Kronthaler U, Stoll G, Dickneite G, Nieswandt B (2010) Factor XIIa inhibitor recombinant human albumin Infestin-4 abolishes occlusive arterial thrombus formation without affecting bleeding. Circulation 121(13):1510–1517. doi:10.1161/CIRCULATIONAHA.109.924761
34. Heemskerk JW, Mattheij NJ, Cosemans JM (2013) Platelet-based coagulation: different populations, different functions. J Thromb Haemost: JTH 11(1):2–16. doi:10.1111/jth.12045
35. Heydenreich N, Nolte MW, Gob E, Langhauser F, Hofmeister M, Kraft P, Albert-Weissenberger C, Brede M, Varallyay C, Gobel K, Meuth SG, Nieswandt B, Dickneite G, Stoll G, Kleinschnitz C (2012) C1-inhibitor protects from brain ischemia-reperfusion injury by combined antiinflammatory and antithrombotic mechanisms. Stroke J Cereb Circ 43(9):2457–2467. doi:10.1161/STROKEAHA.112.660340
36. Hughes M, Lip GY, Guideline Development Group NCGfMoAFiP, Secondary Care NIfH, Clinical E (2008) Stroke and thromboembolism in atrial fibrillation: a systematic review of stroke risk factors, risk stratification schema and cost effectiveness data. Thromb Haemost 99(2):295–304. doi:10.1160/TH07-08-0508
37. Iadecola C, Anrather J (2011) The immunology of stroke: from mechanisms to translation. Nat Med 17(7):796–808. doi:10.1038/nm.2399
38. Kanaji T, Russell S, Ware J (2002) Amelioration of the macrothrombocytopenia associated with the murine Bernard-Soulier syndrome. Blood 100(6):2102–2107. doi:10.1182/blood-2002-03-0997
39. Kellert L, Hametner C, Rohde S, Bendszus M, Hacke W, Ringleb P, Stampfl S (2013) Endovascular stroke therapy: tirofiban is associated with risk of fatal intracerebral hemorrhage and poor outcome. Stroke J Cereb Circ 44(5):1453–1455. doi:10.1161/STROKEAHA.111.000502
40. Kenne E, Renne T (2014) Factor XII: a drug target for safe interference with thrombosis and inflammation. Drug Discov Today 19(9):1459–1464. doi:10.1016/j.drudis.2014.06.024
41. Kleinschnitz C, Stoll G, Bendszus M, Schuh K, Pauer HU, Burfeind P, Renne C, Gailani D, Nieswandt B, Renne T (2006) Targeting coagulation factor XII provides protection from pathological thrombosis in cerebral ischemia without interfering with hemostasis. J Exp Med 203(3):513–518. doi:10.1084/jem.20052458
42. Kleinschnitz C, Pozgajova M, Pham M, Bendszus M, Nieswandt B, Stoll G (2007) Targeting platelets in acute experimental stroke: impact of glycoprotein Ib, VI, and IIb/IIIa blockade on infarct size, functional outcome, and intracranial bleeding. Circulation 115(17):2323–2330. doi:10.1161/CIRCULATIONAHA.107.691279
43. Kleinschnitz C, De Meyer SF, Schwarz T, Austinat M, Vanhoorelbeke K, Nieswandt B, Deckmyn H, Stoll G (2009) Deficiency of von Willebrand factor protects mice from ischemic stroke. Blood 113(15):3600–3603. doi:10.1182/blood-2008-09-180695
44. Kleinschnitz C, Schwab N, Kraft P, Hagedorn I, Dreykluft A, Schwarz T, Austinat M, Nieswandt B, Wiendl H, Stoll G (2010) Early detrimental T-cell effects in experimental cerebral ischemia are neither related to adaptive immunity nor thrombus formation. Blood 115(18):3835–3842. doi:10.1182/blood-2009-10-249078
45. Kleinschnitz C, Kraft P, Dreykluft A, Hagedorn I, Gobel K, Schuhmann MK, Langhauser F, Helluy X, Schwarz T, Bittner S, Mayer CT, Brede M, Varallyay C, Pham M, Bendszus M et al (2013) Regulatory T cells are strong promoters of acute ischemic stroke in mice by inducing dysfunction of the cerebral microvasculature. Blood 121(4):679–691. doi:10.1182/blood-2012-04-426734
46. Kraft P, Gob E, Schuhmann MK, Gobel K, Deppermann C, Thielmann I, Herrmann AM, Lorenz K, Brede M, Stoll G, Meuth SG, Nieswandt B, Pfeilschifter W, Kleinschnitz C (2013) FTY720

ameliorates acute ischemic stroke in mice by reducing thrombo-inflammation but not by direct neuroprotection. Stroke J Cereb Circ 44(11):3202–3210. doi:10.1161/STROKEAHA.113.002880
47. Kraft P, Schuhmann MK, Fluri F, Lorenz K, Zernecke A, Stoll G, Nieswandt B, Kleinschnitz C (2015) Efficacy and safety of platelet glycoprotein receptor blockade in aged and comorbid mice with acute experimental stroke. Stroke J Cereb Circ 46(12):3502–3506. doi:10.1161/STROKEAHA.115.011114
48. Langhauser F, Gob E, Kraft P, Geis C, Schmitt J, Brede M, Gobel K, Helluy X, Pham M, Bendszus M, Jakob P, Stoll G, Meuth SG, Nieswandt B, McCrae KR et al (2012) Kininogen deficiency protects from ischemic neurodegeneration in mice by reducing thrombosis, blood-brain barrier damage, and inflammation. Blood 120(19):4082–4092. doi:10.1182/blood-2012-06-440057
49. Liesz A, Suri-Payer E, Veltkamp C, Doerr H, Sommer C, Rivest S, Giese T, Veltkamp R (2009) Regulatory T cells are key cerebroprotective immunomodulators in acute experimental stroke. Nat Med 15(2):192–199. doi:10.1038/nm.1927
50. Massberg S, Gawaz M, Gruner S, Schulte V, Konrad I, Zohlnhofer D, Heinzmann U, Nieswandt B (2003) A crucial role of glycoprotein VI for platelet recruitment to the injured arterial wall in vivo. J Exp Med 197(1):41–49
51. Momi S, Tantucci M, Van Roy M, Ulrichts H, Ricci G, Gresele P (2013) Reperfusion of cerebral artery thrombosis by the GPIb-VWF blockade with the Nanobody ALX-0081 reduces brain infarct size in guinea pigs. Blood 121(25):5088–5097. doi:10.1182/blood-2012-11-464545
52. Morowski M, Vogtle T, Kraft P, Kleinschnitz C, Stoll G, Nieswandt B (2013) Only severe thrombocytopenia results in bleeding and defective thrombus formation in mice. Blood 121(24):4938–4947. doi:10.1182/blood-2012-10-461459
53. Muller F, Renne T (2008) Novel roles for factor XII-driven plasma contact activation system. Curr Opin Hematol 15(5):516–521. doi:10.1097/MOH.0b013e328309ec85
54. Muller F, Mutch NJ, Schenk WA, Smith SA, Esterl L, Spronk HM, Schmidbauer S, Gahl WA, Morrissey JH, Renne T (2009) Platelet polyphosphates are proinflammatory and procoagulant mediators in vivo. Cell 139(6):1143–1156. doi:10.1016/j.cell.2009.11.001
55. Nieswandt B, Schulte V, Bergmeier W, Mokhtari-Nejad R, Rackebrandt K, Cazenave JP, Ohlmann P, Gachet C, Zirngibl H (2001) Long-term antithrombotic protection by in vivo depletion of platelet glycoprotein VI in mice. J Exp Med 193(4):459–469
56. Nieswandt B, Watson SP (2003) Platelet-collagen interaction: is GPVI the central receptor? Blood 102(2):449–461. doi:10.1182/blood-2002-12-3882
57. Nieswandt B, Kleinschnitz C, Stoll G (2011) Ischaemic stroke: a thrombo-inflammatory disease? J Physiol 589(Pt 17):4115–4123. doi:10.1113/jphysiol.2011.212886
58. Renne T, Pozgajova M, Gruner S, Schuh K, Pauer HU, Burfeind P, Gailani D, Nieswandt B (2005) Defective thrombus formation in mice lacking coagulation factor XII. J Exp Med 202(2):271–281. doi:10.1084/jem.20050664
59. Savage B, Almus-Jacobs F, Ruggeri ZM (1998) Specific synergy of multiple substrate-receptor interactions in platelet thrombus formation under flow. Cell 94(5):657–666
60. Shichita T, Sugiyama Y, Ooboshi H, Sugimori H, Nakagawa R, Takada I, Iwaki T, Okada Y, Iida M, Cua DJ, Iwakura Y, Yoshimura A (2009) Pivotal role of cerebral interleukin-17-producing gammadeltaT cells in the delayed phase of ischemic brain injury. Nat Med 15(8):946–950. doi:10.1038/nm.1999
61. Stegner D, Deppermann C, Kraft P, Morowski M, Kleinschnitz C, Stoll G, Nieswandt B (2013) Munc13-4-mediated secretion is essential for infarct progression but not intracranial hemostasis in acute stroke. J Thromb Haemost: JTH 11(7):1430–1433. doi:10.1111/jth.12293
62. Stegner D, Thielmann I, Kraft P, Frohman MA, Stoll G, Nieswandt B (2013) Pharmacological inhibition of phospholipase D protects mice from occlusive thrombus formation and ischemic stroke – brief report. Arterioscler Thromb Vasc Biol 33(9):2212–2217. doi:10.1161/ATVBAHA.113.302030
63. Stegner D, Haining EJ, Nieswandt B (2014) Targeting glycoprotein VI and the immunoreceptor tyrosine-based activation motif signaling pathway. Arterioscler Thromb Vasc Biol 34(8):1615–1620. doi:10.1161/ATVBAHA.114.303408

64. Stoll G, Kleinschnitz C, Nieswandt B (2008) Molecular mechanisms of thrombus formation in ischemic stroke: novel insights and targets for treatment. Blood 112(9):3555–3562. doi:10.1182/blood-2008-04-144758
65. Stoll G, Kleinschnitz C, Nieswandt B (2010) Combating innate inflammation: a new paradigm for acute treatment of stroke? Ann N Y Acad Sci 1207:149–154. doi:10.1111/j.1749-6632.2010.05730.x
66. Storini C, Rossi E, Marrella V, Distaso M, Veerhuis R, Vergani C, Bergamaschini L, De Simoni MG (2005) C1-inhibitor protects against brain ischemia-reperfusion injury via inhibition of cell recruitment and inflammation. Neurobiol Dis 19(1–2):10–17. doi:10.1016/j.nbd.2004.11.001
67. van Eeuwijk JM, Stegner D et al (2016) The Novel Oral Syk Inhibitor, B1002494, Protects Mice From Arterial Thrombosis and Thromboinflammatory Brain Infarction. Arterioscler Thromb Vasc Biol. 36(6):1247–1253. doi:10.1161/ATVBAHA.115.306883.
68. Varga-Szabo D, Braun A, Kleinschnitz C, Bender M, Pleines I, Pham M, Renne T, Stoll G, Nieswandt B (2008) The calcium sensor STIM1 is an essential mediator of arterial thrombosis and ischemic brain infarction. J Exp Med 205(7):1583–1591. doi:10.1084/jem.20080302
69. Varga-Szabo D, Pleines I, Nieswandt B (2008) Cell adhesion mechanisms in platelets. Arterioscler Thromb Vasc Biol 28(3):403–412. doi:10.1161/ATVBAHA.107.150474
70. Wu D, Vanhoorelbeke K, Cauwenberghs N, Meiring M, Depraetere H, Kotze HF, Deckmyn H (2002) Inhibition of the von Willebrand (VWF)-collagen interaction by an antihuman VWF monoclonal antibody results in abolition of in vivo arterial platelet thrombus formation in baboons. Blood 99(10):3623–3628
71. Yilmaz G, Arumugam TV, Stokes KY, Granger DN (2006) Role of T lymphocytes and interferon-gamma in ischemic stroke. Circulation 113(17):2105–2112. doi:10.1161/CIRCULATIONAHA.105.593046
72. Zhao BQ, Chauhan AK, Canault M, Patten IS, Yang JJ, Dockal M, Scheiflinger F, Wagner DD (2009) von Willebrand factor-cleaving protease ADAMTS13 reduces ischemic brain injury in experimental stroke. Blood 114(15):3329–3334. doi:10.1182/blood-2009-03-213264
73. Zhou Z, Nguyen TC, Guchhait P, Dong JF (2010) Von Willebrand factor, ADAMTS-13, and thrombotic thrombocytopenic purpura. Semin Thromb Hemost 36(1):71–81. doi:10.1055/s-0030-1248726
74. Zhu Z, Fu Y, Tian D, Sun N, Han W, Chang G, Dong Y, Xu X, Liu Q, Huang D, Shi FD (2015) Combination of the immune modulator fingolimod with alteplase in acute ischemic stroke: a pilot trial. Circulation 132(12):1104–1112. doi:10.1161/CIRCULATIONAHA.115.016371

Insights in Megakaryopoiesis and Platelet Biogenesis from Studies of Inherited Thrombocytopenias

14

Kathleen Freson

Abstract

Inherited thrombocytopenias (ITs) are clinically and genetically heterogeneous disorders characterized by defects in megakaryopoiesis and platelet biogenesis. Patients with ITs have a variable degree or even no bleeding symptoms because of low platelet count that can be associated with platelet dysfunction. ITs have been described as an isolated blood disorder but can also be part of a multisystem disorder. Mutations in more than 30 genes have already been shown to be implicated in ITs and will be recapitulated in this review. Because of that complexity, detailed clinical and laboratory investigations are needed for diagnosis of ITs. However, confirmed diagnosis can usually only be made by genetic analysis. In this review paper, a focus is made on the major contributions of next-generation sequencing approaches to the genetic landscape of ITs. Numerous novel, often unexpected, genes have been identified; still more are expected, as about 50 % of the IT cases received not yet a genetic diagnosis. In addition, the phenotypic spectrum associated with a single gene also constantly gains in complexity as illustrated with some examples. All these factors make genotype-phenotype correlations particularly difficult. Finally, we highlight some functional genetic approaches that increasingly appear important to understand the role of novel genes in defective megakaryopoiesis and platelet biogenesis.

K. Freson
Department of Cardiovascular Sciences, Center for Molecular and Vascular Biology, University of Leuven, Leuven, Belgium
e-mail: kathleen.freson@med.kuleuven.be

H. Schulze, J. Italiano (eds.), *Molecular and Cellular Biology of Platelet Formation*,
DOI 10.1007/978-3-319-39562-3_14

14.1 Megakaryopoiesis and Platelet Biogenesis

Megakaryopoiesis starts with the commitment of hematopoietic stem cells to the myeloid lineage and subsequent differentiation into the bipotent megakaryocyte-erythroid progenitor cells. These progenitor cells undergo a complex process of megakaryocyte (MK) commitment, proliferation, and maturation to become proplatelet-forming MKs (--> Chap. 1) [1]. Finally, platelets arise from cytoplasmic fragmentation of proplatelet-forming MKs and play an essential role in primary hemostasis and other processes such as inflammation, angiogenesis, wound healing, or cancer. Different molecular pathways are known to tightly regulate megakaryopoiesis and platelet production as main regulators thrombopoietin (TPO) and diverse lineage-specific transcription factors [2]. TPO is the growth factor necessary for all stages of MK development. TPO signaling goes via its receptor MPL (c-MPL) and is supported by additional growth factors such as interleukin (IL)-3, stem cell factor, IL-1 alpha, IL-6, and IL-11 [1]. Cell type-specific transcription factors such as GATA1, FLI1, RUNX1, and GFI1B play an essential role in MK lineage commitment through regulation of gene expression [3]. Proplatelet-forming MKs arise from immature MKs or megakaryoblasts. During maturation, megakaryoblasts undergo endomitosis and become polyploid [4]. They undergo repeated rounds of DNA replication without cell division and usually result in a very large MK with a single nucleus and DNA contents up to 128N. The MKs start showing a budding lobulated shape filled with granules, cytoskeletal proteins, and demarcation membrane system [5]. MK maturation is further characterized by the progressive formation and appearance of three main types of secretion granules: alpha (α) granules, dense (δ) granules, and lysosomes (--> Chap. 6). Next, the proplatelet-forming MK undergoes a cytoskeletal rearrangement that results in proplatelet formation. Proplatelets are long, branching cytoplasmic extensions, which fragment at their end to finally form platelets. Cytoskeletal proteins are important during platelet formation; microtubules, actin filaments, and other cytoskeletal proteins play an essential role in transporting platelet granules along these proplatelet extensions. These proplatelets extend into the bone marrow sinusoids, where they fragment into platelets (--> Chap. 4) [1, 6]. The fragmentation of the MK cytoplasm into proplatelets is regulated via different processes including apoptosis (--> Chap. 9) [7] though a role of apoptosis for platelet production is still a controversial matter [8] that requires more studies.

14.2 Thrombocytopenia: Phenotype Analysis and Genetics

A normal platelet count ranges between 150 and 450×10^9/L. Defects in megakaryopoiesis with decreased platelet production or increased platelet destruction can lead to abnormal low platelet counts, termed thrombocytopenia. Both inherited (primary) and acquired (secondary) thrombocytopenia do exist. In this review, only inherited thrombocytopenias (ITs) are discussed in detail, while the acquired causes are the topic of Chap. 15. ITs are a heterogeneous group of diseases characterized by variable expressivity of the bleeding tendency because of low platelet counts that can also be associated with abnormal platelet function. ITs have been shown to result from defects in transcription regulation, TPO signaling, cytoskeletal organization, apoptosis, granule trafficking, and receptor signaling [9]. Examples of these defects are discussed in more detail in the next parts.

IT patients with bleeding typically present with mucocutaneous bleeding symptoms such as easy bruising, purpura, gingival bleeding, and sand epistaxis, usually already obvious after birth or during early childhood [3]. Milder cases only develop bleeding problems after trauma or surgery. Spontaneous life-threatening bleeding complications such as intracranial hemorrhage and gastrointestinal or genitourinary bleeding are rare [10]. Female patients can have menorrhagia and bleeding during pregnancy and labor. Diagnostics of ITs is challenging and requires laboratory techniques such as platelet functional and morphological studies that are often additionally hampered by the (very) low platelet count. The first laboratory test for thrombocytopenia screening is the measurement of platelet count with the mean platelet volume and a peripheral blood smear. Different laboratory investigations are possible to study platelet functional defects such as measurement of platelet aggregation tests, ATP secretion, flow cytometry to quantify expression of platelet receptors, platelet adhesion by the platelet function analyzer (PFA-100®, Siemens Healthcare Diagnostics, Marburg, Germany), and electron microscopy that have been standardized to further analyze defects in platelet function and/or morphology [5]. Misdiagnosis can result in inappropriate treatments with serious consequences, as some forms predispose to other illnesses, such as leukemia or kidney failure, etc. Moreover, the platelet count in specific disorders can be increased by medical treatment, and patients can receive drugs instead of platelet transfusions [11, 12]. It is difficult to estimate the prevalence of IT, but numbers are expected to be higher than assumed due to the lack of genetic diagnostic confirmation of the exact type of thrombocytopenia.

ITs are considered as a highly heterogeneous group of diseases that can be inherited in an autosomal recessive, autosomal dominant, or X-linked manner. It is important to examine the family history and assess for other clinical phenotypes, as ITs are not only a blood-specific disorder but it can also occur as a component of a multisystem disorder [10]. The presence of skeletal, renal, dysmorphic, ocular, audiological, neurological, endocrinological, cardiac, and immune problems associated with bleeding is suggestive of multisystem disorders. Sanger sequencing of potential candidate gene defects was the standard approach for genetic diagnosis of ITs. A major step forward in the field of genetics for ITs was made during the last 3 years using new-generation genetic approaches that resulted in the discovery of novel genes such as NBEAL2, RBM8A, and ACTN1-GFI1B-ETV6 for some well-known ITs that cause gray platelet syndrome, thrombocytopenia-absent radius syndrome, and autosomal dominant thrombocytopenias, respectively. In the future, Sanger sequencing for genetic diagnosis will be replaced by a targeted next-generation sequencing (NGS) approach that will allow the simultaneous screening of all known platelet genes that cause thrombocytopenia [13]. Genome-wide and whole-exome sequencing studies are likely to help in the further screening of ITs for which all known genetic causes have been excluded as to date about 50 % of ITs still receive no genetic diagnosis [14]. More than 20 genes have already been found to explain ITs [14], and it is expected that many more will be discovered. In order to facilitate gene discovery by NGS technologies, the analysis of large case collections recruited through consortia of investigators will require the analysis of shared data through systematic phenotype coding in order to facilitate the identification of similarities between unrelated cases. For this purpose, a novel approach using Human Phenotype Ontology (HPO) terms was recently developed to standardize the annotation of detailed clinical and laboratory characteristics of bleeding and platelet disorder cases [10, 15]. HPO coding also allows the description of non-hematological terms to describe multisystem disorders. Cluster analyses are then used to characterize groups of cases with similar HPO terms and variants in the same genes [10]. For this review, an overview of currently known ITs will be presented as being a blood-specific or part of a multisystem disorder (Table 14.1).

14.3 IT as an Isolated Blood Disorder

14.3.1 Transcription Factor GATA1

GATA1 is the founding member of the GATA family of transcription factors and is required for megakaryocyte and erythrocyte differentiation [16]. Hemizygous (X-linked) loss-of-function mutations in N-terminal zinc finger of GATA1 have been identified in patients with macrothrombocytopenia and result in abnormal binding to DNA or its transcriptional cofactor FOG1. GATA1 regulates megakaryocyte proliferation and differentiation, and dysmegakaryopoiesis with hyperproliferation of immature hypolobulated MKs is described in the bone marrow of GATA1-defective patients [16, 17]. The mutations can be associated with dyserythropoiesis, anemia, beta-thalassemia, or congenital erythropoietic porphyria depending on the missense variant [17]. GATA1 defects also result in dysfunctional and large platelets with reduced platelet aggregation responses and a paucity of alpha granules.

14.3.2 Transcription Factor RUNX1

RUNX1 is a member of the RUNT family of transcription factors and is a master regulator of hematopoiesis. Autosomal dominant mutations in the *RUNX1* (or *CBFA2*) gene cause familial platelet disorder with associated myeloid leukemia (FPD-AML) [18]. The disease is characterized by thrombocytopenia, impaired platelet function with abnormal aggregations, and reduced ATP secretion. Depending on whether the mutation is haploinsufficient or dominant negative, patients are at risk to develop leukemia [18]. RUNX1 has been shown to cooperate with FLI1 in the late stages of megakaryopoiesis. Interestingly, it was recently suggested to use retained MYH10 protein expression in patients' platelets as a diagnostic biomarker for underlying RUNX1 and FLI1 defects [19].

Table 14.1 Genetic and clinical characteristics of inherited thrombocytopenias

Gene name(s)	Gene OMIM number(s)	Mode of inheritance	Name of the disease (as in OMIM)	Clinical symptoms
IT as an isolated blood disorder				
GATA1	305371	X-linked	Anemia, with/without neutropenia and/or platelet abnormalities Thrombocytopenia with beta-thalassemia Thrombocytopenia, with or without dyserythropoietic anemia	Dysfunctional and enlarged platelets, reduced aggregation, giant α-granule, paucity of α-granules, abnormal aggregations, and abnormal erythopoiesis
RUNX1 (CBFA2/AML1)	151385	Dominant	Familial platelet disorder – acute myeloid leukemia	Dysfunctional platelets, abnormal platelet aggregation, reduced ATP secretion, δ-SPD, and presence of MYH10. High risk to developed leukemia
GFI1B	604383	Dominant	Bleeding disorder, platelet-type, 17	Macrothrombocytopenia. Dysfunctional platelets, abnormal platelet aggregation, deficient α-granules, and abnormal erythropoiesis with red cell anisopoikilocytosis
ETV6	600618	Dominant	Thrombocytopenia 5	Thrombocytopenia with diverse hematologic neoplasms
MPL	159530	Recessive	Congenital amegakaryocytic thrombocytopenia (CAMT)	Severe thrombocytopenia. Normal ultrastructure, predisposition for aplastic anemia or myelodysplastic syndrome and pancytopenia

(continued)

Table 14.1 (continued)

Gene name(s)	Gene OMIM number(s)	Mode of inheritance	Name of the disease (as in OMIM)	Clinical symptoms
ANKRD26	610855	Dominant (UTR variants)	Thrombocytopenia 2	Impaired pro-platelet formation, small platelets, low levels of GPα2, deficient α-granules, predisposition to acute leukemia
TUBB1	612901	Dominant	Macrothrombocytopenia, TUBB1-related	Normal platelet function and structure, and no clinical bleeding problems
WAS	300392	X-linked	Wiskott-Aldrich syndrome	Dysfunctional platelets, abnormal platelet aggregations, fewer granules, neutropenia associated severe immune dysfunctions with predisposition to infections, eczema, autoimmune disease and malignancies
ACNT1	102575	Dominant	Bleeding disorder, platelet-type, 15	Macrothrombocytopenia. Enlarged platelets with normal subcellular structure and function, and modest bleeding tendency
CYCS	123970	Dominant	Thrombocytopenia 4	Normal platelet function and structure, no clinical bleeding problems
GP1BA *GP1BB* *GP9*	606672 138720 173515	Recessive	Bernard-Soulier Syndrome	Macrothrombocytopenia, bleeding, and platelet dysfunction

GP1BA	606672	Dominant Bolzano variant L57F	Bernard-Soulier syndrome, type A2 (dominant)	Mild or no clinical bleeding, macrothrombocytopenia
VWF	613160	Dominant (mutations in VWF A1 domain)	von Willebrand disease types 2B	Macrothrombocytopenia, bleeding
GP1BA	606672	Dominant (gain-of-function missense mutations)	von Willebrand disease, platelet-type (pseudo-VWD)	Macrothrombocytopenia, bleeding
NBEAL2	614169	Recessive	Gray platelet syndrome	Macrothrombocytopenia, bleeding, platelet dysfunction with absent α-granules. Myelofibrosis.
IT as part of a multisystem disorder				
FLI1	193067	Dominant 11q deletion	Paris-Trousseau thrombocytopenia Jacobsen syndrome	Dysfunctional platelets, giant α-granules, increased bleeding tendency, δ-SPD, presence of MYH10, heart defects, gastrointestinal and genitourinary problems, mental retardation, ophthalmologic, gross and fine motor delays, and infection of respiratory system
RBM8A (Y14)	605313	Recessive	Thrombocytopenia-absent radius syndrome	Hypomegakaryocytic thrombocytopenia, structural abnormalities, bilateral radial aplasia, capillary hemangiomata, facial dysmorphism, gastroenteritis, cow's milk intolerance, renal and cardiac defects, shortness and macrocephaly

(continued)

Table 14.1 (continued)

Gene name(s)	Gene OMIM number(s)	Mode of inheritance	Name of the disease (as in OMIM)	Clinical symptoms
MYH9	160775	Dominant	May-Hegglin syndrome Sebastian syndrome Fechtner syndrome Epstein syndrome	Thrombocytopenia with giant platelets, defect in clot retraction, Döhle-like inclusion in leukocytes, risk for proteinuric nephropathy, cataract, and deafness
FLNA	300017	X-linked	Cardiac valvular dysplasia Short bowel disease Frontometaphyseal dysplasia Periventricular heterotopia Melnick-Needles syndrome Otopalatodigital syndrome	Thrombocytopenia with dysfunctional platelets, abnormal platelet aggregation, secretion and adhesion, giant α-granules, wide-spectrum disease that can include brain malformations with periventricular nodular heterotopia, skeletal dysplasia, cardiac valvular dystrophy, congenital intestinal pseudo-obstruction, and terminal osseous dysplasia
ADCYAP1 (PACAP)	102980	Dominant Trisomy 18p	–	Thrombocytopenia. Dysfunctional platelets, bleeding symptoms, severe mental retardation and endocrinological defects

NPC1	257220	Recessive	Niemann-Pick type C1	Thrombocytopenia. Dysfunctional platelets, abnormal aggregation, abnormal α-granules, deficient δ-granules, accumulation of cholesterol in the endosomal-lysosomal system, anemia and petechial rash
GBA	606463	Recessive	Gaucher disease	Hepatosplenomegaly, skeletal complications and in some forms involves the central nervous system. Bleeding, thrombocytopenia, abnormal platelet function.
STIM1	605921	Dominant Gain-of-function	Stormorken syndrome	Bleeding diathesis, thrombocytopenia, miosis, and tubular myopathy

14.3.3 Transcription Factor GFI1B

GFI1B is a transcription factor that was recently identified by NGS as important for normal human erythropoiesis and megakaryocyte development. A nonsense mutation in the *GFI1B* gene was described in a family with autosomal dominant gray platelet syndrome, because of the deficiency of α-granules in platelets. This mutation resulted in a GFI1B variant that inhibited the transcriptional activity of the nonmutant GFI1B in a dominant-negative manner [20]. Similar as described for GATA1 deficiency, bone marrow megakaryocytes remain immature, and patients can develop myelofibrosis. Autosomal dominant mutations in *GFI1B* were in two other reported cases to result in mild clinical bleeding symptoms with macrothrombocytopenia, red cell anisopoikilocytosis, and platelet dysfunction with strongly reduced numbers of α-granules and decreased platelet aggregation responses to all agonists except ristocetin [21, 22].

14.3.4 Transcription Factor ETV6

Germline missense mutations in *ETV6* gene identified by NGS were shown to cosegregate with the dominant transmission of thrombocytopenia and hematologic malignancy in three unrelated kindreds, defining a new hereditary syndrome featuring thrombocytopenia with susceptibility to diverse hematologic neoplasms [23, 24]. Functional studies show that the mutations abrogate DNA binding, alter subcellular localization, decrease transcriptional repression in a dominant-negative fashion, and impair MK maturation and platelet biogenesis. The identification of germline predisposition to leukemia, myelodysplastic syndrome (MDS), or dyserythropoietic anemia informs the diagnosis and medical management of at-risk individuals. The exact role of ETV6 in megakaryopoiesis and platelet biogenesis is not yet known.

14.3.5 Growth Factor TPO and Its Receptor MPL

TPO is the main growth factor for MK and platelet formation and binds to the MPL receptor to regulate the different steps of megakaryopoiesis. Congenital amegakaryocytic thrombocytopenia (CAMT) is an autosomal recessive disorder caused by mutations in the *MPL* gene. CAMT is characterized by severe hypomegakaryocytic thrombocytopenia during the first year of life and typically develops into pancytopenia leading to bone marrow failure in later childhood and anemia prior to leukopenia. The disease can be fatal unless treated with hematopoietic stem cell transplantation. Since patients have strongly reduced MKs in their bone marrow, plasma TPO levels are high [25]. TPO has been shown to induce endomitosis as well as being responsible for stimulating the expression of characteristic cell surface proteins including integrin alpha2b/beta3 and glycoprotein Ibalpha during MK maturation. TPO is constitutively produced in the liver and stimulates in synergy with other cytokines' MK maturation. TPO acts via different downstream signaling pathways that include mitogen-activated protein kinase (MAPK), Janus kinase-signal transducers and activators of transcription (JAK-STAT), and phosphoinositol-3-kinase (PI3K) [1, 25].

14.3.6 Gene with Undefined Function: ANKRD26

Variants in the 5′ UTR of the *ANKRD26* gene cause autosomal dominant thrombocytopenia [26]. Recently, the underlying causative mechanism was resolved as variants in the 5′ UTR of *ANKRD26* result in loss of RUNX1 and FLI1 transcription factor binding. The interaction with these transcription factors is responsible for silencing the *ANKRD26* gene during later stages of megakaryopoiesis and platelet development. Without silencing, the persistent expression results in increased signaling via the TPO-MPL pathway with impaired proplatelet formation from MKs [26]. The disease causes a mild bleeding tendency with relatively small platelets that have deficient α-granules [26].

14.3.7 Cytoskeletal Regulatory Gene TUBB1

Tubulin beta1 (TUBB1) is the major tubulin isoform within platelet and MK microtubules. The marginal microtubule band maintains normal platelet morphology and is important as linkage between the membrane skeleton and the cytoskeletal actin filaments. α- and β-tubulins form heterodimers that are assembled in microtubules. An autosomal dominant missense mutation in the *TUBB1* gene was found to cause macrothrombocytopenia without bleeding symptoms. A defect in platelet formation was expected, but aggregation and morphology studies for this patient showed normal platelets [27].

14.3.8 Cytoskeletal Regulatory Gene WAS

WASP is encoded by the *WAS* gene that is selectively expressed in hematopoietic cells and plays an essential role as regulator of the actin cytoskeleton to maintain normal platelet function. Mutations in the *WAS* gene cause the X-linked disorder Wiskott-Aldrich syndrome (WAS). The syndrome is characterized by microthrombocytopenia with dysfunctional platelets, but patients also often have primary immunodeficiency. WAS patients often also have neutropenia associated with severe immune dysfunction with susceptibility to infections, eczema, autoimmune diseases, and malignancies [28]. Severely affected patients die without hematopoietic stem cell transplantation.

14.3.9 Cytoskeletal Regulatory Gene ACNT1

Exome sequencing discovered a new pathology related to a defective cytoskeletal organization [29]. They found a missense variant in the ACTN1 gene that causes autosomal dominant macrothrombocytopenia with a modest bleeding tendency. ACTN1 deficiency seems to be the most frequent cause of dominant IT [10]. The *ACTN1* gene encodes for the α-actinin-1 protein and is a member of the actin cross-linking protein family. ACTN1 is mainly expressed in platelets and MKs. The disease is characterized by enlarged platelets with normal subcellular structures and function [29].

14.3.10 Apoptosis Regulatory Gene CYCS

Cytochrome C is involved in the initiation of the intrinsic pathway of apoptosis. It is localized at the inner membrane of mitochondria, and upon release it leads to apoptosome formation and activates the caspase cascade. Autosomal dominant missense mutations were found in the *CYCS* gene in two studies that report dominant IT without platelet dysfunction or bleeding [30, 31]. The cytochrome C variant caused by this mutation was able to enhance apoptotic activity in vitro.

14.3.11 Glycoprotein Receptor GPIb/GPIX/GPV

Bernard-Soulier syndrome (BSS) is characterized by defective platelet adhesion caused by mutations in *GP1BA*, *GP1BB*, or *GP9*, the genes that code for the GPIb/GPIX/GPV platelet glycoprotein receptor. The three subunits are all type 1 transmembrane proteins containing leucine-rich repeat domains. GPIb/GPIX/GPV is the main receptor for the von Willebrand factor (VWF), which is important in clot formation. BSS is an autosomal recessive disorder with moderate to severe macrothrombocytopenia, decreased platelet survival, and often a spontaneous bleeding tendency. The platelets show no response to VWF or ristocetin. The underlying molecular mechanism for the defect in megakaryopoiesis that is linked to defective GPIb/GPIX/GPV signaling is not yet elucidated [32].

Autosomal dominant "Mediterranean" macrothrombocytopenia is caused by a monoallelic c.515C>T transition in the GP1BA gene, which is previously reported as the "Bolzano variant" [33]. The genetic mutation results in an Ala156Val amino acid substitution in the GP1BA gene, which is suspected to cause an abnormal conformation of the receptor and reduces the ability to interact with VWF. The disorder is characterized by mild or no clinical bleeding symptoms and normal platelet function. Platelets are enlarged and have a reduced level of GPIb/GPIX/GPV platelet expression.

14.3.12 Von Willebrand Factor VWD2B and Platelet-Type VWD

Von Willebrand disease type 2B (VWD2B) is an autosomal dominant disorder caused by mutations in the *VWF gene*, while platelet-type VWD is the result of specific dominant mutations in the *GP1BA* gene [34]. Mutations in both genes result in an increased affinity of VWF to platelet GPIb. These gain-of-function mutations are associated with an increased platelet aggregation and increased proteolysis of VWF subunits causing a decrease of large VWF multimers. The disease is characterized by macrothrombocytopenia and bleeding diathesis [34]. The VWD type 2 has typical mutations localized within or near the A1 domain of the *VWF* gene; this domain helps platelet GPIbα-binding [34].

14.3.13 BEACH Domain Protein NBEAL2

Autosomal recessive mutations were found in the *NBEAL2* gene by different NGS approaches as cause for the gray platelet syndrome [35–37]. This disorder causes macrothrombocytopenia with enlarged dysfunctional platelets and the absence of α-granules. Additionally, many patients develop myelofibrosis. The underlying mechanism of NBEAL2 defects in megakaryopoiesis is unknown, but it is expected to regulate MK maturation and the trafficking of α-granules.

14.4 IT as Part of a Multisystem Disorder

14.4.1 Transcription Factor FLI1

The *FLI1* (Friend leukemia virus integration) proto-oncogene is a member of the *ETS* gene family of winged helix-turn-helix transcription factors that bind the purine-rich consensus sequence GGA(A/T). *FLI1* is localized on chromosome 11q24 and is mainly expressed in the hematopoietic lineages and vascular endothelial cells. The Paris-Trousseau syndrome (previously called Jacobsen syndrome) is caused by 11q chromosomal deletions [38, 39]. A large study of 110 patients showed that nearly all patients with 11q deletion have thrombocytopenia and platelet dysfunction with giant α-granules and an increased bleeding tendency [39]. Bone marrow investigation also showed dysmegakaryopoiesis with defects in MK maturation. The disease is also associated with other clinical phenotypes such as gastrointestinal and genitourinary problems, mental retardation, ophthalmologic problems, heart defects, gross and fine motor delays, and infections of the upper respiratory system in at least a large subset in patients depending on the size of the chromosomal deletion [38, 39].

14.4.2 Exon Junction Complex Member RBM8A

Thrombocytopenia with absent radius (TAR) syndrome is characterized by bilateral radial aplasia with preservation of the thumbs. Additional skeletal abnormalities, including more complex upper or lower limb malformations and phocomelia, are also observed. Further, common non-skeletal defects are capillary hemangiomata, facial dysmorphism, gastroenteritis, cow's milk intolerance, renal and cardiac defects, shortness, and macrocephaly. Patients suffering with TAR syndrome have hypomegakaryocytosis and thrombocytopenia [40]. During the first year of life, thrombocytopenia and bleeding tendency are typically severe and might lead to fatal hemorrhages, which diminish in frequency and severity with age.

Interestingly, TAR is associated with high levels of TPO and defective TPO signaling despite that they have a normal *MPL* gene [40]. Recently, exome sequencing revealed that a null mutation in the *RBM8A* gene in combination with two low-frequency single nucleotide polymorphisms causes TAR syndrome [41]. The *RBM8A* gene encodes for the exon junction complex subunit Y14, which performs essential RNA-processing tasks, but a role in defective TPO signaling or MK biology remains completely unknown. The two noncoding single nucleotide polymorphisms are hypomorphic mutations affecting the regulatory regions of the *RBM8A* gene, and patients with TAR syndrome have a lower expression of Y14 in their platelets [41].

14.4.3 Cytoskeletal Regulatory Gene MYH9

Myosin heavy chain (MYH) IIA proteins are abundantly present in platelets. They play an important role in proplatelet formation and contribute to changes in platelet shape after activation. The cytoskeletal protein MYH9 contains a myosin headlike and a tail domain. The headlike domain interacts with actin and hydrolyses ATP. Autosomal dominant mutations in the *MYH9* gene cause MYH9-related diseases such as May-Hegglin, Sebastian, Fechtner, and Epstein syndromes [10, 42, 43]. MYH9-related diseases are characterized by thrombocytopenia with giant platelets. Clinical phenotype variability has been described as patients are also at risk to develop proteinuric nephropathy, cataract, and/or deafness, and the disease can be associated with Döhle-like leukocyte inclusions [10, 42, 43].

14.4.4 Cytoskeletal Regulatory Gene FLNA

Filamin A (FLNA) plays an essential role in stabilization of actin filament networks and anchoring some adhesion receptors of platelet and MK membranes to the cytoskeleton. The *FLNA* gene is located on chromosome Xq28, and this filamin isoform is predominantly expressed in platelets. Mutations in the *FLNA* gene are responsible for a spectrum of disorders, including brain malformations with periventricular nodular heterotopia [44]. Some patients are also described with macrothrombocytopenia with enlarged α-granules [45]. The exact role of FLNA in MK maturation and platelet biogenesis is not known. In addition, platelet dysfunction heterogeneity is described with main findings as reduced secretion, aggregation, and defective platelet adhesion [45].

14.4.5 Apoptosis Regulator PACAP

Patients with a trisomy 18p have three gene copies of the *PACAP* gene (coding for the pituitary adenylyl cyclase-activating peptide) and present with severe bleeding problems, dysfunctional platelets, and mild thrombocytopenia in addition to severe mental retardation and endocrinological defects [46]. PACAP binds to the

Gs-coupled VPAC1 receptor, which is expressed in MKs and platelets. Patients have elevated PACAP plasma concentrations, which cause inactivation of platelets via activation of the cAMP pathway [46]. Stimulation of the VPAC1 signaling pathway results in decreased MK maturation and platelet formation [46, 47]. PACAP is a protective factor against apoptosis in megakaryocytes via stimulation of the NF-*k*B pathway [47].

14.4.6 Lysosomal Proteins NPC1 and GBA

Autosomal recessive mutations in the *NPC1* gene cause the lysosomal storage disease Niemann-Pick type C1 [48]. The disease is characterized by defects in the sphingomyelin/cholesterol metabolism. Additionally in some cases, the disease is associated with thrombocytopenia, anemia, and petechial rash. NPC1 is expressed in platelets and MKs and plays a role in platelet function and formation. Patients have reduced aggregation, diminished P-selectin expression and ATP secretion with morphological abnormal α-granules, and reduced δ-granules [48]. A defective cholesterol metabolism was suggested to interfere with normal megakaryopoiesis and granule formation, but further studies are needed to support this hypothesis.

Gaucher disease is caused by autosomal recessive mutations in the *GBA* (glucosidase, beta, acid) gene and results from deficient activity of the lysosomal enzyme beta-glucosidase or glucocerebrosidase. Gaucher disease is associated with hepatosplenomegaly, cytopenias, and skeletal complications and in some forms involves the central nervous system [49]. Patients can have bleedings that include mucosal and surgical hemorrhages due to thrombocytopenia, abnormal platelet function, reduced production of coagulation factors, and activation of fibrinolysis, but none of these defects have been studied in detail.

14.4.7 Calcium Sensor STIM1

Defects in the store-operated Ca^{2+} release-activated Ca^{2+} (CRAC) channel due to dominant gain-of-function mutations in the *STIM1* gene that codes for the Ca^{2+}-sensing protein stromal interaction molecule 1 lead to thrombocytopenia in patients with the Stormorken syndrome. The CRAC channel is important and regulates critical cellular functions such as cell growth and differentiation, gene expression, and Ca^{2+} homeostasis. Clinical symptoms associated with Stormorken syndrome are bleeding diathesis, thrombocytopenia, miosis, and tubular myopathy [50], but its exact role in megakaryopoiesis is not studied. However, mice expressing an activating EF hand mutant of *Stim1* have macrothrombocytopenia and an associated bleeding disorder [51]. This study showed that basal [Ca2+]i levels were increased in platelets, resulting in a preactivation state with a selective unresponsiveness to ITAM-coupled agonists and increased platelet consumption.

14.5 Functional Genetics to Understand the Role of Gene Defects in Thrombocytopenia

From a mechanistic perspective, ITs can be the result from mutations in genes with diverse functions, although for many the exact function in megakaryocytes and platelets still remains unclear as indicated above for several disorders. Animal (mostly knockout mice and CD41+ transgenic zebrafish) and stem cell approaches have shown to be useful to study the role of such genes in megakaryopoiesis and platelet biogenesis. Bone marrow- and peripheral blood-derived CD34+ hematopoietic stem cells (HSCs) from patients are useful as a tool for ex vivo models of megakaryopoiesis and the production of functional platelets. Such models have delivered important insights in novel molecular mechanisms of megakaryopoiesis, disease modeling, and correcting relevant disease defects. More recently, in addition to HSCs, also inducible pluripotent stem cells (iPS) from patients have been used to uncover novel molecular mechanisms of megakaryopoiesis, disease modeling, and correcting relevant diseases [52–55]. Moreover, efforts to derive megakaryocytes and platelets from pluripotent stem cells foster the opportunity of a revolutionary cellular therapy for the future treatment of multiple platelet-associated diseases.

Conclusion

Though the list of genes responsible for ITs is rapidly growing, about 50 % of IT patients still receive no genetic diagnosis. The identification of molecular defects in patients with IT has improved our understanding of normal megakaryocyte and platelet biology, as well as mechanisms of hemostasis and thrombosis. The utilization of NGS approaches will potentially revolutionize not on the current diagnosis via functional platelet testing of IT with only a blood disease but possibly also the rapid identification of individuals with increased risk of bleeding often hidden but in association with other clinical phenotypes. Future medical DNA sequencing is also expected to give clinicians important information regarding the genetic phenotype of the IT in their patients to improve early diagnosis.

Acknowledgments KF is supported by the Fund for Scientific Research-Flanders (FWO-Vlaanderen, Belgium, G.0B17.13N] and by the Research Council of the University of Leuven (BOF KU Leuven, Belgium, OT/14/098].

References

1. Thon J, Italiano JE (2012) Platelet formation. Semin Hematol 47:220–226
2. Doré LC, Crispino JD (2011) Transcription factor networks in erythroid cell and megakaryocyte development. Blood 118:231–239
3. Tijssen MR, Ghevaert C (2013) Transcription factors in late megakaryopoiesis and related platelet disorders. J Thromb Haemost 11:593–604
4. Patel SR, Hartwih JH, Italiano JE (2014) The biogenesis of platelets from megakaryocyte proplatelets. J Clin Invest 115:3348–3354

5. Schulze H, Korpal M, Hurov J, Kim S-W, Zhang J, Canteley LC, Graf T, Shivadasani RA (2006) Characterization of the megakaryocyte demarcation membrane system and its role in thrombopoiesis. Blood 107:3868–3875
6. Szalai G, LaRue AC, Watson DK (2006) Molecular mechanisms of megakaryopoiesis. Cell Mol Life Sci 63:2460–2476
7. Avanzi MP, Izak M, Oluwadara OE, Mitchell WB (2015) Actin inhibition increases megakaryocyte proplatelet formation through an apoptosis-dependent mechanism. PLoS One 10(4), e0125057
8. Josefsson EC, Burnett DL, Lebois M, Debrincat MA, White MJ, Henley KJ, Lane RM, Moujalled D, Preston SP, O'Reilly LA, Pellegrini M, Metcalf D, Strasser A, Kile BT (2014) Platelet production proceeds independently of the intrinsic and extrinsic apoptosis pathways. Nat Commun 5:3455
9. Freson K, Wijgaerts A, van Geet C (2014) Update on the causes of platelet disorders and functional consequences. Int J Lab Hematol 36(3):313–325
10. Westbury SK, Turro E, Greene D, Lentaigne C, Kelly AM, Bariana TK, Simeoni I, Pillois X, Attwood A, Austin S, Jansen SB, Bakchoul T, Crisp-Hihn A, Erber WN, Favier R, Foad N, Gattens M, Jolley JD, Liesner R, Meacham S, Millar CM, Nurden AT, Peerlinck K, Perry DJ, Poudel P, Schulman S, Schulze H, Stephens JC, Furie B, Robinson PN, van Geet C, Rendon A, Gomez K, Laffan MA, Lambert MP, Nurden P, Ouwehand WH, Richardson S, Mumford AD, Freson K, BRIDGE-BPD Consortium (2015) Human phenotype ontology annotation and cluster analysis to unravel genetic defects in 707 cases with unexplained bleeding and platelet disorders. Genome Med 7(1):36
11. Bolton-Maggs PHB, Chalmers EA, Collins PW, Harrison P, Kitchen S, Liesner RJ, Minford A, Mumford AD, Parapia LA, Perry DJ, Watson SP, Wilde JT, Williams MD (2006) A review of inherited platelet disorders with guidelines for their management on behalf of the UKHCDO. Br J Haematol 135:603–633
12. Balduini CL, Savoia A, Seri M (2013) Inherited thrombocytopenias frequently diagnosed in adults. J Thromb Haemost 11:1006–1009
13. Thrombogenomics. https://haemgen.haem.cam.ac.uk/thrombogenomics. Accessed 17 June 2015
14. Savoia A (2015) Molecular basis of inherited thrombocytopenias. Clin Genet. doi:10.1111/cge.12607 [Epub ahead of print]
15. Köhler S, Doelken SC, Mungall CJ, Bauer S, Firth HV, Bailleul-Forestier I, Black GC, Brown DL, Brudno M, Campbell J, FitzPatrick DR, Eppig JT, Jackson AP, Freson K, Girdea M, Helbig I, Hurst JA, Jähn J, Jackson LG, Kelly AM, Ledbetter DH, Mansour S, Martin CL, Moss C, Mumford A, Ouwehand WH, Park SM, Riggs ER, Scott RH, Sisodiya S, Van Vooren S, Wapner RJ, Wilkie AO, Wright CF, Vulto-van Silfhout AT, de Leeuw N, de Vries BB, Washingthon NL, Smith CL, Westerfield M, Schofield P, Ruef BJ, Gkoutos GV, Haendel M, Smedley D, Lewis SE, Robinson PN (2014) The Human Phenotype Ontology project: linking molecular biology and disease through phenotype data. Nucleic Acids Res 42:D966–D974
16. Freson K, Matthijs G, Thys C, Mariën P, Hoylaerts MF, Vermylen J, Van Geet C (2002) Different substitutions at residue D218 of the X-linked transcription factor GATA1 lead to altered clinical severity of macrothrombocytopenia and anemia and are associated with variable skewed X inactivation. Hum Mol Genet 11:147–152
17. Freson K, Devriendt K, Matthijs G, Van Hoof A, De Vos R, Thys C, Minner K, Hoylaerts MF, Vermylen J, Van Geet C (2001) Platelet characteristics in patients with X-linked macrothrombocytopenia because of a novel GATA1 mutation. Blood 98:85–92
18. Song W-J, Sullivan MG, Legare RD, Hutchings S, Tan X, Kufrin D, Ratajczak J, Resende IC, Haworth C, Hock R, Loh M, Felix C, Roy D-C, Busque L, Kurnit D, Willman C, Gewirtz AM, Speck NA, Bushweller JH, Li FP, Gardiner K, Poncz M, Maris JM, Gilliland DG (1999) Haploinsufficiency of CBFA2 causes familial thrombocytopenia with propensity to develop acute myelogenous leukaemia. Nat Genet 23:166–175
19. Antony-Debré I, Bluteau D, Itzykson R, Baccini V, Renneville A, Boehlen F, Morabito M, Droin N, Deswarte C, Chang Y, Leverger G, Solary E, Vainchenker W, Favier R, Raslova H

(2012) MYH10 protein expression in platelets as a biomarker of RUNX1 and FLI1 alterations. Blood 120:2719–2722
20. Monteferrario D, Bolar NA, Marneth AE, Hebeda KM, Bergevoet SM, Veenstra H, Laros-van Gorkom BA, Mackenzie MA, Khandanpour C, Botezatu L, Fransen E, Van Camp G, Duijnhouwer AL, Salemink S, Willemsen B, Huls G, Preijers F, Van Heerde W, Jansen JH, Kempers MJ, Loeys BL, Van Laer L, Van der Reijden BA (2013) A dominant-negative GFI1B mutation in the gray platelet syndrome. N Engl J Med 370:245–253
21. Stevenson WS, Morel-Kopp MC, Chen Q, Liang HP, Bromhead CJ, Wright S, Turakulov R, Ng AP, Roberts AW, Bahlo M, Ward CM (2013) GFI1B mutation causes a bleeding disorder with abnormal platelet function. J Thromb Haemost 11:2039–2047
22. Chen L, Kostadima M, Martens JH, Canu G, Garcia SP, Turro E, Downes K, Macaulay IC, Bielczyk-Maczynska E, Coe S, Farrow S, Poudel P, Burden F, Jansen SB, Astle WJ, Attwood A, Bariana T, de Bono B, Breschi A, Chambers JC, BRIDGE Consortium, Choudry FA, Clarke L, Coupland P, van der Ent M, Erber WN, Jansen JH, Favier R, Fenech ME, Foad N, Freson K, van Geet C, Gomez K, Guigo R, Hampshire D, Kelly AM, Kerstens HH, Kooner JS, Laffan M, Lentaigne C, Labalette C, Martin T, Meacham S, Mumford A, Nürnberg S, Palumbo E, van der Reijden BA, Richardson D, Sammut SJ, Slodkowicz G, Tamuri AU, Vasquez L, Voss K, Watt S, Westbury S, Flicek P, Loos R, Goldman N, Bertone P, Read RJ, Richardson S, Cvejic A, Soranzo N, Ouwehand WH, Stunnenberg HG, Frontini M, Rendon A (2014) Transcriptional diversity during lineage commitment of human blood progenitors. Science 345(6204):1251033
23. Zhang MY, Churpek JE, Keel SB, Walsh T, Lee MK, Loeb KR, Gulsuner S, Pritchard CC, Sanchez-Bonilla M, Delrow JJ, Basom RS, Forouhar M, Gyurkocza B, Schwartz BS, Neistadt B, Marquez R, Mariani CJ, Coats SA, Hofmann I, Lindsley RC, Williams DA, Abkowitz JL, Horwitz MS, King MC, Godley LA, Shimamura A (2015) Germline ETV6 mutations in familial thrombocytopenia and hematologic malignancy. Nat Genet 47(2):180–185
24. Noetzli L, Lo RW, Lee-Sherick AB, Callaghan M, Noris P, Savoia A, Rajpurkar M, Jones K, Gowan K, Balduini CL, Pecci A, Gnan C, De Rocco D, Doubek M, Li L, Lu L, Leung R, Landolt-Marticorena C, Hunger S, Heller P, Gutierrez-Hartmann A, Xiayuan L, Pluthero FG, Rowley JW, Weyrich AS, Kahr WH, Porter CC, Di Paola J (2015) Germline mutations in ETV6 are associated with thrombocytopenia, red cell macrocytosis and predisposition to lymphoblastic leukemia. Nat Genet 47(5):535–538
25. Germeshausen M, Ballmaier M, Welte K (2006) MPL mutations in 23 patients suffering from congenital amegakaryocytic thrombocytopenia: the type of mutation predicts the course of the disease. Hum Mutat 27:296
26. Bluteau D, Balduini A, Balayn N, Currao M, Nurden P, Deswarte C, Leverger G, Noris P, Perrotta S, Solary E, Vainchenker W, Debili N, Favier R, Raslova H (2014) Thrombocytopenia-associated mutations in the ANKRD26 regulatory region induce MAPK hyperactivation. J Clin Invest 124:580–591
27. Kunishima S, Kobayashi R, Itoh TJ, Hamaguchi M, Saito H (2009) Mutation of the beta1-tubulin gene associated with congenital macrothrombocytopenia affecting microtubule assembly. Blood 113:458–461
28. Notarangelo LD, Miao CH, Ochs HD (2008) Wiskott-Aldrich syndrome. Curr Opin Hematol 15:30–36
29. Kunishima S, Okuno Y, Yoshida K, Shiraishi Y, Sanada M, Muramatsu H, Chiba K, Tanaka H, Miyazaki K, Sakai M, Ohtake M, Kobayashi R, Iguchi A, Niimi G, Otsu M, Takahashi Y, Miyano S, Saito H, Kojima S, Ogawa S (2013) ACTN1 mutations cause congenital macrothrombocytopenia. Am J Hum Genet 92:431–438
30. De Rocco D, Cerqua C, Goffrini P, Russo G, Pastore A, Meloni F, Nicchia E, Moraes CT, Pecci A, Salviati L, Savoia A (1842) Mutations of cytochrome c identified in patients with thrombocytopenia THC4 affect both apoptosis and cellular bioenergetics. Biochim Biophys Acta 2013:269–274
31. Morison IM, Cramer Bordé EM, Cheesman EJ, Cheong PL, Holyoake AJ, Fichelson S, Weeks RJ, Lo A, Davies SM, Wilbanks SM, Fagerlund RD, Ludgate MW, da Silva Tatley FM, Coker MS, Bockett NA, Hughes G, Pippig DA, Smith MP, Capron C, Ledgerwood EC (2008) A mutation of human cytochrome c enhances the intrinsic apoptotic pathway but causes only thrombocytopenia. Nat Genet 40:387–389

32. Ware J, Russel SR, Marchese P, Ruggeri ZM (1993) Point mutation in a leucine-rich repeat of platelet glycoprotein Ib alpha resulting in the Bernard-Soulier syndrome. J Clin Invest 92:1213–1220
33. Noris P, Perrota S, Bottega R, Pecci A, Melazzini F, Civashi E, Russo S, Magrin S, Loffredo G, Di salvo V, Russo G, Casale M, De Rocco D, Grignani C, Cattanea M, Baronci C, Dragani A, Albano V, Jankovic M, Scianguetta S, Savoia A, Balduini CL (2012) Clinical and laboratory features of 103 patients from 42 Italian families with inherited thrombocytopenia derived from the monoallelic Ala156Val mutation of GPIbα (Bolzano mutation). Haematologica 97:82–88
34. Sadler JE, Budde U, Eikenboom JC, Favaloro EJ, Hill FGH, Holmberg L, Ingerslev J, Lee CA, Lillicrap D, Mannucci PM, Mazurier C, Meyer D, Nichols WL, Nishino M, Peake IR, Rodeghiero F, Schneppenheim R, Ruggeri ZM, Srivastava A, Montgomery RR, Federici AB (2006) Update on the pathophysiology and classification of von Willebrand disease: a report of the Subcommittee on von Willebrand Factor. J Thromb Haemost 4:2103–2114
35. Albers CA, Cvejic A, Favier R, Bouwmans EE, Alessi MC, Bertone P, Jordan G, Kettleborough RN, Kiddle G, Kostadima M, Read RJ, Sipos B, Sivapalaratnam S, Smethurst PA, Stephens J, Voss K, Nurden A, Rendon A, Nurden P, Ouwehand WH (2011) Exome sequencing identifies NBEAL2 as the causative gene for gray platelet syndrome. Nat Genet 43(8):735–737
36. Gunay-Aygun M, Falik-Zaccai TC, Vilboux T, Zivony-Elboum Y, Gumruk F, Cetin M, Khayat M, Boerkoel CF, Kfir N, Huang Y, Maynard D, Dorward H, Berger K, Kleta R, Anikster Y, Arat M, Freiberg AS, Kehrel BE, Jurk K, Cruz P, Mullikin JC, White JG, Huizing M, Gahl WA (2011) NBEAL2 is mutated in gray platelet syndrome and is required for biogenesis of platelet α-granules. Nat Genet 43(8):732–734
37. Kahr WH, Hinckley J, Li L, Schwertz H, Christensen H, Rowley JW, Pluthero FG, Urban D, Fabbro S, Nixon B, Gadzinski R, Storck M, Wang K, Ryu GY, Jobe SM, Schutte BC, Moseley J, Loughran NB, Parkinson J, Weyrich AS, Di Paola J (2011) Mutations in NBEAL2, encoding a BEACH protein, cause gray platelet syndrome. Nat Genet 43(8):738–740
38. Hart A, Melet F, Grossfeld P, Chien K, Jones C, Tunnacliffe A, Favier R, Bernstrein A (2000) Fli-1 required for murine vascular and megakaryocytic development and is hemizygously deleted in patients with thrombocytopenia. Immunity 13:167–177
39. Grossfeld PD, Mattina T, Lai Z, Favier R, Jones KL, Cotter F, Jones C (2004) The 11q terminal deletion disorder: a prospective study of 110 cases. Am J Med Genet A 129A:51–61
40. Klopocki E, Schulze H, Strauss G, Ott C-E, Hall J, Trotier F, Fleischhauer S, Greenhalgh L, Newbury-Ecob RA, Neumann LM, Habenicht R, Konig R, Seemanova E, Megarbane A, Ropers H-H, Ullmann R, Horn D, Mundlos S (2007) Complex inheritance pattern resembling autosomal recessive inheritance involving a microdeletion in thrombocytopenia-absent radius syndrome. Am J Hum Genet 80:232–240
41. Albers CA, Paul DS, Schulze H, Freson K, Stephens JC, Smethurst PA, Jolley JD, Cvejic A, Kostadima M, Bertone P, Breuning MH, Debili N, Deloukas P, Favier R, Fiedler J, Hobbs CM, Huang N, Hurles ME, Kiddle G, Krapels I, Nurden P, Ruivenkamp CA, Sambrook JG, Smith K, Stemple DL, Strauss G, Thys C, van Geet C, Newbury-Ecob R, Ouwehand WH, Ghevaert C (2012) Compound inheritance of a low-frequency regulatory SNP and a rare null mutation in exon-junction complex subunit RBM8A causes TAR syndrome. Nat Genet 44:435–439
42. Balduini CL, Pecci A, Savoia A (2011) Recent advances in the understanding and management of MYH9-related inherited thrombocytopenias. Br J Haematol 154:161–174
43. Pecci A, Panza E, Pujol-Moix, Klersy C, Di Bari F, Bozzi V, Gresele P, Lethagen S, Fabris F, Dufour C, Granata A, Doubek M, Pecoraro C, Koivisto PA, Heller PG, Iolascon A, Alvisi P, Schwabe D, De Candia E, Rocca B, Russo U, Ramenghi U, Noris P, Seri M, Balduini CL, Savoia A (2008) Position of nonmuscle myosin heavy chain II1 (NMMHC-IIA) mutations predicts the natural history of MYH9-related disease. Hum Mutat 29:409–417
44. Parrini E, Ramazzotti A, Dobyns WB, Mei D, Moro F, Veggiotti P, Marini C, Brilstra EH, Dalla Bernardina B, Goodwin L, Bodell A, Jones MC, Nangeroni M, Palmeri S, Said E, Sander JW, Striano P, Takahashi Y, Van Maldergem L, Leonardi G, Wright M, Walsh CA, Guerrini R (2006) Periventricular heterotopia: phenotypic heterogeneity and correlation with Filamin A mutations. Brain 129:1892–1906

45. Berrou E, Adam F, Lebret M, Fergelot P, Kauskot A, Coupry I, Jandrot-Perrus M, Nurden A, Favier R, Rosa JP, Goizet C, Nurden P, Bryckaert M (2013) Heterogeneity of platelet functional alterations in patients with filamin A mutations. Arterioscler Thromb Vasc Biol 33:e11–e18
46. Freson K, Peeters K, De Vos R, Wittevrongel C, Thys C, Hoylaerts MF, Vermylen J, Van Geet C (2008) PACAP and its receptor VPAC1 regulate megakaryocyte maturation: therapeutic implications. Blood 111:1885–1893
47. Di Michele M, Peeters K, Loyen S, Thys C, Waelkens E, Overbergh L, Hoylaerts M, Van Geet C, Freson K (2012) Pituitary Adenylate Cyclase-Activating Polypeptide (PACAP) impairs the regulation of apoptosis in megakaryocytes by activating NF-kB: a proteomic study. Mol Cell Proteomics 11(1):M111.007625
48. Louwette S, Régal L, Wittevrongel C, Thys C, Vandeweeghde G, Decuyper E, Leemans P, De Vos R, Van Geet C, Jeaken J, Freson K (2013) NPC1 defect results in abnormal platelet formation and function: studies in Niemann-Pick disease type C1 patients and zebrafish. Hum Mol Genet 22:61–73
49. Motta I, Filocamo M, Poggiali E, Stroppiano M, Dragani A, Consonni D, Barcellini W, Gaidano G, Facchini L, Specchia G, Cappellini MD (2015) Splenomegaly Gaucher Disease study group. A multicentre observational study for early diagnosis of Gaucher disease in patients with Splenomegaly and/or Thrombocytopenia. Eur J Haematol. 96:352–59 [Epub ahead of print]
50. Nesin V, Wiley G, Kousi M, Ong EC, Lehmann T, Nicholi DJ, Suri M, Shahrizaila N, Katasanis N, Gaffney PM, Wierenga KJ, Tsiokas L (2014) Activating mutations in STIM1 and ORAI1 cause overlapping syndromes of tubular myopathy and congenital miosis. Proc Natl Acad Sci U S A 111:4197–4202
51. Grosse J, Braun A, Varga-Szabo D, Beyersdorf N, Schneider B, Zeitlmann L, Hanke P, Schropp P, Mühlstedt S, Zorn C, Huber M, Schmittwolf C, Jagla W, Yu P, Kerkau T, Schulze H, Nehls M, Nieswandt B (2007) An EF hand mutation in Stim1 causes premature platelet activation and bleeding in mice. J Clin Invest 117(11):3540–3550
52. Hirata S, Takayama N, Jono-ohnishi R, Endo H, Nakamura S, Dohda T, Nishi M, Hamazaki Y, Ishii E, Kaneko S, Otsu M, Nakauchi H, Kunishima S, Eto K (2013) Congenital amegakaryocytic thrombocytopenia iPS cells exhibit defective MPL-mediated signaling. J Clin Invest 123:3802–3814
53. Sakurai M, Kunimoto H, Watanabe N, Fukuchi Y, Yuasa S, Yamazaki S, Nishimura T, Sadahira K, Fukuda K, Okano H, Nakauchi H, Morita Y, Matsumura I, Kudo K, Ito E, Ebihara Y, Tsuji K, Harada Y, Harada H, Okamoto S, Nakajima H (2014) Impaired hematopoietic differentiation of RUNX1-mutated induced pluripotent stem cells derived from FPD/AML patients. Leukemia 28(12):2344–2354
54. Antony-Debré I, Manchev VT, Balayn N, Bluteau D, Tomowiak C, Legrand C, Langlois T, Bawa O, Tosca L, Tachdjian G, Leheup B, Debili N, Plo I, Mills JA, French DL, Weiss MJ, Solary E, Favier R, Vainchenker W, Raslova H (2015) Level of RUNX1 activity is critical for leukemic predisposition but not for thrombocytopenia. Blood 125(6):930–940
55. Ingrungruanglert P, Amarinthnukrowh P, Rungsiwiwut R, Maneesri-le Grand S, Sosothikul D, Suphapeetiporn K, Israsena N, Shotelersuk V (2015) Wiskott-Aldrich syndrome iPS cells produce megakaryocytes with defects in cytoskeletal rearrangement and proplatelet formation. Thromb Haemost 113(4):792–805

Acquired Thrombocytopenia

15

Tamam Bakchoul and Andreas Greinacher

Abstract

In this chapter, we will discuss the most frequently acquired disorders which should be considered in the differential diagnosis when a patient presents with a low platelet count. We will particularly focus on the mechanisms of platelet destruction in immune-mediated thrombocytopenias.

In this chapter, we will discuss the most frequently acquired disorders which should be considered in the differential diagnosis when a patient presents with a low platelet count. We will particularly focus on the mechanisms of platelet destruction in immune-mediated thrombocytopenias.

15.1 Physiology of Platelet Production

15.1.1 Normal Platelet Production

Platelets are produced from megakaryocytes in the bone marrow. The exact steps leading from megakaryocytes to mature platelets are discussed in Chap. 1.

The average platelet life span in humans is 7–10 days. After their formation from bone marrow megakaryocytes, platelets are distributed in two compartments: (1) the

T. Bakchoul (✉)
Center for Clinical Transfusion Medicine, University Hospital of Tuebingen, Otfried-Müller-Straße 4/1, Tübingen 72076, Germany
e-mail: tamam.bakchoul@med.uni-tuebingen.de

A. Greinacher
Institut für Immunologie und Transfusionsmedizin, Universitätsmedizin Greifswald, Ferdinand-Sauerbruch-Straße, Greifswald 17475, Germany
e-mail: greinach@uni-greifswald.de

H. Schulze, J. Italiano (eds.), *Molecular and Cellular Biology of Platelet Formation*,
DOI 10.1007/978-3-319-39562-3_15

stationary compartment in the spleen comprising about one third of the whole platelet mass and (2) circulating platelets representing the remaining two thirds. The platelet count in the peripheral blood is the final result of the relative rate between platelet production, their distribution into the two compartments, consumption due to adhesion to vessel wall damage or formation of platelet aggregates, and clearance. Daily production of more than 100×10^9 platelets is necessary to maintain steady state of normal platelet count ($150–350 \times 10^9$/L) in the peripheral blood.

The number of circulating platelets is regulated mainly by thrombopoietin (TPO), a growth factor which is produced in the liver in a constant amount in steady state [31, 56, 57]. TPO binds to megakaryocytes and hematopoietic stem cells via its receptor c-Mpl resulting in megakaryocyte proliferation and differentiation, ultimately resulting in platelet production [56]. c-Mpl is also expressed on mature platelets, which bind and clear TPO from the circulation [114]. As a consequence, the level of free TPO is regulated primarily by the number of circulating platelets, the platelet life span, and the megakaryocyte mass [57]. TPO levels in thrombocytopenic patients vary largely depending on the etiology of the disorder [18]. While patients with aplastic anemia show a decreased megakaryocyte mass despite high levels of free TPO, immune thrombocytopenia patients have low-normal levels of TPO and an expanded megakaryocyte mass [19, 84, 103]. Also the Ashwell-Morell receptor on murine hepatocytes binds platelets that have lost sialic acid residues on their surface as a result of cell aging. This binding activates a JAK-STAT signaling pathway, resulting in increased hepatic TPO mRNA and TPO production [46].

A second more rapid mechanism is platelet release from megakaryocytes via a rupture mechanism induced by interleukin-1α (IL-1α). This yields higher platelet numbers independently of TPO and may occur during situations of acute platelet need [79]. More detailed information on regulation of platelet life span can be found in Chap. 9.

15.1.2 Thrombocytopenia

Thrombocytopenia occurs when platelet loss and/or consumption in the circulation exceed their production from megakaryocytes in the bone marrow. The following five mechanisms can be distinguished although more than one of them often contribute to thrombocytopenia in the individual patient: (1) hemodilution, (2) platelet consumption, (3) decreased platelet production, (4) increased sequestration of platelets, and (5) immune-mediated destruction of platelets.

15.1.2.1 Hemodilution and Platelet Consumption

Acquired thrombocytopenia due to hemodilution is a frequent finding in the early postoperative phase, mostly combined with platelet consumption [100]. A platelet count decline of 30–70 % occurs universally following major surgery. Such hemodilution-associated thrombocytopenia is most prominent after cardiac surgery [109]. At the same time, platelets are consumed in the hemostatic process at the wound site. This platelet count fall is abrupt and occurs within a few hours following surgery, an effect which correlates with the magnitude of tissue trauma. An important characteristic of the post-surgery platelet count is that it usually continues to decline over the next 1 or 2 days, with the postoperative nadir (lowest platelet count value) usually occurring at a median of postoperative day 2, with a range between postoperative days 1–4 [94, 107]. Due to the lower perioperative platelet numbers, more thrombopoietin becomes available to stimulate megakaryocytes, which, however, require several days to produce platelets after receiving the TPO signal. According to the observed kinetic in platelet count in ITP patients and healthy volunteer after application of TPO analogues, platelet count increases after 7–10 days [108]. Until the platelet count increases, platelets are further consumed at an increased rate at the side of tissue injury to provide sufficient hemostasis. Subsequently, there is a rise in the platelet count as response to the increased TPO signal that peaks at approximately day 14, at levels often two to three times the patient's preoperative platelet count value, before it returns to baseline over the next 2 weeks (~day 28) (Fig. 15.1a). This time pattern can be effectively used to differentiate a normal platelet count decrease after major surgery from immune-mediated thrombocytopenias, especially the adverse drug effect of heparin-induced thrombocytopenia (HIT) (Fig. 15.1b; see section "Heparin-Induced Thrombocytopenia").

15.1.2.2 Impaired Platelet Production

An acquired decrease in platelet production can be caused by generalized marrow disorders (most common) resulting in pancytopenia or isolated megakaryocytic disorders.

Nonneoplastic Disorders

Thrombocytopenia usually accompanies nonneoplastic hypoproliferative stem cell disorders such as aplastic anemia, chemotherapy-induced pancytopenia, and nonimmune drug-induced aplasia [75]. Sometimes, isolated thrombocytopenia is a precursor of these disorders. Acquired thrombocytopenia is frequently observed in patients having myelodysplasia (MDS). Isolated, refractory thrombocytopenia might be even the initial presentation in up to 3 % of patients with MDS [48]. Peripheral blood cell morphologic clues pointing to a diagnosis of MDS include red cell macrocytosis, leukocyte left shift, various leukocyte abnormalities (e.g., hypogranular neutrophils, pseudo-Pelger-Huet anomaly, etc.), and hypogranular, enlarged platelets.

Neoplastic Disorders

Primary neoplastic disorders of the bone marrow such as acute myeloid and acute lymphoid leukemia, as well as many chronic lymphoproliferative disorders, such as chronic lymphoid leukemia, hairy-cell leukemia, non-Hodgkin's lymphoma, adult T-cell leukemia and lymphoma, multiple myeloma, and Waldenström macroglobulinemia, are commonly associated with thrombocytopenia [27, 53]. Usually, pancytopenia and the presence of circulating malignant cells point to the diagnosis.

Thrombocytopenia can also be secondary to infiltration of the bone marrow by non-hematologic metastatic malignancies and infectious organisms. Peripheral blood features include normoblasts (nucleated red cells), teardrop red cells, and leukopenia usually with leukocyte left shift [64]. Often thrombocytopenia is a

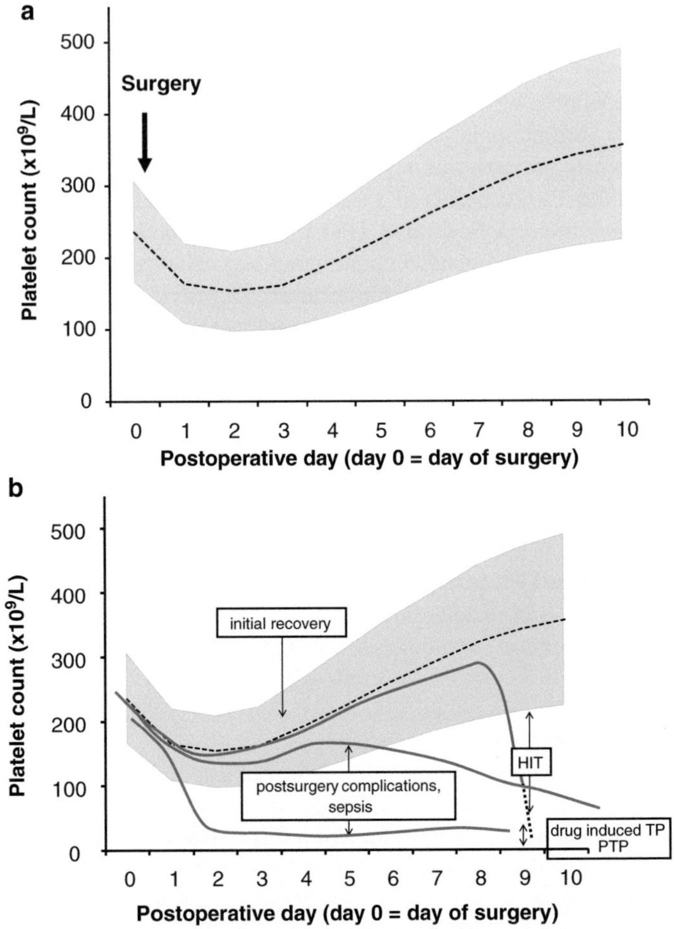

Fig. 15.1 Platelet count courses in patients undergoing major surgery. (**a**) Typical course: the *gray* background area shows the mean +– 2 SD of the platelet count course after cardiopulmonary bypass surgery as an example. (**b**) Platelet count course in patients who developed heparin-induced thrombocytopenia after surgery: the *solid line* indicates a platelet count pattern typical for early- and late-onset complications in a surgical patient. The first decrease in platelet counts is caused by major surgery (compare with (**a**)). After initial start of platelet count recovery, late-onset nonimmune complications cause a gradual decrease of platelet counts. Besides HIT, other immune-mediated thrombocytopenias are posttransfusion purpura (HIT) and drug-dependent thrombocytopenia (quinine type). Both show typically a platelet count decrease to values below 10,000/μL, while in HIT the platelet count nadir is usually above 20,000/μL

composite of impaired megakaryocyte numbers and enhanced platelet consumption due to activation of the clotting system. Especially infiltration of the bone marrow by adenocarcinoma cells is often associated with massive clotting activation and/or disseminated intravascular coagulopathy and consecutive enhanced consumption of platelets.

Nutrition and Alcohol

Impaired platelet production can be caused by vitamin B12 or folic acid deficiency [16]. Vitamin deficiency induces defective DNA synthesis resulting in low ploidy of megakaryocytes that produce a reduced number of large platelets [16]. The combination of large, heterogeneous red cells together with large platelets should prompt exclusion of vitamin B12 or folate deficiency. A transient platelet count decline is also observed approximately 1 week after beginning iron replacement therapy in anemic patients, most likely due to a shift of stem cells to erythropoiesis.

Another example of acquired thrombocytopenia is induced by heavy and prolonged alcohol consumption. Bone marrow aspirates of these patients show reduced megakaryocyte numbers and sometimes vacuolation of normoblasts and promyelocytes [60]. The platelet count rises 2–3 days after cessation of alcohol consumption, usually with "rebound" thrombocytosis.

Viral Infections and Parasites

Viral infections commonly cause acute thrombocytopenia, in part through decreased platelet production [33]. A prominent role for direct megakaryocyte infection has been best shown for HIV infection [118]. However, viral infections can also trigger platelet-reactive autoantibodies which can produce severe immune thrombocytopenia (see section "Immune Thrombocytopenia"). A special mechanism underlies dengue virus infection [13]. The manifestation of dengue in its severe form of hemorrhagic fever only occurs after a second infection. Although the virus is infecting platelets and megakaryocytes, it is likely that an immune-mediated mechanism causes typical dengue-associated hemorrhagic fever with antibodies (cross-)reacting with platelets, which have been induced during the first infection episode [32].

Parasite infections, especially malaria, can induce thrombocytopenia caused by consumption of platelets, which bind to the infected erythrocytes [71]. These platelet-erythrocyte aggregates are likely the cause of cerebral manifestations of malaria.

Isolated Megakaryocytic Disorders

Acquired amegakaryocytic thrombocytopenia is a rare disorder characterized by isolated thrombocytopenia associated with a total or marked reduction in megakaryocytes. Sometimes, subsequent progression to aplastic anemia or MDS occurs. The pathogenesis of the disorder is heterogeneous, with at least three different implicated mechanisms: an intrinsic stem cell defect (defective growth of megakaryocyte precursors), T-cell-mediated suppression of megakaryocyte development, and autoantibodies directed against megakaryocyte precursors. The fact that many patients benefit from immunosuppressive treatment suggests an important role of the immune system in the pathophysiology of this disorder.

15.1.2.3 Platelet Sequestration

Hypersplenism

In a case of hypersplenism, reduction in one or more of cell lines of the peripheral blood is frequently observed due to sequestration within the enlarged spleen. Often, the degree of anemia is not as marked as neutropenia and thrombocytopenia. The proportion of platelets within the enlarged spleen can be as high as ~90 %, leading to thrombocytopenia which usually develops subacutely or chronically, rather than acutely [6, 50]. In patients with advanced cirrhosis, thrombocytopenia is further aggravated by impaired platelet production due to reduced production of TPO and cirrhosis-associated subclinical disseminated intravascular coagulopathy (see section "Disseminated Intravascular Coagulation"). Treatment of splenomegaly-associated thrombocytopenia is usually not required, as the platelet count rarely declines to clinically relevant low levels. If these patients require interventions or chemotherapy, platelet transfusion often does not lead to a sufficient platelet count increase.

Hypothermia

Another mechanism for increased platelet sequestration is hypothermia which is used in various clinical settings to inhibit ischemia-related organ damage [29, 104]. Platelet activation and thrombocytopenia have been described as potential side effects of hypothermia. Data from animal model suggests that the platelet agonist ADP may play a key role in hypothermia-induced platelet activation [82]. Platelet kinetic studies in cooled dogs demonstrated reversible hepatic (potentially via the Ashwell receptor) and splenic platelet sequestration. Hypothermia-mediated thrombocytopenia is transient and corrects with rewarming [82].

15.1.2.4 Increased Platelet Consumption

The loss of platelets during major bleeding is an obvious cause of thrombocytopenia. Especially in trauma patients, this is combined with increased consumption at the site of tissue damage. An interesting debate is whether the high platelet counts in humans 150–400,000/μL have been developed during evolution to compensate for an increased requirement of platelets during massive trauma or whether this rather indicates that platelets have other important roles besides hemostasis, e.g., in innate immunity (Chap. 10). In the non-bleeding patient, reduced platelet survival due to increased platelet consumption is usually caused by platelet activation such as in DIC or sepsis.

Disseminated Intravascular Coagulation

Disseminated intravascular coagulopathy (DIC) is an acquired syndrome characterized by intravascular activation of coagulation with loss of localization, arising from different causes, e.g., sepsis, trauma, organ destruction (e.g., in severe pancreatitis), tumors, severe hepatic failure, and other causes [37]. The complex pathophysiology comprises a massive inflammatory response of the host that provokes increased tissue factor release, which then results in systemic activation of the coagulation system and consecutive hyperfibrinolysis. Consumption of platelets in DIC is thought to be responsible for the associated thrombocytopenia. Clinically, patients present usually with bleeding combined with microthrombi (e.g., digital ischemia). The most severe form occurs during infection with meningococci resulting in Waterhouse-Friderichsen syndrome. Therapy should focus on managing the underlying cause of DIC [86]. Patients with mild DIC and no evidence of bleeding require no further treatment of their clotting system when thrombosis prophylaxis is part of the standard therapy [54, 89]. A platelet transfusion trigger of 50×10^9/L should be considered in DIC although this recommendation is expert opinion based and not supported by valid studies.

Sepsis

Approximately 50 % of patients with systemic bacterial or fungal infections present with thrombocytopenia. Platelet counts in septic patients are generally mildly to moderately reduced, and bleeding as a result of the low platelet count is infrequent [76]. The mechanisms for thrombocytopenia in septicemia in the absence of DIC are uncertain but could include chemokine-induced macrophage ingestion of platelets or direct activation of platelets by endogenous mediators of inflammation or certain microbial products [34, 97, 105]. Prompt recognition and treatment of the infection is most important, as platelet count recovery tends to parallel the resolution of the infection. Prophylactic platelet transfusions are generally not required unless the platelet count falls to less than 10×10^9/L. However, in patients with comorbid clinical features such as concomitant coagulopathy or renal failure, this threshold should be individually assessed. The use of anticoagulants in septic patients is controversial. Heparin may benefit a subset of septic patients with clinical evidence of DIC and microvascular thrombosis as it can decrease further platelet consumption due to excess thrombin generation.

HIV Infection

Many potential mechanisms could explain thrombocytopenia in patients infected with HIV [80, 90]. These include immune platelet destruction, impaired platelet production secondary to HIV infection of megakaryocytes, drug-induced myelosuppression, HIV-associated thrombotic microangiopathy, hypersplenism, and marrow infiltration by tumors or opportunistic infections. Immune mechanisms for platelet destruction include antibodies that cross-react with GPIIb/IIIa complexes (molecular mimicry; see section "Immune Thrombocytopenia"). Anti-HIV chemotherapy (e.g., zidovudine, HAART) often raises the platelet count in patients with HIV-associated thrombocytopenia [81, 83].

Hepatitis C Infection

In chronic hepatitis C infection, thrombocytopenia is frequent. It can be caused by liver cirrhosis (see above) and splenomegaly, more often antibodies induced by the virus cross-react with platelets, thereby mediating immune thrombocytopenia [26, 30]. Until recently, thrombocytopenia often limited treatment of hepatitis C with interferon, as interferon itself my cause thrombocytopenia with critical low platelet counts, which required cessation of treatment [15, 102]. This is less an issue with the new antiviral drugs for treatment of hepatitis C.

Hemophagocytic Lymphohistiocytosis

Secondary or acquired hemophagocytic lymphohistiocytosis (HLH) is associated with lymphomas, infection (especially Epstein-Barr virus, but also HIV), immunosuppressive drugs (e.g., transplant recipients), or autoimmune disorders [49, 51, 117]. In the macrophage activation (or hemophagocytic) syndrome, nonmalignant macrophages phagocytose red cells, granulocytes, and platelets, resulting in pan- or line-specific cytopenia. Thrombocytopenia is frequently present and may be complicated by bleeding especially due to the concomitant hypofibrinogenemia. Cardinal symptoms are fever (due to hypercytokinemia), hepatosplenomegaly (accumulation of macrophages), bi- or pancytopenia (hemophagocytosis in the bone marrow), elevated triglyceride levels (inhibition of lipoprotein lipase by elevated TNF-alpha levels), elevated ferritin levels (increased secretion from the macrophages), and low fibrinogen levels (release of tissue plasminogen activator from activated monocytes).

Bone marrow examination shows numerous well-differentiated macrophages actively phagocytosing hematopoietic cells [49]. Treatment of HLH includes suppression of hyper-inflammation and reduction of activated cells by immunosuppressive therapy [1, 74, 101].

Envenomation

Venomous snake bite frequently produces a profound consumptive coagulopathy that can be accompanied by thrombocytopenia [67, 115]. Although most toxins in snake venom activate the clotting cascade, some of them can activate platelets directly via collagen receptor glycoprotein (GP) VI such as convulxin, via CLEC-2 such as rhodocytin, or via von Willebrand receptor GPIb such as bilinexin [63, 116]. Venom-induced consumptive coagulopathy is usually only complicated by bleeding. However, a subgroup of patients develops features that resemble thrombotic microangiopathy, with renal failure, microangiopathic hemolysis, and thrombocytopenia which may persist longer than the coagulopathy [67]. The most important treatment is to administer antivenom as soon as possible.

Pulmonary Embolism

Thrombocytopenia can accompany pulmonary embolism. The low platelet count presumably results from thrombin-induced platelet activation and/or platelet accretion within the emboli [73, 112]. More severe thrombocytopenia occurs in patients who develop DIC secondary to pulmonary embolism. An important differential diagnosis in patients who develop pulmonary embolism or deep-vein thrombosis associated with thrombocytopenia during treatment with heparin is HIT (see section "Heparin-Induced Thrombocytopenia").

Cardiovascular Disorders and Devices

Valvular heart diseases, especially aortic valve stenosis, can induce variable severity of thrombocytopenia most likely due to increased platelet activation. However, the major cause of increased bleeding in these patients is acquired von Willebrand syndrome due to accelerated clearance of the largest multimers rather than low platelet counts [9, 111]. Artificial heart valve-associated thrombocytopenia is often caused either by infection of the valve or by a paravalvular leak which produces shear stress-induced activation of platelets [62, 106].

Extracorporeal and intravascular devices are frequently responsible for thrombocytopenia in patients with heart diseases. They can induce a decreased platelet count through a variety of mechanisms, including platelet activation at prothrombotic surfaces, high shear stress leading to shear-induced platelet activation, increased thrombin generation, and, finally, mechanical alterations of platelets [98]. For instance, during cardiopulmonary bypass (CPB), there is extensive contact between heparin-anticoagulated blood and the synthetic surfaces of the extracorporeal perfusion device (blood oxygenator). Typically, the platelet count falls in patients undergoing CPB by 33–50 %, initially due to hemodilution, aggravated by losses within the extracorporeal perfusion device, hypothermia, and consumption at the side of bleeding [99]. Another example of device-associated thrombocytopenia is induced by intra-aortic balloon pumps. Approximately half of the patients who received this device show a decrease in platelet count of about 30–40 % from the initial value [88]. Thrombocytopenia is also very frequent in patients with cardiac assist devices, which are designed to support left, right, or biventricular function. All of these devices activate platelets, and several of them mechanically destroy platelets. Although thrombocytopenia is most likely caused by the device itself, improving hemodynamic function frequently alleviates platelet consumption [44].

In addition to the reduction in platelet numbers, activation of platelets and enhanced cleavage of platelet receptors (e.g., by shear-activated ADAMTS 17) induce platelet function defects. They are more relevant for bleeding complications than the moderately reduced platelet count.

Non-cardiovascular Devices

Hemodialysis, extracorporeal oxygenation (ECMO), chronic veno-venous renal replacement therapy, and plasmapheresis (with or without immunoadsorption) can all induce a mild to moderate decrease in platelet count [2]. A more pronounced decrease may result from increased thrombin generation due to insufficient anticoagulation. Therefore, anticoagulant regimen should be monitored and reassessed in patients in whom platelet counts decrease more than expected. However, in case of an abrupt decline in platelet count by more than 50 % within the second week after start of extracorporeal treatment devices, HIT should be considered, especially if new clots occur in the device [113].

Microangiopathic Hemolysis

The main feature of microangiopathic hemolytic anemias is Coombs test-negative hemolysis (red cell fragments, increased lactate dehydrogenase, reticulocytosis) and thrombocytopenia, with a variable component of neurologic or renal dysfunction and fever [95]. Some disorders are well characterized, e.g., thrombotic thrombocytopenic purpura (TTP) associated with autoantibodies against ADAMTS13 (a disintegrin and metalloprotease with thrombospondin-1-like repeats) or hereditary deficiency of this enzyme [36]. The platelet count in TTP often falls below 30,000 platelets/μL [23]. The other two syndromes are the hemolytic-uremic syndrome (HUS) associated with toxin-associated diarrheal syndrome (typical form) or hereditary abnormalities of complement (atypical form, aHUS) and the HELLP syndrome [52, 87]. The HELLP syndrome is a pregnancy-related complication with hemolysis, elevated liver enzymes, and low platelet counts.

Distinguishing between these different disorders may be difficult because of extensive overlap in symptoms. Patients with TTP generally present acutely or subacutely with fatigue and malaise, with variable neurologic symptoms that may range from mild personality changes to paresthesia. Renal insufficiency may or may not be present. In contrast, typical HUS follows infection with enteropathogenic E. coli and often is preceded by bloody diarrhea and abdominal pain. Renal insufficiency is usually the most prominent component of typical HUS. aHUS presents in a similar manner like TTP, but it may demonstrate a more chronic presentation with progressive renal insufficiency, low grade of anemia, and thrombocytopenia. Neurologic defects are less common in aHUS than in TTP. aHUS is often caused by hereditary or (rarely) acquired defects in the complement system. The HELLP syndrome is always associated with pregnancy; most important is differentiation between TTP and HELLP in pregnant women with acute severe thrombocytopenia [14].

Plasma exchange is the standard of care for TTP [93]. The exchange of 1.5 plasma volumes per day is a standard initial treatment. However, patients with an inadequate response may benefit from larger exchange volumes or more frequent exchanges. Plasma exchange should be continued daily until the platelet count and fragmented cells reach normal levels, LDH normalizes, and symptoms have resolved. TTP patients may also benefit from the use of B-cell depletion medication such as anti-CD20 (rituximab), especially those patients who do not respond rapidly to plasma exchange [92]. Platelet transfusion is not sufficient to treat immune-mediated TTP. Treatment of typical HUS is generally supportive; it was long assumed that the use of antibiotics may lead to increased toxin release and worse outcome; however, during an epidemic outbreak of typical HUS, antibiotic treatment was associated with reduced morbidity [28, 40]. A benefit for plasma exchange in typical HUS has not been demonstrated, while immunoadsorption using protein G columns (IgG depletion) resulted in rapid improvement of severe neurological symptoms and renal dysfunction [43]. Response rates to plasma exchange in patients with aHUS are also not as robust as in TTP. Eculizumab, an antibody blocking complement C5, has shown efficacy in patients with aHUS [59]. Eculizumab should be initiated promptly in patients with a high likelihood for aHUS, i.e., thrombotic microangiopathy without severe ADAMTS13 deficiency, with no Shiga toxin-producing E. coli infection, or who do not respond to plasma exchange.

15.1.2.5 Increased Platelet Destruction

Compared to platelet consumption, increased platelet destruction indicates reduced platelet life span that is caused by platelet-reactive antibodies, such as autoantibodies, alloantibodies, or drug-dependent antibodies. The involved antibodies usually target platelet surface GPs. In contrast to most of the other causes of thrombocytopenia discussed above, platelet counts decrease significantly, and severe thrombocytopenia ($<20 \times 10^9$/L) is often present [23]. With the exception of TTP, mucocutaneous bleeding is the predominant symptom in patients with platelet counts below 10–20,000/L.

Immune Thrombocytopenia

Immune thrombocytopenia (ITP, previously referred to as idiopathic thrombocytopenic purpura) is an autoimmune disorder characterized by isolated thrombocytopenia [21, 85]. The incidence of ITP in adults is approximately 2–4 per 100,000/year

[35, 55, 78, 91]. The low platelet count in ITP is thought to be caused by increased destruction of platelets opsonized by antiplatelet autoantibodies [22, 47, 68, 69]. Antiplatelet autoantibodies can induce a decrease in platelet count by different mechanisms, or any combination of the following: Fc-dependent and Fc-independent phagocytosis of antibody-coated platelets, complement activation, and direct lysis of platelets. ITP can also be caused in part by immune-mediated impairment of platelet production. T-cell cytotoxicity, abnormalities in megakaryocytopoiesis, and impaired platelet production have been shown as contributing factors in ITP [12, 68].

The clinical course of ITP is variable and to some extent age dependent. While the great majority of children with ITP will have a brief and uneventful course with spontaneous remission, most adult patients experience chronic ITP which can be associated with clinically significant bleeding. However, the risk of severe bleeding and especially intracranial hemorrhage is low in younger patients [21, 25], but increases with age and in the presence of comorbidities. Patients older than 60 years with chronic ITP and symptomatic bleeding have the highest risk for intracranial hemorrhage.

The diagnosis of ITP rests on a consistent clinical history, physical examination, and exclusion of other causes of thrombocytopenia. Leukocyte counts and hemoglobin are characteristically normal unless thrombocytopenic bleeding has resulted in anemia. Examination of the peripheral blood film should be performed to exclude pseudothrombocytopenia, microangiopathic hemolytic anemia (fragmented red cells), or abnormalities suggestive of other disorders. The mean platelet volume may be increased in patients with ITP. However, ITP patients always show a heterogeneous platelet population with up to ~40 % enlarged platelets. If more than 60 % large or even giant platelets (less than two platelets fit into one red blood cell) are present, hereditary macrothrombocytopenia (see Chap. 14) is much more likely the underlying cause of thrombocytopenia. Bone marrow examination is not required routinely and is generally not useful for diagnosing ITP, but should be performed to exclude other causes of thrombocytopenia when atypical features such as unexplained anemia, lymphadenopathy, or splenomegaly are present. Megakaryocyte number is typically normal or increased in the marrow of patients with ITP. The role of laboratory investigations for platelet-reactive autoantibodies in the diagnostic workup of ITP is a matter of debate [70]. Because of low sensitivity reported in prospective studies, the routine use of platelet autoantibody testing has not been endorsed in guidelines; however, recent studies showed an important impact of the glycoprotein specificity on the patient's responsiveness to particular treatments, especially IVIG [17]. The clinical value of antibody testing in ITP is currently best established in patients with several potential causes for a low platelet count, e.g., in case of underlying CLL. Due to the high specificity of glycoprotein-specific platelet autoantibody tests, a positive test makes ITP very likely.

The main therapeutic goal in ITP is to minimize the risk of severe bleeding. This can be achieved by preventing platelet destruction and managing the patient toward a rapid and persistent rise in platelet count. As discussed above, ITP represents a heterogeneous spectrum of clinical presentations and pathophysiologic events creating the need for individualized patient management. Treatment protocols are not restricted to a particular group of drugs. Conventional therapeutic options include corticosteroids, intravenous immunoglobulins (IVIG), and anti-D-immunoglobulin (Anti-D) as first-line treatments [65, 85]. Corticosteroids suppress nonspecifically T- and B-cell responses and typically require continuing drug exposure for usually three to six cycles to achieve a significant improvement in platelet count or reduced bleeding symptoms. While high-dose dexamethasone administered at 2- to 4-week intervals has been reported to be associated with durable responses in some patients (and may be more effective than daily prednisone) [66], no significant difference exists in the long-term response rates between the various drugs of this group. Anti-D is an effective treatment for ITP patients who express the RhD antigen on their red blood cells. The mechanism of action of anti-D in ITP remains highly speculative, and it should be used with caution due to the risk of undesired reactions related to intravascular hemolysis. Second-line treatments include rituximab and splenectomy to inhibit antibody production by depletion of B cells and to prevent accelerated platelet destruction, respectively [5, 77, 96]. TPO receptor agonists increase platelet production through activation of TPO receptors on megakaryocytes and hematopoietic progenitor cells: two TPO receptor agonists were shown to be effective in ITP treatment, romiplostim and eltrombopag [18, 58] (see Chap. 18). Only recently, the temporary use of TPO agonists was reported to induce prolonged remission in some patients [38]. It is intriguing, yet unproven, to speculate that overexpression of platelet antigens via increased production might restore self-tolerance. Still, a word of caution should be said. Despite the well-documented short-term efficacy and safety, except for bone marrow fibrosis [39], long-term safety and efficacy data 5 years of treatment, are still limited [72].

Drug-Induced Immune Thrombocytopenia

More than 300 drugs have been implicated in drug-induced immune thrombocytopenia (DITP). A systematic review of individual patient data found that the most commonly reported drugs with a definite or probable causal relation to thrombocytopenia were quinidine, quinine, rifampin, and trimethoprim-sulfamethoxazole. The most common drug involved in DITP is, however, heparin. As it is caused by a different mechanism, we will discuss heparin-induced thrombocytopenia (HIT) separately.

DITP may be induced by several pathogenic mechanisms [7]:

1. *Quinine-type mechanism*: The classic drug-dependent antibodies (DDAbs) attach tightly to platelets only in the presence of the sensitizing drug and most often target GPIIb/IIIa or GPIb/IX/V. The mechanisms by which quinine-type DDAbs form and react with platelets have been intensely studied [20]. A model for the interaction between drug and platelets was proposed suggesting patients have autoantibodies with only weak reactivity for platelet GP epitopes in the absence of drug, but that this reactivity is significantly enhanced through bridging interactions by drug binding between both the protein and antibody [8]; the resultant sandwich entraps a drug at the antigen-antibody interface and greatly increases antibody binding energy for the GP autoepitope.

2. *Neoepitope mechanism* (such as GPIIb/IIIa inhibitors, tirofiban, and eptifibatide): Interactions between drug and membrane proteins are speculated to induce conformational changes in the protein structure that elicit DDAbs in certain individuals. With both mechanisms, when the drug is cleared from the circulation and the conformation of the platelet protein reverts to its native state, DDAbs, although still present, can no longer bind and destroy the platelets. Rarely, antibodies are triggered by GPIIb/IIIa inhibitors which cause pronounced thrombocytopenia about one week after cessation of the drug. These antibodies cross-react with the GP in its native form.
3. *Hapten mechanism*: Small molecules (low molecular weight, <5000 Da, e.g., penicillin and some cephalosporins) require a covalent coupling to a larger carrier protein, mostly GPIIb/IIIa, to elicit drug-specific antibodies, which then bind to the small molecule drug rather than to the platelet protein.
4. *Autoantibody mechanism*: These antibodies are induced after drug exposure (especially gold therapy), but are not dependent on the presence of the drug for their binding to platelets.
5. *Drug-specific mechanism* (such as abciximab): This drug is a chimeric (mouse-human) monoclonal antibody Fab fragment specific for GPIIIa and used primarily to prevent platelet aggregate formation. Drug-specific antibodies that appear to recognize murine sequences in the complimentary determining region 3 (CDR3) of abciximab are responsible for this type of DITP.

The diagnosis of DITP is based on five clinical criteria [4]: (1) exposure to the candidate drug preceded thrombocytopenia; (2) recovery from thrombocytopenia was complete and sustained after discontinuing candidate drug; (3) candidate drug was the only drug used before the onset of thrombocytopenia, or other drugs were continued or reintroduced after discontinuation of the candidate drug with a sustained normal platelet count; (4) other causes for thrombocytopenia were excluded; and (5) reexposure to the candidate drug resulted in recurrent thrombocytopenia. Evidence is considered *definite* if criteria 1, 2, 3, and 4 are met; *probable* if criteria 1, 2, and 3 are met; *possible* if criterion 1 only is met; and *unlikely* if criterion 1 is not met.

DITP often occurs in hospitalized patients who are taking multiple medications and who have comorbidities that can also cause thrombocytopenia. Therefore, relating thrombocytopenia to a particular drug depending solely on clinical information is difficult. Specialized laboratory testing for antibodies that bind to platelets in the presence of drugs or a drug metabolite has been developed [3]. Although such assays are only available for a limited number of drugs and drug metabolites, they may provide useful confirmation of DITP.

DITP is characteristically severe with median nadir platelet counts of $<20 \times 10^9$/L and a high risk of hemorrhage. Treatment for DITP involves discontinuation of the offending drug (in a case of multiple medications, all drugs started within the last 2 weeks should be stopped (especially antibiotics) and replaced). The platelet count usually starts to recover after four to five half-lives of the responsible drug or drug metabolite. High doses of IVIG can be given to patients with severe thrombocytopenia and bleeding as well as to those at high risk of bleeding. Platelet transfusion is generally ineffective as long as the drug or its metabolite(s) is present in plasma. Rare forms of DITPs are induced by food ingredients, such as quinine containing beverages or walnuts. Diagnosis of DITP requires usually a high grade of clinical suspicion and careful "detective" workup to identify the causative drug.

Heparin-Induced Thrombocytopenia

Antibodies against multimolecular complexes of platelet factor 4 (PF4) and heparin are responsible for heparin-induced thrombocytopenia (HIT) [41]. Binding of HIT antibodies to Fc receptors on monocytes and platelets causes intravascular cellular activation leading to elevated levels of circulating microparticles and an intensely prothrombotic state. The incidence of HIT varies widely between patient groups, with reported incidences of 0.2–5.0 % [41]. The risk of HIT associated with low molecular weight heparin (LMWH) is five- to tenfold lower than with unfractionated heparin (UFH), and the incidence of HIT is approximately threefold greater in surgical than in medical patients. Despite the occurrence of thrombocytopenia, bleeding is extremely rare in HIT. In contrast, thrombosis develops in 40–50 % of unmanaged patients with HIT.

The clinical symptoms include a decrease in platelet counts by >50 % and/or new thromboembolic complications [45]. As a general rule, the clinical symptoms manifest between days 5 and 14 after initiation of heparin. An exception is rapid-onset HIT, in which patients with recent heparin exposure (usually within the last 30 days) and preexisting HIT antibodies may manifest clinical HIT within hours of heparin re-exposure. Uncommonly, HIT may develop 2–3 weeks after heparin exposure (delayed-onset HIT) [110]. In these patients, the antibodies have gained autoreactivity, i.e., they recognize PF4 bound to endogenous glycosaminoglycans on platelets and therefore activate platelets even in the absence of heparin. Rarely, HIT can manifest as an autoimmune disorder without any exposure to heparin, so-called spontaneous or autoimmune HIT [42]. The autoantibodies in these patients are often triggered by a preceding infection or major surgery which may be explained by a cross-reactivity of the antibodies to PF4/bacteria polyanion complexes or endogenous PF4/DNA complexes. The resulting antibodies are high titer, and, in contrast to typical HIT antibodies, they can persist for months.

Two approaches can help the treating physician to confirm or rule out HIT: (a) systematic assessment of the clinical presentation using scoring systems that determine the pretest probability of HIT and (b) in vitro demonstration of anti-PF4/heparin antibodies [11].

Several clinical scoring systems have been developed to assist with determining the pretest probability of HIT. The most commonly used is the 4Ts system (Thrombocytopenia, Timing, Thrombosis, and oTher; see Table 15.1) [61]. This system has been shown to have a high negative predictive value (i.e., a low score is useful in ruling out HIT), but its effectiveness is limited by modest interobserver agreement and a relatively low positive predictive value. Another system is the HIT expert probability score [24], but the clinical experience with this system is not as extensive. While a low 4T score (<4 points) makes HIT unlikely, in case of missing

Table 15.1 The 4Ts scoring system to evaluate the pretest risk for HIT

4Ts	2 points	1 point	0 point
Thrombocytopenia	Platelet count fall >50 % and platelet nadir ≥20	Platelet count fall 30–50 % or platelet nadir 10–19	Platelet count fall <30 % or platelet nadir <10
Timing of platelet count fall	Clear onset between days 5 and 10 or platelet fall ≤1 day[a]	Onset after day 10 or fall ≤1 day[a]	Platelet count fall <4 days or >14 days after exposure
Thrombosis or other sequelae	New thrombosis[b] (confirmed)	Suspected thrombosis (not proven)	None
OTher causes for thrombocytopenia	None apparent	Possible	Definite

The resulting clinical probability score is divided into high (6–8 points), intermediate (4–5 points), and low (≤3 points)
[a]In case of prior heparin exposure within 30–100 days ago)
[b]Also skin necrosis, acute systemic reaction post-intravenous unfractionated heparin (UFH) bolus, progressive or recurrent thrombosis, non-necrotizing (erythematous) skin lesions

information (e.g., start of heparin) or comorbidity also causing thrombocytopenia, the 4Ts score may be falsely low. This is the reason that HIT may be the underlying cause in 2–3 % of patients with low score. In these situations, laboratory testing for heparin-dependent antibodies is a mainstay of the diagnosis.

Two different classes of assays are available: immunoassays, to detect binding of anti-PF4/heparin antibodies, and functional laboratory assays, to investigate the capability of antibodies in the patient serum/plasma to activate platelets in the presence of heparin [10].

Two main types of immunoassays are available: enzyme-linked immunosorbent assays (ELISAs) and particle-based assays. The sensitivity of the PF4/heparin ELISA approaches 100 %, and thus a negative test is useful in excluding HIT. Difficulties concerning its use include long turnaround time in institutions in which it is not performed daily and its low specificity and positive predictive value, particularly in the post-cardiac surgery setting. Several approaches can be performed to increase specificity: (1) considering the level of positivity (optical density), (2) the use of an ELISA that detects only anti-PF4/heparin IgG, and (3) the addition of a confirmatory step performed in the presence of high heparin concentrations.

The clinically most relevant antibodies against PF4/heparin complexes are those that have the capability to activate platelets. This can be investigated using functional assays using platelet of normal donors in the presence of patient sera/plasma and heparin. The readout is platelet activation, and the specificity can be increased by inhibition of platelet activation by adding heparin in excess (100 IU/mL, proving the heparin dependency) and by showing Fcγ receptor IIa-dependent activation using a blocking monoclonal antibody (clone IV.3). Although functional assays have improved specificity compared with the ELISA, they are technically challenging,

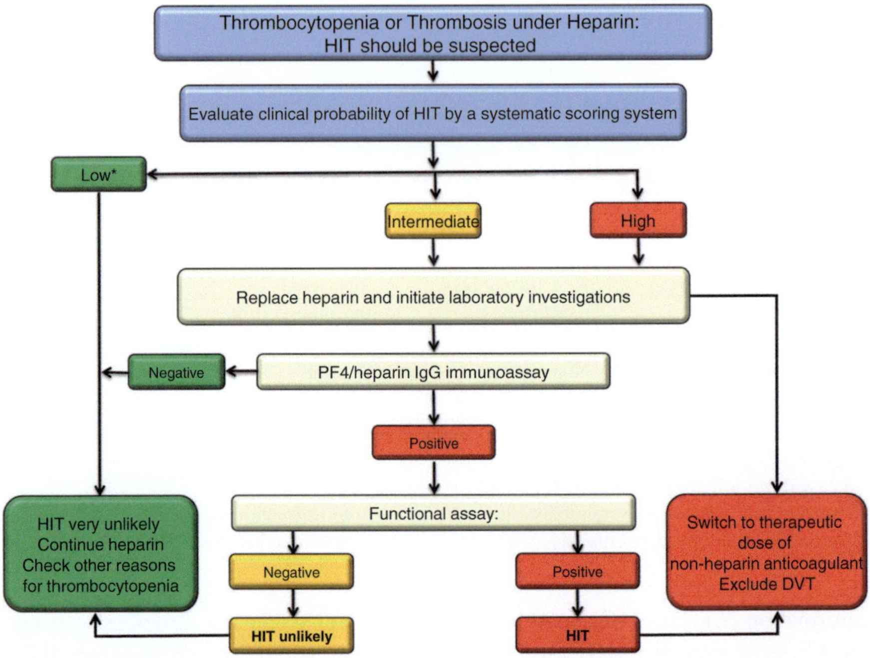

Fig. 15.2 A suggested approach to diagnosis and initial management of patients with suspected HIT. This approach to the diagnosis and initial management of patients with suspected HIT is based on clinical assessment supported by complementary laboratory investigations. As indeterminate results may be occasionally obtained using laboratory tests, reevaluating the clinical probability of HIT in an individual patient may be helpful to overcome some diagnostic uncertainty. * The 4T score might be falsely low in case of missing data on preceding platelet counts or start of heparin, or the patient might have comorbidities causing thrombocytopenia. In these situations, laboratory tests may show clinically relevant PF4/heparin antibodies despite a low 4T score, and a PF4/heparin ELISA is justified despite a low score

requiring washed donor platelets and, for the SRA, radioisotopes. Because of these considerations, it is recommended to use an algorithm including the clinical feature and laboratory investigation in the management of HIT-suspected cases (Fig. 15.2). By at least intermediate probability for HIT immunoassays for PF4/heparin antibodies should be performed. Depending on the obtained results (OD higher than 1.0, inhibition by high concentration of heparin), functional assays should be also performed. The diagnosis of HIT can only be confirmed by in vitro demonstration of platelet-activating anti-PF4/heparin antibodies using functional assays.

The cornerstone of the management of HIT is immediate discontinuation of heparin when the disease is suspected, even before laboratory test results are reported [10]. Anticoagulation using one of the non-heparin anticoagulants in therapeutic dose should be initiated even in patients with no thrombosis, because of the ongoing high risk of thrombosis, and continued until platelet counts have normalized (i.e., until a stable plateau is reached for two consecutive days). Patients with HIT and thrombosis should receive at least 3 months of therapeutic dose anticoagulation. Warfarin must be avoided in acute HIT. It may be started once the platelet count has reached a stable plateau, but only overlapping with an alternative parenteral anticoagulant.

HIT antibodies are transient and typically disappear within 3 months. When patients are re-exposed thereafter to heparin, only rarely an anamnestic immune response will occur, and if so, it requires again at least 5 days until HIT develops. This allows controlled reexposure to heparin during cardiac surgery in patients with a history of HIT. Before and after surgery, however, alternative anticoagulants should be used.

15.1.3 Future Perspectives

As outlined in this chapter, a decreased platelet count is a sensitive indication for disturbed homeostasis. As it may be the first symptom of severe disease, acquired thrombocytopenia requires prompt clinical attention and further diagnostic work-up. Gaining new insights into pathological states of the immune and coagulation systems is of major relevance to improve the quality of patient care. Recent data make it highly likely that endogenous coagulation and platelet proteins are involved in pathogen host defense and may act as a danger label for the immune system. Further understanding of these mechanisms bears the possibility to identify new strategies for anti-infection treatment, but also for prevention of thrombocytopenia as a side effect of infection and autoimmune disease.

References

1. Al-Lamki Z, Wali YA, Pathare A, Ericson KG, Henter JI (2003) Clinical and genetic studies of familial hemophagocytic lymphohistiocytosis in Oman: need for early treatment. Pediatr Hematol Oncol 20(8):603–609
2. Anderson HL 3rd, Cilley RE, Zwischenberger JB, Bartlett RH (1986) Thrombocytopenia in neonates after extracorporeal membrane oxygenation. ASAIO Trans 32(1):534–537
3. Arnold DM, Curtis BR, Bakchoul T, I. Subcommittee on Platelet (2015) Recommendations for standardization of laboratory testing for drug-induced immune thrombocytopenia: communication from the SSC of the ISTH. J Thromb Haemost 13(4):676–678
4. Arnold DM, Kukaswadia S, Nazi I, Esmail A, Dewar L, Smith JW, Warkentin TE, Kelton JG (2013) A systematic evaluation of laboratory testing for drug-induced immune thrombocytopenia. J Thromb Haemost 11(1):169–176
5. Arnold DM, Nazi I, Santos A, Chan H, Heddle NM, Warkentin TE, Kelton JG (2010) Combination immunosuppressant therapy for patients with chronic refractory immune thrombocytopenic purpura. Blood 115(1):29–31
6. Aster RH (1966) Pooling of platelets in the spleen: role in the pathogenesis of "hypersplenic" thrombocytopenia. J Clin Invest 45(5):645–657
7. Aster RH (1999) Drug-induced immune thrombocytopenia: an overview of pathogenesis. Semin Hematol 36(1 Suppl 1):2–6
8. Aster RH, Bougie DW (2007) Drug-induced immune thrombocytopenia. N Engl J Med 357(6):580–587
9. Avila ML, Lee KJ, Bouskill V, Rand ML, James P, Carcao M (2015) Acquired von Willebrand syndrome in paediatric patients with congenital heart disease: challenges in the diagnosis and management of this rare condition. Haemophilia 21(1):e89–e92

10. Bakchoul T, Greinacher A (2012) Recent advances in the diagnosis and treatment of heparin-induced thrombocytopenia. Ther Adv Hematol 3(4):237–251
11. Bakchoul T, Zollner H, Greinacher A (2014) Current insights into the laboratory diagnosis of HIT. Int J Lab Hematol 36(3):296–305
12. Ballem PJ, Segal GM, Stratton JR, Gernsheimer T, Adamson JW, Slichter SJ (1987) Mechanisms of thrombocytopenia in chronic autoimmune thrombocytopenic purpura. Evidence of both impaired platelet production and increased platelet clearance. J Clin Invest 80(1):33–40
13. Basanta Otero P, Gonzalez Villalonga C, Orbeal Aldama L (1983) Platelet autoantibodies in Dengue hemorrhagic fever. Acta Haematol 70(2):141–142
14. Battinelli EM (2014) TTP and pregnancy. Blood 123(11):1624–1625
15. Benci A, Caremani M, Tacconi D (2003) Thrombocytopenia in patients with HCV-positive chronic hepatitis: efficacy of leucocyte interferon-alpha treatment. Int J Clin Pract 57(1): 17–19
16. Bessman JD (1984) The relation of megakaryocyte ploidy to platelet volume. Am J Hematol 16(2):161–170
17. Brighton TA, Evans S, Castaldi PA, Chesterman CN, Chong BH (1996) Prospective evaluation of the clinical usefulness of an antigen-specific assay (MAIPA) in idiopathic thrombocytopenic purpura and other immune thrombocytopenias. Blood 88(1):194–201
18. Bussel JB, Provan D, Shamsi T, Cheng G, Psaila B, Kovaleva L, Salama A, Jenkins JM, Roychowdhury D, Mayer B, Stone N, Arning M (2009) Effect of eltrombopag on platelet counts and bleeding during treatment of chronic idiopathic thrombocytopenic purpura: a randomised, double-blind, placebo-controlled trial. Lancet 373(9664):641–648
19. Chang M, Suen Y, Meng G, Buzby JS, Bussel J, Shen V, van de Ven C, Cairo MS (1996) Differential mechanisms in the regulation of endogenous levels of thrombopoietin and interleukin-11 during thrombocytopenia: insight into the regulation of platelet production. Blood 88(9):3354–3362
20. Christie DJ, Aster RH (1982) Drug-antibody-platelet interaction in quinine- and quinidine-induced thrombocytopenia. J Clin Invest 70(5):989–998
21. Cines DB, Liebman HA (2009) The immune thrombocytopenia syndrome: a disorder of diverse pathogenesis and clinical presentation. Hematol Oncol Clin N Am 23(6):1155–1161
22. Cines DB, McMillan R (2007) Pathogenesis of chronic immune thrombocytopenic purpura. Curr Opin Hematol 14(5):511–514
23. Cuker A (2015) ITP and TTP: interpreting evidence in light of patient values. Blood 125(10): 1514–1515
24. Cuker A, Arepally G, Crowther MA, Rice L, Datko F, Hook K, Propert KJ, Kuter DJ, Ortel TL, Konkle BA, Cines DB (2010) The HIT Expert Probability (HEP) Score: a novel pre-test probability model for heparin-induced thrombocytopenia based on broad expert opinion. J Thromb Haemost 8(12):2642–2650
25. D'Orazio JA, Neely J, Farhoudi N (2013) ITP in children: pathophysiology and current treatment approaches. J Pediatr Hematol Oncol 35(1):1–13
26. Dimitroulis D, Valsami S, Stamopoulos P, Kouraklis G (2012) Immunological HCV-associated thrombocytopenia: short review. Clin Dev Immunol 2012:378653
27. Dubansky AS, Boyett JM, Falletta J, Mahoney DH, Land VJ, Pullen J, Buchanan G (1989) Isolated thrombocytopenia in children with acute lymphoblastic leukemia: a rare event in a Pediatric Oncology Group Study. Pediatrics 84(6):1068–1071
28. Ducker C, Dautel P, Wagner K, Przewozna J, Oehlerking S, Repenthin J, Bruning R, Meyer TN, Faiss S (2011) Clinical symptoms, treatment and outcome of EHEC and EHEC-HUS patients treated as in-patients. Dtsch Med Wochenschr 136(36):1770–1776
29. Easterbrook PJ, Davis HP (1985) Thrombocytopenia in hypothermia: a common but poorly recognised complication. Br Med J (Clin Res Ed) 291(6487):23
30. El Barbary MA, Saad AE, Attia FM, Mandour MI, Haidara MA, Dallak MM, Isenovic ER (2010) Thrombocytopenia in patients with chronic hepatitis C: a possible role of HCV on platelet progenitor cell maturation. Angiology 61(3):304–313

31. Emmons RV, Reid DM, Cohen RL, Meng G, Young NS, Dunbar CE, Shulman NR (1996) Human thrombopoietin levels are high when thrombocytopenia is due to megakaryocyte deficiency and low when due to increased platelet destruction. Blood 87(10):4068–4071
32. Falconar AK (1997) The dengue virus nonstructural-1 protein (NS1) generates antibodies to common epitopes on human blood clotting, integrin/adhesin proteins and binds to human endothelial cells: potential implications in haemorrhagic fever pathogenesis. Arch Virol 142(5):897–916
33. Flaujac C, Boukour S, Cramer-Borde E (2010) Platelets and viruses: an ambivalent relationship. Cell Mol Life Sci 67(4):545–556
34. Francois B, Trimoreau F, Vignon P, Fixe P, Praloran V, Gastinne H (1997) Thrombocytopenia in the sepsis syndrome: role of hemophagocytosis and macrophage colony-stimulating factor. Am J Med 103(2):114–120
35. Frederiksen H, Schmidt K (1999) The incidence of idiopathic thrombocytopenic purpura in adults increases with age. Blood 94(3):909–913
36. Furlan M, Robles R, Solenthaler M, Lammle B (1998) Acquired deficiency of von Willebrand factor-cleaving protease in a patient with thrombotic thrombocytopenic purpura. Blood 91(8):2839–2846
37. Gando S, Wada H, Thachil J, Scientific, D. I. C. o. t. I. S. o. T. Standardization Committee on and Haemostasis (2013) Differentiating disseminated intravascular coagulation (DIC) with the fibrinolytic phenotype from coagulopathy of trauma and acute coagulopathy of trauma-shock (COT/ACOTS). J Thromb Haemost 11(5):826–835
38. Ghadaki B, Nazi I, Kelton JG, Arnold DM (2013) Sustained remissions of immune thrombocytopenia associated with the use of thrombopoietin receptor agonists. Transfusion 53(11): 2807–2812
39. Ghanima W, Geyer JT, Lee CS, Boiocchi L, Imahiyerobo AA, Orazi A, Bussel JB (2014) Bone marrow fibrosis in 66 patients with immune thrombocytopenia treated with thrombopoietin-receptor agonists: a single-center, long-term follow-up. Haematologica 99(5):937–944
40. Goldwater PN, Bettelheim KA (2012) Treatment of enterohemorrhagic Escherichia coli (EHEC) infection and hemolytic uremic syndrome (HUS). BMC Med 10:12
41. Greinacher A (2009) Heparin-induced thrombocytopenia. J Thromb Haemost 7(Suppl 1):9–12
42. Greinacher A (2014) Me or not me? The danger of spontaneity. Blood 123(23):3536–3538
43. Greinacher A, Friesecke S, Abel P, Dressel A, Stracke S, Fiene M, Ernst F, Selleng K, Weissenborn K, Schmidt BM, Schiffer M, Felix SB, Lerch MM, Kielstein JT, Mayerle J (2011) Treatment of severe neurological deficits with IgG depletion through immunoadsorption in patients with Escherichia coli O104:H4-associated haemolytic uraemic syndrome: a prospective trial. Lancet 378(9797):1166–1173
44. Greinacher A, Selleng K (2010) Thrombocytopenia in the intensive care unit patient. Hematol Am Soc Hematol Educ Program 2010:135–143
45. Greinacher A, Warkentin TE (2006) Recognition, treatment, and prevention of heparin-induced thrombocytopenia: review and update. Thromb Res 118(2):165–176
46. Grozovsky R, Begonja AJ, Liu K, Visner G, Hartwig JH, Falet H, Hoffmeister KM (2015) The Ashwell-Morell receptor regulates hepatic thrombopoietin production via JAK2-STAT3 signaling. Nat Med 21(1):47–54
47. Harrington WJ, Minnich V, Hollingsworth JW, Moore CV (1951) Demonstration of a thrombocytopenic factor in the blood of patients with thrombocytopenic purpura. J Lab Clin Med 38(1):1–10
48. Hellstrom-Lindberg E, Kanter-Lewensohn L, Nichol J, Ost A (1999) Spontaneous and cytokine-induced thrombocytopenia in myelodysplastic syndromes: serum thrombopoietin levels and bone marrow morphology. Scandinavian MDS Group, Sweden and Norway. Br J Haematol 105(4):966–973
49. Henter JI, Horne A, Arico M, Egeler RM, Filipovich AH, Imashuku S, Ladisch S, McClain K, Webb D, Winiarski J, Janka G (2007) HLH-2004: diagnostic and therapeutic guidelines for hemophagocytic lymphohistiocytosis. Pediatr Blood Cancer 48(2):124–131
50. Jandl JH, Aster RH, Forkner CE, Fisher AM, Vilter RW (1967) Splenic pooling and the pathophysiology of hypersplenism. Trans Am Clin Climatol Assoc 78:9–27

51. Janka GE (2007) Familial and acquired hemophagocytic lymphohistiocytosis. Eur J Pediatr 166(2):95–109
52. Kappler S, Ronan-Bentle S, Graham A (2014) Thrombotic microangiopathies (TTP, HUS, HELLP). Emerg Med Clin North Am 32(3):649–671
53. Kim MJ, Park PW, Seo YH, Kim KH, Seo JY, Jeong JH, Park MJ, Ahn JY (2014) Comparison of platelet parameters in thrombocytopenic patients associated with acute myeloid leukemia and primary immune thrombocytopenia. Blood Coagul Fibrinolysis 25(3):221–225
54. Kobayashi T, Terao T, Maki M, Ikenoue T (2001) Diagnosis and management of acute obstetrical DIC. Semin Thromb Hemost 27(2):161–167
55. Kurata Y, Fujimura K, Kuwana M, Tomiyama Y, Murata M (2011) Epidemiology of primary immune thrombocytopenia in children and adults in Japan: a population-based study and literature review. Int J Hematol 93(3):329–335
56. Kuter DJ (1996) The physiology of platelet production. Stem Cells 14(Suppl 1):88–101
57. Kuter DJ, Beeler DL, Rosenberg RD (1994) The purification of megapoietin: a physiological regulator of megakaryocyte growth and platelet production. Proc Natl Acad Sci U S A 91(23): 11104–11108
58. Kuter DJ, Bussel JB, Lyons RM, Pullarkat V, Gernsheimer TB, Senecal FM, Aledort LM, George JN, Kessler CM, Sanz MA, Liebman HA, Slovick FT, de Wolf JT, Bourgeois E, Guthrie TH Jr, Newland A, Wasser JS, Hamburg SI, Grande C, Lefrere F, Lichtin AE, Tarantino MD, Terebelo HR, Viallard JF, Cuevas FJ, Go RS, Henry DH, Redner RL, Rice L, Schipperus MR, Guo DM, Nichol JL (2008) Efficacy of romiplostim in patients with chronic immune thrombocytopenic purpura: a double-blind randomised controlled trial. Lancet 371(9610):395–403
59. Lapeyraque AL, Malina M, Fremeaux-Bacchi V, Boppel T, Kirschfink M, Oualha M, Proulx F, Clermont MJ, Le Deist F, Niaudet P, Schaefer F (2011) Eculizumab in severe Shiga-toxin-associated HUS. N Engl J Med 364(26):2561–2563
60. Lindenbaum J, Lieber CS (1969) Hematologic effects of alcohol in man in the absence of nutritional deficiency. N Engl J Med 281(7):333–338
61. Lo GK, Juhl D, Warkentin TE, Sigouin CS, Eichler P, Greinacher A (2006) Evaluation of pretest clinical score (4 T's) for the diagnosis of heparin-induced thrombocytopenia in two clinical settings. J Thromb Haemost 4(4):759–765
62. Loeffelbein F, Funk D, Nakamura L, Zieger B, Grohmann J, Siepe M, Kroll J, Stiller B (2014) Shear-stress induced acquired von Willebrand syndrome in children with congenital heart disease. Interact Cardiovasc Thorac Surg 19(6):926–932
63. Lu Q, Clemetson JM, Clemetson KJ (2005) Snake venoms and hemostasis. J Thromb Haemost 3(8):1791–1799
64. Makoni SN, Laber DA (2004) Clinical spectrum of myelophthisis in cancer patients. Am J Hematol 76(1):92–93
65. Matzdorff A, Giagounidis A, Greinacher A, Hiller E, Kiefel V, Muller-Beissenhirtz H, Ostermann H, Rummel M, Sachs UJ, Salama A (2010) Diagnosis and therapy of autoimmune thrombocytopenia. Recommendations of a joint Expert Group of DGHO, DGTI, DTH. Onkologie 33(Suppl 3):2–20
66. Mazzucconi MG, Fazi P, Bernasconi S, De Rossi G, Leone G, Gugliotta L, Vianelli N, Avvisati G, Rodeghiero F, Amendola A, Baronci C, Carbone C, Quattrin S, Fioritoni G, D'Alfonso G, Mandelli F, E. d. A. T. W. P. Gruppo Italiano Malattie (2007) Therapy with high-dose dexamethasone (HD-DXM) in previously untreated patients affected by idiopathic thrombocytopenic purpura: a GIMEMA experience. Blood 109(4):1401–1407
67. McCleary RJ, Kini RM (2013) Snake bites and hemostasis/thrombosis. Thromb Res 132(6): 642–646
68. McMillan R (2005) The role of antiplatelet autoantibody assays in the diagnosis of immune thrombocytopenic purpura. Curr Hematol Rep 4(2):160–165
69. McMillan R (2007) The pathogenesis of chronic immune thrombocytopenic purpura. Semin Hematol 44(4 Suppl 5):S3–S11

70. McMillan R, Wang L, Tani P (2003) Prospective evaluation of the immunobead assay for the diagnosis of adult chronic immune thrombocytopenic purpura (ITP). J Thromb Haemost 1(3): 485–491
71. McMorran BJ, Wieczorski L, Drysdale KE, Chan JA, Huang HM, Smith C, Mitiku C, Beeson JG, Burgio G, Foote SJ (2012) Platelet factor 4 and Duffy antigen required for platelet killing of Plasmodium falciparum. Science 338(6112):1348–1351
72. Mitchell WB, Bussel JB (2015) Thrombopoietin receptor agonists: a critical review. Semin Hematol 52(1):46–52
73. Monreal M, Lafoz E, Casals A, Ruiz J, Arias A (1991) Platelet count and venous thromboembolism. A useful test for suspected pulmonary embolism. Chest 100(6):1493–1496
74. Mougiakakos D, Machaczka M, Jitschin R, Klimkowska M, Entesarian M, Bryceson YT, Henter JI, Sander B, Le Blanc K (2012) Treatment of familial hemophagocytic lymphohistiocytosis with third-party mesenchymal stromal cells. Stem Cells Dev 21(17):3147–3151
75. Nahirniak S, Slichter SJ, Tanael S, Rebulla P, Pavenski K, Vassallo R, Fung M, Duquesnoy R, Saw CL, Stanworth S, Tinmouth A, Hume H, Ponnampalam A, Moltzan C, Berry B, Shehata N, G. International Collaboration for Transfusion Medicine (2015) Guidance on platelet transfusion for patients with hypoproliferative thrombocytopenia. Transfus Med Rev 29(1):3–13
76. Neame PB, Kelton JG, Walker IR, Stewart IO, Nossel HL, Hirsh J (1980) Thrombocytopenia in septicemia: the role of disseminated intravascular coagulation. Blood 56(1):88–92
77. Neunert C, Lim W, Crowther M, Cohen A, Solberg L Jr, Crowther MA, H. American Society of (2011) The American Society of Hematology 2011 evidence-based practice guideline for immune thrombocytopenia. Blood 117(16):4190–4207
78. Neylon AJ, Saunders PW, Howard MR, Proctor SJ, Taylor PR, G. Northern Region Haematology (2003) Clinically significant newly presenting autoimmune thrombocytopenic purpura in adults: a prospective study of a population-based cohort of 245 patients. Br J Haematol 122(6):966–974
79. Nishimura S, Nagasaki M, Kunishima S, Sawaguchi A, Sakata A, Sakaguchi H, Ohmori T, Manabe I, Italiano JE Jr, Ryu T, Takayama N, Komuro I, Kadowaki T, Eto K, Nagai R (2015) IL-1alpha induces thrombopoiesis through megakaryocyte rupture in response to acute platelet needs. J Cell Biol 209(3):453–466
80. Passos AM, Treitinger A, Spada C (2010) An overview of the mechanisms of HIV-related thrombocytopenia. Acta Haematol 124(1):13–18
81. Piketty C, Gilquin J, Kazatchkine MD (1994) Successful treatment of HIV-related thrombocytopenia with didanosine (ddI). J Acquir Immune Defic Syndr 7(5):521–522
82. Pina-Cabral JM, Ribeiro-da-Silva A, Almeida-Dias A (1985) Platelet sequestration during hypothermia in dogs treated with sulphinpyrazone and ticlopidine – reversibility accelerated after intra-abdominal rewarming. Thromb Haemost 54(4):838–841
83. Pizzigallo E, Sisto A, Iezzi A (1996) Efficacy of a combined zidovudine plus didanosine therapy in one case of HIV-related thrombocytopenia non-responder to zidovudine alone or to anti-Rho immunoglobulin administration. AIDS 10(14):1741–1742
84. Porcelijn L, Folman CC, de Haas M, Kanhai HH, Murphy MF, von dem Borne AE, Bussel JB (2002) Fetal and neonatal thrombopoietin levels in alloimmune thrombocytopenia. Pediatr Res 52(1):105–108
85. Provan D, Stasi R, Newland AC, Blanchette VS, Bolton-Maggs P, Bussel JB, Chong BH, Cines DB, Gernsheimer TB, Godeau B, Grainger J, Greer I, Hunt BJ, Imbach PA, Lyons G, McMillan R, Rodeghiero F, Sanz MA, Tarantino M, Watson S, Young J, Kuter DJ (2010) International consensus report on the investigation and management of primary immune thrombocytopenia. Blood 115(2):168–186
86. Richey ME, Gilstrap LC 3rd, Ramin SM (1995) Management of disseminated intravascular coagulopathy. Clin Obstet Gynecol 38(3):514–520
87. Riedl M, Orth-Holler D, Wurzner R (2014) An update on the thrombotic microangiopathies hemolytic uremic syndrome (HUS) and thrombotic thrombocytopenic purpura (TTP). Semin Thromb Hemost 40(4):413–415

88. Roy SK, Howard EW, Panza JA, Cooper HA (2010) Clinical implications of thrombocytopenia among patients undergoing intra-aortic balloon pump counterpulsation in the coronary care unit. Clin Cardiol 33(1):30–35
89. Saito H (1993) Thrombotic microangiopathy (DIC, TTP and HUS) – recent advances in pathogenesis and management. Nihon Rinsho 51(1):5–10
90. Scaradavou A (2002) HIV-related thrombocytopenia. Blood Rev 16(1):73–76
91. Schoonen WM, Kucera G, Coalson J, Li L, Rutstein M, Mowat F, Fryzek J, Kaye JA (2009) Epidemiology of immune thrombocytopenic purpura in the General Practice Research Database. Br J Haematol 145(2):235–244
92. Scully M (2012) Rituximab in the treatment of TTP. Hematology 17(Suppl 1):S22–S24
93. Scully M, Hunt BJ, Benjamin S, Liesner R, Rose P, Peyvandi F, Cheung B, Machin SJ, H. British Committee for Standards in (2012) Guidelines on the diagnosis and management of thrombotic thrombocytopenic purpura and other thrombotic microangiopathies. Br J Haematol 158(3):323–335
94. Selleng S, Malowsky B, Strobel U, Wessel A, Ittermann T, Wollert HG, Warkentin TE, Greinacher A (2010) Early-onset and persisting thrombocytopenia in post-cardiac surgery patients is rarely due to heparin-induced thrombocytopenia, even when antibody tests are positive. J Thromb Haemost 8(1):30–36
95. Shenkman B, Einav Y (2014) Thrombotic thrombocytopenic purpura and other thrombotic microangiopathic hemolytic anemias: diagnosis and classification. Autoimmun Rev 13(4–5):584–586
96. Shih A, Nazi I, Kelton JG, Arnold DM (2014) Novel treatments for immune thrombocytopenia. Presse Med 43(4 Pt 2):e87–e95
97. Stephan F, Thioliere B, Verdy E, Tulliez M (1997) Role of hemophagocytic histiocytosis in the etiology of thrombocytopenia in patients with sepsis syndrome or septic shock. Clin Infect Dis 25(5):1159–1164
98. Thiele T, Selleng K, Selleng S, Greinacher A, Bakchoul T (2013) Thrombocytopenia in the intensive care unit-diagnostic approach and management. Semin Hematol 50(3):239–250
99. Thielmann M, Bunschkowski M, Tossios P, Selleng S, Marggraf G, Greinacher A, Jakob H, Massoudy P (2010) Perioperative thrombocytopenia in cardiac surgical patients – incidence of heparin-induced thrombocytopenia, morbidities and mortality. Eur J Cardiothorac Surg 37(6):1391–1395
100. Triulzi DJ, Ness PM (1995) Intraoperative hemodilution and autologous platelet rich plasma collection: two techniques for collecting fresh autologous blood. Transfus Sci 16(1):33–44
101. Trottestam H, Horne A, Arico M, Egeler RM, Filipovich AH, Gadner H, Imashuku S, Ladisch S, Webb D, Janka G, Henter JI, Histiocyte S (2011) Chemoimmunotherapy for hemophagocytic lymphohistiocytosis: long-term results of the HLH-94 treatment protocol. Blood 118(17):4577–4584
102. Uygun A, Kadayifci A, Ercin N, Bagci S, Dagalp K (2000) Interferon treatment for thrombocytopenia associated with chronic HCV infection. Int J Clin Pract 54(10):683–684
103. van den Oudenrijn S, Bruin M, Folman CC, Bussel J, de Haas M, von dem Borne AE (2002) Three parameters, plasma thrombopoietin levels, plasma glycocalicin levels and megakaryocyte culture, distinguish between different causes of congenital thrombocytopenia. Br J Haematol 117(2):390–398
104. Vella MA, Jenner C, Betteridge DJ, Jowett NI (1988) Hypothermia-induced thrombocytopenia. J R Soc Med 81(4):228–229
105. Venkata C, Kashyap R, Farmer JC, Afessa B (2013) Thrombocytopenia in adult patients with sepsis: incidence, risk factors, and its association with clinical outcome. J Intensiv Care 1(1):9
106. Waldow HC, Westhoff-Bleck M, Widera C, Templin C, von Depka M (2014) Acquired von Willebrand syndrome in adult patients with congenital heart disease. Int J Cardiol 176(3):739–745
107. Wandt H, Schafer-Eckart K, Greinacher A (2014) Platelet transfusion in hematology, oncology and surgery. Dtsch Arztebl Int 111(48):809–815

108. Wang B, Nichol JL, Sullivan JT (2004) Pharmacodynamics and pharmacokinetics of AMG 531, a novel thrombopoietin receptor ligand. Clin Pharmacol Ther 76(6):628–638
109. Warkentin TE, Greinacher A (2003) Heparin-induced thrombocytopenia and cardiac surgery. Ann Thorac Surg 76(2):638–648
110. Warkentin TE, Kelton JG (2001) Delayed-onset heparin-induced thrombocytopenia and thrombosis. Ann Intern Med 135(7):502–506
111. Warkentin TE, Moore JC, Anand SS, Lonn EM, Morgan DG (2003) Gastrointestinal bleeding, angiodysplasia, cardiovascular disease, and acquired von Willebrand syndrome. Transfus Med Rev 17(4):272–286
112. Weitz JI, Hudoba M, Massel D, Maraganore J, Hirsh J (1990) Clot-bound thrombin is protected from inhibition by heparin-antithrombin III but is susceptible to inactivation by antithrombin III-independent inhibitors. J Clin Invest 86(2):385–391
113. Welp H, Ellger B, Scherer M, Lanckohr C, Martens S, Gottschalk A (2014) Heparin-induced thrombocytopenia during extracorporeal membrane oxygenation. J Cardiothorac Vasc Anesth 28(2):342–344
114. Wendling F, Tambourin P (1991) The oncogene V-MPL, a putative truncated cytokine receptor which immortalizes hematopoietic progenitors. Nouv Rev Fr Hematol 33(2):145–146
115. White J (1990) Thrombocytopenia after snake envenomation. Med J Aust 152(8):445–446
116. Wisner A, Braud S, Bon C (2001) Snake venom proteinases as tools in hemostasis studies: structure-function relationship of a plasminogen activator purified from Trimeresurus stejnegeri venom. Haemostasis 31(3–6):133–140
117. Zhang L, Zhou J, Sokol L (2014) Hereditary and acquired hemophagocytic lymphohistiocytosis. Cancer Control 21(4):301–312
118. Zucker-Franklin D, Cao YZ (1989) Megakaryocytes of human immunodeficiency virus-infected individuals express viral RNA. Proc Natl Acad Sci U S A 86(14):5595–5599

Megakaryocytes and Platelet Production During Viral Infection 16

Günther Schönrich and Martin J. Raftery

Abstract

It is now recognized that megakaryocytes and platelets have further functions beyond hemostasis and thrombosis. In particular they have been shown to play a role in infection, both in viral pathogenesis and in the host defense against viral infections. Although unproven it is very likely that viral immune-evasion mechanisms also function in platelets. Platelets can also serve as vehicles that bind and spread viruses within the infected organism. The intricate interplay between viruses on one side and megakaryocytes and platelets on the other is poorly understood. In fact many viruses that are pathogenic for humans cause thrombocytopenia, a drop in platelet numbers, and diminished platelet function resulting in thrombosis and bleeding. These include important viral pathogens with a worldwide distribution such as human immunodeficiency virus (HIV) and viral hemorrhagic fever (VHF) viruses. The mechanisms of HIV-induced thrombocytopenia have been studied quite extensively. In contrast, the mechanisms underlying thrombocytopenia after infection with VHF viruses are less well known. We comparatively describe what is known about how two quite different viruses, HIV and VHF-inducing hantaviruses, cause thrombocytopenia.

16.1 Introduction

In addition to hemostasis and thrombosis, it has become apparent that megakaryocytes and platelets can also react to pathogens and play an important role in both innate and adaptive immunity [37, 61] (Chap. 10). In fact, thrombosis/hemostasis

G. Schönrich (✉) • M.J. Raftery
Institute of Medical Virology, Helmut-Ruska-Haus, Charité – Universitätsmedizin Berlin, Charitéplatz 1, Berlin 10117, Germany
e-mail: guenther.schoenrich@charite.de

H. Schulze, J. Italiano (eds.), *Molecular and Cellular Biology of Platelet Formation*,
DOI 10.1007/978-3-319-39562-3_16

and immune function of platelets overlap to a certain extent although the response of platelets to immune versus thrombotic stimulation can be quite different [19, 53]. They have been shown to contribute to the host defense against viral infections and viral pathogenesis [4, 20, 69], and it is probable that viral immune-evasion mechanisms function in platelets. Quite often platelet-virus and megakaryocyte-virus interactions result in loss of platelets or diminished platelet function thereby contributing to virus-associated thrombosis and bleeding [25]. Viruses have been reported to hide within platelets in order to disseminate within the organism in a Trojan horse strategy. Numerous viruses modulate the function of megakaryocytes, and some viruses replicate in megakaryocytes or even platelets [63]. Platelets normally assist antiviral responses; however, in the context of specific infectious pathogens, they may also cause pathology. In order to understand virus-associated thrombocytopenia, the causative mechanisms need to be explored in detail for each individual virus.

Worldwide the most prominent viral pathogens known to be associated with platelet abnormalities are human immunodeficiency virus (HIV) and viral hemorrhagic fever (VHF) viruses. In humans HIV causes severe immune suppression that – if untreated – leads to acquired immunodeficiency syndrome (AIDS). HIV belongs to the subgroup *Lentivirinae* of the virus family *Retroviridae* and is a small and very compact enveloped virus. Besides regulatory proteins, it encodes structural proteins such as the envelope proteins gp120 and gp41. For entry into host cells, fusion of the viral envelope with the membrane of the host cells is required. For this purpose HIV interacts not only with CD4, its main receptor, but also with co-receptors such as CXCR4 and CCR5 [30]. Whereas only approximately 10 % of all HIV-infected individuals develop chronic thrombocytopenia, 60 % of patients with AIDS do so [22]. In some cases thrombocytopenia represents the first manifestation of HIV infection and may result in severe bleeding later on. In accordance with an antiviral function of platelets, platelet counts inversely correlate with viral load [54].

VHF comprises a group of distinct zoonotic pathogens that all typically cause fever and bleeding [71]. Platelets play a crucial role in the pathogenesis of VHF, which is defined as a syndrome by acute fever, vascular hyperpermeability, and reduced numbers of platelets [48]. VHF has pathological aspects in common with sepsis and systemic inflammatory response syndrome (SIRS) such as platelet dysfunction. In severe cases, disseminated intravascular coagulation (DIC), shock, and death are observed. Although VHF viruses belong to different virus families (*Flaviviridae*, *Bunyaviridae*, *Arenaviridae*, *Filoviridae* and *Rhabdoviridae*), they share some common features such as a single-stranded RNA genome and replication in the cytoplasm. Whereas humans represent dead-end hosts that are incidentally infected, the natural host range includes rodents and bats, with arthropods serving as vectors for some of these viruses. Consequently, their geographic distribution is determined by the habitats of their natural host. Intriguingly, all VHF viruses target certain cell types including professional antigen-presenting cells, i.e., dendritic cells (DCs) and macrophages. Other target cells are endothelial cells and cells of the hematopoietic system such as megakaryocytes and platelets. The VHF-inducing hantaviruses belong to the virus family *Bunyaviridae*. They are transmitted to humans by aerosols derived from persistently infected but asymptomatic natural hosts, i.e., small mammals such as rodents [18]. Hantavirus species in Europe and Asia cause hemorrhagic fever with renal syndrome (HFRS), whereas hantavirus species in the Americas are associated with hantavirus cardiopulmonary syndrome (HCPS). However, kidney failure and cardiopulmonary dysfunction can occur in both clinical syndromes although to different extents indicating that HFRS and HCPS are pathogenically closely related [15]. The case fatality rate of human hantavirus infection varies from less than 1 % to more than 50 % [31, 36]. Immunological mechanisms play a major role in hantavirus-induced disruption of the endothelial barrier [58, 59]. As with most other types of VHF, there are no specific antivirals or approved vaccines available for hantaviruses.

This book chapter is a comparative pandect of the current knowledge on how two very different viruses, HIV, a very prominent and extensively researched pathogen, and VHF-inducing hantaviruses, which are much less well known, decrease platelet numbers and function.

16.2 Viral Interference with Platelet Production

The platelet counts in healthy individuals range from 150,000 to 450,000 platelets per microliter. Symptoms such as petechiae, purpura, epistaxis, or bruising are first signs of abnormally low numbers of platelets, which is called thrombocytopenia. There are two different types of pathogenic cascades that lead to thrombocytopenia in the peripheral blood of virus-infected individuals. There can be either increased platelet destruction or decreased platelet production. Decreased platelet production may be a consequence of viral infection and subsequent elimination of hematopoietic cells such as hematopoietic progenitor cells (HPCs) in the bone marrow. In addition viruses may target the bone marrow megakaryocytes which, when mature, give rise to platelets. Finally, viral infection can modulate the microenvironment in the bone marrow in such a way that it no longer supports growth and survival of HPCs.

16.2.1 Viral Infection of HPCs

HPCs represent a heterogeneous cell population that comprises pluripotent CD34+ hematopoietic stem cells (HSCs) as well as more differentiated but still multipotent progenitor cells (MPP) derived from HSCs. In turn MPPs develop into common lymphoid progenitors (CLPs) or common myeloid progenitors (CMPs). The latter give rise to the megakaryocyte lineage that finally produces platelets (Chap. 1). Viruses may target HPCs that are subsequently eliminated thereby decreasing hematopoiesis in general, leading to pancytopenia. For example, deficiencies in all hematopoietic lineages are observed in some HIV-infected patients resulting not only in reduced platelet production but also in anemia and neutropenia [57]. Indeed depletion of HPCs is observed in HIV-infected individuals, in particular CD34+ CD4+ cells despite showing no sign of HIV infection [7]. Additionally, in vitro growth of HPCs derived from HIV-infected individuals is impaired [39, 72]. The fact that HPCs express the HIV receptors CD4, CCR5, and CXCR4 suggests that they are susceptible to HIV infection. However, only a low fraction of HPCs can be infected with HIV in vitro and only under certain experimental conditions [22]. Moreover, HIV is only detected in HPCs in a low percentage of HIV-infected patients [46]. Thus, it is unlikely that HIV infection of HPCs plays a major role in elimination of bone marrow-derived progenitor cells and the pancytopenia resulting thereof.

Other mechanisms may contribute to pancytopenia observed during HIV infection. For example, HIV-encoded proteins such as gp120 and trans-activator of transcription (Tat) protein sensitize even uninfected T cells to CD95-mediated apoptosis [9, 68]. The HIV-1-encoded gp120 molecule, a glycoprotein inserted in the viral envelope that is required for HIV entry into the host cells, is shed in relatively large amounts in the spleen and lymph nodes in vivo [56]. HIV-encoded Tat serves multiple functions in the viral life cycle and plays a critical role in viral replication. A soluble form of Tat is released from productively infected cells and is able to enter neighboring cells. HIV gp120 also induces apoptosis in HPCs [8, 41, 73]. Thus, soluble HIV-encoded proteins can also affect proliferation and survival in uninfected HPCs thereby contributing to decreased platelet production in HIV patients.

16.2.2 Viral Infection of Megakaryocytic Cells

It is possible that viruses infect more differentiated cells such as cells of the megakaryocytic lineage which, when mature, are responsible for the release of high numbers of platelets into the circulation. In fact megakaryocytes express the CD4 receptor for HIV [35]. Similarly, CD61 (β3 integrin), CD18 (β2 integrin), and CD55 (decay accelerating factor, DAF) which represent important receptors for pathogenic hantaviruses are also expressed on megakaryocytic cells [40, 50, 52]. It has been found that megakaryocytes isolated from HIV-infected individuals harbor viral genomes [75]. Moreover, HIV is able to replicate in megakaryocytic cells after infection in vitro [14, 55].

Pathogenic hantaviruses also target megakaryocytic cells, whereas apathogenic hantaviruses do not infect this cell type [40]. Intriguingly, after induction of differentiation, hantavirus replication in megakaryocytic cells is increased without changes in cell survival or differentiation, in contrast to HIV. This argues against the possibility that pathogenic hantaviruses affect platelet production by directly eliminating or functionally impairing differentiating megakaryocytic cells. However, expression of HLA class I molecules, the target structures for cytotoxic T cells (CTLs), is increased. This could lead to more efficient elimination of hantavirus-infected megakaryocytic cells by CTLs. On the other hand, increased expression of HLA class I molecules on megakaryocytic cells may inhibit attack by NK cells. Clinical observations argue against elimination of hantavirus-infected megakaryocytes by immune cells [16]. The levels of thrombopoietin (TPO), the main regulator of thrombopoiesis, as well as the immature platelet fraction and the mean platelet volume all point toward increased thrombopoiesis during HFRS. Thus, despite its presence in the bone marrow, pathogenic hantaviruses do not seem to interfere with platelet production, and hantavirus-induced thrombocytopenia is thus most likely caused by increased platelet destruction. However, the underlying mechanisms are unclear at the moment.

16.2.3 Viral Modulation of the Microenvironment in the Bone Marrow

Platelet production is not only decreased after viral infection and elimination of hematopoietic cells. It is also possible that viruses alter the microenvironment in the bone marrow in such a way that growth and development of hematopoietic cells is dysregulated (Chap. 4). Besides HPCs stromal cells at all stages of differentiation are found in the bone marrow. They represent a mixture of different cell types that migrate from the blood to the bone marrow such as endothelial cells, fibroblasts, osteoblasts, and osteoclasts. In addition, immune cells such as macrophages, DCs, and lymphocytes (both T and B) are found. These stromal cells produce important hematopoietic growth factors that regulate survival, proliferation, and differentiation of HSCs. Stromal cells from healthy individuals that have been infected with HIV in vitro are unable to support the growth of uninfected HPCs [6, 23, 60]. HIV infects bone marrow microvascular endothelial cells, an important element of the microenvironment in the bone marrow [47]. As pathogenic hantaviruses have a strong tropism for endothelial cells, they probably also infect endothelial cells in the bone marrow, although this has not yet been demonstrated. In addition to endothelial cells, HIV infects other cell types of the bone marrow stroma including macrophages, fibroblasts, and T cells. The virus-associated disruption of the normal cytokine milieu generated by the microenvironment in the bone marrow strongly contributes to the pancytopenia observed in HIV-infected patients. In contrast to HIV, pathogenic hantaviruses do not cause pancytopenia and, therefore, are unlikely to modulate the microenvironment in the bone marrow in such a way that proliferation and differentiation of HPCs is disturbed.

16.3 Virus-Induced Platelet Destruction

Virus-induced platelet destruction in the periphery is another mechanism besides decreased platelet production that can result in thrombocytopenia. Thrombocytopenic purpura in early HIV infection results from autoimmune destruction based on molecular mimicry [10, 26, 32], whereas at the late stage, thrombocytopenia is thought to be due to a defective thrombopoiesis. Molecular mimicry describes the situation in which a host protein shares epitope(s) with a protein encoded by an invading microbe. In essence, antibodies directed against HIV epitopes cross-react with glycoproteins on platelets and cause lysis. For example, antibodies directed against a variable epitope region within HIV-1 protein Nef also recognized GPIIIa49-66, an immunodominant epitope of the β3 (glycoprotein IIIa [GPIIIa]) integrin [38]. Interestingly, the anti-GPIIIa49-66 antibody isolated from HIV-infected individuals can inhibit megakaryocyte differentiation in vitro through β3 integrin signaling [49].

Signs of peripheral platelet destruction are observed in hantavirus-infected individuals [16] and in a humanized mouse model of hantavirus infection [33]. However, the underlying mechanism is unknown. Neutrophils can clear activated platelets from the circulation [42]. Stimulation of both neutrophils and platelets further increases the number of neutrophil-platelet adhesive events and results in large numbers of platelets binding to each single neutrophil [12]. Blocking of the β2 integrin complex CD11b/CD18 on neutrophils strongly reduces the percentage of neutrophils coated with platelets. Intriguingly, after blockade of both P-selectin and CD11b/18, the stimulated increase in the percentage of neutrophils binding platelets and the number of platelets bound per neutrophil is further reduced. Thus, sequestration of platelets in hantavirus-infected patients could be due to the following scenario: Hantaviruses bind to human platelets via β3 integrin [21]. Subsequently, the hantavirus-coated platelets could be bridged to phagocytic cells such as macrophages or neutrophils through β2 integrins, a receptor for pathogenic hantaviruses on leukocytes [52]. These bridged platelets could then be phagocytosed. In line with such a mechanism, in a mouse model of severe fever with thrombocytopenia syndrome (SFTS), which is caused by another member of the *Bunyaviridae*, SFTS virus adhered to platelets and facilitated phagocytosis and clearance of platelets by splenic macrophages [29]. However, the precise mechanism of hantaviruses-induced platelet destruction has yet to be elucidated.

16.4 Activation of Platelets During Viral Infections

Virus-induced activation of platelets per se can result in platelet consumption, i.e., disseminated intravascular coagulation, or in platelet sequestration in the spleen and liver followed by their clearance [28]. The latter involves adhesion of activated platelets to neutrophils and their subsequent internalization. Adhesion is dependent on CD62P (P-selectin), a marker of platelet activation, and on the β2 integrin complex CD11b/CD18, which allows more efficient binding to neutrophils, resulting in phagocytosis of bound platelets [43]. Adhesion of activated platelets causes neutrophil degranulation [43] or triggers the release of neutrophil extracellular traps [45].

How viruses activate platelets is mostly unclear. Platelets may be activated indirectly, i.e., by cytokines released after stimulation of other immune cells that sense pathogens, or directly by viral pathogen-associated molecular patterns that stimulate platelet-associated pattern recognition receptors (PRRs). In line with the latter view, platelets harbor many PRRs, mainly Toll-like receptors, similar to core immune cells such as DCs and macrophages but unlike erythrocytes [1, 3, 62, 74]. In fact platelets may represent prime pathogen sensors, especially in the light of their abundance and ubiquitous distribution. Lack of a nucleus would seem to preclude their function in platelets; however, it has been shown that platelets transfer PRRs to special compartments resulting in novel PRR organization and PRR signaling [65]. Moreover, platelets express FcγRIIa, which functions as a receptor for the Fc-portion of IgG and mediates platelet activation after binding IgG-coated pathogens [2]. Taken together, direct virus-induced platelet activation can be a conditio sine qua non for the survival of the infected host but at the same time may also contribute to pathogenesis [34].

Intriguingly, HIV can activate infected as well as uninfected platelets [13]. The soluble form of HIV-encoded Tat, which is released from productively infected cells, has been identified as a trigger of platelet activation [67]. Moreover, activated platelets are detected in HIV-infected individuals especially in AIDS patients. After antiretroviral therapy, almost complete normalization of enhanced platelet activation occurred [27, 44]. In the case of HIV, activated platelets have been demonstrated to perform antiviral functions. For example, activated platelets release the chemokine CXCL4 which in turn suppresses spread of HIV in cultured T cells and acts as a broad spectrum antiviral against HIV [5, 64].

It has been postulated that pathogenic hantaviruses prevent activation of platelets by binding and stabilizing inactive β3 integrin [21]. Moreover, platelets derived from patients with HFRS show decreased platelet aggregation in vitro [17]. Similarly, circulating platelets derived from HFRS patients showed decreased function as compared to circulating platelets from follow-up patients [16]. On the other hand, increased levels of soluble P-selectin and GPVI were observed in hantavirus-infected patients indicative of platelet activation in vivo [16]. It is unlikely, however, that hantavirus-exposed platelets are per se resistant to activation by direct or indirect mechanisms. Thus, it is possible that during hantavirus infection, platelets become refractory after stimulation in vivo and cannot subsequently be fully restimulated in vitro.

16.5 Platelets as Vehicles of Viral Dissemination

HIV binds to platelets through interaction of its gp120 with fibronectin on the surface of platelets [51]. Platelets express HIV co-receptors including CXCR4 [13]. In addition, dendritic cell-specific ICAM-grabbing non-integrin (DC-SIGN), a C-type lectin receptor that is expressed on both macrophages and dendritic cells, is detectable on platelets [11].

DC-SIGN acts as an attachment molecule for many viruses [30]. Moreover, after ligation of DC-SIGN by HIV on DCs, viral particles are internalized and transferred into an intracellular compartment, in which the virus remains infectious for several days. From this safe storage compartment, the virus can be relocated to the cell surface to productively infect susceptible cells such as CD4 T cells [66]. Such a reversible storage mechanism may also act in other DC-SIGN expressing cell types such as platelets and for other DC-SIGN binding viruses.

Platelets loaded with viral particles in safe storage compartments may act as carriers that help to disseminate viruses within the infected organisms. For example, analysis of platelets derived from HIV-infected patients has indeed demonstrated that HIV is internalized by platelets, thereby sheltering the virus particles from effector components of the host immune system [76]. Internalization exclusively occurs within platelets with an activated phenotype [70]. Whether pathogenic hantaviruses attach to DC-SIGN has not been reported. However, it is known that pathogenic hantaviruses bind to platelets via a β3 integrin-dependent mechanism [21, 24]. At the time being, it is unclear whether pathogenic hantaviruses are also internalized by platelets. Both HIV and pathogenic hantaviruses, however, do not productively replicate in platelets.

16.6 Concluding Remarks

Many life-threatening viral infections result in decreased peripheral blood platelet counts. The underlying mechanisms are multiple, however, and need to be elucidated for each viral infection. Thrombocytopenia plays a crucial role in the pathogenesis of VHF. Thus, it is of utmost importance to understand how VHF viruses, which belong to different virus families, decrease platelet numbers, especially as there are no specific antiviral drugs available for VHF treatment. It is likely but not yet proven that platelets are involved in host defense against VHF viruses. This would imply that platelets are activated prior to their consumption, destruction, or sequestration to fulfill their tasks as immune cells. In this light VHF-associated thrombocytopenia would be the consequence and end point of a massive immune response involving platelets which overlaps with the other important platelet functions such as hemostasis and thrombosis.

References

1. Andonegui G, Kerfoot SM, McNagny K, Ebbert KV, Patel KD, Kubes P (2005) Platelets express functional toll-like receptor-4. Blood 106:2417–2423
2. Arman M, Krauel K, Tilley DO, Weber C, Cox D, Greinacher A, Kerrigan SW, Watson SP (2014) Amplification of bacteria-induced platelet activation is triggered by FcgammaRIIA, integrin alphaIIbbeta3, and platelet factor 4. Blood 123:3166–3174
3. Aslam R, Speck ER, Kim M, Crow AR, Bang KW, Nestel FP, Ni H, Lazarus AH, Freedman J, Semple JW (2006) Platelet Toll-like receptor expression modulates lipopolysaccharide-induced thrombocytopenia and tumor necrosis factor-alpha production in vivo. Blood 107:637–641
4. Assinger A (2014) Platelets and infection – an emerging role of platelets in viral infection. Front Immunol 5:649
5. Auerbach DJ, Lin Y, Miao H, Cimbro R, Difiore MJ, Gianolini ME, Furci L, Biswas P, Fauci AS, Lusso P (2012) Identification of the platelet-derived chemokine CXCL4/PF-4 as a broad-spectrum HIV-1 inhibitor. Proc Natl Acad Sci U S A 109:9569–9574
6. Bahner I, Kearns K, Coutinho S, Leonard EH, Kohn DB (1997) Infection of human marrow stroma by human immunodeficiency virus-1 (HIV-1) is both required and sufficient for HIV-1-induced hematopoietic suppression in vitro: demonstration by gene modification of primary human stroma. Blood 90:1787–1798
7. Banda NK, Simon GR, Sipple JD, Terrell KL, Archer P, Shpall EJ, Akkina RK, Myers AM, Harrison GS (1999) Depletion of CD34+ CD4+ cells in bone marrow from HIV-1-infected individuals. Biol Blood Marrow Transplant 5:162–172
8. Banda NK, Tomczak JA, Shpall EJ, Sipple J, Akkina RK, Steimer KS, Hami L, Curiel TJ, Singer HG (1997) HIV-gp120 induced cell death in hematopoietic progenitor CD34+ cells. Apoptosis 2:61–68
9. Berndt C, Mopps B, Angermuller S, Gierschik P, Krammer PH (1998) CXCR4 and CD4 mediate a rapid CD95-independent cell death in CD4(+) T cells. Proc Natl Acad Sci U S A 95:12556–12561

10. Bettaieb A, Fromont P, Louache F, Oksenhendler E, Vainchenker W, Duedari N, Bierling P (1992) Presence of cross-reactive antibody between human immunodeficiency virus (HIV) and platelet glycoproteins in HIV-related immune thrombocytopenic purpura. Blood 80:162–169
11. Boukour S, Masse JM, Benit L, Dubart-Kupperschmitt A, Cramer EM (2006) Lentivirus degradation and DC-SIGN expression by human platelets and megakaryocytes. J Thromb Haemost 4:426–435
12. Brown KK, Henson PM, Maclouf J, Moyle M, Ely JA, Worthen GS (1998) Neutrophil-platelet adhesion: relative roles of platelet P-selectin and neutrophil beta2 (DC18) integrins. Am J Respir Cell Mol Biol 18:100–110
13. Chabert A, Hamzeh-Cognasse H, Pozzetto B, Cognasse F, Schattner M, Gomez RM, Garraud O (2015) Human platelets and their capacity of binding viruses: meaning and challenges? BMC Immunol 16:26
14. Chelucci C, Federico M, Guerriero R, Mattia G, Casella I, Pelosi E, Testa U, Mariani G, Hassan HJ, Peschle C (1998) Productive human immunodeficiency virus-1 infection of purified megakaryocytic progenitors/precursors and maturing megakaryocytes. Blood 91:1225–1234
15. Clement J, Maes P, Van RM (2014) Hemorrhagic fever with renal syndrome in the new, and hantavirus pulmonary syndrome in the old world: paradi(se)gm lost or regained? Virus Res 187:55–58
16. Connolly-Andersen AM, Sundberg E, Ahlm C, Hultdin J, Baudin M, Larsson J, Dunne E, Kenny D, Lindahl TL, Ramstrom S, Nilsson S (2015) Increased thrombopoiesis and platelet activation in hantavirus-infected patients. J Infect Dis 212:1061
17. Cosgriff TM, Lee HW, See AF, Parrish DB, Moon JS, Kim DJ, Lewis RM (1991) Platelet dysfunction contributes to the haemostatic defect in haemorrhagic fever with renal syndrome. Trans R Soc Trop Med Hyg 85:660–663
18. Elliott RM, Schmaljohn CS (2013) In: Knipe D, Howley P (eds) Fields Virology, 7th ed. Bunyaviridae. Lippincott Williams & Wilkins, Philadelphia, PA, USA, pp 1244–1282
19. Engelmann B, Massberg S (2013) Thrombosis as an intravascular effector of innate immunity. Nat Rev Immunol 13:34–45
20. Flaujac C, Boukour S, Cramer-Borde E (2010) Platelets and viruses: an ambivalent relationship. Cell Mol Life Sci 67:545–556
21. Gavrilovskaya IN, Gorbunova EE, Mackow ER (2010) Pathogenic hantaviruses direct the adherence of quiescent platelets to infected endothelial cells. J Virol 84:4832–4839
22. Gibellini D, Clo A, Morini S, Miserocchi A, Ponti C, Re MC (2013) Effects of human immunodeficiency virus on the erythrocyte and megakaryocyte lineages. World J Virol 2:91–101
23. Gill V, Shattock RJ, Scopes J, Hayes P, Freedman AR, Griffin GE, Gordon-Smith EC, Gibson FM (1997) Human immunodeficiency virus infection impairs hemopoiesis in long-term bone marrow cultures: nonreversal by nucleoside analogues. J Infect Dis 176:1510–1516
24. Goeijenbier M, Meijers JC, Anfasa F, Roose JM, van de Weg CA, Bakhtiari K, Henttonen H, Vaheri A, Osterhaus AD, van Gorp EC, Martina BE (2015) Effect of Puumala hantavirus infection on human umbilical vein endothelial cell hemostatic function: platelet interactions, increased tissue factor expression and fibrinolysis regulator release. Front Microbiol 6:220
25. Goeijenbier M, van Wissen M, van de Weg C, Jong E, Gerdes VE, Meijers JC, Brandjes DP, van Gorp EC (2012) Review: viral infections and mechanisms of thrombosis and bleeding. J Med Virol 84:1680–1696
26. Hohmann AW, Booth K, Peters V, Gordon DL, Comacchio RM (1993) Common epitope on HIV p24 and human platelets. Lancet 342:1274–1275
27. Holme PA, Muller F, Solum NO, Brosstad F, Froland SS, Aukrust P (1998) Enhanced activation of platelets with abnormal release of RANTES in human immunodeficiency virus type 1 infection. FASEB J 12:79–89
28. Hottz ED, Oliveira MF, Nunes PC, Nogueira RM, Valls-de-Souza R, Da Poian AT, Weyrich AS, Zimmerman GA, Bozza PT, Bozza FA (2013) Dengue induces platelet activation, mitochondrial dysfunction and cell death through mechanisms that involve DC-SIGN and caspases. J Thromb Haemost 11:951–962

29. Jin C, Liang M, Ning J, Gu W, Jiang H, Wu W, Zhang F, Li C, Zhang Q, Zhu H, Chen T, Han Y, Zhang W, Zhang S, Wang Q, Sun L, Liu Q, Li J, Wang T, Wei Q, Wang S, Deng Y, Qin C, Li D (2012) Pathogenesis of emerging severe fever with thrombocytopenia syndrome virus in C57/BL6 mouse model. Proc Natl Acad Sci U S A 109:10053–10058
30. Jolly CL, Sattentau QJ (2013) Attachment factors. Adv Exp Med Biol 790:1–23
31. Jonsson CB, Figueiredo LT, Vapalahti O (2010) A global perspective on hantavirus ecology, epidemiology, and disease. Clin Microbiol Rev 23:412–441
32. Karpatkin S, Nardi M, Lennette ET, Byrne B, Poiesz B (1988) Anti-human immunodeficiency virus type 1 antibody complexes on platelets of seropositive thrombocytopenic homosexuals and narcotic addicts. Proc Natl Acad Sci U S A 85:9763–9767
33. Kobak L, Raftery MJ, Voigt S, Kuhl AA, Kilic E, Kurth A, Witkowski P, Hofmann J, Nitsche A, Schaade L, Kruger DH, Schonrich G (2015) Hantavirus-induced pathogenesis in mice with a humanized immune system. J Gen Virol 96:1258–1263
34. Koupenova M, Vitseva O, MacKay CR, Beaulieu LM, Benjamin EJ, Mick E, Kurt-Jones EA, Ravid K, Freedman JE (2014) Platelet-TLR7 mediates host survival and platelet count during viral infection in the absence of platelet-dependent thrombosis. Blood 124:791–802
35. Kouri YH, Borkowsky W, Nardi M, Karpatkin S, Basch RS (1993) Human megakaryocytes have a CD4 molecule capable of binding human immunodeficiency virus-1. Blood 81:2664–2670
36. Krüger DH, Figueiredo LT, Song JW, Klempa B (2015) Hantaviruses – globally emerging pathogens. J Clin Virol 64:128–136
37. Leslie M (2010) Cell biology. Beyond clotting: the powers of platelets. Science 328: 562–564
38. Li Z, Nardi MA, Karpatkin S (2005) Role of molecular mimicry to HIV-1 peptides in HIV-1-related immunologic thrombocytopenia. Blood 106:572–576
39. Louache F, Henri A, Bettaieb A, Oksenhendler E, Raguin G, Tulliez M, Vainchenker W (1992) Role of human immunodeficiency virus replication in defective in vitro growth of hematopoietic progenitors. Blood 80:2991–2999
40. Lütteke N, Raftery MJ, Lalwani P, Lee MH, Giese T, Voigt S, Bannert N, Schulze H, Krüger DH, Schönrich G (2010) Switch to high-level virus replication and HLA class I upregulation in differentiating megakaryocytic cells after infection with pathogenic hantavirus. Virology 405:70–80
41. MacEneaney OJ, Connick E, DeSouza CA (2011) Effects of HIV-1 gp120 and protease inhibitors on apoptotic susceptibility of CD34+ hematopoietic progenitor cells. J Acquir Immune Defic Syndr 56:e49–e50
42. Manfredi AA, Rovere-Querini P, Maugeri N (2010) Dangerous connections: neutrophils and the phagocytic clearance of activated platelets. Curr Opin Hematol. 17(1):3–8. doi:10.1097/MOH.0b013e3283324f97. Review. PMID: 19770653
43. Maugeri N, Rovere-Querini P, Evangelista V, Covino C, Capobianco A, Bertilaccio MT, Piccoli A, Totani L, Cianflone D, Maseri A, Manfredi AA (2009) Neutrophils phagocytose activated platelets in vivo: a phosphatidylserine, P-selectin, and {beta}2 integrin-dependent cell clearance program. Blood 113:5254–5265
44. Mayne E, Funderburg NT, Sieg SF, Asaad R, Kalinowska M, Rodriguez B, Schmaier AH, Stevens W, Lederman MM (2012) Increased platelet and microparticle activation in HIV infection: upregulation of P-selectin and tissue factor expression. J Acquir Immune Defic Syndr 59:340–346
45. McDonald B, Urrutia R, Yipp BG, Jenne CN, Kubes P (2012) Intravascular neutrophil extracellular traps capture bacteria from the bloodstream during sepsis. Cell Host Microbe 12:324–333
46. Moses A, Nelson J, Bagby GC Jr (1998) The influence of human immunodeficiency virus-1 on hematopoiesis. Blood 91:1479–1495
47. Moses AV, Williams S, Heneveld ML, Strussenberg J, Rarick M, Loveless M, Bagby G, Nelson JA (1996) Human immunodeficiency virus infection of bone marrow endothelium reduces induction of stromal hematopoietic growth factors. Blood 87:919–925

48. Paessler S, Walker DH (2013) Pathogenesis of the viral hemorrhagic fevers. Annu Rev Pathol 8:411–440
49. Pan R, Wang J, Nardi MA, Li Z (2011) The inhibition effect of anti-GPIIIa49-66 antibody on megakaryocyte differentiation. Thromb Haemost 106:484–490
50. Philippeaux MM, Vesin C, Tacchini-Cottier F, Piguet PF (1996) Activated human platelets express beta2 integrin. Eur J Haematol 56:130–137
51. Pugliese A, Savarino A, Cantamessa C, Torre D (1996) Influence of fibronectin on HIV-1 infection and capability of binding to platelets. Cell Biochem Funct 14:291–296
52. Raftery MJ, Lalwani P, Krautkrämer E, Peters T, Scharffetter-Kochanek K, Krüger R, Hofmann J, Seeger K, Krüger DH, Schönrich G (2014) beta2 integrin mediates hantavirus-induced release of neutrophil extracellular traps. J Exp Med 211:1485–1497
53. Rex S, Beaulieu LM, Perlman DH, Vitseva O, Blair PS, McComb ME, Costello CE, Freedman JE (2009) Immune versus thrombotic stimulation of platelets differentially regulates signalling pathways, intracellular protein-protein interactions, and alpha-granule release. Thromb Haemost 102:97–110
54. Rieg G, Yeaman M, Lail AE, Donfield SM, Gomperts ED, Daar ES (2007) Platelet count is associated with plasma HIV type 1 RNA and disease progression. AIDS Res Hum Retroviruses 23:1257–1261
55. Sakaguchi M, Sato T, Groopman JE (1991) Human immunodeficiency virus infection of megakaryocytic cells. Blood 77:481–485
56. Santosuosso M, Righi E, Lindstrom V, Leblanc PR, Poznansky MC (2009) HIV-1 envelope protein gp120 is present at high concentrations in secondary lymphoid organs of individuals with chronic HIV-1 infection. J Infect Dis 200:1050–1053
57. Scadden DT, Zon LI, Groopman JE (1989) Pathophysiology and management of HIV-associated hematologic disorders. Blood 74:1455–1463
58. Schönrich G, Kruger DH, Raftery MJ (2015) Hantavirus-induced disruption of the endothelial barrier: neutrophils are on the payroll. Front Microbiol 6:222
59. Schönrich G, Rang A, Lutteke N, Raftery MJ, Charbonnel N, Ulrich RG (2008) Hantavirus-induced immunity in rodent reservoirs and humans. Immunol Rev 225:163–189
60. Schwartz GN, Kessler SW, Szabo JM, Burrell LM, Francis ML (1995) Negative regulators may mediate some of the inhibitory effects of HIV-1 infected stromal cell layers on erythropoiesis and myelopoiesis in human bone marrow long term cultures. J Leukoc Biol 57:948–955
61. Semple JW, Italiano JE Jr, Freedman J (2011) Platelets and the immune continuum. Nat Rev Immunol 11:264–274
62. Shiraki R, Inoue N, Kawasaki S, Takei A, Kadotani M, Ohnishi Y, Ejiri J, Kobayashi S, Hirata K, Kawashima S, Yokoyama M (2004) Expression of toll-like receptors on human platelets. Thromb Res 113:379–385
63. Simon AY, Sutherland MR, Pryzdial EL (2015) Dengue virus binding and replication by platelets. Blood 126:378–385
64. Solomon TT, Gnirss K, Rahe-Meyer N, Kiene M, Kramer-Kuhl A, Behrens G, Munch J, Pohlmann S (2013) Platelet activation suppresses HIV-1 infection of T cells. Retrovirology 10:48
65. Thon JN, Peters CG, Machlus KR, Aslam R, Rowley J, Macleod H, Devine MT, Fuchs TA, Weyrich AS, Semple JW, Flaumenhaft R, Italiano JE Jr (2012) T granules in human platelets function in TLR9 organization and signaling. J Cell Biol 198:561–574
66. van Kooyk Y, Geijtenbeek TB (2003) DC-SIGN: escape mechanism for pathogens. Nat Rev Immunol 3:697–709
67. Wang J, Zhang W, Nardi MA, Li Z (2011) HIV-1 Tat-induced platelet activation and release of CD154 contribute to HIV-1-associated autoimmune thrombocytopenia. J Thromb Haemost 9:562–573
68. Westendorp MO, Frank R, Ochsenbauer C, Stricker K, Dhein J, Walczak H, Debatin KM, Krammer PH (1995) Sensitization of T cells to CD95-mediated apoptosis by HIV-1 Tat and gp120. Nature 375:497–500
69. Yeaman MR (2014) Platelets: at the nexus of antimicrobial defence. Nat Rev Microbiol 12:426–437

70. Youssefian T, Drouin A, Masse JM, Guichard J, Cramer EM (2002) Host defense role of platelets: engulfment of HIV and Staphylococcus aureus occurs in a specific subcellular compartment and is enhanced by platelet activation. Blood 99:4021–4029
71. Zapata JC, Cox D, Salvato MS (2014) The role of platelets in the pathogenesis of viral hemorrhagic fevers. PLoS Negl Trop Dis 8:e2858
72. Zauli G, Re MC, Davis B, Sen L, Visani G, Gugliotta L, Furlini G, La PM (1992) Impaired in vitro growth of purified (CD34+) hematopoietic progenitors in human immunodeficiency virus-1 seropositive thrombocytopenic individuals. Blood 79:2680–2687
73. Zauli G, Vitale M, Gibellini D, Capitani S (1996) Inhibition of purified CD34+ hematopoietic progenitor cells by human immunodeficiency virus 1 or gp120 mediated by endogenous transforming growth factor beta 1. J Exp Med 183:99–108
74. Zhang S, Zhang S, Hu L, Zhai L, Xue R, Ye J, Chen L, Cheng G, Mruk J, Kunapuli SP, Ding Z (2015) Nucleotide-binding oligomerization domain 2 receptor is expressed in platelets and enhances platelet activation and thrombosis. Circulation 131:1160–1170
75. Zucker-Franklin D, Cao YZ (1989) Megakaryocytes of human immunodeficiency virus-infected individuals express viral RNA. Proc Natl Acad Sci U S A 86:5595–5599
76. Zucker-Franklin D, Seremetis S, Zheng ZY (1990) Internalization of human immunodeficiency virus type I and other retroviruses by megakaryocytes and platelets. Blood 75:1920–1923

Diagnosis of Inherited and Acquired Platelet Disorders

17

Tamam Bakchoul and Harald Schulze

Abstract

In this chapter we will focus on current laboratory methods to identify platelet function tests to identify the underlying (genetic) defect, allowing to apply the best possible treatment options. There are many reviews written on platelet function assays, especially on drug-induced defects or on drug monitoring, as used in many point-of-care applications [1–3]. Here, we will focus less on the methods underlying the distinct platelet function assays, but rather discuss methods to identify inherited thrombocytopathies.

17.1 Aim of This Chapter

In this chapter we will focus on current laboratory methods to identify platelet function tests to identify the underlying (genetic) defect, allowing to apply the best possible treatment options. There are many reviews written on platelet function assays, especially on drug-induced defects or on drug monitoring, as used in many point-of-care applications [1–3]. Here, we will focus less on the methods underlying the distinct platelet function assays, but rather discuss methods to identify inherited thrombocytopathies.

T. Bakchoul
Center for Clinical Transfusion Medicine, Tübingen, Germany

H. Schulze (✉)
Experimental Biomedicine, University Hospital Würzburg,
Josef-Schneider-Str. 2, D15, Würzburg 97080, Germany
e-mail: harald.schulze@uni-wuerzburg.de

H. Schulze, J. Italiano (eds.), *Molecular and Cellular Biology of Platelet Formation*,
DOI 10.1007/978-3-319-39562-3_17

17.2 Overview: Prerequisites and Preanalytics

The platelet concentration in peripheral blood is an essential parameter and is provided by all automated blood cell counters. Even with small counters, red blood cells, leukocytes, and platelets are measured. The latter cell type, however, is prone to false results, mostly, as platelets become activated when in contact with an injured vessel, which is intrinsic to a typical blood withdrawal at the antecubital vein. In addition, fully pre-evacuated tubes like Vacutainers® force the withdrawn blood against the tube wall before resuspension in anticoagulants. This becomes of special interest when blood of neonatal or young pediatric patients is analyzed, as blood is typically withdrawn by dropping blood into these tubes from cranial veins or veins on the back of hands or feet. While it is well established that platelet numbers and function are different in neonates, newborn, and children [4] and Chap. 12, the lack of adequate healthy controls makes it difficult to interpret results in these patients.

Classical anticoagulants like EDTA or citrate, prefilled as concentrated solutions into the collecting tubes, might interact with platelet surface receptors and can cause agglutination and activation. This is not uncommon when EDTA tubes are used. Pseudo-thrombocytopenia can readily be excluded, when a blood smear from that tube is stained and platelet aggregates are present.

In addition, when platelet function is to be analyzed, preanalytics become even more critical [5, 6]. Tubes should not be stored in the cold (fridge), as this causes receptor clustering and can lead to preactivation of platelets [7, 8]. This activation can be masked, when tubes are transported across campus or towns and subjected to extreme changes in temperature.

When screening tests like the PFA-100, VerifyNow, Cone or Plate Analyzer, etc., imply a bleeding defect, especially due a prolonged closure time, non-platelet-mediated causes have to be excluded. Many drugs cause a platelet function defect. Indeed, an “aspirin-like defect” attributes an observed phenotype to the intake of this medication. Of note, physicians need to specifically ask for intake of acetyl salicylic acid (ASA), indomethacin, or other non-steroidal anti-inflammatory drugs, as patients often do not consider these medications as “drugs.” Platelets are anucleated and therefore not able of protein resynthesis. Drugs like ASA irreversibly inhibit enzymes for the rest of the platelet’s life span. Thus, for optimal platelet function testing, patients should avoid these drugs for at least 7–10 days.

In addition, some foods or food supplements have an impact on platelet function. Lists with commonly used food are available and should be shown in case of an unknown bleeding cause.

In addition, any coagulation-based bleeding diathesis needs to be excluded. While the complete lack of most clotting factor is readily recognized by using routine assays, more complex alterations of several factors simultaneously requires more substantiated testing.

Bleeding can also be due to altered vWF expression or function which needs to be excluded. Multimer analysis is typically performed in specialized laboratories [9, 10].

Bleeding scores are used to evaluate and grade the severity. Standardized sheets can be highly informative but require compliance by both physician and patient. Detailed questions allow a better differential diagnosis but also request more time, in an often tightly packed clinical day. In Germany, bleeding scores are adopted from questionnaires for diagnosis of vWD or the pQB [11]. The human phenotype ontology (HPO) coding has recently been used to standardize hematological and non-hematological features [12]. This score is supposed to help in identifying patients sharing similar phenotypes for future studies [13]. However, a full HPO coding is quite time consuming and still dependent on the reporting physician or scientist. Nevertheless, standardized bleeding scores will be essential in the future to identify similarities in databases and registries within countries or in international databases like the BRIDGE-BPD consortium [13].

17.3 Investigations for Platelet Count and Morphology

The initial laboratory investigation should include an automated complete blood count and peripheral blood smear analysis, which can guide further laboratory investigations. As indicated above, pseudo-thrombocytopenia must be excluded. Furthermore, the overall appearance of the platelet morphology, especially the size and the staining pattern, is readily recognized, even by less experienced physicians and investigators.

17.3.1 Blood Count Analyzer

Automated impedance and optical platelet-counting methods provide accurate platelet counts within a broad range. However, less accuracy can be produced when platelet counts are $<50 \times 10^9/L$ or most importantly when platelets are significantly large. In these situations, manual counting using a hemocytometer or immuno-counting using flow cytometer should be considered [14]. It is important to remember that neither the absence nor the presence of thrombocytopenia does rule out an additional platelet dysfunction. Some inherited platelet disorders are characterized by both decreased platelet numbers and abnormal function [15].

17.3.2 The Mean Platelet Volume (MPV)

Most automated cell counters provide MPV of measured platelets in whole blood samples. Of note, MPV is strongly dependent on the used machine, and this parameter should always be communicated with the associated reference range. MPV is further influenced by the type of blood sample collection (citrated or EDTA anticoagulant) and storage conditions (duration and temperature) [14]. Assuming that these are standardized, obtaining an accurate MPV can still be problematic for samples with macrothrombocytopenia, as very large platelets might not be detected in the correct gate. Evaluation of the platelet diameter using Wright's or May-Grünwald-Giemsa staining of blood smear is very helpful in this case. Of note, storage of blood samples anticoagulated with EDTA for longer than 24 may increase platelet volume.

17.3.3 Evaluation of Blood Smears

Platelet morphology can be easily assessed using a Wright- or May-Grünwald-Giemsa-stained blood smear. This method provides additional information about platelet number, size, clumping, and granularity, as well as the morphology of erythrocytes and granulocytes. The platelet diameter can be quantified using ocular micrometer or computer-assisted image analysis. A mean platelet diameter (MPD) >3.7 μm indicates an inherited thrombocytopenia with giant platelets, e.g., biallelic Bernard-Soulier syndrome and MYH9-related disorders (Fig. 17.1a). In contrast, an MPD <2.6 μm is suggestive for thrombocytopenia with X-chromosomal-linked thrombocytopenia (XLT), and Wiskott-Aldrich syndrome (WAS). Therefore, MPD can be helpful to distinguish inherited thrombocytopenia from ITP. The intensity of platelet staining is also important for the interoperation of platelet morphology. Pale appearance can be caused by a reduced number of alpha granules and indicates alpha storage pool deficiencies (e.g., gray platelet syndrome). Also, the white blood cell morphology requires attention, because small blue staining inclusion bodies in polymorphonuclear leukocytes suggest strongly for MYH9-related thrombocytopenia.

17.3.4 Immunofluorescence Microscopy

The diagnosis of hereditary thrombocytopenias and platelet function disorders is overall challenging, and recent studies suggest that about half of patients with a manifest bleeding diathesis remain without a substantiated diagnosis, despite extensive laboratory platelet function testings [16]. Especially in pediatric patients, the required blood volume is often problematic. Immunofluorescence microscopy is a promising technology that facilitates the diagnosis of hereditary platelet disorders by combining basic and widely available preanalytical methods with centralized cutting-edge morphological techniques.

Blood smears are prepared by standard method, air-dried, and shipped by regular mail. In the reference laboratory, blood smears are fixed and permeabilized before incubation with a panel of antibodies against specific proteins affected in various platelet disorders. The choice of the antigen panel to be tested is driven by a few basic elements: (a) platelet size and possible other information derived from examination of blood cell morphology and (b) key features of patients' personal and family history. Antibody binding is then detected by fluorophore-labeled secondary antibodies. Blood smears are examined by immunofluorescence microscopy, and the expression of the protein of interest is assessed in comparison to a normal control sample, stained in parallel. Using this method, expression of the functionally relevant membrane glycoprotein complexes GPIIb/IIIa or GPIb/V/IX can be investigated for the diagnosis of Glanzmann's thrombasthenia and Bernard-Soulier syndrome, respectively (Fig. 17.1b). In addition, abnormal expression of alpha or delta granule markers may direct the follow-up step in the diagnostic algorithm to test for alpha or delta storage pool diseases, respectively. Finally, subcellular localization of non-muscle myosin heavy chain IIA in granulocytes provides an important finding for the diagnosis of MYH9 disorders [17], and ß1-tubulin-related macrothrombocytopenia is typically characterized by the absence of this ß-tubulin isoform from the platelet marginal

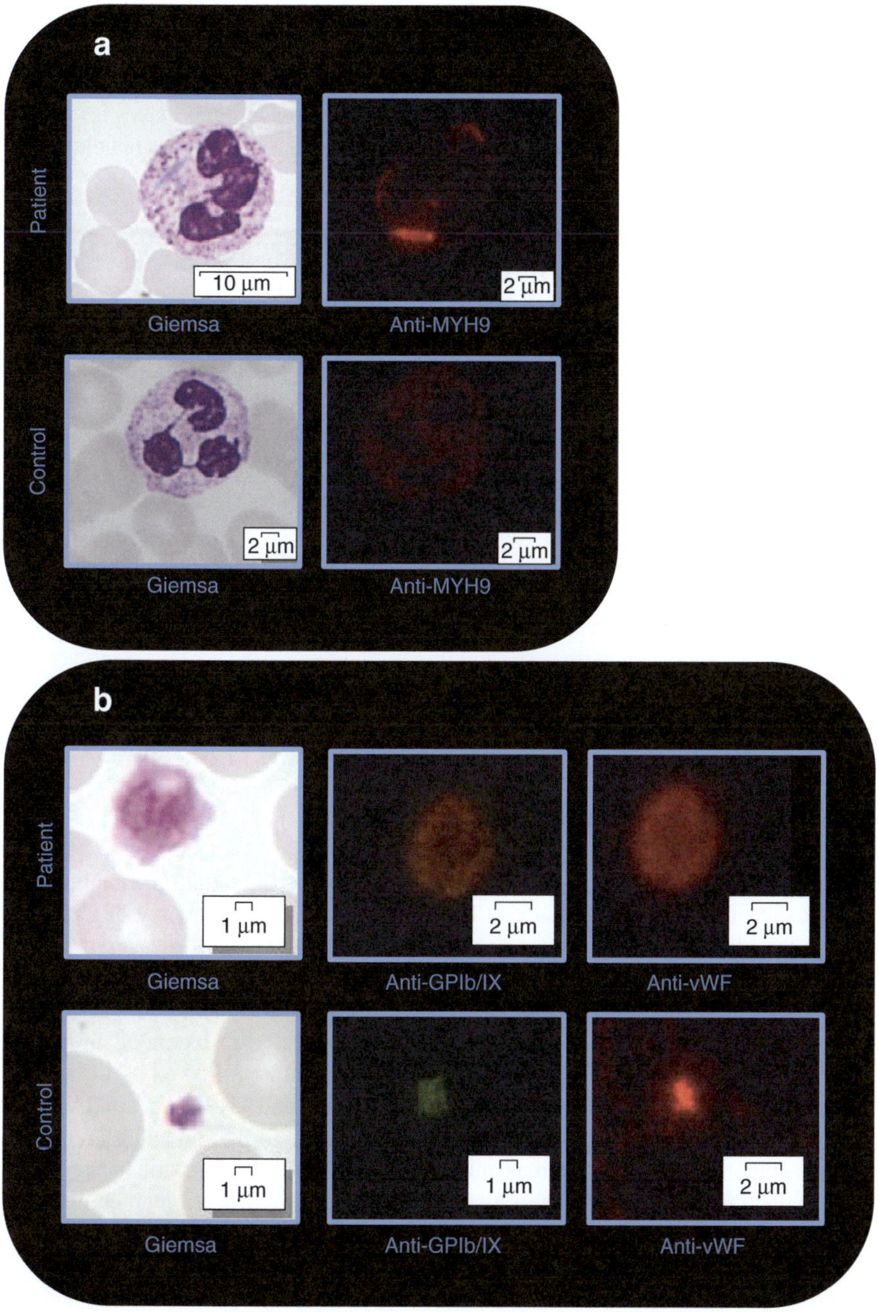

Fig. 17.1 Light microscopy and immunofluorescence features of hereditary thrombocytopenias. (**a**) The slide shows the blood smear of a patient with MYH9-related thrombocytopenia. In the granulocytes inclusion bodies consisting of non-muscle myosin heavy chain IIA, stain light blue. The inclusion bodies are much better visualized by immunofluorescence (*red*). (**b**) Platelet from a patient with Bernard-Soulier Syndrome is shown. For immunofluorescence, platelets stained positive for von Willebrand factor (*red*) but not for glycoprotein Ibα (*green*)

zone [18]. The biggest advantage of this method is that the preanalytical requirements for making a blood smear are minimal and widely available. Subsequent techniques, namely, fixation procedures, platelet permeabilization, and use of appropriate antibodies, can be performed at specialized centers. This approach makes it feasible to provide testing for a large proportion of known platelet disorders to neonates and young children due to the minimal sample requirements.

17.4 Platelet Function Assays

Platelet function defects can be assessed using several distinct laboratory tests. Many of these assays determine one or more of the biological functions: adhesion, secretion, or aggregation, either as an endpoint or as a dynamic measurement.

17.4.1 Screening Assays

Several methods have been introduced for the use as screening tests in routine laboratories such as PFA-100 and PFA-200, VerifyNow, Plateletworks, Impact-CPA, Placar PRT-7000, thromboelastography (TEG), and thromboelastometry (Rotem). While these assays are easy to perform with very short turnaround time, they lack the specificity and are very labile and sensitive for hematocrit, platelet count, etc. In addition, vWD should also be considered as a potential reason for prolonged closure time in PFA-100.

17.4.2 Light Transmission Aggregometry (LTA)

This method measures in a real-time fashion platelet aggregation enabling the assessment of several key functions. The change in light transparency of a rapidly stirred sample of citrated platelet-rich plasma (PRP) at 37 °C is measured photometrically and recorded. The curves derived by LTA can be divided into three major stages (Fig. 17.2) [1]: shape change, the addition of agonists leads to a change in platelet form from a discoid to a more spherical shape, with extended filopodia, resulting in a transient, small decrease in light transmission, followed by [2] a "first wave," reversible aggregation (fibrinogen-dependent), and [3] the "second wave," which indicates an irreversible aggregation due to TxA2 formation and secretion of granule contents. Typically, the maximum increase in light transmission (% aggregation) is documented. Platelet agglutination stimulated by ristocetin, which changes the conformation of plasma vWF allowing it to bind to GPIbα, can also be measured and allows the identification of Bernard-Soulier syndrome or subtypes of vWD. LTA is currently considered to be the best practice ("gold standard") for individual clinical laboratories and an achievable goal. However, this method is significantly impacted by differences in sample collection and preparation. Standardization of LTA has recently been addressed in published guidelines and recommendations [19]. Moreover, due to the overall large volume required, it is not applicable for young children or patients with a profound thrombocytopenia. Standardizing preanalytical and analytical variables can improve the consistency of results (Table 17.1).

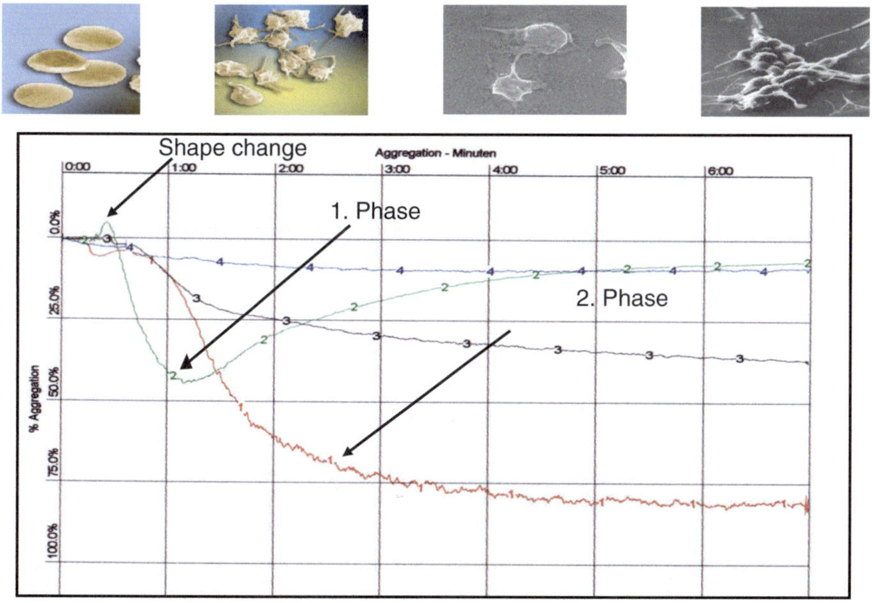

Fig. 17.2 Light Transmission Aggregometry (LTA). The figure illustrates an example for aggregation curves from a patient with storage pool disease (*green*) compared to healthy control (*red curve*). Aggregation was induced by ADP. Blue line is the curve in absence of any agonist. Normal curves derived by LTA can be divided into three major stages: shape change, first and second wave

Table 17.1 Selected suggestions to improve the performance of light transmission aggregometry

Preanalytical	No medications that affect platelet function for 7–10 days
	Blood sampling after a short period of rest, refrained from smoking or caffeine
	Blood collection by atraumatic venipuncture with minimal tourniquet pressure into buffered citrate anticoagulant
	Do not use first few mLs of the drawn blood
Preparation	PRP: centrifuging of whole blood samples at 200 g for 10 min without brake
	PRP should be allowed to rest at room temperature for 15 min before testing
Analysis	Testing should be completed within 4 h of blood collection
	After addition of the agonist, the aggregation response should be monitored for a minimum of 3–5 min
	A standard panel with agonist concentrations should be chosen
	Basic lists should include at least ADP, collagen, epinephrine, arachidonic acid, and ristocetin
	In case of abnormal finding, an extended agonist panel should be used to retest with higher concentrations or investigate additional agonists such as thromboxane mimetics, thrombin receptor activation peptides, or convulxin
Interpretation	Aggregation curves should be examined for the quality of the baseline, the presence of shape change, the length of the lag phase, the slope of the aggregation curve, the maximum percent aggregation, and the presence of deaggregation

17.4.3 Whole Blood Aggregometry (WBA)

Platelet function is measured as a change in electrical impedance between two electrodes due to platelet adhesion and aggregation in response to agonists. Adherent platelets increase the impedance, which is displayed as an aggregation curve. The most important advantages of WBA are that less amounts of blood are required and that there is less manipulation of the sample. Fully automated aggregometers with disposable electrodes are also available (Multiplate analyzer). Agonist responsiveness differs between LTA and WBA, and direct comparison studies of the two methodologies are still required [20, 21].

17.4.4 ATP Release

Recent studies indicate that LTA alone is not sensitive enough to identify all patients with secretion defects [22]. Therefore, measurement of released nucleotides from platelet dense granules is usually performed as an adjunct to aggregation studies using a lumino-aggregometer, which detects light emitted from freshly released ATP reacting with the bioluminescent reagent luciferin/luciferase [23, 24]. A decreased ATP release indicates abnormalities of platelet dense granule number, content, or secretion dynamic. However, abnormal results do not differentiate between abnormalities of dense granule per se and disorders of secretion. Semiquantification of platelet dense granules can be determined by immunofluorescence microscopy to identify patients with decreased ATP release secondary to decreased numbers of dense granules. Platelet counts and size play an important role in this assay. Alternative methods for the measurement of total nucleotide content and release can be performed by luminometry or high-performance liquid chromatography. Uptake and release of mepacrine or radiolabeled serotonin can also be used to evaluate dense granule secretion.

17.4.5 Flow Cytometry

This method is essential to confirm or rule out the diagnosis of Bernard-Soulier syndrome or Glanzmann's thrombasthenia. Specific monoclonal antibodies are used to quantitate the appropriate membrane glycoprotein receptor density (GPIb/IX/V and GPIIb/IIIa, respectively). Flow cytometry can also be employed to identify quantitative abnormalities of receptors for collagen (GPVI, integrin $\alpha2\beta1$), fibronectin (integrin $\alpha5\beta1$), or laminin (integrin $\alpha6\beta1$) or to measure the exposure of anionic phospholipids like phosphatidylserine as a pro-coagulative surface by adding fluorophore-labeled annexin V [24, 25].

The activation of platelets by standardized agonists can also be measured by flow cytometry. Most guidelines recommend ADP and TRAP with one suboptimal and an optimal concentration each [6, 26]. Further agonists like collagen-related peptide or the thromboxane analog U46619 (alone or in combination with ADP) are typically used when the initial screening was not informative. Three readouts are readily possible: (a) the fluorophore-labeled antibody PAC-1 recognizes only the activated GPIIb/IIIa receptor and provides information on inside-out signaling of this major integrin. (b) CD62P (P-selectin) is expressed on the membrane of alpha granules and almost absent on the surface of resting platelets. Agonists typically cause a release of granules and this increase is measured. (c) Expression of the lysosomal marker protein CD63 is a marker for lysosomal and granule storage or release defects.

The storage pool disorders can often be identified by a defective loading or release of the dye mepacrine which binds preferentially to dense granules but also to uncharacterized membranes. The decrease is measured after an indicated time (endpoint) or followed kinetically over 5–10 min. Similarly, signal transduction defects that affect a reduced calcium influx can be measured by loading platelets with dyes that bind selectively to divalent cations like Fluo-3 or Fura-2. In general, these assays are mostly restricted to specialized laboratory, although the basic flow cytometric assay is a robust assay. However, recording of data is often more complicated, as several procedures (MFI values, percentage positive cells over threshold or a semiquantitative comparison to standardized beads) are used that are difficult to compare.

17.5 Antibody Binding Assays

17.5.1 Laboratory Assays for Platelet-Reactive Allo- and Autoantibodies

Laboratory investigations for antibody-mediated thrombocytopenia involve identification and classification of antibodies present in patient sera or attached to patient platelets. Table 17.2 summarizes the available types of platelet antibody testing and applications in disorders such as neonatal alloimmune thrombocytopenia, post-transfusion purpura, multiple platelet transfusion refractoriness, immune thrombocytopenia, and drug-induced thrombocytopenia. For more information of the pathophysiology and clinical manifestation of these diseases, please refer to Chap. 15.

17.5.2 Laboratory Assays for Heparin-Induced Thrombocytopenia (HIT)

As HIT can often not be excluded on clinical grounds, laboratory testing for heparin-dependent antibodies is a mainstay of the diagnosis (see Chap. 15). Two different classes of assays are currently available: immunoassays that detect binding of anti-PF4/heparin antibodies and functional laboratory assays that investigate the capability of antibodies in the patient's serum/plasma to activate platelets in the presence of heparin.

Table 17.2 Methods used to test for platelet antibodies

Test	Description	Application	Advantages	Disadvantages
Indirect platelet antibody tests				
Flow cytometry	Uses whole platelet from donors with known HPAs	Identify platelet specific antibodies in patient serum or plasma	Detects some antibodies that cannot be detected by glycoprotein-specific tests More sensitive for labile and low density antigens	Not recommended for ITP Results reported qualitatively Does not identify specific antibody target
MAIPA	Glycoprotein specific (specific for antibodies to HPAs) Uses captured platelet glycoproteins from lysed platelets as source of known HPAs	Preferred for antibody identification in NAIT (using maternal serum)	No reactivity from non-HPA antibodies (HLA, blood group)	Clinical utility of antibody titer is controversial Commercial kits detect only common platelet antibodies Low titer and low avidity antibodies may not be detected
Direct platelet antibody tests				
Flow cytometry	Uses whole platelet from patient	Identify antibodies on patient platelets for diagnosis of ITP	Easy to perform High sensitivity	Detects any platelet associated antibody Low specificity, often positive in non-ITP
MAIPA	Detects glycoprotein-specific antibody by eluting bound antibody from patient platelets or using solubilized patient platelets with bound autoantibodies	Identify antibodies bound to platelet specific glycoproteins for diagnosis of ITP	More sensitive and specific than indirect tests for autoantibody detection	Sensitivity is compromised by severe thrombocytopenia

17.5.3 Immunoassays

In order to detect the pathogen antibodies responsible for HIT, the target antigen (complexes of PF4 and heparin or polyanion) is bound onto a solid phase, e.g., microtiter plate wells in enzyme immune absorbent assays (EIA assays) or microparticles in bead-based assays, to which patient's serum or plasma is added. In EIAs, a color change is used to detect anti-PF4/heparin antibodies binding in a semiquantitative fashion. The intensity of color change is proportional to the amount of bound antibodies and measured as optical density (OD). Although ELISAs have excellent negative predictive value to rule out HIT, their specificity is lower (40–80 %), depending on the assay and patient cohorts, accounting for the high rate of false-positive results. Several strategies have been suggested to reduce overdiagnosis (false-positive results) of HIT. These include (1) the use of ELISAs that exclusively detect IgG antibodies, (2) the implementation of an inhibition step using a high heparin concentration, and (3) the consideration of the OD magnitude. These approaches have been shown to increase of the predictive capability of the IgG ELISA for the presence of platelet-activating (clinically relevant) antibodies. However, this advantage varied among the studies, probably due to differences in study design and patient demography (for more details, see review [27]).

Another drawback of most immunoassays is that they cannot detect antibodies directed against heparin-dependent antigens other than PF4, with the exception of those which use platelet lysate. Some of the platelet-activating antibodies against other chemokines such as interleukin-8 (IL-8) or neutrophil-activating peptide 2 (NAP-2) may induce symptoms resembling clinical HIT [28].

Particle-based immunoassays detect anti-PF4/heparin antibodies either through the agglutination of particles, which can be evaluated visually in the gel centrifugation assay PaGIA, the membrane filter assay "HealthTEST Heparin/Platelet factor 4 Antibody Assay," or using lateral flow technology as in the lateral flow immunoassay for the detection of HIT antibodies (LFI-HIT). Some particle-based immunoassays are suitable for automated laboratories such as HemosIL® HIT-Ab and HemosIL® AcuStar HIT-Ab and HemosIL® AcuStar HIT-IgG. A summary of the performance characteristics of the immunoassays is presented in Table 17.3.

17.5.4 Functional Assays

The clinically most relevant antibodies against PF4/heparin complexes are those which have the capability to activate platelets. This can be investigated using functional assays that are based on assessing platelet activation of normal donors in the presence of patient sera/plasma and heparin. Two groups of functional assays can be distinguished: (i) assays using whole blood or platelet-rich plasma (PRP) and (ii) assays using washed platelets. The readout for all assays is platelet activation. The specificity of all functional assays can be increased by inhibition of platelet activation after adding heparin in excess (100 IU/mL; proving heparin dependency) and by showing Fcγ receptor IIa-dependent activation using a blocking monoclonal antibody (clone IV.3). Assays that use washed platelet are currently considered the "gold standard" for the diagnosis of platelet-activating antibodies. Most commonly the heparin-induced platelet activation (HIPA) assay and the SRA [38, 39] are used. Both assays employ washed platelets and detect their activation by heparin-dependent antibodies by visually observing formation of platelet aggregates or release of serotonin, respectively, and have similar operating characteristics. Although both functional assays are generally considered very specific for

Table 17.3 Performance characteristics of immunoassays used for the diagnosis of HIT [29–37]

Test system	Principle	Detected antibodies	Sensitivity	Specificity	NPV	PPV	Advantages	Drawbacks
PaGIA	Beads, gel centrifugation	Polyspecific	91–94	81–88	99.5–99.1	35.1–36.6	One sample, short turnaround time	Visual interpretation Difficult to quantify
HealthTEST	Beads, membrane filter	Polyspecific	Non-informative	30–35			One sample, easy to perform, no special equipment	Non-informative
LFI-HIT Melinea	Beads, lateral flow technology	IgG	Visual: 100	93	100	54.0	One sample, short turnaround time, no special equipment	Serum samples are required
			Scanner: 97	93	99.7	52.4		
HemosIL® HIT-Ab	Beads, agglutination	Polyspecific	100	76	100	45.5	Automated and quantifiable, standardization makes results comparable between laboratories	
HemosIL® AcuStar HIT-Ab	Beads, chemiluminescent detection	Polyspecific	98	82	99.7	41.8	Automated and quantifiable, standardization makes results comparable between laboratories	

HemosIL® AcuStar HIT-IgG	Beads, chemiluminescent detection	IgG	96.2	96.5	99.5	78.1	Automated and quantifiable, standardization makes results comparable between laboratories	
Stago Diagnostica	Enzyme-linked immunosorbent assays	IgG (polyspecific)	98	90				Time and cost intensive
Immucor GTI Diagnostics	Enzyme-linked immunosorbent assays	IgG (polyspecific)	100	89	100	43		Time and cost intensive
			100	81	100	28		
HYPHEN BioMed	Enzyme-linked immunosorbent assays	IgG (polyspecific)	100	87	100	39	Detects heparin-dependent antigens other than PF4	Time and cost intensive
In-house ELISAs	Enzyme-linked immunosorbent assays	Separately IgG, IgM, IgA	98–100	86–88	99.7	47.2	Require in-house quality control measures	

diagnosing HIT, they have several disadvantages and limitations. These assays are time intensive, which limits their usefulness in upfront clinical decision-making. They also require highly skilled laboratory technicians as well as platelets from healthy donors. Moreover, the SRA relay on the use of radioisotopes, which most clinical laboratories are now trying to avoid due to regulations and safety issues.

17.6 Target Gene Sequencing, Panel-Based Sequencing and Genome-Wide Studies

The complete sequencing of the human genome has led to two features affecting platelet function analysis: first, sequencing methods have become cheap due to the development of several independent next-generation sequencing (NGS) platforms. Second, a reference sequence of the human genome is available for virtually all known genes. With an increasing annotation of coding genes, as well as noncoding regions (including intronic sequences, promoters, enhancers, and regulatory microRNAs), genetic variants that are causative for disorders are identified and deposited in variant databases like the dbSNP or the ExAC variant databases. With an increasing input from population studies like the UK10K project, where nearly 10,000 individuals are sequenced by whole exome sequencing (WES) or whole genome sequencing (WGS) [40] and whole genome association studies (WGAS), pathogenic and non-pathogenic variants can be attributed to a low or high platelet number of mean platelet volume [41]. Variants are put to their relative occurrence in a certain population (typically Caucasians) and connected with a minor allele frequency (MAF) that somehow defines a single nucleotide polymorphism (SNP) with a MAF of more than 1 % or a "classical" mutation with a MAF of less than 1 %, although these numbers are a matter of debate. Of note, new variants in functional genes related to megakaryopoiesis and platelet formation are now described with a speed that exceeds by far the possibility to perform functional assays and to correlate variants and functions or clinical "phenotypes." In the light that about one-third of genetic mutations described in the current literature might not be pathogenic, there is a high risk of making quick conclusions from sequencing results to the genetic cause underlying a rare bleeding defect. This becomes additionally problematic when genetic counseling in families occurs or risks are evaluated that are based on reported, but not necessarily causative variants.

Nevertheless, the unambiguous identification of a genetic mutation that causes a bleeding disease like Glanzmann's thrombasthenia is helpful for prognosis of bleeding tendency or optimized therapeutic options. Identifying mutations in MYH9 additionally gives information of an increased risk for cataract, deafness, or kidney problems. Finally, mutations in genes like ANKRD26, ETV6, or RUNX1 that explain a defect in platelet number or function also lead to a predisposition for developing leukemia [42–44]. Thus, these "accidental" findings might bear the burden for the clinician that the patient might not want to be confronted with it at this moment or in general. This is best regulated by the informed consent sheets that the patient has to read, understand, and sign prior to a complex analysis and which is typically required by all local ethical boards.

Thus, the speed of advancement might not be determined by the new technology platforms or the management of "big data" or their analysis, but by a careful performance of experiments that characterize genetic variants by functional assays and compliant reporting of groups worldwide of whether a new (or a reported) variant is pathogenic of benign.

Panel-based sequencing is currently performed to restrict the number of analyzed genes to a more or less characterized minimal dataset. The current chip by the ThromboGenomics consortium in the UK comprises 75 genes with known function in coagulation and platelet production or function. The application of a large cohort of healthy donors has led to identification of 114 novel missense variants in the GPIIb/IIIa genes compared to 111 variants reported in the literature so far [45]. The authors highlight the challenges in predicting the clinical significance of these newly identified variants. It is expected that an updated ThromboGenomics chip will go from a research-based analysis of coagulation and platelet genes into the routine diagnosis at hemostaseologic centers in the UK.

Taken together, there is a need for updating current regulations to integrate clinical data and information with functional testing and genetic data from state-of-the art technologies. Thus, these revolutionized methods are currently still a challenge and their implementation requires exchange of data and experiences between basic scientists, geneticists, and finally the clinicians who report and discuss the consequences of an identified genetic variant with the patient.

References

1. O'Brien SH (2012) An update on pediatric bleeding disorders: bleeding scores, benign joint hypermobility, and platelet function testing in the evaluation of the child with bleeding symptoms. Am J Hematol 87(Suppl 1):S40–S44
2. Dovlatova N (2015) Current status and future prospects for platelet function testing in the diagnosis of inherited bleeding disorders. Br J Haematol 170(2):150–161
3. Gross L, Aradi D, Sibbing D (2016) Platelet function testing in patients on antiplatelet medications. Semin Thromb Hemost 17, Epub ahead of print
4. Andres O, Schulze H, Speer CP (2015) Platelets in neonates: central mediators in haemostasis, antimicrobial defence and inflammation. Thromb Haemost 113(1):3–12
5. Favaloro EJ, Lippi G, Franchini M (2010) Contemporary platelet function testing. Clin Chem Lab Med 48(5):579–598
6. Knofler R, Eberl W, Schulze H, Bakchoul T, Bergmann F, Gehrisch S et al (2014) Diagnosis of inherited diseases of platelet function. Interdisciplinary S2K guideline of the Permanent Paediatric Committee of the Society of Thrombosis and Haemostasis Research (GTH e. V.). Hamostaseologie 34(3):201–212
7. Jansen AJ, Josefsson EC, Rumjantseva V, Liu QP, Falet H, Bergmeier W et al (2012) Desialylation accelerates platelet clearance after refrigeration and initiates GPIbalpha metalloproteinase-mediated cleavage in mice. Blood 119(5):1263–1273
8. Hoffmeister KM (2011) The role of lectins and glycans in platelet clearance. J Thromb Haemost: JTH 9(Suppl 1):35–43
9. Schneppenheim R (2011) The pathophysiology of von Willebrand disease: therapeutic implications. Thromb Res 128(Suppl 1):S3–S7
10. von Lillicrap D (2013) Willebrand disease: advances in pathogenetic understanding, diagnosis, and therapy. Hematol/Educ Program Am Soc Hematol Am Soc Hematol Educ Program 2013:254–260

11. Marcus PD, Nire KG, Grooms L, Klima J, O'Brien SH (2011) The power of a standardized bleeding score in diagnosing paediatric type 1 von Willebrand's disease and platelet function defects. Haemophilia: Off J World Fed Hemophilia 17(2):223–227
12. Robinson PN, Kohler S, Bauer S, Seelow D, Horn D, Mundlos S (2008) The Human Phenotype Ontology: a tool for annotating and analyzing human hereditary disease. Am J Hum Genet 83(5):610–615
13. Westbury SK, Turro E, Greene D, Lentaigne C, Kelly AM, Bariana TK et al (2015) Human phenotype ontology annotation and cluster analysis to unravel genetic defects in 707 cases with unexplained bleeding and platelet disorders. Genome Med 7(1):36
14. Briggs C, Harrison P, Machin SJ (2007) Continuing developments with the automated platelet count. Int J Lab Hematol 29(2):77–91
15. Nurden AT, Nurden P (2014) Congenital platelet disorders and understanding of platelet function. Br J Haematol 165(2):165–178
16. Quiroga T, Goycoolea M, Panes O, Aranda E, Martinez C, Belmont S et al (2007) High prevalence of bleeders of unknown cause among patients with inherited mucocutaneous bleeding. A prospective study of 280 patients and 299 controls. Haematologica 92(3):357–365
17. Pecci A, Noris P, Invernizzi R, Savoia A, Seri M, Ghiggeri GM et al (2002) Immunocytochemistry for the heavy chain of the non-muscle myosin IIA as a diagnostic tool for MYH9-related disorders. Br J Haematol 117(1):164–167
18. Navarro-Nunez L, Teruel R, Anton AI, Nurden P, Martinez-Martinez I, Lozano ML et al (2011) Rare homozygous status of P43 beta1-tubulin polymorphism causes alterations in platelet ultrastructure. Thromb Haemost 105(5):855–863
19. Cattaneo M, Cerletti C, Harrison P, Hayward CP, Kenny D, Nugent D, et al (2013) Recommendations for the standardization of light transmission aggregometry: a consensus of the working party from the Platelet Physiology Subcommittee of SSC/ISTH. J Thromb Haemost: JTH Journal of Thrombosis and Haemostasis 11(6):1183–1189
20. McGlasson DL, Fritsma GA (2009) Whole blood platelet aggregometry and platelet function testing. Semin Thromb Hemost 35(2):168–180
21. Peerschke EI, Castellone DD, Stroobants AK, Francis J (2014) Reference range determination for whole-blood platelet aggregation using the Multiplate analyzer. Am J Clin Pathol 142(5):647–656
22. Pai M, Wang G, Moffat KA, Liu Y, Seecharan J, Webert K et al (2011) Diagnostic usefulness of a lumi-aggregometer adenosine triphosphate release assay for the assessment of platelet function disorders. Am J Clin Pathol 136(3):350–358
23. Cattaneo M (2009) Light transmission aggregometry and ATP release for the diagnostic assessment of platelet function. Semin Thromb Hemost 35(2):158–167
24. Cattaneo M, Hayward CP, Moffat KA, Pugliano MT, Liu Y, Michelson AD (2009) Results of a worldwide survey on the assessment of platelet function by light transmission aggregometry: a report from the platelet physiology subcommittee of the SSC of the ISTH. J Thromb Haemost: JTH 7(6):1029
25. Israels SJ, Kahr WH, Blanchette VS, Luban NL, Rivard GE, Rand ML (2011) Platelet disorders in children: a diagnostic approach. Pediatr Blood Cancer 56(6):975–983
26. Gresele P, Subcommittee on Platelet Physiology of the International Society on T, Hemostasis (2015) Diagnosis of inherited platelet function disorders: guidance from the SSC of the ISTH. J Thromb Haemost: JTH 13(2):314–322
27. Bakchoul T, Zollner H, Greinacher A (2014) Current insights into the laboratory diagnosis of HIT. Int J Lab Hematol 36(3):296–305
28. Prechel M, Hilby AM, Walenga JM (2008) Cross reactivity of heparin-induced PF4 antibodies with heparin-bound IL-8. Blood 112(11):3429
29. Althaus K, Hron G, Strobel U, Abbate R, Rogolino A, Davidson S, et al (2013)_ Evaluation of automated immunoassays in the diagnosis of heparin induced thrombocytopenia. Thromb Res 131(3):e85–e90
30. Arepally G, Reynolds C, Tomaski A, Amiral J, Jawad A, Poncz M et al (1995) Comparison of PF4/heparin ELISA assay with the 14C-serotonin release assay in the diagnosis of heparin-induced thrombocytopenia. Am J Clin Pathol 104(6):648–654

31. Bakchoul T, Giptner A, Bein G, Santoso S, Sachs UJ (2011) Performance characteristics of two commercially available IgG-specific immunoassays in the assessment of heparin-induced thrombocytopenia (HIT). Thromb Res 127(4):345–348
32. Bakchoul T, Giptner A, Najaoui A, Bein G, Santoso S, Sachs UJ (2009) Prospective evaluation of PF4/heparin immunoassays for the diagnosis of heparin-induced thrombocytopenia. J Thromb Haemost 7(8):1260–1265
33. Lindhoff-Last E, Gerdsen F, Ackermann H, Bauersachs R (2001) Determination of heparin-platelet factor 4-IgG antibodies improves diagnosis of heparin-induced thrombocytopenia. Br J Haematol 113(4):886–890
34. Pouplard C, Gueret P, Fouassier M, Ternisien C, Trossaert M, Regina S et al (2007) Prospective evaluation of the '4Ts' score and particle gel immunoassay specific to heparin/PF4 for the diagnosis of heparin-induced thrombocytopenia. J Thromb Haemost 5(7): 1373–1379
35. Sachs UJ, von HJ, Santoso S, Bein G, Bakchoul T (2011) Evaluation of a new nanoparticle-based lateral-flow immunoassay for the exclusion of heparin-induced thrombocytopenia (HIT). Thromb Haemost 106(6):1197–11202
36. Whitlatch NL, Kong DF, Metjian AD, Arepally GM, Ortel TL (2010) Validation of the high-dose heparin confirmatory step for the diagnosis of heparin-induced thrombocytopenia. Blood 116(10):1761–1766
37. Whitlatch NL, Perry SL, Ortel TL (2008) Anti-heparin/platelet factor 4 antibody optical density values and the confirmatory procedure in the diagnosis of heparin-induced thrombocytopenia. Thromb Haemost 100(4):678–684
38. Greinacher A, Michels I, Kiefel V, Mueller-Eckhardt C (1991) A rapid and sensitive test for diagnosing heparin-associated thrombocytopenia. Thromb Haemost 66(6):734–736
39. Sheridan D, Carter C, Kelton JG (1986) A diagnostic test for heparin-induced thrombocytopenia. Blood 67(1):27–30
40. Consortium UK, Walter K, Min JL, Huang J, Crooks L, Memari Y et al (2015) The UK10K project identifies rare variants in health and disease. Nature 526(7571):82–90
41. Gieger C, Radhakrishnan A, Cvejic A, Tang W, Porcu E, Pistis G et al (2011) New gene functions in megakaryopoiesis and platelet formation. Nature 480(7376):201–208
42. Noris P, Perrotta S, Seri M, Pecci A, Gnan C, Loffredo G et al (2011) Mutations in ANKRD26 are responsible for a frequent form of inherited thrombocytopenia: analysis of 78 patients from 21 families. Blood 117(24):6673–6680
43. Noetzli L, Lo RW, Lee-Sherick AB, Callaghan M, Noris P, Savoia A et al (2015) Germline mutations in ETV6 are associated with thrombocytopenia, red cell macrocytosis and predisposition to lymphoblastic leukemia. Nat Genet 47(5):535–538
44. Jongmans MC, Kuiper RP, Carmichael CL, Wilkins EJ, Dors N, Carmagnac A et al (2010) Novel RUNX1 mutations in familial platelet disorder with enhanced risk for acute myeloid leukemia: clues for improved identification of the FPD/AML syndrome. Leukemia 24(1):242–246
45. Buitrago L, Rendon A, Liang Y, Simeoni I, Negri A, ThromboGenomics C et al (2015) alphaIIbbeta3 variants defined by next-generation sequencing: predicting variants likely to cause Glanzmann thrombasthenia. Proc Natl Acad Sci U S A 112(15):E1898–E1907

Part IV

Future Perspectives for Platelet Biogenesis

Cellular Therapies for Wiskott-Aldrich Syndrome

18

Christian J. Braun and Christoph Klein

Abstract

Wiskott-Aldrich syndrome (WAS) is an X-linked inherited rare disease caused by mutations of the *WAS* gene and affecting various subsets of the hematopoietic system. Affected patients experience severe bleedings, autoimmunity, immunodeficiency, and an increased risk of hematopoietic malignancies. Since several decades, patients with WAS have been treated successfully using allogeneic hematopoietic stem cell transplantation – this procedure may however be associated with significant morbidity as well as mortality, in particular in case of HLA-disparity. Hematopoietic stem cell gene therapy has been developed as an alternative, but oncogene activation secondary to insertional mutagenesis can lead to oncogenesis. More recent therapeutic approaches include the use of self-inactivating lentiviral vectors, promising a better safety-to-risk ratio. Therapeutic efficacy and safety are currently being assessed. New methods of genomic engineering employing zinc finger nucleases, TALENs, and CRISPR-Cas as tools may offer new perspectives to site-specific genomic repair of disease-causing mutations.

C.J. Braun
Dr. von Hauner Children's Hospital, Ludwig-Maximilians-Universität München, Lindwurmstraße 4, Munich 80337, Germany

Koch Institute for Integrative Cancer Research, Massachusetts Institute of Technology (MIT), Cambridge, MA, USA

C. Klein (✉)
Dr. von Hauner Children's Hospital, Ludwig-Maximilians-Universität München, Lindwurmstraße 4, Munich 80337, Germany
e-mail: Christoph.Klein@med.uni-muenchen.de

H. Schulze, J. Italiano (eds.), *Molecular and Cellular Biology of Platelet Formation*, DOI 10.1007/978-3-319-39562-3_18

18.1 Wiskott-Aldrich Syndrome (WAS)

Wiskott-Aldrich syndrome (WAS) is an X-linked recessive inherited disease of the hematopoietic system caused by mutations in the *WAS* gene [24]. Even though the clinical triad of bleeding symptoms (due to microthrombocytopenia), eczema (due to autoimmunity), and severe recurring infections (due to immunodeficiency) is usually pathognomonic for severe WAS, this combination only occurs in about 30 % of all WAS patients [69]. The life expectation of patients with severe clinical symptoms is about 10 years of age due to hemorrhage, overwhelming infections, or serious infections. The Bavarian pediatrician Alfred Wiskott was the first who published a case study correctly describing the clinical presentation of three young brothers with classical symptoms of WAS as a novel "hereditary thrombopathy" [81]. The American pediatrician Robert Aldrich demonstrated the X-linked mode of inheritance in a family with many affected males over multiple generations [2].

The clinical severity of WAS can range from an isolated thrombocytopenia with mild or no eczema and infections (the so-called X-linked thrombocytopenia, XLT) to the classic WAS [51]. The so-called X-linked neutropenia is rare and presents clinically as an isolated neutropenia without microthrombocytopenia [25].

WAS is a rare disease with about four cases per million live male births in the United States [56].

18.2 From Gene Mutations to Functional Impairments of Hematopoietic Cells

18.2.1 WASP Is a Key Regulator of Actin Polymerization

The expression of the *WAS* gene is restricted to cells of the hematopoietic system [55]. The gene product WASP is involved in the transduction of signals from cell surface receptors to the actin cytoskeleton and in the initiation of subsequent actin polymerization [70]. The so-called VCA region of WASP (V: verprolin-homology, C: cofilin-homology, A: acidic region) can bind to an actin monomer and to the Arp2/3 complex [73] – Arp2/3 activation by WASP is thought to initiate actin polymerization [58, 74]. The EVH1/WH1 domain of WASP can bind to so-called WIP proteins (a family of "WASP-interacting proteins") [60, 80]. Interaction with WIP proteins is thought to modulate WASP activity. The EVH1/WH1 domain is a hotspot for genetic mutations [80]. The GTPase-binding domain (GBD) of WASP interacts with the VCA domain and, as a result, can interfere with the activation of the Arp2/3 complex and with actin polymerization [54]. In detail, binding of GTP-Cdc42 to GBD leads to protein conformation changes of WASP and increases its binding affinity to Arp2/3 [35, 64]. Correlation of *WAS* gene mutations and a clinical phenotype is not always clear, but generally patients without any gene expression or with a mostly truncated protein expression are affected by a more severe WAS phenotype than patients with full-length mutated protein expression [32, 33]. Functional impairment or subnormal expression levels of WASP lead to a disturbed polymerization of actin [6].

18.2.2 A Disturbed Actin Polymerization Leads to Broad Functional Impairments of Hematopoietic Cells

WASP is expressed only in the hematopoietic system (in all cell lineages except erythrocytes) – this includes not only mature cells of adults but already the earliest hematopoietic progenitor cells during fetal development [41, 55]. Female carriers of *WAS* gene mutations experience a nonrandom inactivation of their X chromosomes [38]. Sometimes new mutations in the hematopoietic cells of WAS patients restore WASP expression – this can occur in more than 10 % of patients' peripheral blood cells, and sometimes even multiple different reversion mutants can be detected to the same patient [23], which underlines the evolutionary pressure against *WAS* mutations and the cellular advantage of expressing a functional WASP.

Multiple functions of innate immune system cells are affected by *WAS* mutations. Firstly, functional actin polymerization (mediated in part by WASP) is required for the formation of the so-called phagocytic cup, an actin-dense structure at the plasma membrane of phagocytes that form prior to the ingestion of foreign material [79]. Hematopoietic cells of the innate immune system therefore exhibit a severely impaired phagocytosis if WASP is not functional. Secondly, polymerization of actin is necessary for cellular movement mediated by the formation of podosomes [16], lamellipodia, and filopodia [5, 77] – contributing to the cellular migration defect observed in WAS patients [9]. Cytotoxic activity of NK cells is impaired in WAS patients due to problems assembling the NK cell immunological synapse [49, 50]. WASP-deficient neutrophils show a reduced integrin-dependent activation of degranulation and respiratory burst [86].

Furthermore, cellular and humoral functions of the adaptive immunity cells are impaired in WAS. In T cells, WASP is important for the formation of the T-cell immunological synapse [26] leading to an impairment of T-cell receptor (TCR)-mediated signaling and T-cell proliferation. In B cells, certain subsets are more affected by mutations in *WAS* than others: especially circulating mature B cells and splenic marginal zone precursors/B cells deplete selectively [43, 77]. Furthermore, the B-cell receptor (BCR)-mediated signaling is impaired, which probably weakens B-cell activation. WAS patients typically show very high levels of IgE [32, 62].

18.2.3 Platelet Disturbances in WAS

Microthrombocytopenia is a very characteristic and persistent feature for WAS and can usually be found (even if the general disease phenotype is very mild). Despite this prominence many parts of the pathophysiology still remain unclear. Interestingly, mutant WASP is usually not expressed in platelets (even if it is expressed in other cells of the hematopoietic system), most likely as a result of diminished protein stability [66]. Platelet numbers and also platelet size rise after splenectomy. This evidences that destruction of WASP-negative platelets in the spleen contributes to the disease phenotype [39]. Interestingly WASP deficiency in mice (as opposed to humans) only leads to a very mild phenotype of microthrombocytopenia – most likely because smaller mouse platelets might not be cleared as effectively by the spleen [68, 85]. Isotope studies however suggested that insufficient platelet production in WAS patients might contribute to thrombocytopenia [65]. There is some evidence that megakaryocytes, which normally reside in the extravascular space of the bone marrow, depend on the transendothelial projection of cellular processes into the blood vessel lumen for effective platelet formation [15, 75]. It is also possible that megakaryocytes fail to respond appropriately to activating intercellular signals due to actin polymerization defects [3, 31].

18.2.4 WAS in Autoimmunity and Cancer Formation

Autoimmunity (like colitis, vasculitis, autoimmunity-mediated cytopenias) is a very common symptom in severe WAS [18]. The actual pathophysiological mechanism for this is still not absolutely clear. There is some evidence from mouse experiments that WASP may be necessary for normal maturation of T cells (including the negative selection of autoreactive T cells) [85].

The development of malignancies is a relatively frequent event in WAS (13 % of 154 patients with a mean age of 9.5 years [69]). Lymphomas are the most frequent malignancy. Impairment of tumor surveillance mechanisms (especially NK cells) and susceptibility to EBV infections (EBV-associated Burkitt lymphoma is frequent in WAS patients) probably contribute to this susceptibility [14]. Whether WASP deficiency directly affects the maintenance of genomic integrity is currently under investigation.

18.3 Clinical Diagnostics

Since WAS is an X-linked, recessive disease, it mostly affects males (there are only a few reported cases of affected heterozygous females with a nonrandom inactivation of the healthy X chromosome [8]). The diagnosis of WAS or XLT should therefore be considered primarily in males showing classical symptoms (eczema, bleeding, severe and recurrent infections, autoimmunity, lymphoma).

18.3.1 Laboratory Diagnostics

Microthrombocytopenia can be diagnosed in peripheral blood smears; blood cell counts confirm thrombocytopenia and low mean platelet volume. Serum immunoglobulin levels typically show an elevated IgE, while IgM, IgA, and IgG are typically low (but can also be normal or even elevated). Vaccination titers show diminished responses even after multiple rounds of immunizations in particular development of anti-polysaccharide antibodies is decreased, autoantibodies against neutrophils and platelets can sometimes be detected, and T-cell proliferation assays show reduced responses to T-cell receptor induced signals. WASp expression can be assayed using flow cytometry or Western blot assays [14].

18.3.2 Genetic Analysis

Sanger sequencing of the *WAS* gene can provide the ultimate proof for the diagnosis of WAS. It typically includes sequencing of all exons including exon-intron boundaries (which make the discovery of splice site mutations possible). Genetic testing can also be used for identification of female carriers, for prenatal testing [67], and for preimplantation genetic diagnosis [63] (depending on legal and ethical restrictions).

18.4 Clinical Classifications

After the diagnosis of WAS has been made, a clinical scoring system can help to decide about different treatment alternatives. Zhou et al introduced a scoring system based on clinical features in 1997 [87]. It ranges between 1 (very mild symptoms) and 5 (severe symptoms of WAS) and depends on the presence of clinical symptoms and thrombocytopenia (Table 18.1). A WAS score of 3 and higher is considered "classic WAS." Recent study data suggests that very young patients with a higher WAS score may have a high risk of early mortality and might benefit from early hematopoietic stem cell transplantation (HSCT) [40].

Table 18.1 Clinical WAS score (According to Zhu et al. [87])

	Autoimmune disease	Platelets	Eczema	Infections	Malignancies
1	None	Thrombocytopenia and small platelets	None	None	None
2	None	Thrombocytopenia and small platelets	Mild and transient	Minor (with and without)	None
3	None	Thrombocytopenia and small platelets	Persistent but manageable (and/or infections)	Recurrent infections (and/or eczema)	None
4	None	Thrombocytopenia and small platelets	Persistent, hard to control (and/or infections)	Frequent life-threatening infections (and/or eczema)	None
5	And/or malignancies	Thrombocytopenia and small platelets	Persistent, hard to control (and/or infections, autoimmune disease, malignancies)	Frequent life-threatening infections (and/or eczema, autoimmune disease, malignancies)	And/or autoimmune disease

18.5 Therapy

18.5.1 Supportive Therapy

18.5.1.1 Thrombocytopenia and Bleeding

Thrombocytopenia and bleeding are the most frequent symptoms of WAS and XLT. Splenectomy is a classical supportive treatment [32, 48] – platelet numbers usually improve following this procedure (unless platelet autoantibodies prevent this). Previously, splenectomy has been a classical supportive treatment option [32, 48]. However, splenectomy is associated with an increased risk of sepsis due to encapsulated organisms such as *S. pneumoniae* and *H. influenzae* and is currently rarely done. Splenectomized patients should therefore be vaccinated prior to the procedure, and a lifelong antibiotic prophylaxis has to be considered [22]. Intravenous immunoglobulin (IVIG) administration does not seem to improve platelet counts significantly [42]. Irradiated and cytomegalovirus-negative platelet transfusions can support WAS patients during severe bleeding episodes and if at high risk of bleeding (e.g., perioperative) [14].

18.5.1.2 Infections

The most important prophylaxis for the prevention of severe infections is the regular supplementation with IVIG for every patient with recurring infections, abnormally low immunoglobulin levels, and vaccination nonresponses. Some patients with a history of severe bacterial infections might benefit from antibiotic prophylaxis (especially for patients with recurring sinus and lung infections). Live vaccinations have to be avoided, but patients should be immunized with conjugated and unconjugated "dead" vaccines. Manifest infections should be treated early and with broad-spectrum antibiotics [14].

18.5.1.3 Autoimmunity

WAS patients suffering from autoimmunity often benefit from immunosuppression: Corticosteroids are widely utilized (but can have severe side effects like growth retardation, weight gain, glaucoma, muscle weakness, stomach ulcers, diabetes, body fat distribution changes, opportunistic infections). Some patients profit from Some patients benefit from immunosuppressive drugs.

18.5.2 Stem Cell Transplantation

Allogeneic hematopoietic stem cell transplantation can provide a definitive cure for patients with WAS. If no HLA-matched sibling donor (MSD) or HLA-matched family donor (MRD) is available, search for non-related HLA-matched donors (MUDs) should be started. If an HLA-identical sibling donor, an MRD, or an MUD is available, HSCT should strongly be considered, since HSCT from HLA matched donors now results in high survival rates [46]. In the absence of an

HLA-compatible donor, transplantation of WAS patients from T-cell-depleted partially HLA-mismatched family donors (MMFD) can be a treatment alternative. However, this approach has historically resulted in a much lower survival than matched donor transplantation [27, 28, 30, 36, 46, 53], mostly because of non-engraftment, graft rejection, and severe graft-versus-host disease (GvHD). Mixed chimerism after transplantation is frequently associated with persistent symptoms of WAS (especially low lymphocyte counts, autoimmunity, and thrombocytopenia) [46]. Age at HSCT seems to be an important predictor of outcome: WAS patients undergoing unrelated donor HSCT at an age younger than 5 years have comparable outcomes to children transplanted from matched siblings donors [52].

18.5.3 Molecular Therapy for WAS

18.5.3.1 Primary Immunodeficiencies (PIDs): At the Forefront of Clinical Gene Therapy

Hematopoietic stem cell gene therapy (GT) is an emerging option for curative treatment in WAS and other primary immunodeficiencies (PIDs). Gene therapy for PIDs can potentially compensate for some of the biggest shortcomings in HSCT: GvHD, serious side effects from strong cytotoxic conditioning regiments, and potentially costs. PIDs are at the forefront for clinical application of GT for multiple reasons (a more detailed discussion can be found in [59]): (1) PIDs are very well studied, deep knowledge is available about the molecular pathophysiology, and spontaneous somatic reversion of the underlying mutation in certain cellular subsets demonstrated growth advantage and reversion of disease, (2) even small numbers of engrafted cells can be enough to achieve symptom control due to a proliferative advantage of corrected cells, and (3) hematopoietic cells can be grown and manipulated in vitro. Clinical trials for GT in PIDs have been performed for X-linked severe combined immunodeficiency (SCID-X1), adenosine deaminase deficiency (SCID-ADA), chronic granulomatous disease (CGD), and WAS.

18.5.3.2 Gene Transfer Technologies

Retroviral vectors for gene transfer have been essential for the development of clinical gene therapy: production (even in larger batches) is relatively easy, due to their stable integration into the host genome, they promise long-term transgene expression, hematopoietic cells can efficiently be transduced, and – due to safety precautions and a split genome – standard retroviral vectors for gene therapy have been shown to be non-replicating. There is a traditional separation of so-called *cis* and *trans* elements within the genetic structure of a retrovirus. While *cis* regions are biologically active as nucleic acids (e.g., LTRs, long-terminal repeats, and identical sequences of DNA that repeat many times and are important for the regulation of viral integration into the host genome, transcriptional activity, viral packaging, and reverse transcription), *trans* regions (e.g., env, a protein that forms the viral envelope) are coding for viral proteins. Most retroviral gene vectors have split genomes with a clear separation of *cis* and *trans* regions on different DNA plasmids, therefore preventing the production of replication competent viral particles. The ability of retroviral vector to integrate into the host genome can affect the expression of neighboring genes (insertional mutagenesis (IM)). The most frequent IM is the transactivation of an endogenous oncogene by the inserted viral LTR – either by directly acting as a transcriptional enhancer/promoter, by activating an endogenous enhancer/promoter that then consecutively activates the expression of a cellular oncogene, or by forming a transgene/oncogene chimeric transcript. Silencing of a cellular (tumor suppressor) transcript less frequently contributes to oncogenesis (since usually biallelic silencing is required for loss of the tumor suppressor).

Self-inactivating (SIN) retroviral vectors were developed to reduce the possibility of oncogene transactivation by introducing deletions in the 3′ LTR covering the LTR promoter and enhancer [4, 57, 76, 82, 83]. Upon reverse transcription and genomic integration, the SIN carries the SIN LTR at both the 3′ and at the 5′ flank of the transgene [83]. Transgene transcription is driven by an internal promoter (which typically leads to weaker expression levels if compared to standard LTRs and can be cell specific) [47]. This then theoretically minimizes the risk of cellular (oncogene) transactivation around the insertion site and therefore IM. *Lentiviruses* are more complex than *retroviruses* and contain additional open reading frames that encode for additional genes (such as *rev*, which mediates nuclear export of viral mRNA, and *tat*, which acts as a viral transcription factor). Some of these additional proteins are considered factors that subsequently moved away from the actual viral backbone to helper plasmids in order to increase biosafety and to make room for larger transgenes [37]. The overall insertion site pattern of lentiviral vectors is different from that of γ-retroviral vectors. Lentiviral vectors appear to have a somewhat broader integration spectrum into actively transcribed genes than γ-retroviral vectors [1] and furthermore seem to lack most of the transcription start site (TSS) affinity of γ-retrovirus and somewhat randomly integrate throughout the entire gene [57]. The use of SIN lentiviral vectors still carries the risk of growth factor transactivation as demonstrated recently in a gene therapy trial for ß-thalassemia in which a growth-promoting gene (*HMGA2*) was activated transcriptionally after IM rendering HMGA2 mRNA insensitive to degradation by let-7 microRNAs – the general belief however is that the risk of leukomogenesis is much lower in SIN lentiviral GT than in γ-retroviral GT (Fig. 18.1).

18.5.3.3 Clinical Trials for Wiskott-Aldrich Syndrome

WAS patients have been enrolled into GT trials using standard γ-retrovirus [11, 12] as well as recently SIN *Lentivirus* [1], whereas the time of observation after GT is much longer for the classic γ-retrovirus group. A strong increase in the proportion of WASP-positive peripheral lymphoid cells was seen in patients after γ-retrovirus GT, which confirms the hypothesis of a proliferative advantage of gene-corrected cells [10]. An overall increase of platelet counts was noted as

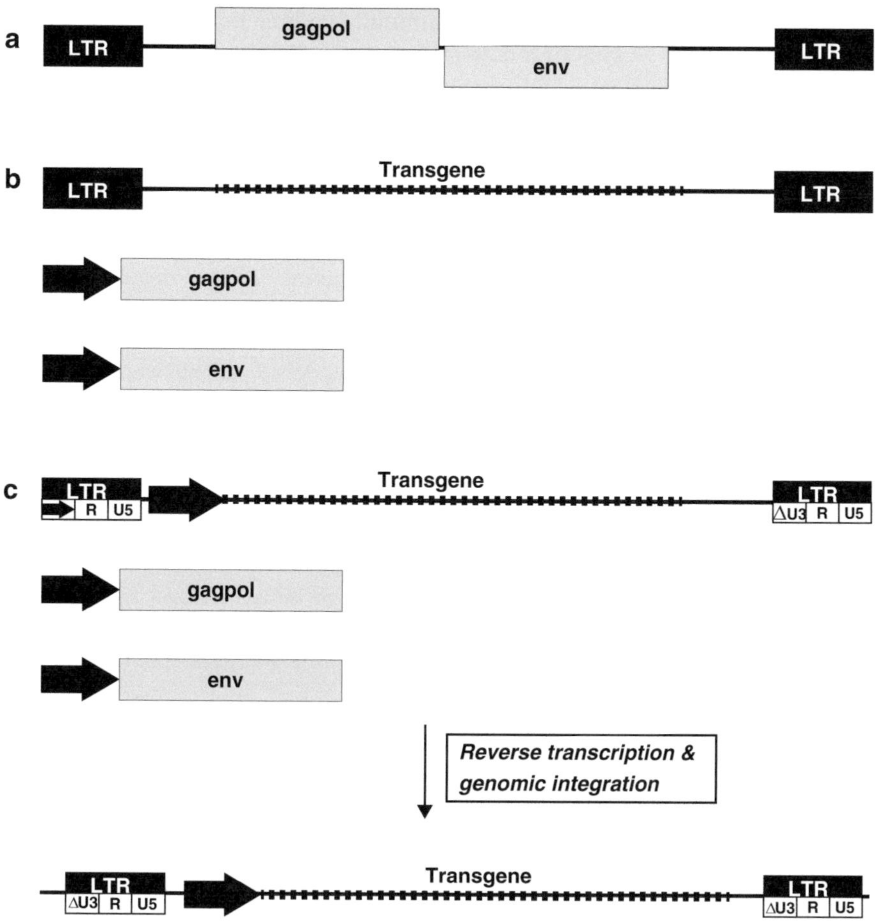

Fig. 18.1 (**a**) Retrovirus, (**b**) split genome gene vector, (**c**) self-inactivating (SIN) gene vector

well, along with a normalization of platelet size and a cessation of the bleeding diathesis. IVIG substitution could be discontinued in some patients, and T-cell proliferative responses improved along with a normalization of T-cell receptor (TCR) Vß repertoire usage. Many patients had not been able to form immunological synapses between NK cells and tumor cells in vitro pre-GT – this improved significantly over time after GT. Furthermore, clinical signs of autoimmunity (like eczema, autoimmune cytopenias, and colitis) resolved after therapy. Severe infections were less frequent after therapy. Analysis of the clonal dynamics of hematopoietic repopulation was carried out using standard and nonrestrictive (nr) linear amplification-mediated polymerase chain reaction (LAM-PCR) high-throughput sequencing of peripheral blood and also bone marrow samples [29]. Shortly after infusion of genetically modified HSC GT a highly polyclonal repopulation was observed. Of note, a fluctuating pattern of clones with targeted oncogenes contributed to hematopoiesis, initially without malignant transformation. Years after GT, a total of 8 patients developed leukemia. T cell acute lymphoblastic leukemia (T-ALL) was associated with retroviral vector integration in close proximity to the T-ALL oncogene *LMO2* locus. Two patients additionally developed acute myeloid leukemia (AML) shortly after or even during their maintenance therapy for their T-ALL, now with dominant insertion sites close or within *MN1* and *MDS1* (both genes had been associated in AML before). One patient was diagnosed with primary AML. These data clearly show that GT for WAS can be very effective in improving the symptoms of WAS, but also that the use of classic γ-retrovirus vectors is associated with a high degree of genotoxic side effects [11, 13].

Results from the first SIN lentiviral trial for GT in WAS patients were published in 2013 for three patients with a median observation time of 25 months [1]. All three patients showed increasing expression levels of WASP in their peripheral blood lymphocytes and platelets (even though platelet counts at the time of publication had not reached physiological levels yet); eczema as well as an existing bleeding diathesis improved in all patients. Two patients experienced severe infections within the first 6 months after GT, but overall frequency and severity of infections decreased over time. Fluctuating viral insertion patterns were noted, without evidence for clonal outgrowth. These results are promising in view of efficacy and toxicity, but long term long-term observations are still pending for WAS patients treated with SIN lentiviral GT.

18.6 Gene Engineering: True Mutational Correction Instead of Mere Overexpression?

Despite all advanced toward safer and more effective gene therapy using retro- and lentiviral systems, the random insertion of viral DNA into the genome will probably always be associated with some risk of genotoxicity. This is why at least in theory the actual repair of a disease-causing mutation would be much more desirable than the mere overexpression of a transgene. This could potentially reestablish a fully physiological expression of the targeted gene: accessible for epigenetic regulations, expressed from a natural promoter, and without affecting other regions of the genome. The development of three new techniques for genome engineering has provoked a lot of hope that a targeted genome repair may soon be feasible.

18.6.1 Zinc Finger Nucleases (ZFNs)

ZFNs are artificially produced enzymes that are capable of cutting genomic DNA. These enzymes are fusion products of a zinc finger DNA-binding domain and a DNA cleavage domain. Zinc finger domains can be designed in such a way that they only bind a defined DNA sequence, which is then cut by the attached zinc finger nuclease. A pair of ZFNs targeted to the same region of the genome is able to produce double-strand breaks (DSBs). A provided DNA strand, which shares homology with the targeted DNA sequence, can act as a template for homologous recombination (HR) – part of the cellular DNA repair machinery (Fig. 18.2a). Enzymes involved in HR search for DNA sequences similar to the region affected by the DSB. If such a template is found, the DSB is filled up with the DNA template sequence (regardless of whether this fragment actually contains the original sequence or not) [17, 34].

18.6.2 Transcription Activator-Like Effector Nuclease (TALENs)

TALENS are artificial fusion enzymes between a transcription activator-like (TAL) effector DNA-binding domain and a DNA cleavage domain. TAL domains were initially derived from *Xanthomonas* bacteria. These proteins are able to recognize and subsequently bind certain nucleotide sequences in DNA. Targeting a genomic region with multiple TALENs can lead to DSBs that – in close similarity to ZFNs – can be used to edit the genome and potentially repair disease-causing mutations [7, 19, 20, 45] (Fig. 18.2b).

18.6.3 Clustered Regularly Interspaced Short Palindromic Repeats (CRISPR)

CRISPRs are DNA loci containing short repetitions of base sequences. Together with CRISPR-associated genes (Cas-genes), they can be found in many bacteria and most archaea. This naturally occurring system has been adapted for genome engineering (Fig. 18.2c). A short RNA molecule (the so-called guide RNA) is used to target the bacterial Cas9 endonuclease to specific genomic loci of choice. Cutting of Cas9 produces DSBs; these can then be exploited (again in analogy to ZFNs and TALENs) to alter genomic DNA sequences [21, 61, 78].

18.7 Induced Pluripotent Stem Cells (iPSCs): A Better Disease Model System and a Therapeutic Approach?

A pluripotent stem cell has the potency to differentiate into any cell of the human body. The best-characterized pluripotent stem cell is the embryonic stem cell (ESC). ESCs can be derived from the inner mass of an embryo and then be kept in a culture dish. Since the use of human ESCs cells is very limited by ethical consideration, it has been a major breakthrough when cells with a similar phenotype were first derived from adult somatic cells. These so-called induced pluripotent stem cells (iPSCs) can be obtained by a forced expression of four transcription factors (Oct4,

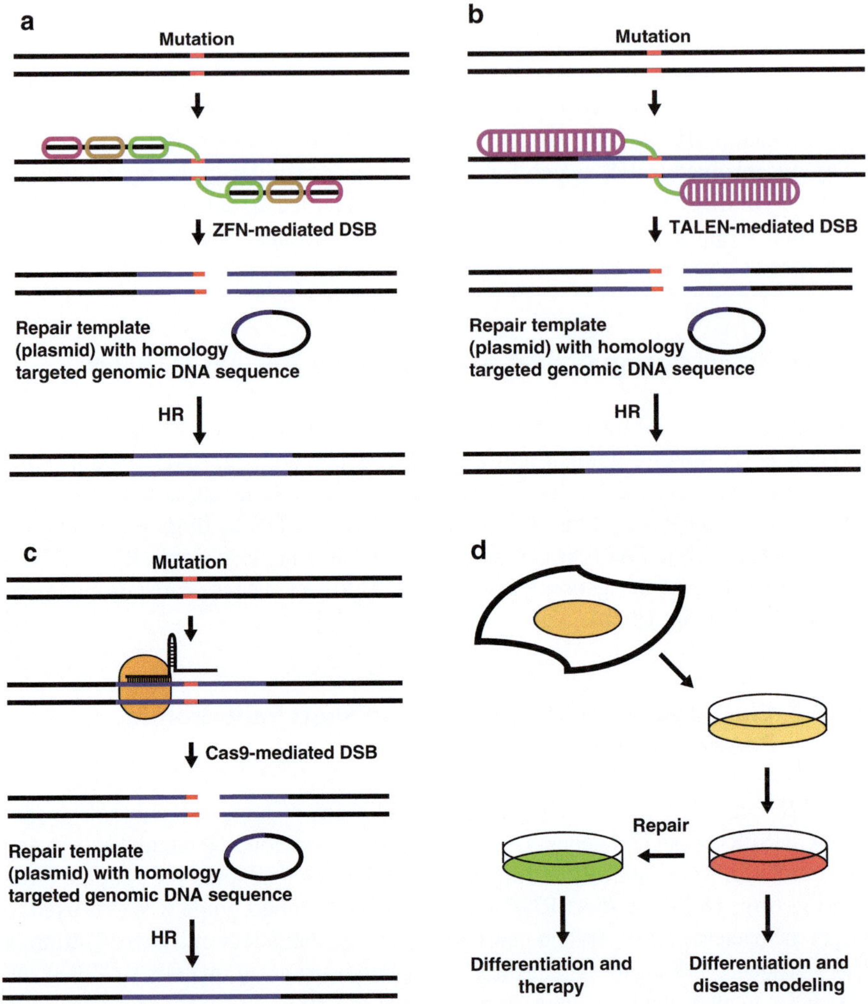

Fig. 18.2 Methods of genome engineering (**a**) ZFNs, (**b**) TALENs, (**c**) CRISPR-Cas9, (**d**) and the use of induced pluripotent stem cells (iPSCs) for research and therapy (Modified from Mikkers et al. [44])

Sox2, Klf4, and c-Myc) in differentiated cells [71, 72]. iPSC lines derived from patients can be dedifferentiated into somatic tissue cells and be used to model human disease [84]. Furthermore, disease-causing mutations in patient-derived iPSCs can be repaired making use of HR, re-differentiated, for example, into HSCs and used HSCT (depending on the underlying disease) (Fig. 18.2d). This would then help to overcome the bottleneck of too few healthy donors for HSCT and also the problem of GvHD. Limitations of this approach however include general safety concerns of genetic reprogramming and weak efficacy of deriving transplantable HSCs from iPSCs [44].

References

1. Aiuti A, Biasco L, Scaramuzza S et al (2013) Lentiviral hematopoietic stem cell gene therapy in patients with Wiskott-Aldrich syndrome. Science 341:1233151. doi:10.1126/science.1233151
2. Aldrich RA, Steinberg AG, Campbell DC (1954) Pedigree demonstrating a sex-linked recessive condition characterized by draining ears, eczematoid dermatitis and bloody diarrhea. Pediatrics 13:133–139
3. Avecilla ST, Hattori K, Heissig B et al (2004) Chemokine-mediated interaction of hematopoietic progenitors with the bone marrow vascular niche is required for thrombopoiesis. Nat Med 10:64–71. doi:10.1038/nm973
4. Avedillo Díez I, Zychlinski D, Coci EG et al (2011) Development of novel efficient SIN vectors with improved safety features for Wiskott-Aldrich syndrome stem cell based gene therapy. Mol Pharm 8:1525–1537. doi:10.1021/mp200132u
5. Biyasheva A, Svitkina T, Kunda P et al (2004) Cascade pathway of filopodia formation downstream of SCAR. J Cell Sci 117:837–848. doi:10.1242/jcs.00921
6. Blanchoin L, Amann KJ, Higgs HN et al (2000) Direct observation of dendritic actin filament networks nucleated by Arp2/3 complex and WASP/Scar proteins. Nature 404:1007–1011. doi:10.1038/35010008
7. Boch J (2011) TALEs of genome targeting. Nat Biotechnol 29:135–136. doi:10.1038/nbt.1767
8. Boonyawat B, Dhanraj S, Al Abbas F et al (2013) Combined de-novo mutation and non-random X-chromosome inactivation causing Wiskott-Aldrich syndrome in a female with thrombocytopenia. J Clin Immunol 33:1150–1155. doi:10.1007/s10875-013-9927-9
9. Bouma G, Burns SO, Thrasher AJ (2009) Wiskott-Aldrich Syndrome: immunodeficiency resulting from defective cell migration and impaired immunostimulatory activation. Immunobiology 214:778–790. doi:10.1016/j.imbio.2009.06.009
10. Boztug K, Dewey RA, Klein C (2006) Development of hematopoietic stem cell gene therapy for Wiskott-Aldrich syndrome. Curr Opin Mol 8(5):390–395
11. Boztug K, Schmidt M, Schwarzer A et al (2010) Stem-cell gene therapy for the Wiskott-Aldrich syndrome. N Engl J Med 363:1918–1927. doi:10.1056/NEJMoa1003548
12. Braun CJ, Witzel M, Paruzynski A, Boztug K (2014a) Gene therapy for Wiskott-Aldrich syndrome—long-term reconstitution and clinical benefits, but increased risk for leukemogenesis. Rare 2(1):e947749
13. Braun CJ, Boztug K, Paruzynski A et al (2014) Gene therapy for Wiskott-Aldrich syndrome – long-term efficacy and genotoxicity. Sci Transl Med 6:227ra33. doi:10.1126/scitranslmed.3007280
14. Buchbinder D, Nugent DJ, Fillipovich AH (2014) Wiskott-Aldrich syndrome: diagnosis, current management, and emerging treatments. Appl Clin Genet 7:55–66. doi:10.2147/TACG.S58444
15. Burns S, Cory GO, Vainchenker W, Thrasher AJ (2004) Mechanisms of WASp-mediated hematologic and immunologic disease. Blood 104:3454–3462. doi:10.1182/blood-2004-04-1678
16. Calle Y, Antón IM, Thrasher AJ, Jones GE (2008) WASP and WIP regulate podosomes in migrating leukocytes. J Microsc 231:494–505. doi:10.1111/j.1365-2818.2008.02062.x
17. Carroll D (2011) Genome engineering with zinc-finger nucleases. Genetics 188:773–782. doi:10.1534/genetics.111.131433
18. Catucci M, Castiello MC, Pala F et al (2012) Autoimmunity in wiskott-Aldrich syndrome: an unsolved enigma. Front Immunol 3:209. doi:10.3389/fimmu.2012.00209
19. Cermak T, Doyle EL, Christian M et al (2011) Efficient design and assembly of custom TALEN and other TAL effector-based constructs for DNA targeting. Nucleic Acids Res 39:e82. doi:10.1093/nar/gkr218
20. Christian M, Cermak T, Doyle EL et al (2010) Targeting DNA double-strand breaks with TAL effector nucleases. Genetics 186:757–761. doi:10.1534/genetics.110.120717
21. Cong L, Ran FA, Cox D et al (2013) Multiplex genome engineering using CRISPR/Cas systems. Science 339:819–823. doi:10.1126/science.1231143
22. Davidson RN, Wall RA (2001) Prevention and management of infections in patients without a spleen. Clin Microbiol Infect. doi:10.1046/j.1198-743x.2001.00355.x

23. Davis BR, Dicola MJ, Prokopishyn NL et al (2008) Unprecedented diversity of genotypic revertants in lymphocytes of a patient with Wiskott-Aldrich syndrome. Blood 111:5064–5067. doi:10.1182/blood-2007-06-095299
24. Derry JM, Ochs HD, Francke U (1994) Isolation of a novel gene mutated in Wiskott-Aldrich syndrome. Cell 79:following 922. doi: 10.1016/0092-8674(94)90528-2
25. Devriendt K, Kim AS, Mathijs G et al (2001) Constitutively activating mutation in WASP causes X-linked severe congenital neutropenia. Nat Genet 27:313–317. doi:10.1038/85886
26. Dustin ML, Cooper JA (2000) The immunological synapse and the actin cytoskeleton: molecular hardware for T cell signaling. Nat Immunol 1:23–29. doi:10.1038/76877
27. Filipovich A (2001) Impact of donor type on outcome of bone marrow transplantation for Wiskott-Aldrich syndrome: collaborative study of the international bone marrow transplant registry and the national marrow donor program. Blood. doi:10.1182/blood.V97.6.1598
28. Friedrich W, Schütz C, Schulz A, Benninghoff U (2009) Results and long-term outcome in 39 patients with Wiskott–Aldrich syndrome transplanted from HLA-matched and-mismatched donors. Immunol Res 44(1-3):18–24
29. Gabriel R, Eckenberg R, Paruzynski A et al (2009) Comprehensive genomic access to vector integration in clinical gene therapy. Nat Med 15:1431–1436. doi:10.1038/nm.2057
30. Gennery AR, Slatter MA, Grandin L, Taupin P (2010) Transplantation of hematopoietic stem cells and long-term survival for primary immunodeficiencies in Europe: entering a new century, do we do better? J Allergy Clin Immunol 126(3):602–10.e1–11
31. Haddad E, Zugaza JL, Louache F (2001) The interaction between Cdc42 and WASP is required for SDF-1–induced T-lymphocyte chemotaxis. Blood 97(1):33–38. doi:10.1182/blood.V97.1.33
32. Imai K, Morio T, Zhu Y et al (2004) Clinical course of patients with WASP gene mutations. Blood 103:456–464. doi:10.1182/blood-2003-05-1480
33. Jin Y, Mazza C, Christie JR et al (2004) Mutations of the Wiskott-Aldrich Syndrome Protein (WASP): hotspots, effect on transcription, and translation and phenotype/genotype correlation. Blood 104:4010–4019. doi:10.1182/blood-2003-05-1592
34. Kim YG, Cha J (1996) Hybrid restriction enzymes: zinc finger fusions to Fok I cleavage domain. Proc Natl Acad Sci U S A 93(3):1156–1160. doi:10.1073/pnas.93.3.1156
35. Kim AS, Kakalis LT, Abdul-Manan N et al (2000) Autoinhibition and activation mechanisms of the Wiskott-Aldrich syndrome protein. Nature 404:151–158. doi:10.1038/35004513
36. Kobayashi R, Ariga T, Nonoyama S et al (2006) Outcome in patients with Wiskott-Aldrich syndrome following stem cell transplantation: an analysis of 57 patients in Japan. Br J Haematol 135:362–366. doi:10.1111/j.1365-2141.2006.06297.x
37. Kootstra NA, Verma IM (2003) Gene therapy with viral vectors. Annu Rev Pharmacol Toxicol 43:413–439. doi:10.1146/annurev.pharmtox.43.100901.140257
38. Lacout C, Haddad E, Sabri S et al (2003) A defect in hematopoietic stem cell migration explains the nonrandom X-chromosome inactivation in carriers of Wiskott-Aldrich syndrome. Blood 102:1282–1289. doi:10.1182/blood-2002-07-2099
39. Lum LG, Tubergen DG, Corash L, Blaese RM (1980) Splenectomy in the management of the thrombocytopenia of the Wiskott-Aldrich syndrome. N Engl J Med 302:892–896. doi:10.1056/NEJM198004173021604
40. Mahlaoui N, Pellier I, Mignot C et al (2013) Characteristics and outcome of early-onset, severe forms of Wiskott-Aldrich syndrome. Blood 121:1510–1516. doi:10.1182/blood-2012-08-448118
41. Marshall CJ, Thrasher AJ (2001) The embryonic origins of human haematopoiesis. Br J Haematol. doi:10.1046/j.1365-2141.2001.02537.x
42. Mathew P, Conley ME (1995) Effect of intravenous gamma globulin (IVIG) on the platelet count in patients with Wiskott-Aldrich syndrome. Pediatr Allergy Immunol 6:91–94. doi:10.1111/j.1399-3038.1995.tb00265.x
43. Meyer-Bahlburg A, Becker-Herman S, Humblet-Baron S et al (2008) Wiskott-Aldrich syndrome protein deficiency in B cells results in impaired peripheral homeostasis. Blood 112:4158–4169. doi:10.1182/blood-2008-02-140814
44. Mikkers H, Pike-Overzet K, Staal FJ (2012) Induced pluripotent stem cells and severe combined immunodeficiency: merely disease modeling or potentially a novel cure? Pediatr Res 71:427–432. doi:10.1038/pr.2011.65

45. Miller JC, Tan S, Qiao G et al (2011) A TALE nuclease architecture for efficient genome editing. Nat Biotechnol 29:143–148. doi:10.1038/nbt.1755
46. Moratto D, Giliani S, Bonfim C et al (2011) Long-term outcome and lineage-specific chimerism in 194 patients with Wiskott-Aldrich syndrome treated by hematopoietic cell transplantation in the period 1980–2009: an international collaborative study. Blood 118:1675–1684. doi:10.1182/blood-2010-11-319376
47. Moreno-Carranza B, Gentsch M, Stein S et al (2009) Transgene optimization significantly improves SIN vector titers, gp91phox expression and reconstitution of superoxide production in X-CGD cells. Gene Ther 16:111–118. doi:10.1038/gt.2008.143
48. Nathan DG (1980) Splenectomy in the Wiskott-Aldrich syndrome. N Engl J Med. 17;302(16):916–7
49. Orange JS (2008) Formation and function of the lytic NK-cell immunological synapse. Nat Rev Immunol 8:713–725. doi:10.1038/nri2381
50. Orange JS, Ramesh N, Remold-O'Donnell E et al (2002) Wiskott-Aldrich syndrome protein is required for NK cell cytotoxicity and colocalizes with actin to NK cell-activating immunologic synapses. Proc Natl Acad Sci U S A 99:11351–11356. doi:10.1073/pnas.162376099
51. Oshima K, Imai K, Albert MH et al (2014) Hematopoietic stem cell transplantation for X-linked thrombocytopenia with mutations in the WAS gene. J Clin Immunol. doi:10.1007/s10875-014-0105-5
52. Ozsahin H, Cavazzana-Calvo M, Notarangelo LD et al (2008) Long-term outcome following hematopoietic stem-cell transplantation in Wiskott-Aldrich syndrome: collaborative study of the European society for immunodeficiencies and European group for blood and marrow transplantation. Blood 111:439–445. doi:10.1182/blood-2007-03-076679
53. Pai SY, DeMartiis D, Forino C, Cavagnini S (2006) Stem cell transplantation for the Wiskott–Aldrich syndrome: a single-center experience confirms efficacy of matched unrelated donor transplantation. Bone Marrow Transplant. 38(10):671–679. doi:10.1038/sj.bmt.1705512
54. Panchal SC, Kaiser DA, Torres E et al (2003) A conserved amphipathic helix in WASP/Scar proteins is essential for activation of Arp2/3 complex. Nat Struct Biol 10:591–598. doi:10.1038/nsb952
55. Parolini O, Berardelli S, Riedl E et al (1997) Expression of Wiskott-Aldrich syndrome protein (WASP) gene during hematopoietic differentiation. Blood 90:70–75
56. Perry GS, Spector BD, Schuman LM et al (1980) The Wiskott-Aldrich syndrome in the United States and Canada (1892–1979). J Pediatr 97:72–78
57. Persons DA (2010) Lentiviral vector gene therapy: effective and safe? Mol Ther 18:861–862. doi:10.1038/mt.2010.70
58. Pollard TD, Borisy GG (2003) Cellular motility driven by assembly and disassembly of actin filaments. Cell 112:453–465
59. Qasim W, Gaspar HB, Thrasher AJ (2009) Progress and prospects: gene therapy for inherited immunodeficiencies. Gene Ther 16:1285–1291. doi:10.1038/gt.2009.127
60. Ramesh N, Antón IM, Hartwig JH, Geha RS (1997) WIP, a protein associated with wiskott-aldrich syndrome protein, induces actin polymerization and redistribution in lymphoid cells. Proc Natl Acad Sci U S A 94:14671–14676
61. Ran FA, Hsu PD, Lin C-YY et al (2013) Double nicking by RNA-guided CRISPR Cas9 for enhanced genome editing specificity. Cell 154:1380–1389. doi:10.1016/j.cell.2013.08.021
62. Recher M, Burns SO, Miguel A, Volpi S (2012) B cell–intrinsic deficiency of the Wiskott-Aldrich syndrome protein (WASp) causes severe abnormalities of the peripheral B-cell compartment in mice. Blood 119(12):2819–2828. doi:10.1182/blood-2011-09-379412
63. Rechitsky S, Kuliev A, Tur-Kaspa I et al (2004) Preimplantation genetic diagnosis with HLA matching. Reprod BioMed Online 9:210–221
64. Rohatgi R, Ma L, Miki H et al (1999) The interaction between N-WASP and the Arp2/3 complex links Cdc42-dependent signals to actin assembly. Cell 97:221–231
65. Sabri S, Foudi A, Boukour S et al (2006) Deficiency in the Wiskott-Aldrich protein induces premature proplatelet formation and platelet production in the bone marrow compartment. Blood 108:134–140. doi:10.1182/blood-2005-03-1219

66. Shcherbina A, Rosen FS, Remold-O'Donnell E (1999) WASP levels in platelets and lymphocytes of wiskott-aldrich syndrome patients correlate with cell dysfunction. J Immunol 163:6314–6320
67. Siminovitch KA (2003) Prenatal diagnosis and genetic analysis of Wiskott-Aldrich syndrome. Prenat Diagn 23:1014–1016. doi:10.1002/pd.745
68. Snapper SB, Rosen FS, Mizoguchi E, Cohen P (1998) Wiskott-Aldrich syndrome protein-deficient mice reveal a role for WASP in T but not B cell activation. Immunity 9:81–91
69. Sullivan KE, Mullen CA, Blaese RM (1994) A multiinstitutional survey of the Wiskott-Aldrich syndrome. J Pediatr 125(6 Pt 1):876–85
70. Symons M, Derry J, Karlak B et al (1996) Wiskott–Aldrich syndrome protein, a novel effector for the GTPase CDC42Hs, is implicated in actin polymerization. Cell 84:723–734
71. Takahashi K, Tanabe K, Ohnuki M et al (2007) Induction of pluripotent stem cells from adult human fibroblasts by defined factors. Cell 131:861–872. doi:10.1016/j.cell.2007.11.019
72. Takahashi K, Yamanaka S (2006) Induction of pluripotent stem cells from mouse embryonic and adult fibroblast cultures by defined factors. Cell 126:663–676. doi:10.1016/j.cell.2006.07.024
73. Takenawa T, Suetsugu S (2007) The WASP–WAVE protein network: connecting the membrane to the cytoskeleton. Nat Rev Mol Cell Biol. doi:10.1038/nrm2069
74. Takenawa T, Miki H (2001) WASP and WAVE family proteins: key molecules for rapid rearrangement of cortical actin filaments and cell movement. J Cell Sci 114:1801–1809
75. Tavassoli M, Aoki M (1981) Migration of entire megakaryocytes through the marrow--blood barrier. Br J Haematol 48:25–29
76. Thornhill SI, Schambach A, Howe SJ et al (2008) Self-inactivating gammaretroviral vectors for gene therapy of X-linked severe combined immunodeficiency. Mol Ther 16:590–598. doi:10.1038/sj.mt.6300393
77. Thrasher AJ, Burns SO (2010) WASP: a key immunological multitasker. Nat Rev Immunol 10:182–192. doi:10.1038/nri2724
78. Trevino AE, Zhang F (2014) Genome editing using Cas9 nickases. Methods Enzymol 546:161–174. doi:10.1016/B978-0-12-801185-0.00008-8
79. Tsuboi S, Meerloo J (2007) Wiskott-Aldrich syndrome protein is a key regulator of the phagocytic cup formation in macrophages. J Biol Chem 282:34194–34203. doi:10.1074/jbc.M705999200
80. Volkman BF, Prehoda KE, Scott JA et al (2002) Structure of the N-WASP EVH1 domain-WIP complex: insight into the molecular basis of Wiskott-Aldrich Syndrome. Cell 111:565–576
81. Wiskott, A. (1937) Familiärer, angeborener Morbus Werlhofii? Monatsschrift für Kinderheilkunde. 68:212–216
82. Witzel M, Braun C, Boztug K, Klein C (2013) Gene therapy for Wiskott–Aldrich syndrome. Expert Opin Orphan Drugs. doi:10.1517/21678707.2013.830068
83. Xu W, Russ JL, Eiden MV (2012) Evaluation of residual promoter activity in γ-retroviral self-inactivating (SIN) vectors. Mol Ther 20:84–90. doi:10.1038/mt.2011.204
84. Yamanaka S (2010) Patient-specific pluripotent stem cells become even more accessible. Cell Stem Cell 7:1–2. doi:10.1016/j.stem.2010.06.009
85. Zhang J, Shehabeldin A, da Cruz L (1999) Antigen receptor–induced activation and cytoskeletal rearrangement are impaired in Wiskott-Aldrich syndrome protein–deficient lymphocytes. J Exp Med 190(9):1329–1342
86. Zhang H, Schaff UY, Green CE et al (2006) Impaired integrin-dependent function in Wiskott-Aldrich syndrome protein-deficient murine and human neutrophils. Immunity 25:285–295. doi:10.1016/j.immuni.2006.06.014
87. Zhu Q, Watanabe C, Liu T et al (1997) Wiskott-Aldrich syndrome/X-linked thrombocytopenia: WASP gene mutations, protein expression, and phenotype. Blood 90:2680–2689

Generating Blood from iPS Cells

19

Peter Karagiannis, Hiroshi Endo, and Koji Eto

Abstract

Blood transfusions are the oldest contemporary form of cell therapy and remain the standard treatment for anemia and thrombocytopenia. Due to the enucleate nature of the cells, patients that require erythrocyte or platelet transfusions must rely on a constant supply of blood donors. Aging populations and other demographic trends, however, make this dependency unsustainable. Consequently, investigators have sought other sources for these cell types. Despite only being discovered 10 years ago, induced pluripotent stem cells (iPSCs) have quickly gained the imagination of researchers seeking new clinical treatments for an assortment of diseases, including hematopathologies. The appeal of iPSCs is that they permit the conversion of any cell in the body to another type. This cell reprogramming makes it possible to conduct autologous transplantation by taking cells such as fibroblasts from the patient and converting them to erythrocytes or platelets before reintroduction. iPSCs also allow for investigations on hematopoiesis that were previously impossible, as they enable each intermediate step in the development of a blood cell to be controlled and therefore studied in greater detail. In this chapter, we review the basics about cell reprogramming to the iPSC state and the application of this technology to generating erythrocytes and platelets ex vivo.

19.1 Introduction

There is perhaps no more anticipation about new medical treatments than those of stem cell therapies. These treatments have the promise to correct diseased cells and supply healthy ones indefinitely. The excitement over stem cell therapies has

P. Karagiannis • H. Endo • K. Eto (✉)
Center for iPS Cell Research and Application, Kyoto University,
53 Kawahara-cho, Shogoin, Sakyo-ku, Kyoto 606-8507, Japan
e-mail: kojieto@cira.kyoto-u.ac.jp

H. Schulze, J. Italiano (eds.), *Molecular and Cellular Biology of Platelet Formation*,
DOI 10.1007/978-3-319-39562-3_19

intensified with the discovery of embryonic stem cells (ESCs) and more recently induced pluripotent stem cells (iPSCs) [1, 2]. Yet stem cell therapies date well before these stem cells, including those of hematopoietic stem cells (HSCs) [3]. HSCs have the potential to differentiate into all hematopoietic and lymphatic lineages, thus providing sustenance to patients whose bodies fail to supply sufficient or healthy blood. This method is so standard that globally nearly 100 million blood units are collected annually from donors, a total that exceeds the amount used in transfusions [4, 5]. Despite the abundance of blood supplies, however, there is fear that blood donations are not sustainable. Already, many countries, especially poorer ones, are undersupplied [6]. Even developed countries are not immune to this worry, with some anticipating a blood donor shortage as high as 20 % within a generation [7]. In addition, transfusions are not without risk, as immune rejections due to their allogeneic nature can trigger morbidity and mortality [8]. Autologous transplantations are therefore preferred, but these are rarely viable for several reasons. First, the HSC constituency of peripheral blood is only 1 % (and only 0.1 % in cord blood), which implies that an extraordinary amount of blood from the patient must be harvested [9]. While HSCs can be expanded, they cannot be expanded indefinitely. Furthermore, for patients with genetic diseases such as sickle cell anemia and thalassemia, genome editing of the HSCs is required before their reintroduction, but this task is not trivial. Thus, alternative sources of blood cells are being explored.

Pluripotent stem cells (PSCs), which include ESCs and iPSCs, are one example of an alternative source. Unlike HSCs, which are the progenitors of all hematopoietic cells, PSCs have the potential to differentiate into any cell of the body, putting them above HSCs in the developmental hierarchy. Furthermore, PSCs are far easier to expand and to have their genomes edited [10]. Finally, while ESCs can only be used for allogeneic transplantation, because they are derived from the blastocyst stage and thus unavailable from adult donors, iPSCs, on the other hand, can be harvested from adults, including, most importantly, the patient, which means that unlike ESCs, they have autologous transplantation potential.

The creation of iPSCs is arguably the most poignant demonstration of cell reprogramming. In effect, their invention demonstrated that any cell in the human body could be reprogrammed to the pluripotent state from which it can then be differentiated into a desired lineage. In other words, by mastering the mechanisms of cell reprogramming, any cell in the body can be refashioned into another, including blood cells. Additionally, PSCs can be expanded indefinitely, which means their supply is not exhausted with use. The implications of this property for blood donations, especially of enucleated cell types such as erythrocytes and platelets, are immense. However, while in principle one can generate iPSCs from a patient, the reality makes this strategy infeasible both in terms of time and cost [11]. Accordingly, a great deal of international collaboration is underway to build iPSC banks that accumulate cell lines of homozygous HLA haplotypes [12]. Matching HLAs between donors and recipient should mitigate many of the concerns about transplant rejection seen currently in cell therapies like blood transfusion.

The challenge for researchers then is to develop effective protocols that both acquire iPSCs and differentiate them into the desired cell type. Here we consider the derivation of the two most common blood cells transfused, erythrocytes and platelets, from iPSCs, by reviewing how differentiated cells are reverted to the pluripotent state and from there redirected to the hematopoietic lineage. To accomplish this goal, researchers are learning intricate details about hematopoiesis, including the iterations that lead a cell to its final hematopoietic fate.

19.2 iPS Cells and Development

The origin of any somatic cell in the body can be traced to the zygote. Upon its formation, the zygote immediately commences a series of events that gradually narrows the number of potential lineages until the cell reaches its final differentiated state. In Waddington's famed landscape, the fate of the cell is visualized as rolling down a valley, passing through lesser degrees of lineage potential until its end point, with gravity preventing the cell from returning to a more primitive state (Fig. 19.1). This image is an effective visual of the then contemporary view that differentiation was irreversible. John Gurdon's Nobel Prize winning work changed that perception, however, as he showed that the nuclear transfer of the somatic cell could reprogram an oocyte [13]. This work is the foundation for animal cloning, including Dolly and other mammal cells, and for the invention of iPSCs [14, 15].

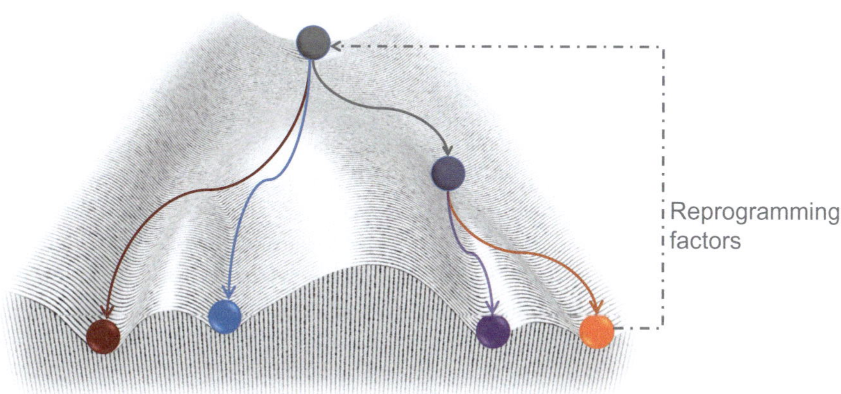

Fig. 19.1 In Waddington's landscape, development is represented as a valley in which a cell (*circles*) descends to its final state. During the descent, the number of cell types into which a cell can differentiate reduces. The landscape assumes that development is irreversible. The induction of reprogramming factors shows, however, that a cell can return to the top of the valley from which it can descend to a different fate

Cell reprogramming took another leap with the isolation of ESCs from mouse embryo and eventually from human embryo (mESCs and hESCs, respectively). When injected into a blastocyst, ESCs can differentiate into the three germ layers, demonstrating their pluripotency. Pluripotency allows for the genome of these cells to be included into offspring, inviting the creation of chimeric mice, which has since opened the door to new models for disease pathogenesis, body development, and cell therapies. In contrast, HSCs possess only multipotency, which limits their differentiation potential to hematopoietic lineages. However, because their isolation requires the destruction of the embryo, hESCs are wrought in controversy, and ambiguous policies have discouraged many scientists from considering their use [16].

The isolation of human iPSCs (hiPSCs) circumvents many of these problems. The first iPSCs were made using mouse cells (miPSCs). Takahashi and Yamanaka reverted the somatic state of mouse fibroblasts to the pluripotent state by inducing four transcription factors, Oct4, Sox2, Klf4, and c-Myc (OSKM) [17]. A year later, they and another group did the same with human fibroblasts and the same OSKM concoction [1, 18]. Importantly, iPSCs are functionally equivalent to ESCs and thus can be used for essentially the same applications. Modifying the genome of iPSCs and then differentiating and transplanting the modified cells into mouse models are a proven effective cell therapy for several mutant-based diseases. Coincidently, the first proof of concept of iPSC-based cell therapy was demonstrated using a sickle cell anemia mouse model, as Hanna et al. reprogrammed fibroblasts into iPSCs that carry the sickle cell anemia mutant [19]. The authors then partitioned the iPSCs into two groups and modified the globin genes in one. The edited and nonedited iPSCs were differentiated into HSCs and transplanted into the diseased mice. Mice that received the corrected cells showed recovery, while those that did not remained ill. A similar strategy using hiPSCs has also been demonstrated [20, 21]. Within a few years of the first iPSC report, an exhaustive number of other somatic cells from several mammalian species have been reprogrammed to the pluripotent state (see Stadtfeld and Hochedlinger [22] for a list), including human peripheral and cord blood [23]. Additionally, a large number of iPSC-based human disease models have been created, including many for hematopathologies [24, 25].

One striking feature about iPSCs is the remarkable simplicity with which they can be made, as only the induction of OSKM is sufficient. Since the first iPSC reports, it has been demonstrated that OSKM is not necessary for reprogramming, suggesting redundancy in the pluripotency network. For example, Yu et al. managed to reprogram fibroblasts to the pluripotent state when substituting Klf4 and c-Myc with Nanog and Lin28, and Montserrat et al. showed that Oct4 can be replaced with Gata3 for the reprogramming [18, 26]. These are only a few examples of the many concoctions that can reprogram a cell to the pluripotent state.

Along with the different reprogramming factors, there are a number of induction methods that can be used for the reprogramming (Table 19.1). The original OSKM protocol involved the integration of retroviruses into the host chromatin. This approach risks mutagenesis and is therefore unsuitable for clinical treatment. Investigators have sought ways to excise the insertion, such as by piggyBac

Table 19.1 Brief list of reprogramming vehicles

Method	Examples
Integration	Lentivirus, retrovirus
Nonintegration, DNA based	Episomes and other plasmids
Nonintegration, RNA based	Sendai virus
Nonintegration, nucleotide free	Small molecules such as inhibitors

transposon, but these methods are laborious [27]. Accordingly, nonintegrating measures to induce the reprogramming have been considered. Such examples include episomal plasmids, which offer the best efficiency-to-safety ratio, Sendai virus, which is RNA based, or the use of synthesized RNA or proteins [28–33]. Hou et al. managed to reprogram cells using only small molecules [34]. Small molecules are attractive because of their cost and relatively easy standardization, which are crucial considerations for the industrialization of iPSC technology [35]. These nonintegrating methods all have their advantages and disadvantages, but none outmatch the reprogramming efficiency of retroviruses.

Despite Waddington's landscape, it would be naïve to think that reprogramming is a simple reversal of the events that differentiate a pluripotent cell to its final state. mESCs are equivalent to cells in the blastocyst, which then proceed to the epiblast at which point gastrulation begins with the formation of the primitive streak. For the primitive streak to branch to the three germ lines, ectoderm, endoderm, and mesoderm, an epithelial-to-mesenchymal transition (ETM) must occur. The early phase of reprogramming depends on the reversal of this transition, a mesenchymal-to-epithelial transition (MET), and the initial gene targets of the reprogramming factors are regulators of MET [36–38]. Polo et al. decomposed the reprogramming process into two waves [39]. In the first wave, the expression of genes associated with pluripotency, metabolism, and cytoskeleton organization is upregulated, while those associated with development are downregulated. It is only in the second wave, which appears several days after the first, when genes that promote stem cell maintenance and embryonic development are upregulated. To be fully reprogrammed to the pluripotent state, a cell must complete these two waves. The success rate of reprogramming is relatively paltry, with only <1 % reaching a stable pluripotent state [15]. Interestingly, Tanabe et al. showed that a much higher percentage of cells initiate the reprogramming process, suggesting that many cells undergo partial reprogramming but fail completion [40]. This observation reflects the theory that reprogramming can be divided into sequential stochastic and deterministic phases [41]. Cells in the stochastic phase are heterogeneous, indicating that disparate mechanisms may be at play, but once they enter the deterministic phase, gene expressions synchronize, beginning with the expression of Nanog and Oct4 [41].

It is assumed that the transition between the two phases depends on a rate-limiting step that once passed essentially assures cells will not fall refractory to the reprogramming factors. This rate-limiting step may be attributed to epigenetic changes in the cell. Chromatin remodeling is extensive during the reprogramming process, and Oct4 and Nanog are both hypermethylated until the late stages of reprogramming [39, 42, 43]. Papp and Plath thus postulated demethylation of these genes is the defining event that transitions cells to the deterministic phase [44]. Consistent with this conclusion, it has been reported that vitamin C, a small molecule that regulates histone demethylation, enhances the reprogramming efficiency [45–47]. Another possible epigenetic factor is chromatin modifications, as these occur with the initiation of reprogramming [42].

One challenge in identifying this rate-limiting step is that cell lines, be they ESCs or iPSCs, have variations in their transcriptome and epigenome. These differences may indicate that different stem cells may use different mechanisms to exercise their stemness [48]. One possible explanation for these differences, at least in the case of iPSCs, is that reprogramming to the pluripotent state must follow different routes due to the variable nature of different somatic cells even of the same type.

19.3 iPSCs to Somatic Cells

Reprogramming cells into iPSCs is only one half of the challenge in generating erythrocytes and platelets ex vivo. Once iPSCs are acquired, they must be perturbed toward the desired hematopoietic lineage. Thus, the conversion of iPSCs to erythrocytes and platelets requires comprehensive understanding of hematopoiesis.

All hematopoietic cells are progeny of the mesoderm. Before proceeding to the three germ layers, murine ESCs form the epiblast. Epiblast cells are also pluripotent, and epiblast stem cells (EpiSCs) have been generated in the lab [49, 50]. Although both ESCs and EpiSCs are pluripotent, their pluripotency differs, leading to the categorization of two states: naïve pluripotency and primed pluripotency [51]. Primed pluripotent cells have less pluripotency, as they show lineage bias, which causes a heterogeneous nature in their development. Thus, a population of primed PSCs is, due to its heterogeneity, more apt to respond variably to common reprogramming induction. In contrast, a naïve PSC population, because it is homogeneous, should result in a more homogeneous response. This difference is important when considering clinical application, as any cell therapy must have predictable behavior. Furthermore, this difference is pertinent when considering murine and human PSCs: murine PSCs, including mESCs and miPSCs, show properties indicative of the naïve pluripotent state, whereas human PSCs have attributes of the primed state.

The primary factors for mesoderm induction are BMP4, Wnt3, and FGF2 [52–55]. As an attribute to how the naïve or primed state influences the differentiation protocol, BMP4 promotes the differentiation of EpiSCs but stabilizes the undifferentiated state of ESCs. FGF2 makes cells sensitive to VEGF, which is thought to stimulate the hematopoietic lineage [56, 57]. HSCs are marked as CD34+ cells; however these cells are not specific to the hematopoietic lineage. Vodyanik et al. have recommended CD34+43+ cells instead [58].

The classical model of hematopoiesis begins with HSCs at the apex (Fig. 19.2, solid arrows). In the next iteration, the lymphoid and myeloid pathways separate. The differentiation of megakaryocytes and erythrocytes is assumed to be the product of a bipotent progenitor, megakaryocyte-erythrocyte progenitor (MEP), which is

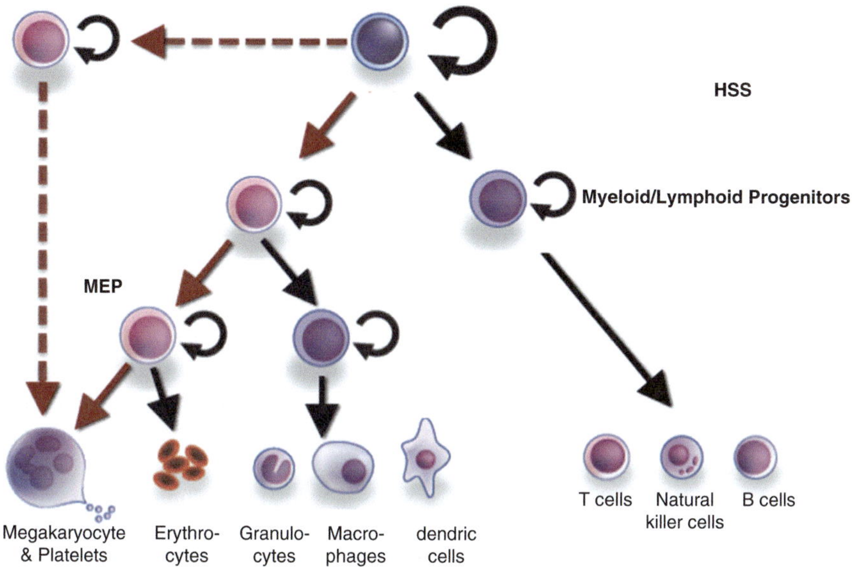

Fig. 19.2 In the classical model of hematopoiesis (*solid arrows*), megakaryocytes are derived from the bipotent progenitor MEP. MEP is preceded by a series of intermediate steps. *Red lines* show the megakaryocyte lineage; *black lines* show other lineages. Several studies have indicated, however, that there exists a megakaryocyte progenitor with self-renewing capability that is a direct intermediate of HSC. The existence of this progenitor may provide a simpler and efficient pathway for differentiation of megakaryocytes in the lab (*dotted arrows*). *Circular arrows* indicate self-renewing capability

marked by a large increase in the expression of GATA1 and NF-E2 [59]. Human MEP cells are equivalent to CD41a+CD235a+ [58, 60, 61]. This model is appealing in that it shows clearly bifurcations for all hematopoietic lineages. At the same time, these bifurcations demonstrate the large risk of heterogeneous populations when deriving megakaryocytes and erythrocytes, since it would be very easy for a subpopulation of HSCs (or any other starting point) to diverge toward the wrong lineage. In the ideal differentiation protocol, the original cell would be derived into the desired cell type in just one step.

Later studies have demonstrated that the classical model is an overly simplistic description. Hematopoiesis has been deconstructed into two sequential phases, primitive hematopoiesis and definitive hematopoiesis. HSCs emerge through definitive hematopoiesis, which happens at E10.5, 3 days after primitive hematopoiesis [62]. The earlier phase occurs in the yolk sac and generates nucleated erythrocytes, erythroblasts and progenitors of definitive erythrocytes (or definitive type of erythroblasts), and megakaryocytes [63]. Definitive hematopoiesis occurs in multiple regions, but the HSCs will converge at the fetal liver where they are expanded before settling at their final destination [64]. This phase will also see the appearance of enucleated erythrocytes, which have different globins from the aforementioned nucleated ones [65]. Several studies have shown that HSCs and definitive erythrocytes bud off endothelial cells, indicating that an endothelial-hematopoietic transition is crucial for hematopoiesis to occur [66, 67]. These endothelial cells are referred to as "hemogenic endothelium" and are recognized by the expression of the transcription factor RUNX1 [68]. Like erythrocytes, megakaryocytes have been observed in both phases, but of different ploidy [69]. Megakaryocytes generated during primitive hematopoiesis are only diploids and unlikely to be the primary source of platelets in the adult body [70], whereas fully mature megakaryocytes with maximal platelet generation capability are polyploidy.

These insights have suggested that the different hematopoietic lineages cannot be partitioned easily. Doulatov et al. showed that myeloid lineages can appear in cell populations that should, according to the classical model, be lymphoid committed [71]. More interesting, Morita et al. showed that HSCs themselves could be divided into subpopulations that have different potentials for the erythrocyte and megakaryocyte lineages [72]. Sanjuan-Pla went further showing that one subpopulation is primed for platelet-specific gene expressions, suggesting it could be possible to differentiate megakaryocytes directly from HSCs [73]. Indeed, these cells were cloned by Yamamoto et al., who reported a megakaryocyte population that had self-renewing ability and were the daughter cells of HSCs [74]. This direct lineage commitment suggests that HSCs can be differentiated into megakaryocytes that autonomously expand using a minimal number of steps in the culture medium, thus significantly increasing the probability of a homogenous population of unipotent platelet progenitors (Fig. 19.2, dashed arrows). This progenitor population may therefore make an ideal source for platelet generation.

19.4 Thrombopoiesis

Even before the creation of iPSCs, there was extensive effort to generate platelets ex vivo. Patients suffering from thrombocytopenia will normally be treated with blood transfusions. However, the body recycles its platelet population every 10 days, which demands patients receive transfusions regularly [75]. In addition, platelets must be stored at room temperature, which gives them extraordinarily high risk of bacterial contamination and thus a shelf life of only a few days (in Japan the limit is 4 days) [76]. Depending on donors then is perpetually taxing, since donors must be recruited continuously.

One possible solution would be to maintain platelets in a progenitor state that has a much longer shelf life. All platelets are the progeny of megakaryocytes, and megakaryocytes that produce functional platelets are marked as CD41a+CD42b+. The primary factor in this development is thrombopoietin (TPO), as it directs MEP to the megakaryocyte lineage [77]. The importance of TPO is recognized by the fact that the first report on in vitro platelet generation was made soon after its discovery [78]. Knockout of TPO or its receptor, c-MPL, is associated with severe thrombocytopenia and also a significant decrease in mature platelets [79].

Platelets are the end products of membrane protrusions from megakaryocytes, called proplatelets, which extend into the sinusoidal vessels where they are eviscerated by blood flow. A single megakaryocyte will generate thousands of platelets before its membrane is exhausted and the remaining cell body is degraded [80]. To produce such a large number of platelets, megakaryocytes must produce an extraordinary amount of membrane. It has been observed that megakaryocytes undergo not mitosis, but endomitosis, a phenomenon that increases both cell volume, such that a single megakaryocyte can become 0.1 mm wide, and the surface area/volume ratio, which explains the additional membrane for platelet construction [81]. Endomitosis is accompanied by an increase in ploidy, as mature megakaryocytes will have up to 128 N DNA content [82]. The ploidy is believed to positively correlate with the number of platelets generated from a single megakaryocyte [83, 84]. Accordingly, investigators have sought ways to promote endomitosis and ploidy when pursuing ex vivo platelet generation.

Crucial for the cell to enter endomitosis is inhibition of myosin II. This inhibition is achieved by RUNX1, which prevents formation of the cleavage furrow [85]. Two other molecules may also be promising targets to enhance endomitosis; GEF-H1 and ECT2 are downregulated sequentially during this phenomenon; GEF-H1 during the 2 N-4 N transition and ECT2 thereafter [86]. Endomitosis in megakaryocytes is accompanied by a new intracellular structure, the invaginated membrane system (IMS). The IMS acts as the membrane reservoir for the platelet intermediate, proplatelets [81]. Although it exists throughout the cell, the IMS begins its formation in one region of the megakaryocyte and spreads using WASP-WAVE signaling, which initiates actin assembly [87–89]. This initial region is marked by GPIbα of the GPIb-V-IX complex, which also marks the shifting of the megakaryocyte from the osteoblast niche, where it is expanded, to the perivascular niche, where it is matured, a shift promoted by the cytokine SDF1 and its receptor CXCR4 [90, 91]. It is in the perivascular niche where megakaryocytes begin to display proplatelets in correspondence with several extracellular matrix proteins, including von Willebrand factor (vWF), fibrinogen, fibronectin, and VCAM-1. Fibrinogen, fibronectin, and VCAM-1 promote proplatelet formation, while vWF is thought vital for platelet function [92–97].

Proplatelets can be imagined as a long chain of swellings linked by a thin chain. Once proplatelets enter the bloodstream, blood flow will eviscerate them at these links, with each swelling having the necessary components for a functional platelet [98]. Proplatelets are dynamics structures, with cytoskeletal proteins constantly transporting material to cause constant rearrangement with the believed aim of maximizing platelet production [99, 100]. A megakaryocyte will continue to extend proplatelets until it has no more membrane and hence be degraded. Thus, once megakaryocytes are fully matured, the mechanism for platelet production can be viewed as a reductive process until the proper size of a platelet is acquired. How the system knows when to stop rearranging its membrane constituents once the size of a platelet is acquired is unknown.

Proplatelets are directed to the sinusoidal vessels by sphingosine 1-phosphate (S1P) and its receptor, S1pr1. Deficiencies in either of these molecules have no effect on megakaryocyte development, including IMS or proplatelet formation. Instead, deficiency results in proplatelets that extend randomly and do not localize toward the bloodstream [101]. However, the S1P gradient only attracts the proplatelets to the vessels; it does not allow the proplatelets to breach the endothelial wall and enter the vessels. For this purpose, podosomes on the membrane surface are used. These podosomes are also dependent on WASP signaling [102].

The above is a brief description of the key events that occur in order for scientists to perturb an HSC into platelet-generating megakaryocytes. Research has shown that megakaryocytes derived from PSCs, such as human iPSCs, share more in common with megakaryocytes from cord blood than they do with megakaryocytes from peripheral blood, meaning they are generally smaller in size and shed fewer platelets [84, 103–105]. While these characteristics may ostensibly seem disadvantageous, because much of our understanding about thrombopoiesis in humans comes from cord blood sources, it suggests that cord blood-based studies on the maturation mechanisms of megakaryocytes can be extrapolated to iPSC-based systems.

Because human ESCs predate human iPSCs by almost 10 years and that the differentiation processes are essentially identical, the initial studies on ESC-derived megakaryocytes provide important insights for later experiments that used iPSCs. Gaur et al. were the first to derive megakaryocytes from ESCs, but with very modest success, as the megakaryocytes were only of 32 N ploidy and the expansion of the stem cells was very poor, with less than one megakaryocyte being acquired on average per ESC used [106]. Using TPO, SCF, and one additional cytokine, IL-11, Lu et al. reported the derivation of functional platelets from ESCs, but again the number was unacceptably low for clinical purposes [107]. Takayama et al. discovered that a saclike structure generated during the culturing with VEGF could create a more suitable environment for megakaryocyte expansion and maturation, as they found that these structures had higher numbers of megakaryocytes that secreted functional platelets than did other regions in the culture [108]. These studies have provided a basis for the derivation of functional platelets from human iPSCs.

The work by Yamamoto et al. [74] described above indicated the existence of megakaryocytes with self-renewing capacity. Searching for these cells, Nakamura et al. reported the differentiation of hiPSCs to immortalized megakaryocytes, which are megakaryocytes that have extended the normal life span of approximately 2 months to nearly half a year [109]. Key to this discovery was the dynamic control of c-Myc expression (Fig. 19.3). Although c-Myc is among the original OSKM factors, it is actually not part of the core pluripotency network [110, 111]. Rather, it is believed to act as an amplifier of gene expression and considered essential only in the first wave of the reprogramming [112–114]. It has also been associated with partial reprogramming [115, 116]. In megakaryocyte development, c-Myc overexpression has been observed to increase expansion, but at the cost of maturation, as these cells have smaller ploidy [117]. Therefore, Nakamura et al. sought to temporally control the c-Myc expression to exploit its expansion synergy without compromising maturation.

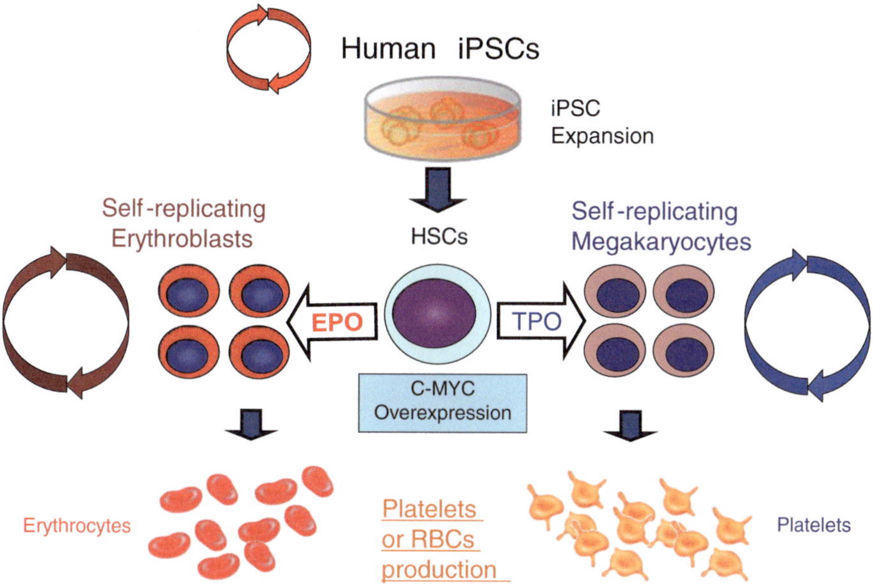

Fig. 19.3 iPSC systems have helped uncover self-replicating erythroblasts and megakaryocytes that are direct intermediates of HSCs. The key to the maintenance of either progenitor is the temporally controlled overexpression of c-Myc. These systems have the potential to respectively provide sufficient numbers of erythrocytes and platelets as alternatives to blood donations for transfusions. *Round arrows* indicate cells with self-renewing capability

To do so, they used a destabilization domain vector system to control the c-Myc expression. This system also incorporated BMI1, a polycomb complex component, which was used to control the expressions of ARF and INK4A, and BCL-Xl, which was used to suppress the expression of caspases. ARF and INK4A are two apoptotic factors that have been shown to affect MK maturity, while caspase-dependent apoptosis is stimulated by c-Myc [118]. Importantly, BCL-XL had to be added in series to the culture, 2–3 weeks after introducing c-Myc and BMI1, as mixing all three together did not result in megakaryocytes that shed functional platelets. Suppressing the expression of these molecules was accompanied by an increase in the expressions of GATA1 and NF-E2. Feng et al. showed that small molecules could also be used to regulate the c-Myc expression and achieve similar platelet output [119].

Lee et al. have estimated that platelet generation ex vivo for adequate therapy will require expanding one HSC into several hundred megakaryocytes and from there acquiring several hundred platelets from one megakaryocyte [120]. iPSC-derived megakaryocytes, including immortalized megakaryocytes, however, fall well short of these targets. Furthermore, the platelets that are shed typically tend to be larger and are less responsive. Therefore, researchers have designed bioreactors that recapitulate the natural microenvironment to enhance megakaryocyte expansion and maturation.

One method is to switch the culturing from two dimensions to three, which by increasing the surface area of the culture could increase yields since proplatelets would have more endothelium with which to interact. Bioreactors that incorporate flow and proteins representative of the osteoblast and perivascular niches show benefits to ex vivo generation. For example, coating the surface with factors such as TPO or vWF promote both MK expansion and maturation and was shown to increase the number of platelets generated [121, 122]. The osteoblast niche is a relatively hypoxic environment, thus lowering the oxygen content could help expand MKs [123]. It has been speculated that hypoxia suppresses the expression of VCAM1, which is part of the signaling pathway regulated by SDF-1, and thus prevents mobilization to the perivascular niche for maturation [124]. Aryl hydrocarbon receptors have been shown to regulate HSC expansion too [125]. StemRegenin 1 is a small molecule that antagonizes the receptor and can increase expansion significantly [126]. Other factors that could improve platelet output are the chemical physical composition of the bioreactor, mild hyperthermia, and metalloproteinase inhibitors. Scaffolds made of aminated nanofibers were seen to expand HSCs, as did the elasticity of the substrate used for culturing [127, 128]. Hyperthermia was found to increase the MK yield, although not the MK ploidy and the molecular mechanism responsible remains unknown [129]. Di Buduo et al. recapitulated the bone marrow niche using an artificial scaffold made from silk. This setup produced functional platelets and modeled multiple steps in thrombopoiesis, including alterations that lead to disease states [130].

Finally, the most important factor in increasing the platelet count may be shear stress. Again, the final step to platelet formation depends on blood flow shearing proplatelets into smaller and smaller size. Incorporating shear stress into the bioreactor design was found to significantly increase both the number of platelets formed and rate of production [131]. Interestingly, the angle of the flow may be an important parameter, as different angles were found to have different platelet outputs, with one study concluding 60° as best [132].

19.5 Erythropoiesis

The count of erythrocytes in the body exceeds that of platelets and any other cell type. Erythrocytes have a longer life span than platelets, reaching 120 days in humans. They also have some unique properties that complicate their ex vivo generation. For one, the development of erythrocytes involves the transition from a nucleated to enucleated state [133]. This transition requires a number of morphological changes and is completed by the activation of macrophages that degrade organelles [134–137]. Antithetical to platelet maturation, which sees an enormous increase in the size of megakaryocytes, erythrocyte maturation is designed to shrink the size of the cell in order to optimize travel in blood, with fully mature erythrocytes being about 35 % smaller than when they first enter the bloodstream [138].

Although hematopoiesis is proving much more complicated than the hierarchy seen in the classical model, which shows that erythrocytes and megakaryocytes are derived from a common bipotent progenitor, it is nevertheless clear that erythropoiesis and thrombopoiesis share much in common. In the case of erythropoiesis, erythropoietin (EPO) is the primary cytokine and acts through its receptor EPOR. EPO stimulates antiapoptotic factors, including BCL-Xl, which has an important role in heme synthesis along with the aforementioned derivation of immortalized megakaryocytes, while at the same time EPO inhibits pro-apoptotic factors [139, 140]. Strengthening the overlap between the two processes, it has also been found that the TPO-c-Mpl signaling axis has a critical role in erythrocyte development [141].

How or even if EPO leads to enucleated erythrocytes is unclear, but this question can be explored by differentiating stem cells. Miharada et al. showed that including mifepristone, an antagonist of glucocorticoid receptors that promote the expansion of HSCs, into an EPO culture increases the yield of enucleated cells [142, 143]. The first derivation of erythrocytes using hPSCs comes from Lu et al. and Lapillonne et al., who, respectively, started with hESCs and hiPSCs. These cells were shown to be safe and effective for transfusions into humans [144, 145]. However, the amount of cells transfused was only 1 % that of a normal transfusion, and the erythrocytes showed a much higher proportion of globulins consistent with fetal blood, not adult blood. Although anemia does not occur in adult patients with fetal hemoglobin, fetal blood does have a higher affinity for oxygen and is thus less efficient at oxygen delivery [4, 146].

Adult hemoglobin is composed of two α–globins and two β–globin subunits (HbA), whereas fetal hemoglobin is composed of two α–globins and two γ–globin subunits (HbF). The fetus benefits from the higher oxygen affinity to acquire the mother's oxygen [147]. Changes in globin expressions are hallmarks of different stages of erythropoiesis. Why proper switching does not occur efficiently in erythrocytes derived from ESCs or iPSCs despite enucleation is not clear, but remains a persistent problem in the field [148–150]. Sankaran et al. reported that deficiency of BCL11A causes a switching failure to occur in mice [151]. Consistent with that report, Ochi et al. found that erythroid cells derived from PSCs have lower expression of BCL11A [152]. To compensate for this problem, the incorporation of organic compounds into erythrocytes has been proposed to manipulate the hemoglobin affinity, but these methods are highly experimental [153].

It should not be surprising based on the late bifurcation in erythrocyte and platelet lineage in the classic model that there exists immortalized erythroblasts, which mirrors the aforementioned immortalized megakaryocytes. Kurita et al. prepared immortalized erythroid progenitor cells that differentiate into enucleated erythrocytes [154]. However, the function of these cells was not tested. On the other hand, our lab was able to generate immortalized erythroblasts using iPSCs based on the same c-Myc strategy we used to prepare immortalized megakaryocytes (Fig. 19.3) [155]. Here too, temporal control of c-Myc was key, as was the simultaneous transduction of c-Myc and BCL-Xl. Unexpectedly, whereas TPO was insufficient to produce immortalized megakaryocytes with the simultaneous transduction of c-Myc and BCL-Xl, EPO was sufficient to produce immortalized erythroblasts. The erythrocytes produced from these cells showed good circulation when transfused into mice.

Conclusion

Despite the success of blood transfusions, there remains a critical proportion of the patient population that does not benefit from these treatments, and the expected shortage of blood donors in the near future demands alternative blood sources be considered. Scientists are therefore investigating ways to prepare blood cells in the lab by mastering the mechanisms driving hematopoiesis. Recent discoveries have shown that subpopulations of HSCs show much earlier markings of megakaryocyte and erythrocyte lineage than previously thought. Isolating these subpopulations may offer an effective way to generating platelets and erythrocytes ex vivo, which could provide a steady source of blood cells that is independent of blood donations. The remarkable ability of PSCs to proliferate indefinitely and retain the capacity to differentiate into any cell type gives them incredible potential for such work. iPSCs especially are attractive, because the source of these cells is readily abundant and can be made using patient cells, which makes them both an excellent tool for disease modeling and autologous transplantation. Moreover, iPSCs are the foundation of immortalized erythroblasts and immortalized megakaryocytes, two cell types with self-renewal capacity. Although the level of self-renewal does not match that of PSCs, these unipotent progenitors exclusively lead to erythrocyte and platelet lineage, respectively, abrogating fears about contamination, which makes them ideal candidates for the generation of clinically relevant erythrocytes and platelets. As the numbers generated of these two cells approach levels for adequate patient care, there is great optimism that these iPSC-based technologies will become the standard source for blood transfusions.

References

1. Takahashi K, Tanabe K, Ohnuki M, Narita M, Ichisaka T, Tomoda K, Yamanaka S (2007) Induction of pluripotent stem cells from adult human fibroblasts by defined factors. Cell 131:861–872
2. Thomson JA, Itskovitz-Eldor J, Shapiro SS, Waknitz MA, Swiergiel JJ, Marshall VS, Jones JM (1998) Embryonic stem cell lines derived from human blastocysts. Science (New York, NY) 282:1145–1147
3. Greenwalt TJ (1997) A short history of transfusion medicine. Transfusion 37:550–563
4. Migliaccio AR, Whitsett C, Papayannopoulou T, Sadelain M (2012) The potential of stem cells as an in vitro source of red blood cells for transfusion. Cell Stem Cell 10:115–119
5. Global Database on Blood Safety: Summary Report 2011. http://www.who.int/bloodsafety/global_databaseGDBS_Summary_Report_2011.pdf
6. Lund TC, Hume H, Allain JP, McCullough J, Dzik W (2013) The blood supply in Sub-Saharan Africa: needs, challenges, and solutions. Transf Apheres Sci Off J World Apher Assoc Off J Eur Soc 49:416–421
7. Japan Ministry of Health LaW (2015) Simulation of future number of blood donors. [Internet]. [23 Jan 2015]. Available from: http://www.mhlw.go.jp/stf/shingi2/0000070616.html. Japanese

8. Spitzer TR, Dey BR, Chen YB, Attar E, Ballen KK (2012) The expanding frontier of hematopoietic cell transplantation. Cytometry B Clin Cytom 82:271–279
9. Xie X, Li Y, Pei X (2014) From stem cells to red blood cells: how far away from the clinical application? Sci China Life Sci 57:581–585
10. Lengerke C, Daley GQ (2010) Autologous blood cell therapies from pluripotent stem cells. Blood Rev 24:27–37
11. Taylor CJ, Peacock S, Chaudhry AN, Bradley JA, Bolton EM (2012) Generating an iPSC bank for HLA-matched tissue transplantation based on known donor and recipient HLA types. Cell Stem Cell 11:147–152
12. Turner M, Leslie S, Martin NG, Peschanski M, Rao M, Taylor CJ, Trounson A, Turner D, Yamanaka S, Wilmut I (2013) Toward the development of a global induced pluripotent stem cell library. Cell Stem Cell 13:382–384
13. Gurdon JB (1962) The developmental capacity of nuclei taken from intestinal epithelium cells of feeding tadpoles. J Embryol Exp Morphol 10:622–640
14. Wilmut I, Schnieke AE, McWhir J, Kind AJ, Campbell KH (1997) Viable offspring derived from fetal and adult mammalian cells. Nature 385:810–813
15. Yamanaka S (2012) Induced pluripotent stem cells: past, present, and future. Cell Stem Cell 10:678–684
16. Wilmut I (2011) Consternation and confusion following EU patent judgment. Cell Stem Cell 9:498–499
17. Takahashi K, Yamanaka S (2006) Induction of pluripotent stem cells from mouse embryonic and adult fibroblast cultures by defined factors. Cell 126:663–676
18. Yu J, Vodyanik MA, Smuga-Otto K, Antosiewicz-Bourget J, Frane JL, Tian S, Nie J, Jonsdottir GA, Ruotti V, Stewart R, Slukvin II, Thomson JA (2007) Induced pluripotent stem cell lines derived from human somatic cells. Science (New York, NY) 318:1917–1920
19. Hanna J, Wernig M, Markoulaki S, Sun CW, Meissner A, Cassady JP, Beard C, Brambrink T, Wu LC, Townes TM, Jaenisch R (2007) Treatment of sickle cell anemia mouse model with iPS cells generated from autologous skin. Science (New York, NY) 318:1920–1923
20. Li HL, Fujimoto N, Sasakawa N, Shirai S, Ohkame T, Sakuma T, Tanaka M, Amano N, Watanabe A, Sakurai H, Yamamoto T, Yamanaka S, Hotta A (2015) Precise correction of the dystrophin gene in duchenne muscular dystrophy patient induced pluripotent stem cells by TALEN and CRISPR-Cas9. Stem Cell Rep 4:143–154
21. Porteus MH (2015) Genome editing of the blood: opportunities and challenges. Curr Stem Cell Rep 1(1):1–8
22. Stadtfeld M, Hochedlinger K (2010) Induced pluripotency: history, mechanisms, and applications. Genes Dev 24:2239–2263
23. Okita K, Yamakawa T, Matsumura Y, Sato Y, Amano N, Watanabe A, Goshima N, Yamanaka S (2013) An efficient nonviral method to generate integration-free human-induced pluripotent stem cells from cord blood and peripheral blood cells. Stem Cells 31:458–466
24. Focosi D, Amabile G, Di Ruscio A, Quaranta P, Tenen DG, Pistello M (2014) Induced pluripotent stem cells in hematology: current and future applications. Blood Cancer J 4:e211
25. Inoue H, Nagata N, Kurokawa H, Yamanaka S (2014) iPS cells: a game changer for future medicine. EMBO J 33:409–417
26. Montserrat N, Nivet E, Sancho-Martinez I, Hishida T, Kumar S, Miquel L, Cortina C, Hishida Y, Xia Y, Esteban CR, Izpisua Belmonte JC (2013) Reprogramming of human fibroblasts to pluripotency with lineage specifiers. Cell Stem Cell 13:341–350
27. Woltjen K, Michael IP, Mohseni P, Desai R, Mileikovsky M, Hamalainen R, Cowling R, Wang W, Liu P, Gertsenstein M, Kaji K, Sung HK, Nagy A (2009) piggyBac transposition reprograms fibroblasts to induced pluripotent stem cells. Nature 458:766–770
28. Fusaki N, Ban H, Nishiyama A, Saeki K, Hasegawa M (2009) Efficient induction of transgene-free human pluripotent stem cells using a vector based on Sendai virus, an RNA virus that does not integrate into the host genome. Proc Jpn Acad Ser B Phys Biol 85:348–362

29. Kim D, Kim CH, Moon JI, Chung YG, Chang MY, Han BS, Ko S, Yang E, Cha KY, Lanza R, Kim KS (2009) Generation of human induced pluripotent stem cells by direct delivery of reprogramming proteins. Cell Stem Cell 4:472–476
30. Okita K, Matsumura Y, Sato Y, Okada A, Morizane A, Okamoto S, Hong H, Nakagawa M, Tanabe K, Tezuka K, Shibata T, Kunisada T, Takahashi M, Takahashi J, Saji H, Yamanaka S (2011) A more efficient method to generate integration-free human iPS cells. Nat Methods 8:409–412
31. Stadtfeld M, Apostolou E, Akutsu H, Fukuda A, Follett P, Natesan S, Kono T, Shioda T, Hochedlinger K (2010) Aberrant silencing of imprinted genes on chromosome 12qF1 in mouse induced pluripotent stem cells. Nature 465:175–181
32. Warren L, Manos PD, Ahfeldt T, Loh YH, Li H, Lau F, Ebina W, Mandal PK, Smith ZD, Meissner A, Daley GQ, Brack AS, Collins JJ, Cowan C, Schlaeger TM, Rossi DJ (2010) Highly efficient reprogramming to pluripotency and directed differentiation of human cells with synthetic modified mRNA. Cell Stem Cell 7:618–630
33. Yu J, Hu K, Smuga-Otto K, Tian S, Stewart R, Slukvin II, Thomson JA (2009) Human induced pluripotent stem cells free of vector and transgene sequences. Science (New York, NY) 324:797–801
34. Hou P, Li Y, Zhang X, Liu C, Guan J, Li H, Zhao T, Ye J, Yang W, Liu K, Ge J, Xu J, Zhang Q, Zhao Y, Deng H (2013) Pluripotent stem cells induced from mouse somatic cells by small-molecule compounds. Science (New York, NY) 341:651–654
35. Silva M, Daheron L, Hurley H, Bure K, Barker R, Carr AJ, Williams D, Kim HW, French A, Coffey PJ, Cooper-White JJ, Reeve B, Rao M, Snyder EY, Ng KS, Mead BE, Smith JA, Karp JM, Brindley DA, Wall I (2015) Generating iPSCs: translating cell reprogramming science into scalable and robust biomanufacturing strategies. Cell Stem Cell 16:13–17
36. Li R, Liang J, Ni S, Zhou T, Qing X, Li H, He W, Chen J, Li F, Zhuang Q, Qin B, Xu J, Li W, Yang J, Gan Y, Qin D, Feng S, Song H, Yang D, Zhang B, Zeng L, Lai L, Esteban MA, Pei D (2010) A mesenchymal-to-epithelial transition initiates and is required for the nuclear reprogramming of mouse fibroblasts. Cell Stem Cell 7:51–63
37. O'Malley J, Skylaki S, Iwabuchi KA, Chantzoura E, Ruetz T, Johnsson A, Tomlinson SR, Linnarsson S, Kaji K (2013) High-resolution analysis with novel cell-surface markers identifies routes to iPS cells. Nature 499:88–91
38. Samavarchi-Tehrani P, Golipour A, David L, Sung HK, Beyer TA, Datti A, Woltjen K, Nagy A, Wrana JL (2010) Functional genomics reveals a BMP-driven mesenchymal-to-epithelial transition in the initiation of somatic cell reprogramming. Cell Stem Cell 7:64–77
39. Polo JM, Anderssen E, Walsh RM, Schwarz BA, Nefzger CM, Lim SM, Borkent M, Apostolou E, Alaei S, Cloutier J, Bar-Nur O, Cheloufi S, Stadtfeld M, Figueroa ME, Robinton D, Natesan S, Melnick A, Zhu J, Ramaswamy S, Hochedlinger K (2012) A molecular roadmap of reprogramming somatic cells into iPS cells. Cell 151:1617–1632
40. Tanabe K, Nakamura M, Narita M, Takahashi K, Yamanaka S (2013) Maturation, not initiation, is the major roadblock during reprogramming toward pluripotency from human fibroblasts. Proc Natl Acad Sci U S A 110:12172–12179
41. Buganim Y, Faddah DA, Cheng AW, Itskovich E, Markoulaki S, Ganz K, Klemm SL, van Oudenaarden A, Jaenisch R (2012) Single-cell expression analyses during cellular reprogramming reveal an early stochastic and a late hierarchic phase. Cell 150:1209–1222
42. Koche RP, Smith ZD, Adli M, Gu H, Ku M, Gnirke A, Bernstein BE, Meissner A (2011) Reprogramming factor expression initiates widespread targeted chromatin remodeling. Cell Stem Cell 8:96–105
43. Mikkelsen TS, Hanna J, Zhang X, Ku M, Wernig M, Schorderet P, Bernstein BE, Jaenisch R, Lander ES, Meissner A (2008) Dissecting direct reprogramming through integrative genomic analysis. Nature 454:49–55
44. Papp B, Plath K (2013) Epigenetics of reprogramming to induced pluripotency. Cell 152:1324–1343
45. Esteban MA, Wang T, Qin B, Yang J, Qin D, Cai J, Li W, Weng Z, Chen J, Ni S, Chen K, Li Y, Liu X, Xu J, Zhang S, Li F, He W, Labuda K, Song Y, Peterbauer A, Wolbank S, Redl H, Zhong

M, Cai D, Zeng L, Pei D (2010) Vitamin C enhances the generation of mouse and human induced pluripotent stem cells. Cell Stem Cell 6:71–79
46. Stadtfeld M, Apostolou E, Ferrari F, Choi J, Walsh RM, Chen T, Ooi SS, Kim SY, Bestor TH, Shioda T, Park PJ, Hochedlinger K (2012) Ascorbic acid prevents loss of Dlk1-Dio3 imprinting and facilitates generation of all-iPS cell mice from terminally differentiated B cells. Nat Genet 44:398–405, S1–2
47. Wang W, Yang J, Liu H, Lu D, Chen X, Zenonos Z, Campos LS, Rad R, Guo G, Zhang S, Bradley A, Liu P (2011) Rapid and efficient reprogramming of somatic cells to induced pluripotent stem cells by retinoic acid receptor gamma and liver receptor homolog 1. Proc Natl Acad Sci U S A 108:18283–18288
48. Cahan P, Daley GQ (2013) Origins and implications of pluripotent stem cell variability and heterogeneity. Nat Rev Mol Cell Biol 14:357–368
49. Brons IG, Smithers LE, Trotter MW, Rugg-Gunn P, Sun B, de Sousa Lopes SM C, Howlett SK, Clarkson A, Ahrlund-Richter L, Pedersen RA, Vallier L (2007) Derivation of pluripotent epiblast stem cells from mammalian embryos. Nature 448:191–195
50. Tesar PJ, Chenoweth JG, Brook FA, Davies TJ, Evans EP, Mack DL, Gardner RL, McKay RD (2007) New cell lines from mouse epiblast share defining features with human embryonic stem cells. Nature 448:196–199
51. Nichols J, Smith A (2009) Naive and primed pluripotent states. Cell Stem Cell 4:487–492
52. Conlon FL, Lyons KM, Takaesu N, Barth KS, Kispert A, Herrmann B, Robertson EJ (1994) A primary requirement for nodal in the formation and maintenance of the primitive streak in the mouse. Development 120:1919–1928
53. Liu P, Wakamiya M, Shea MJ, Albrecht U, Behringer RR, Bradley A (1999) Requirement for Wnt3 in vertebrate axis formation. Nat Genet 22:361–365
54. Saxton TM, Pawson T (1999) Morphogenetic movements at gastrulation require the SH2 tyrosine phosphatase Shp2. Proc Natl Acad Sci U S A 96:3790–3795
55. Winnier G, Blessing M, Labosky PA, Hogan BL (1995) Bone morphogenetic protein-4 is required for mesoderm formation and patterning in the mouse. Genes Dev 9:2105–2116
56. Carmeliet P, Ferreira V, Breier G, Pollefeyt S, Kieckens L, Gertsenstein M, Fahrig M, Vandenhoeck A, Harpal K, Eberhardt C, Declercq C, Pawling J, Moons L, Collen D, Risau W, Nagy A (1996) Abnormal blood vessel development and lethality in embryos lacking a single VEGF allele. Nature 380:435–439
57. Flamme I, Breier G, Risau W (1995) Vascular endothelial growth factor (VEGF) and VEGF receptor 2 (flk-1) are expressed during vasculogenesis and vascular differentiation in the quail embryo. Dev Biol 169:699–712
58. Vodyanik MA, Thomson JA, Slukvin II (2006) Leukosialin (CD43) defines hematopoietic progenitors in human embryonic stem cell differentiation cultures. Blood 108:2095–2105
59. Akashi K, Traver D, Miyamoto T, Weissman IL (2000) A clonogenic common myeloid progenitor that gives rise to all myeloid lineages. Nature 404:193–197
60. Klimchenko O, Mori M, Distefano A, Langlois T, Larbret F, Lecluse Y, Feraud O, Vainchenker W, Norol F, Debili N (2009) A common bipotent progenitor generates the erythroid and megakaryocyte lineages in embryonic stem cell-derived primitive hematopoiesis. Blood 114:1506–1517
61. Sturgeon CM, Ditadi A, Awong G, Kennedy M, Keller G (2014) Wnt signaling controls the specification of definitive and primitive hematopoiesis from human pluripotent stem cells. Nat Biotechnol 32:554–561
62. Slukvin II (2013) Hematopoietic specification from human pluripotent stem cells: current advances and challenges toward de novo generation of hematopoietic stem cells. Blood 122:4035–4046
63. Murry CE, Keller G (2008) Differentiation of embryonic stem cells to clinically relevant populations: lessons from embryonic development. Cell 132:661–680
64. Cumano A, Godin I (2007) Ontogeny of the hematopoietic system. Annu Rev Immunol 25:745–785
65. Nakajima-Takagi Y, Osawa M, Iwama A (2014) Manipulation of hematopoietic stem cells for regenerative medicine. Anat Rec 297:111–120

66. Boisset JC, van Cappellen W, Andrieu-Soler C, Galjart N, Dzierzak E, Robin C (2010) In vivo imaging of haematopoietic cells emerging from the mouse aortic endothelium. Nature 464:116–120
67. Zovein AC, Hofmann JJ, Lynch M, French WJ, Turlo KA, Yang Y, Becker MS, Zanetta L, Dejana E, Gasson JC, Tallquist MD, Iruela-Arispe ML (2008) Fate tracing reveals the endothelial origin of hematopoietic stem cells. Cell Stem Cell 3:625–636
68. Chen MJ, Yokomizo T, Zeigler BM, Dzierzak E, Speck NA (2009) Runx1 is required for the endothelial to haematopoietic cell transition but not thereafter. Nature 457:887–891
69. Tober J, Koniski A, McGrath KE, Vemishetti R, Emerson R, de Mesy-Bentley KK, Waugh R, Palis J (2007) The megakaryocyte lineage originates from hemangioblast precursors and is an integral component both of primitive and of definitive hematopoiesis. Blood 109:1433–1441
70. Potts KS, Sargeant TJ, Markham JF, Shi W, Biben C, Josefsson EC, Whitehead LW, Rogers KL, Liakhovitskaia A, Smyth GK, Kile BT, Medvinsky A, Alexander WS, Hilton DJ, Taoudi S (2014) A lineage of diploid platelet-forming cells precedes polyploid megakaryocyte formation in the mouse embryo. Blood 124:2725–2729
71. Doulatov S, Notta F, Eppert K, Nguyen LT, Ohashi PS, Dick JE (2010) Revised map of the human progenitor hierarchy shows the origin of macrophages and dendritic cells in early lymphoid development. Nat Immunol 11:585–593
72. Morita Y, Ema H, Nakauchi H (2010) Heterogeneity and hierarchy within the most primitive hematopoietic stem cell compartment. J Exp Med 207:1173–1182
73. Sanjuan-Pla A, Macaulay IC, Jensen CT, Woll PS, Luis TC, Mead A, Moore S, Carella C, Matsuoka S, Bouriez Jones T, Chowdhury O, Stenson L, Lutteropp M, Green JC, Facchini R, Boukarabila H, Grover A, Gambardella A, Thongjuea S, Carrelha J, Tarrant P, Atkinson D, Clark SA, Nerlov C, Jacobsen SE (2013) Platelet-biased stem cells reside at the apex of the haematopoietic stem-cell hierarchy. Nature 502:232–236
74. Yamamoto R, Morita Y, Ooehara J, Hamanaka S, Onodera M, Rudolph KL, Ema H, Nakauchi H (2013) Clonal analysis unveils self-renewing lineage-restricted progenitors generated directly from hematopoietic stem cells. Cell 154:1112–1126
75. Deutsch VR, Tomer A (2013) Advances in megakaryocytopoiesis and thrombopoiesis: from bench to bedside. Br J Haematol 161:778–793
76. Williamson LM, Devine DV (2013) Challenges in the management of the blood supply. Lancet 381:1866–1875
77. Kaushansky K, Lok S, Holly RD, Broudy VC, Lin N, Bailey MC, Forstrom JW, Buddle MM, Oort PJ, Hagen FS, Roth GJ, Papayannopoulou T, Foster DC (1994) Promotion of megakaryocyte progenitor expansion and differentiation by the c-Mpl ligand thrombopoietin. Nature 369:568–571
78. Choi ES, Nichol JL, Hokom MM, Hornkohl AC, Hunt P (1995) Platelets generated in vitro from proplatelet-displaying human megakaryocytes are functional. Blood 85:402–413
79. Solar GP, Kerr WG, Zeigler FC, Hess D, Donahue C, de Sauvage FJ, Eaton DL (1998) Role of c-mpl in early hematopoiesis. Blood 92:4–10
80. Avanzi MP, Mitchell WB (2014) Ex vivo production of platelets from stem cells. Br J Haematol 165:237–247
81. Machlus KR, Italiano JE Jr (2013) The incredible journey: from megakaryocyte development to platelet formation. J Cell Biol 201:785–796
82. Zimmet J, Ravid K (2000) Polyploidy: occurrence in nature, mechanisms, and significance for the megakaryocyte-platelet system. Exp Hematol 28:3–16
83. Ferrer-Marin F, Liu ZJ, Gutti R, Sola-Visner M (2010) Neonatal thrombocytopenia and megakaryocytopoiesis. Semin Hematol 47:281–288
84. Mattia G, Vulcano F, Milazzo L, Barca A, Macioce G, Giampaolo A, Hassan HJ (2002) Different ploidy levels of megakaryocytes generated from peripheral or cord blood CD34+ cells are correlated with different levels of platelet release. Blood 99:888–897
85. Lordier L, Bluteau D, Jalil A, Legrand C, Pan J, Rameau P, Jouni D, Bluteau O, Mercher T, Leon C, Gachet C, Debili N, Vainchenker W, Raslova H, Chang Y (2012) RUNX1-induced silencing of non-muscle myosin heavy chain IIB contributes to megakaryocyte polyploidization. Nat Commun 3:717

86. Gao Y, Smith E, Ker E, Campbell P, Cheng EC, Zou S, Lin S, Wang L, Halene S, Krause DS (2012) Role of RhoA-specific guanine exchange factors in regulation of endomitosis in megakaryocytes. Dev Cell 22:573–584
87. Eckly A, Heijnen H, Pertuy F, Geerts W, Proamer F, Rinckel JY, Leon C, Lanza F, Gachet C (2014) Biogenesis of the demarcation membrane system (DMS) in megakaryocytes. Blood 123:921–930
88. Eto K, Nishikii H, Ogaeri T, Suetsugu S, Kamiya A, Kobayashi T, Yamazaki D, Oda A, Takenawa T, Nakauchi H (2007) The WAVE2/Abi1 complex differentially regulates megakaryocyte development and spreading: implications for platelet biogenesis and spreading machinery. Blood 110:3637–3647
89. Schulze H, Korpal M, Hurov J, Kim SW, Zhang J, Cantley LC, Graf T, Shivdasani RA (2006) Characterization of the megakaryocyte demarcation membrane system and its role in thrombopoiesis. Blood 107:3868–3875
90. Avecilla ST, Hattori K, Heissig B, Tejada R, Liao F, Shido K, Jin DK, Dias S, Zhang F, Hartman TE, Hackett NR, Crystal RG, Witte L, Hicklin DJ, Bohlen P, Eaton D, Lyden D, de Sauvage F, Rafii S (2004) Chemokine-mediated interaction of hematopoietic progenitors with the bone marrow vascular niche is required for thrombopoiesis. Nat Med 10:64–71
91. Guerriero R, Mattia G, Testa U, Chelucci C, Macioce G, Casella I, Samoggia P, Peschle C, Hassan HJ (2001) Stromal cell-derived factor 1alpha increases polyploidization of megakaryocytes generated by human hematopoietic progenitor cells. Blood 97:2587–2595
92. Balduini A, Pallotta I, Malara A, Lova P, Pecci A, Viarengo G, Balduini CL, Torti M (2008) Adhesive receptors, extracellular proteins and myosin IIA orchestrate proplatelet formation by human megakaryocytes. J Thromb Haemost JTH 6:1900–1907
93. Eto K, Murphy R, Kerrigan SW, Bertoni A, Stuhlmann H, Nakano T, Leavitt AD, Shattil SJ (2002) Megakaryocytes derived from embryonic stem cells implicate CalDAG-GEFI in integrin signaling. Proc Natl Acad Sci U S A 99:12819–12824
94. Larson MK, Watson SP (2006) Regulation of proplatelet formation and platelet release by integrin alpha IIb beta3. Blood 108:1509–1514
95. Matsunaga T, Fukai F, Kameda T, Shide K, Shimoda H, Torii E, Kamiunten A, Sekine M, Yamamoto S, Hidaka T, Kubuki Y, Yokokura S, Uemura M, Matsuoka A, Waki F, Matsumoto K, Kanaji N, Ishii T, Imataki O, Dobashi H, Bandoh S, Shimoda K (2012) Potentiated activation of VLA-4 and VLA-5 accelerates proplatelet-like formation. Ann Hematol 91:1633–1643
96. Poirault-Chassac S, Nguyen KA, Pietrzyk A, Casari C, Veyradier A, Denis CV, Baruch D (2013) Terminal platelet production is regulated by von Willebrand factor. PLoS One 8:e63810
97. Takizawa H, Eto K, Yoshikawa A, Nakauchi H, Takatsu K, Takaki S (2008) Growth and maturation of megakaryocytes is regulated by Lnk/Sh2b3 adaptor protein through crosstalk between cytokine- and integrin-mediated signals. Exp Hematol 36:897–906
98. Thon JN, Montalvo A, Patel-Hett S, Devine MT, Richardson JL, Ehrlicher A, Larson MK, Hoffmeister K, Hartwig JH, Italiano JE Jr (2010) Cytoskeletal mechanics of proplatelet maturation and platelet release. J Cell Biol 191:861–874
99. Patel SR, Hartwig JH, Italiano JE Jr (2005) The biogenesis of platelets from megakaryocyte proplatelets. J Clin Invest 115:3348–3354
100. Richardson JL, Shivdasani RA, Boers C, Hartwig JH, Italiano JE Jr (2005) Mechanisms of organelle transport and capture along proplatelets during platelet production. Blood 106:4066–4075
101. Zhang L, Orban M, Lorenz M, Barocke V, Braun D, Urtz N, Schulz C, von Bruhl ML, Tirniceriu A, Gaertner F, Proia RL, Graf T, Bolz SS, Montanez E, Prinz M, Muller A, von Baumgarten L, Billich A, Sixt M, Fassler R, von Andrian UH, Junt T, Massberg S (2012) A novel role of sphingosine 1-phosphate receptor S1pr1 in mouse thrombopoiesis. J Exp Med 209:2165–2181
102. Schwertz H, Koster S, Kahr WH, Michetti N, Kraemer BF, Weitz DA, Blaylock RC, Kraiss LW, Greinacher A, Zimmerman GA, Weyrich AS (2010) Anucleate platelets generate progeny. Blood 115:3801–3809

103. Bluteau O, Langlois T, Rivera-Munoz P, Favale F, Rameau P, Meurice G, Dessen P, Solary E, Raslova H, Mercher T, Debili N, Vainchenker W (2013) Developmental changes in human megakaryopoiesis. J Thromb Haemost JTH 11:1730–1741
104. Fuchs DA, McGinn SG, Cantu CL, Klein RR, Sola-Visner MC, Rimsza LM (2012) Developmental differences in megakaryocyte size in infants and children. Am J Clin Pathol 138:140–145
105. Miyazaki R, Ogata H, Iguchi T, Sogo S, Kushida T, Ito T, Inaba M, Ikehara S, Kobayashi Y (2000) Comparative analyses of megakaryocytes derived from cord blood and bone marrow. Br J Haematol 108:602–609
106. Gaur M, Kamata T, Wang S, Moran B, Shattil SJ, Leavitt AD (2006) Megakaryocytes derived from human embryonic stem cells: a genetically tractable system to study megakaryocytopoiesis and integrin function. J Thromb Haemost JTH 4:436–442
107. Lu SJ, Li F, Yin H, Feng Q, Kimbrel EA, Hahm E, Thon JN, Wang W, Italiano JE, Cho J, Lanza R (2011) Platelets generated from human embryonic stem cells are functional in vitro and in the microcirculation of living mice. Cell Res 21:530–545
108. Takayama M, Fujita R, Suzuki M, Okuyama R, Aiba S, Motohashi H, Yamamoto M (2010) Genetic analysis of hierarchical regulation for Gata1 and NF-E2 p45 gene expression in megakaryopoiesis. Mol Cell Biol 30:2668–2680
109. Nakamura S, Takayama N, Hirata S, Seo H, Endo H, Ochi K, Fujita K, Koike T, Harimoto K, Dohda T, Watanabe A, Okita K, Takahashi N, Sawaguchi A, Yamanaka S, Nakauchi H, Nishimura S, Eto K (2014) Expandable megakaryocyte cell lines enable clinically applicable generation of platelets from human induced pluripotent stem cells. Cell Stem Cell 14:535–548
110. Chen X, Xu H, Yuan P, Fang F, Huss M, Vega VB, Wong E, Orlov YL, Zhang W, Jiang J, Loh YH, Yeo HC, Yeo ZX, Narang V, Govindarajan KR, Leong B, Shahab A, Ruan Y, Bourque G, Sung WK, Clarke ND, Wei CL, Ng HH (2008) Integration of external signaling pathways with the core transcriptional network in embryonic stem cells. Cell 133:1106–1117
111. Kim J, Woo AJ, Chu J, Snow JW, Fujiwara Y, Kim CG, Cantor AB, Orkin SH (2010) A Myc network accounts for similarities between embryonic stem and cancer cell transcription programs. Cell 143:313–324
112. Lin CY, Loven J, Rahl PB, Paranal RM, Burge CB, Bradner JE, Lee TI, Young RA (2012) Transcriptional amplification in tumor cells with elevated c-Myc. Cell 151:56–67
113. Nie Z, Hu G, Wei G, Cui K, Yamane A, Resch W, Wang R, Green DR, Tessarollo L, Casellas R, Zhao K, Levens D (2012) c-Myc is a universal amplifier of expressed genes in lymphocytes and embryonic stem cells. Cell 151:68–79
114. Sridharan R, Tchieu J, Mason MJ, Yachechko R, Kuoy E, Horvath S, Zhou Q, Plath K (2009) Role of the murine reprogramming factors in the induction of pluripotency. Cell 136:364–377
115. Nakagawa M, Koyanagi M, Tanabe K, Takahashi K, Ichisaka T, Aoi T, Okita K, Mochiduki Y, Takizawa N, Yamanaka S (2008) Generation of induced pluripotent stem cells without Myc from mouse and human fibroblasts. Nat Biotechnol 26:101–106
116. Wernig M, Meissner A, Cassady JP, Jaenisch R (2008) c-Myc is dispensable for direct reprogramming of mouse fibroblasts. Cell Stem Cell 2:10–12
117. Thompson A, Zhang Y, Kamen D, Jackson CW, Cardiff RD, Ravid K (1996) Deregulated expression of c-myc in megakaryocytes of transgenic mice increases megakaryopoiesis and decreases polyploidization. J Biol Chem 271:22976–22982
118. Juin P, Hunt A, Littlewood T, Griffiths B, Swigart LB, Korsmeyer S, Evan G (2002) c-Myc functionally cooperates with Bax to induce apoptosis. Mol Cell Biol 22:6158–6169
119. Feng Q, Shabrani N, Thon JN, Huo H, Thiel A, Machlus KR, Kim K, Brooks J, Li F, Luo C, Kimbrel EA, Wang J, Kim KS, Italiano J, Cho J, Lu SJ, Lanza R (2014) Scalable generation of universal platelets from human induced pluripotent stem cells. Stem Cell Rep 3:817–831
120. Lee EJ, Godara P, Haylock D (2014) Biomanufacture of human platelets for transfusion: rationale and approaches. Exp Hematol 42:332–346

121. Pallotta I, Lovett M, Kaplan DL, Balduini A (2011) Three-dimensional system for the in vitro study of megakaryocytes and functional platelet production using silk-based vascular tubes. Tissue Eng C Methods 17:1223–1232
122. Sullenbarger B, Bahng JH, Gruner R, Kotov N, Lasky LC (2009) Prolonged continuous in vitro human platelet production using three-dimensional scaffolds. Exp Hematol 37:101–110
123. Lasky LC, Sullenbarger B (2011) Manipulation of oxygenation and flow-induced shear stress can increase the in vitro yield of platelets from cord blood. Tissue Eng C Methods 17:1081–1088
124. Pallotta I, Lovett M, Rice W, Kaplan DL, Balduini A (2009) Bone marrow osteoblastic niche: a new model to study physiological regulation of megakaryopoiesis. PLoS One 4:e8359
125. Smith BW, Rozelle SS, Leung A, Ubellacker J, Parks A, Nah SK, French D, Gadue P, Monti S, Chui DH, Steinberg MH, Frelinger AL, Michelson AD, Theberge R, McComb ME, Costello CE, Kotton DN, Mostoslavsky G, Sherr DH, Murphy GJ (2013) The aryl hydrocarbon receptor directs hematopoietic progenitor cell expansion and differentiation. Blood 122:376–385
126. Boitano AE, Wang J, Romeo R, Bouchez LC, Parker AE, Sutton SE, Walker JR, Flaveny CA, Perdew GH, Denison MS, Schultz PG, Cooke MP (2010) Aryl hydrocarbon receptor antagonists promote the expansion of human hematopoietic stem cells. Science (New York, NY) 329:1345–1348
127. Chua KN, Chai C, Lee PC, Ramakrishna S, Leong KW, Mao HQ (2007) Functional nanofiber scaffolds with different spacers modulate adhesion and expansion of cryopreserved umbilical cord blood hematopoietic stem/progenitor cells. Exp Hematol 35:771–781
128. Holst J, Watson S, Lord MS, Eamegdool SS, Bax DV, Nivison-Smith LB, Kondyurin A, Ma L, Oberhauser AF, Weiss AS, Rasko JE (2010) Substrate elasticity provides mechanical signals for the expansion of hemopoietic stem and progenitor cells. Nat Biotechnol 28:1123–1128
129. Pineault N, Boucher JF, Cayer MP, Palmqvist L, Boyer L, Lemieux R, Proulx C (2008) Characterization of the effects and potential mechanisms leading to increased megakaryocytic differentiation under mild hyperthermia. Stem Cells Dev 17:483–493
130. Di Buduo CA, Wray LS, Tozzi L, Malara A, Chen Y, Ghezzi CE, Smoot D, Sfara C, Antonelli A, Spedden E, Bruni G, Staii C, De Marco L, Magnani M, Kaplan DL, Balduini A (2015) Programmable 3D silk bone marrow niche for platelet generation ex vivo and modeling of megakaryopoiesis pathologies. Blood 125:2254–2264
131. Thon JN, Mazutis L, Wu S, Sylman JL, Ehrlicher A, Machlus KR, Feng Q, Lu S, Lanza R, Neeves KB, Weitz DA, Italiano JE, Jr. (2014) Platelet bioreactor-on-a-chip. Blood 124(12):1857–67
132. Nakagawa Y, Nakamura S, Nakajima M, Endo H, Dohda T, Takayama N, Nakauchi H, Arai F, Fukuda T, Eto K (2013) Two differential flows in a bioreactor promoted platelet generation from human pluripotent stem cell-derived megakaryocytes. Exp Hematol 41:742–748
133. Palis J (2014) Primitive and definitive erythropoiesis in mammals. Front Physiol 5:3
134. Keerthivasan G, Small S, Liu H, Wickrema A, Crispino JD (2010) Vesicle trafficking plays a novel role in erythroblast enucleation. Blood 116:3331–3340
135. Konstantinidis DG, Pushkaran S, Johnson JF, Cancelas JA, Manganaris S, Harris CE, Williams DA, Zheng Y, Kalfa TA (2012) Signaling and cytoskeletal requirements in erythroblast enucleation. Blood 119:6118–6127
136. Wang J, Ramirez T, Ji P, Jayapal SR, Lodish HF, Murata-Hori M (2012) Mammalian erythroblast enucleation requires PI3K-dependent cell polarization. J Cell Sci 125:340–349
137. Yoshida H, Kawane K, Koike M, Mori Y, Uchiyama Y, Nagata S (2005) Phosphatidylserine-dependent engulfment by macrophages of nuclei from erythroid precursor cells. Nature 437:754–758
138. d'Onofrio G, Chirillo R, Zini G, Caenaro G, Tommasi M, Micciulli G (1995) Simultaneous measurement of reticulocyte and red blood cell indices in healthy subjects and patients with microcytic and macrocytic anemia. Blood 85:818–823
139. Koury MJ (2014) Abnormal erythropoiesis and the pathophysiology of chronic anemia. Blood Rev 28:49–66

140. Hafid-Medheb K, Augery-Bourget Y, Minatchy MN, Hanania N, Robert-Lezenes J (2003) Bcl-XL is required for heme synthesis during the chemical induction of erythroid differentiation of murine erythroleukemia cells independently of its antiapoptotic function. Blood 101:2575–2583
141. Hirata S, Takayama N, Jono-Ohnishi R, Endo H, Nakamura S, Dohda T, Nishi M, Hamazaki Y, Ishii E, Kaneko S, Otsu M, Nakauchi H, Kunishima S, Eto K (2013) Congenital amegakaryocytic thrombocytopenia iPS cells exhibit defective MPL-mediated signaling. J Clin Invest 123:3802–3814
142. Kolbus A, Blazquez-Domingo M, Carotta S, Bakker W, Luedemann S, von Lindern M, Steinlein P, Beug H (2003) Cooperative signaling between cytokine receptors and the glucocorticoid receptor in the expansion of erythroid progenitors: molecular analysis by expression profiling. Blood 102:3136–3146
143. Miharada K, Hiroyama T, Sudo K, Nagasawa T, Nakamura Y (2006) Efficient enucleation of erythroblasts differentiated in vitro from hematopoietic stem and progenitor cells. Nat Biotechnol 24:1255–1256
144. Giarratana MC, Rouard H, Dumont A, Kiger L, Safeukui I, Le Pennec PY, Francois S, Trugnan G, Peyrard T, Marie T, Jolly S, Hebert N, Mazurier C, Mario N, Harmand L, Lapillonne H, Devaux JY, Douay L (2011) Proof of principle for transfusion of in vitro-generated red blood cells. Blood 118:5071–5079
145. Lu SJ, Feng Q, Park JS, Vida L, Lee BS, Strausbauch M, Wettstein PJ, Honig GR, Lanza R (2008) Biologic properties and enucleation of red blood cells from human embryonic stem cells. Blood 112:4475–4484
146. Bank A (2006) Regulation of human fetal hemoglobin: new players, new complexities. Blood 107:435–443
147. Sankaran VG, Xu J, Orkin SH (2010) Advances in the understanding of haemoglobin switching. Br J Haematol 149:181–194
148. Dias J, Gumenyuk M, Kang H, Vodyanik M, Yu J, Thomson JA, Slukvin II (2011) Generation of red blood cells from human induced pluripotent stem cells. Stem Cells Dev 20:1639–1647
149. Ledran MH, Krassowska A, Armstrong L, Dimmick I, Renstrom J, Lang R, Yung S, Santibanez-Coref M, Dzierzak E, Stojkovic M, Oostendorp RA, Forrester L, Lako M (2008) Efficient hematopoietic differentiation of human embryonic stem cells on stromal cells derived from hematopoietic niches. Cell Stem Cell 3:85–98
150. Ma F, Ebihara Y, Umeda K, Sakai H, Hanada S, Zhang H, Zaike Y, Tsuchida E, Nakahata T, Nakauchi H, Tsuji K (2008) Generation of functional erythrocytes from human embryonic stem cell-derived definitive hematopoiesis. Proc Natl Acad Sci U S A 105:13087–13092
151. Sankaran VG, Xu J, Ragoczy T, Ippolito GC, Walkley CR, Maika SD, Fujiwara Y, Ito M, Groudine M, Bender MA, Tucker PW, Orkin SH (2009) Developmental and species-divergent globin switching are driven by BCL11A. Nature 460:1093–1097
152. Ochi K, Takayama N, Hirose S, Nakahata T, Nakauchi H, Eto K (2014) Multicolor staining of globin subtypes reveals impaired globin switching during erythropoiesis in human pluripotent stem cells. Stem Cells Transl Med 3:792–800
153. Keyhanian S, Ebrahimifard M, Zandi M (2014) Investigation on artificial blood or substitute blood replace the natural blood. Iran J Pediatr Hematol Oncol 4:72–77
154. Kurita R, Suda N, Sudo K, Miharada K, Hiroyama T, Miyoshi H, Tani K, Nakamura Y (2013) Establishment of immortalized human erythroid progenitor cell lines able to produce enucleated red blood cells. PLoS One 8:e59890
155. Hirose S, Takayama N, Nakamura S, Nagasawa K, Ochi K, Hirata S, Yamazaki S, Yamaguchi T, Otsu M, Sano S, Takahashi N, Sawaguchi A, Ito M, Kato T, Nakauchi H, Eto K (2013) Immortalization of erythroblasts by c-MYC and BCL-XL enables large-scale erythrocyte production from human pluripotent stem cells. Stem Cell Rep 1:499–508

Strategies for the Gene Modification of Megakaryopoiesis and Platelets

20

Lisette Johana Latorre Rey and Ute Modlich

Abstract

Platelets are the central players controlling blood hemostasis. Deregulation of their function leads to various bleeding disorders in human patients with sometimes life-threatening symptoms. To develop therapeutic approaches for these patients, we have to understand the mechanisms behind the diseases.

20.1 Gene Modifications in Megakaryocytes and Platelets

Platelets are the central players controlling blood hemostasis. Deregulation of their function leads to various bleeding disorders in human patients with sometimes life-threatening symptoms. To develop therapeutic approaches for these patients, we have to understand the mechanisms behind the diseases.

Murine models can be employed to study the physiological role of disease-related proteins. One of the widely used models is the gene knockout. Alternatively, genes can be introduced into the genome as transgenic cassettes or can be inserted into the locus of a cellular gene by knock-in approaches. Thanks to new gene targeting methods that developed in the last few years, gene modification can be introduced much faster than the classical homologous recombination. However, germline modification can be laborious and time consuming. In many cases, scientists wish for faster models and target the adult cells rather than developing a new mouse model with the appropriate modification. However, as platelets are anuclear cells and do not contain a DNA genome, their genetic modification is difficult. Therefore, the gene has to be introduced at earlier stages such as the MK progenitors or the hematopoietic stem cells (HSC). For the full reconstitution of the hematopoietic

L.J.L. Rey (✉) • U. Modlich
Research Group for Gene Modification in Stem Cells, LOEWE Centre for Cell and Gene Therapy Frankfurt, Paul-Ehrlich-Institute, Langen, Germany
e-mail: ute.modlich@pei.de

H. Schulze, J. Italiano (eds.), *Molecular and Cellular Biology of Platelet Formation*,
DOI 10.1007/978-3-319-39562-3_20

system after transplantation, long-term HSC have to be modified. In addition, the modification has to be stable and transmitted faithfully onto the daughter cells after cell division. Therefore, in general, integrating vectors, such as the retroviral vectors, will be employed for these approaches.

In this chapter we will introduce the different gene modification strategies and models that can be employed to modify megakaryopoiesis and platelets. Many of these approaches are widely used in biological science; however, we will elucidate their use with a special emphasis on their performance in platelet biology.

20.2 Germline Modifications: Knockout Models

To study the function of genes and pathways in development or disease, model systems with gene deletion (gene knockout) are used. The knockout approach is based on the removal of an important segment of the gene that leads to the disruption of the open reading frame, with the consequent abolition of the expression of a functional gene product. The deletion can be permanent (constitutive) or inducible (conditional) when the effect is desired in a temporal and tissue-specific manner. The knockout technology has allowed for the development of numerous mouse strains and cell lines in which it is possible to study gene functions, signaling pathways, and disease modeling. These genetic modifications are introduced mainly at the embryonic stem cells level. Huge efforts have been invested with the aim of engineering more accurate models since the first knockouts were published.

Modifications introduced in embryonic stem (ES) cells can generate a genetically modified mouse due to their capacity to colonize the germ line and to develop a complete organism (mouse) postimplantation in a foster mother [162]. Mainly male ES cell lines are modified, since the XY karyotype appears to be more stable in culture and generate many more offspring than female karyotype [103, 116]. During the modification and prior to implantation, ES cells are most commonly cultured on mitotically inactive feeder cells (MEFs) in order to maintain their germline properties after genetic manipulation, although there are some feeder-independent ES lines that show high germline competence [107, 215].

Genetic modification of ES cells in classical knockouts is performed by homologous recombination [29, 177, 211]; by taking advantage of the cellular repair machinery, it is possible to align and target a donor engineered plasmid with the corresponding region of homology within the genome, inducing recombination and the replacement of the wild-type genetic locus, by (usually) a drug resistance gene, leading to the removal of an essential DNA region of a gene, disrupting the gene's open reading frame [73] (Fig. 20.1a). Targeting constructs usually contain the coding sequence of a positive drug selection marker to enrich for the population that has successfully recombined with the target sequence. This resistance gene is mostly the neomycin resistance gene (neo^{R}), which is an aminoglycoside phosphotransferase that interferes with protein synthesis in eukaryotic cells. In the targeting plasmid, the recombinant DNA segment is flanked by two arms which are homologous with the target sequence of the gene to be deleted [180]. Although only 2 kb are required, 6–14 kb of homology regions are typically used [119]. However, in cases where the mutation to be introduced leads to the replacement of an essential exon

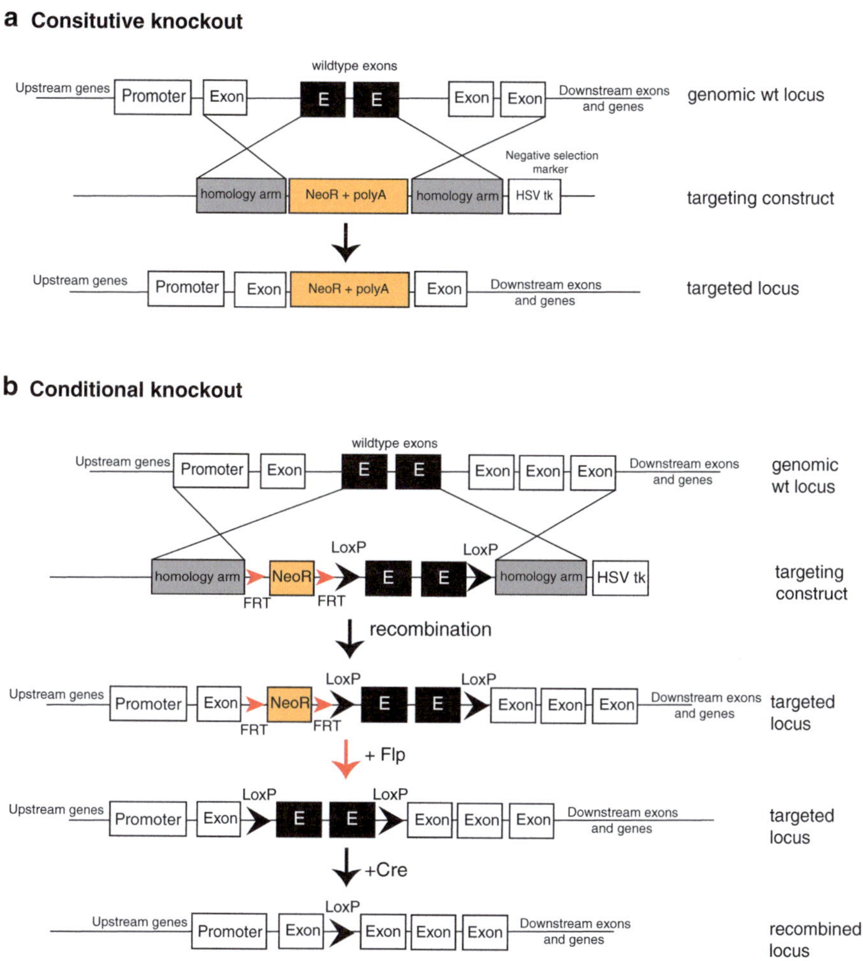

Fig. 20.1 Homologous recombination. (**a**) Constitutive knockout. The targeting vector contains an internal cassette including the neomycin-resistant gene and a polyA sequence (shown in *yellow*), flanked by two homology arms (*gray boxes*) which are able to recombine with the targeted wild-type locus within the genome. The HSV-tk is included as a negative selection marker. After recombination, the neoR cassette is inserted in the targeted locus replacing the wild-type sequence and therefore disrupting the coding sequence of the wild-type gene. (**b**) Conditional knockout. In the first step, the targeting vector is used to generate an embryonic stem cell which contains, flanked with two loxP sites, the coding sequence of the exons to be deleted (shown in *black boxes*); the promoter-neo resistance gene cassette (NeoR, *yellow box*) is flanked by Frt sites. The neoR can be removed by the administration of the Flp recombinase. After the generation of the mice with the engineered genomic locus, these mice can be bred to mice expressing the Cre recombinase. These mice can express Cre in all tissues, in a tissue-specific or inducible manner

by the neo^R without including a polyadenylation signal, downstream exons of the gene could be also transcribed and spliced. To avoid this problem, neo^R is often transcribed in the direction opposite to the wild-type gene. In these cases, a strong promoter drives the expression of the neo^R. The inverted orientation of the neo^R also ensures that the strong promoter will have no transcriptional effect on any genes downstream of the knockout [73, 134].

To increase the accuracy of selection, in most cases the donor plasmids also include a negative selection marker outside of the homology arms in order to discard random integrated clones (Mansour et al. 1988) (Fig. 20.1a). For example, the *herpes simplex virus* thymidine kinase (HSV-tk) can be used for negative selection, as it phosphorylates drugs like ganciclovir or FIAU (1-[2′-deoxy-2′-fluoro- β-D-arabinofuranosyl]-5-iodouracil) into cytotoxic compounds in HSV-tk-positive cells, thereby eliminating random integrated recombinant clones [73, 134].

Many methods can be used for the introduction of the DNA into the ES cells, including calcium phosphate precipitation, electroporation, and nuclear microinjection [116]. However, regardless of the low rate of stable recombinants obtained by electroporation (10^{-5}–10^{-2}), it is the most often used technique, because large amounts of cells can be transfected. In contrast, nuclear microinjection has the advantages of being the more efficient method (10–20 % of the cells become stably transformed) [116] and leading to a better targeting ratio (1:15 targeted recombinants to random integration [73]). To identify clones with the desired recombination event, screenings are needed, usually by Southern blot and PCR analysis. Clones with random integration of the complete vector are excluded upon drug treatment. Once a stem cell with the designated genetic alteration is identified, the ES cell clone is expanded and injected into a blastocyst and implanted into a foster mother. The mice obtained from the mating of chimeric mice with the wild-type mice will be either wild type or heterozygous for the targeted mutation, and these heterozygous mice must be bred to homozygosity in the descendent [13].

20.3 Conditional Knockout Systems

In some cases, constitutive knockout mouse models die at early fetal-embryonic stages or offspring are not viable because the permanent deletion of the respective gene may be detrimental to early stages in development. To overcome this problem, gene deletion has to take place at later stages of development or even in the adult mouse. To achieve this, knockout systems were developed in which a spatial-temporal targeting is possible. These new knockout systems are based on the work of Sternberg and Hamilton who, by studying the bacteriophage Pl site-specific recombination, identified the two important elements for the recombination event in a bacterial-free system [183]: a site called *loxP*, a 34 base-pair sequence containing two 13 base-pair inverted repeats, separated by an eight base-pair spacer region [81], which must be present in both sites of the recombining cassette. The other element identified was an enzyme that mediates the recombination event codified by the PI gene called *Cre* (*cyclization recombination or cause recombination*). Cre recombinase is a protein of 343 amino acids with two distinct domains. The amino terminal domain contains residues 20–129 and it is formed by five alpha helixes; two of which make contact with the loxP sites. The carboxy-terminal domain of the enzyme consists of 132–341 amino acids and harbors the active site of the enzyme [68]. Under this premise, any cassette or gene flanked by two loxP sites and in the presence of the Cre enzyme would recombine and be excised [42] (Fig. 20.1b). In this approach, cassettes for which recombining is desirable very often contain a selection marker and mark the germline-modified mice with floxed alleles (currently phenotypically wild type). These mice have to be bred to Cre-expressing mice to allow the generation of the knockout and open up the possibility of tissue-specific and inducible knockouts. To achieve this, the Cre has to be expressed in a tissue-specific manner, e.g., expression of the Cre is controlled by a tissue-specific promoter or knocked into the locus of a tissue-specific gene, or in an inducible manner when expressed by an inducible promoter.

Another attractive feature of the Cre system is that several types of recombination events can be produced depending on whether the *loxP* sites are placed in the same DNA molecule (recombination in cis) or in different molecules (recombination in trans). Besides, the *loxP* site orientation is important, since it is possible to generate insertions, inversions, translocations, and single point mutations. If the two *loxP* sites have the same orientation, the DNA region situated between these *loxP* sites is deleted during recombination. If the orientation of the two *loxP* sites is opposed to each other, recombination leads to the inversion of the region comprised by the two sites; when the *loxP* sites are both integrated in different places in the genome, recombination in trans induces chromosomal rearrangements. However, these rearrangements are not so easy to generate and therefore require accurate selection methods [13]. A rare event, but also possible, is the generation of insertions, if one *loxP* site is integrated in the genome and the other is carried by a circular plasmid; in this case, an insertion of the sequence carried by the plasmid can be generated. To increase the frequency of such an event, mutant *lox* sites can be used [11].

Although less often used than the Cre recombination system, other recombinases are characterized. In this context, O'Gorman and collaborators in 1991 described a recombination system based on the site-specific recombinase, FLP-1 from *Saccharomyces cerevisiae* [146], which can also mediate recombination events in embryonic stem cells [26, 47]. The Flp recombinase (flippase) consists of a 13-kDa NH_2-terminal domain (p13) and a larger 32-kDa COOH-terminal domain (p32) with the major determinants for DNA binding and was originally isolated by Broach and Hicks [24]. The nucleotide sequence of the flippase recognition target (*FRT*) site is 34-bp long. The site contains two palindromic sequences of 13 bp each, separated by a central and asymmetric sequence of 8 bp, or spacer, which defines the orientation of the site. The recombinase binds the palindromic sequences, whereas the spacer is the site of DNA break, exchange, and ligation [40, 196].

Similarly to the Cre-LoxP recombinase system, the position and orientation of the two *FRT* sites will also determine the outcome of the FLP-mediated recombination. FLP excises DNA fragments as a circular molecule, when the DNA target region is located between two *FRT* sites having the same orientation. The excision reaction is reversible; however, the reintegration of the excised circle into the linear product is kinetically less favorable than excision of the circle, and excision products may accumulate with time [196]. Inversions of DNA segments can be generated, when they are located between two *FRT* sites in opposite orientation. FLP can also mediate recombination between two *FRT* sites located on different or nonhomologous chromosomes; if this is the case, FLP will exchange the chromosomal sequences located downstream of the sites, leading to a reciprocal translocation. To insert a circle into a linear chromosome or to generate an inversion event, *FRT* variants can also be engineered to allow a single and irreversible single-stranded recombination (SSR) event [104].

By sequencing the 7-kb *immC* region from four P1-related phages, Sauer and McDermott identified a novel DNA tyrosine recombinase from the phage D6, which is closely related to the P1 Cre recombinase, including the capacity of recombination in mammalian cells, but differs from Cre in their DNA recognition site [166]. In this case, the Dre recombinase presents DNA specificity for a 32-bp DNA site, called *rox,* and no host bacterial proteins are required for efficient Dre-mediated site-specific DNA recombination. Later on Anastassiadis and collaborators developed a more sophisticated Dre system, called "Dre-PBD," which provided an efficient and tightly controlled system for the inducible expression of Dre recombinase activity in mammalian cells. In their study, they generated a Dre recombinase fused to the PBD (progesterone ligand-binding domain) in a targeting construct, using a PBD that had been truncated at the C-terminus in order to induce the recombination event in a RU486-dependent manner in both the fibroblast and ES reporter cells [6, 7].

20.4 Spatial-Temporal and Tissue-Specific Knockouts to Model Platelet Disease

Cre expression can be tissue specific or temporally controlled based on the expression from inducible promoters (e.g., tet-responsive promoter) or by the generation of ligand-dependent chimeric Cre recombinases (e.g., tamoxifen inducible) (Fig. 20.2).

One inducible model was developed by Feil and collaborators [51, 52] who fused the Cre recombinase to the mutated human estrogen receptor (ER) ligand-binding domain (LBD), and therefore, expression of Cre can be regulated by the addition of tamoxifen [52]. Tamoxifen is administered to the mice either by injection or via their food. This inducible system, for example, has been used to study the conditional ablation of Gata-1 [70]. Another model is based on the combination of the Cre-*loxP* system with tetracycline-mediated regulation of gene expression (Fig. 20.2b). This approach allowed the spatially and temporally controlled expression of the Cre recombinase. In this inducible model, Utomo and collaborators, for example, used the *retinoblastoma* (RB) or the *whey acidic protein* (WAP) promoters to drive expression of the rtTA (for tissue-specific expression), while the Cre expression was controlled by the tetO-hCMV promoter (for temporal induction). Without doxycycline, rtTA is inactive and unable to induce transcription of Cre. In the presence of doxycycline, rtTA binds to the tetO-hCMV promoter leading to Cre expression [197]. Making use of the transgenic mouse expressing the rtTA from the CMV

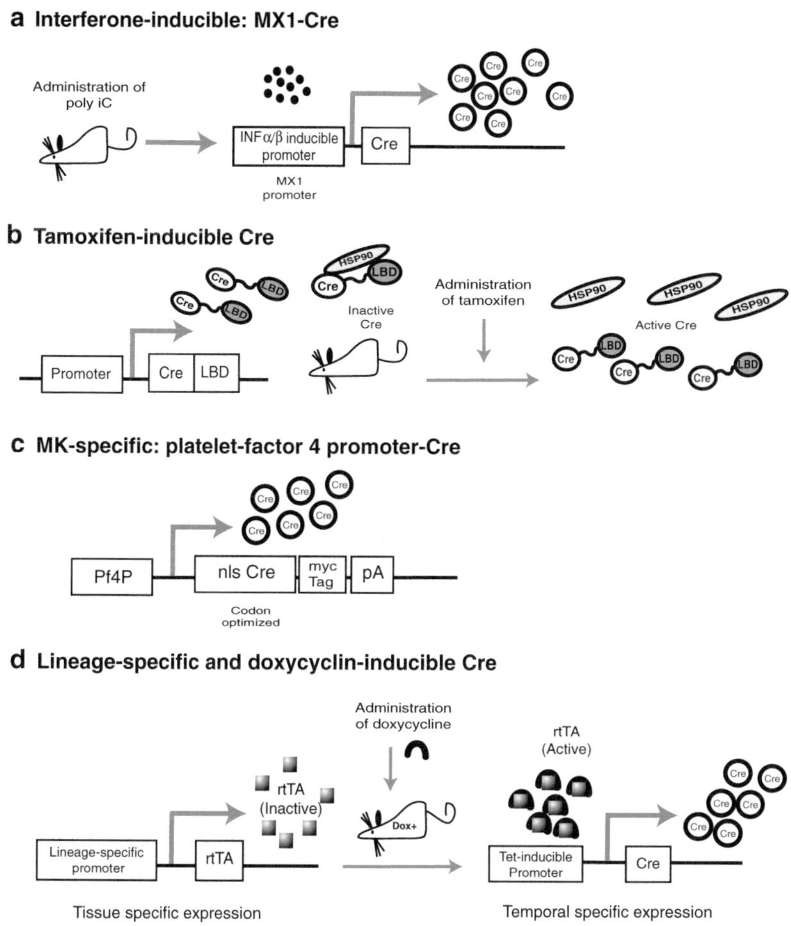

Fig. 20.2 Cre-inducible systems. (**a**) Interferon inducible: MX1-Cre. The interferon-inducible MX1 promoter is used to drive the expression of the Cre. The promoter is induced after the administration of the pl-pC (synthetic double-stranded RNA polyinosinic-polycytidylic acid) which activates the cells' immune response and induced interferon. Mice carrying the floxed targeted gene will only develop the targeted mutation, once the Cre is expressed after the administration of the pl-pC [102]. (**b**) Tamoxifen-inducible Cre. The Cre recombinase is fused with the estrogen receptor ligand-binding domain (LBD), and to avoid permanent induction by the endogenous steroids, an internal mutation is introduced in the LBD domain to keep the affinity just for synthetic hormones. One modification of this model was performed by Feil and colleagues, where, by the introduction of the point mutation G521R, the induction of the LBD domain became sensitive to tamoxifen. In a non-induced state, the Cre is expressed but remains inactive by the binding of the LBD to a heat shock protein (HSP90). After administration of tamoxifen, LBD is not bound anymore, and Cre becomes active to mediate the recombination event [52]. (**c**) Megakaryocyte-specific platelet factor 4 promoter Cre. In this model, the rat platelet factor 4 promoter drives the expression of the Cre, which has been fused to a nuclear localization signal and a myc-tag. The coding sequence is followed by a polyadenylation signal (polyA). Because of the use of a promoter fragment of a megakaryocyte- specific gene, Cre is expected to be expressed only in the megakaryocyte lineage [190]. (**d**) Lineage-specific and doxycycline-inducible Cre. This approach allows the spatial and temporal control of the Cre expression by the use of a tissue-specific promoter driving the expression of the reverse tetracycline-controlled transactivator protein (rtTA), which remains inactive until doxycycline is administrated. Once doxycycline is present, it binds to the rtTA, allowing by this the induction of the Tet-inducible promoter and therefore the expression of Cre [197]

promoter, the efficient expression of a miR downregulating the Bcl-XL antiapoptotic gene induced thrombocytopenias in mice [187]. Bcl-XL is known to be essential for maintaining platelet survival [98] The rtTA can also be expressed in a MK-specific fashion [144]. In one mouse model, the rtTA is expressed by a 1.1-kb fragment of the rat Pf4 promoter previously described by Ravid and colleagues [159] (Fig. 20.2d). Using the Pf4-rtTa mouse, e.g., the role of Survivin in megakaryopoiesis and platelets was studied [118].

In another inducible approach developed by Kühn and colleagues, the expression of the Cre recombinase was placed under the control of the Mx1 promoter. The *myxovirus resistance 1* (Mx1) promoter can be transiently activated in many tissues upon administration of interferon (IFN-α and IFN-β) [83] or by a synthetic double-stranded RNA polyinosinic-polycytidylic acid, also called pI-pC or poly(I:C), which induces the cells' response to viral infections and, in doing so, induces type 1 interferons. Therefore, those mice carrying a floxed target gene and a Mx1-Cre transgene will only develop the targeted modification after the administration of interferons or the poly(I:C) compound [102] (Fig. 20.2a).

Besides the transgenic approaches, Cre can also be delivered by viral vectors. Anton and Graham developed a replication-deficient human adenovirus type 5 vector expressing the Cre recombinase gene under the control of the human cytomegalovirus immediate-early promoter (CMV, vector named AdCre). Transient expression of Cre after coinfection of 293 cells with the AdCre and adenoviral vector carrying the luciferase cDNA whose expression was blocked by the presence of an external floxed DNA sequence turned on luciferase expression after excision of this floxed sequence [9]. Following this initial demonstration of the concept, AdCre has been exploited for in vivo Cre delivery into different tissues [4], or for ex vivo delivery, for example, into T cells [218].

The most widely used application of the Cre-inducible system for the study of gene function in platelets is the transgenic mouse expressing Cre under the control of the murine platelet factor 4 (Pf4) promoter generated by Tiedt et al. [190] (Fig. 20.2c). In this mouse, a codon-optimized Cre was inserted in the position of the first exon and expressed from the murine Pf4 promoter that was included as transgenic cassette into the mouse genome. The upstream and downstream sequence consists of about 100 kb in total, also containing four reading frames of further genes (*Cxcl5*, *Cxcl7 (Ppbp, pro-platelet basic protein),*Cxcl3, and *Cxcl15*). Functionality and specificity of the Cre recombinase in the Pf4-Cre mouse were tested by breeding to the Rosa26 LacZ reporter strain [179] and to the *β1*-flox/flox mouse [63]. Tiedt et al. demonstrate the megakaryocyte-specific LacZ expression and the prevention of the lethal phenotype of the complete (constitutive) integrin β1 knockout by the lineage-specific deletion of the integrin β1 only in megakaryopoiesis. In contrast, Calaminus and collaborators analyzed the specificity of the Pf4-Cre mouse by intercrossing to a reporter line (Rosa26-tdRFP, Pf4-Cre) and reported significant levels of recombination in both fetal liver and bone marrow HSCs. They also demonstrate that Cre activity under the Pf4 promoter extends to other blood cell lineages (myeloid and lymphoid) and up to 50 % in the HSC compartment [28]. However, in a very similar approach by Ng et al., analyzing Pf4-Cre mice crossed with mice carrying a floxed Rosa EYFP locus they did not detect recombination outside the megakaryocytic lineage [143]. Most recently, Pertuy and colleagues [154] detected intestinal epithelial cells in the distal colon to have recombined their floxed genomic locus under the Pf4-Cre conditions. Furthermore, 80 % of the Apc flox/flox;Pf4-cre mice developed APC depletion-induced tumors. In the blood, they only very rarely detected recombined cells (~0.3 %). Positive cells were found as a subset of tissue-infiltrating leukocytes, which were positive for F4/80 and therefore part of the monocyte/macrophage lineage. The activation of immune cells from the spleen with PMA also increased the number of cells that underwent Cre-mediated recombination, as measured by the GFP reporter expression.

In addition to Pf4, CD41 (integrin alpha 2b, glycoprotein IIb) is also a suitable marker for adult megakaryocytes and platelets, and a CD41-Cre transgenic mouse was generated. In a first approach, ~5-kb sequence of the murine *GPIIb* gene (2.5 kb of the upstream and 2.8 kb containing the first exon and intron sequences) was used to express nls-Cre in transgenic mice [49] and crossed with a Rosa26 reporter mouse. In agreement with CD41 expression during the hematopoietic development, recombination could be detected in the yolk sac (day 8.5), AGM and dorsal aorta (day 9.5), and the fetal liver which at day 12.5 contained ~60 % recombined hematopoietic cells. Surprisingly, in the adult hematopoiesis in 2-month-old mice, recombined cells contribute with below 5 %. This was a very surprising observation and could be explained in two ways: (1) only a subset of the CD41-positive fetal HSC give rise to the definitive adult HSC, and/or (2) a second wave of fetal HSC arises that does not undergo a CD41-positive stage. In the same study, a Cre knock-in into the first exon of the endogenous alpha 2b locus was performed which unexpectedly did not mark the embryonic and fetal hematopoiesis to the same extent. In an independent approach to generate a CD41-Cre mouse [145], using a 2.7-kb upstream fragment of the murine alpha 2b gene to drive expression of Cre with a nuclear localization signal (nls-Cre), a restriction of recombination events to the adult megakaryocytic lineage was reported; however, fetal hematopoiesis was not analyzed in depth in this study. Another transgenic mouse utilized a 2.7-kb fragment of the alpha 2b promoter to drive the thymidine kinase (tk) gene which could delete tk-positive cells by the application of ganciclovir. In this study, thrombocytopenia could be induced by ganciclovir underlining the functionality of the CD41 promoter fragment in adult megakaryopoiesis [194]. In the meantime, also a CD41 knock-in mouse was generated expressing the yellow fluorescent protein (YFP) from the endogenous CD41 locus and confirming the specific expression of CD41 in megakaryopoiesis in the adult mouse [222].

20.5 Transgenic Models

In the transgenic approaches, genes are introduced into the genome for overexpression. In general, they are not at their physiological location in the genome, and expression is not controlled by their physiological promoter. In some cases a promoter fragment of the same gene is used. Therefore, mice are provided by an additional gene copy, sometimes even with multiple gene copies. Constructs for generating transgenic mice usually consist of a promoter/enhancer region with transcriptional start site, introns (for improved transgene expression in mouse lines), protein-coding sequence (cDNA) from the gene of interest, and a polyA sequence. The protein-coding sequence (cDNA) has to contain the translation start site, the ATG start codon, sometimes with a Kozak consensus sequence placed in front. The entire transgene element is usually excised from the vector backbone before zygote injection. The inclusion of protein tag sequences to the transgene, like VSV-g, 6xHis, HA, V5, or Myc epitopes, can be helpful to detect the transgene expression as specific antibodies to these epitopes are available. This strategy allows differentiating endogenous versus transgenic expression.

Transgenic mice are generated by microinjecting the transgenic construct into a fertilized oocyte, zygote, or spermatogoneal stem cells, by the use of retro- and lentivirus vectors [32, 84, 94] or by transfecting a transgenic construct into the mouse embryonic stem (ES) cells and then injected into the mouse blastocysts [199]. Transgenic approaches can also be used to analyze tissue-specific or developmental stage-specific gene expression by introducing reporter genes, such as β-galactosidase (*lacZ*) or green fluorescent protein (*GFP*), under the control of a specific gene promoter, whose expression can be followed or tracked during development [36].

One extensively used approach when studying megakaryopoiesis is the transgenic mouse with megakaryocyte-specific transgene expression driven by the platelet factor 4 promoter (Pf4) developed by Ravid and colleagues as was already introduced in this chapter, Sect. 20.4 (rat Pf4 promoter [159]). The targeting plasmid was constructed by fusing 1,104 bases of the 5′ upstream sequence of the rat Pf4 promoter, the cap site to +20 bp, Kozak sequence, and the coding DNA. In contrast to what has been shown in more recent studies using the murine Pf4 promoter consisting of a longer sequence [28], transgenic mice generated with the rat Pf4 promoter expressed the β-galactosidase gene exclusively in the platelets and megakaryocytes, but not in the erythrocytes and leukocytes blood cell lineages in the adult mouse. Analysis in other tissues also revealed the lack of transgene expression in the brain, heart, intestine, kidney, liver, lung, and skeletal muscle in the adult mouse.

Few years later, in 1996, Thompson and collaborators expressed a c-Myc ERT fusion from the Pf4 promoter [188]. They found the overexpression of c-Myc in megakaryocytes of transgenic mice to increase the fraction of proliferating megakaryocytes, but not to change the erythroid or myelomonocytic lineages. When they tested whether the changes in megakaryocytes differentiation were associated with changes in Cyclin A and D3 expression (since Cyclin A is essential for the entry into S phase and Cyclin D3 is essential for the development of polyploid megakaryocytes), they observed that Cyclin D3 was downregulated, while the level of Cyclin A was slightly upregulated, concluding that cultured megakaryocytes overexpressing c-Myc were induced to proliferate, but have a limited potential to fully differentiate.

Transgenic mouse models have also been used to study hemophilia, e.g., hemophilia A. Yarovoi and collaborators generated transgenic FVIII expressing mice, with the aim to correct the FVIII circulating levels in hemophilic mice, by targeting the expression of coagulation FVIII into the alpha granules of the platelets [214]. This is an alternative approach to avoid the immune responses and development of inhibitors against recombinant FVIII in hemophilic patients. They used the GP1b alpha promoter to drive the expression of B-domain-deleted FVIII fused with the alpha granule sorting signal targeting of the von Willebrand factor (vWF). They were able to show that in the transgenic mouse model, targeted FVIII can be ectopically expressed not just in developing megakaryocytes but also in circulating platelets within their alpha granules. This report is the first evidence that platelets can be potentially used as carrier cells for the transport of therapeutic proteins, since from these modified platelets, the transgenic coagulation factor VIII was released and corrected the bleeding time in $FVIII^{null}$ mice. In another model, Kufrin and colleagues generated transgenic mice using 1.2 kb from the murine Pf4 promoter to express the urokinase plasminogen activator (uPA) [101]. Transgenic mice presented with altered platelet biology and bleeding diathesis, similarly to what is observed in patients with Quebec platelet disorder; mice were also resistant to developing occlusive arterial thrombi in a $FeCl_3$-induced carotid artery thrombosis model. As these mice did not develop bleeding symptoms under steady state conditions, the antithrombotic effect was local after platelet activation. The transfusion of urokinase-expressing platelets into wild-type mice also prevented thrombus formation in the transfused mice. This model demonstrates the possibility to genetically modify megakaryocytes to alter platelet function from a pro-thrombotic to anti-thrombotic phenotype.

20.6 Mouse Transgenesis with the Sleeping Beauty Transposon System (SBTS)

Another alternative for the generation of transgenic mice is the modifications of ES cells by the sleeping beauty transposon system. In contrast to retroviral vectors which have the problem to be recognized as foreign genetic material and are often silenced [89, 110, 113], transposon-mediated transgenesis encounters this problem less frequently.

Transposons were discovered in 1940 by Barbara McClintock who later on would earn the Nobel Prize for the discovery. These elements are known as DNA segments which are able to move or "jump" within the genome by a cut and paste mechanism. Ubiquitously found in nature, these elements are part of the genome of basically all living organisms; but they were not used as a vehicle for gene delivery, until 1997, when Ivics and collaborators demonstrated that, by eliminating the inactivating mutations of one ancient transposon system, *sleeping beauty*, a Tc1-like transposon from fish, they could reactivate the transposase, and transposition in human cells was possible [86].

The SBTS technology offers a nonviral system for gene delivery as an alternative for transient and long-term expression. The transposable elements (*sleeping beauty*, but also others such as piggyBac transposon derived from the cabbage looper moth or the *frog prince transposon* derived from the leopard frog) contain a single gene encoding the transposase, which is flanked by terminal inverted repeats (IR/DRs), each one containing two binding sites for the transposase. For gene therapy purposes, the transposase gene within the element can be replaced by a therapeutic gene. The sleeping beauty transposon in a plasmid provides only transient expression of a transgene from a promoter, unless the transgene is transposed for integration into the host genome. Therefore, for long-term expression, the SBTS consists of two components: a transposon containing a gene expression cassette and the coding sequence for the transposase enzyme which is usually provided in trans on a different plasmid than the transposon. The transposase binds the IR/DR domains flanking the transgene and forms a synaptic complex, in which the transposon forms a loop and the inverted repeats are paired. Finally, the transposon is excised from the donor plasmid and integrates into the genome. In doing so, the SBTS works on a cut and paste mechanism [88].

There are several methods for delivery of the transposon system into a cell in culture (with electroporation or transfection) or in a tissue (e.g., by hydrodynamic injection) [12]. For animal transgenesis, pronuclear microinjection of sleeping beauty transposons is performed. In recent papers, Ivics and Izsvác describe a method for the use of the SBTS system by the co-injection of the synthetic mRNA encoding the hyperactive transposase SB100X, together with a circular DNA plasmid carrying the transposon into the pronuclei of fertilized oocytes of rodents [87].

20.7 Knock-In Mouse Models

Knock-in models are based on the targeted insertion of a new gene in a particular locus, by using homologous recombination in embryonic stem cells. In this case, instead of a loss of function by deleting an important segment of a gene, a gain of function from a new coding sequence is desired. In addition, targeted mutations can also be achieved by the replacement of the wild-type sequence by a mutated version within a targeting construct. If the knock-in of a mutant gene was to be inducible, a stop cassette flanked by loxP recombination sites is inserted in front of the mutant gene of interest, thereby preventing its expression. By expression of the recombinase (Cre, Flp, Dre), the stop cassette is excised and the mutant cDNA is expressed [158].

One well-known example in platelet research is the expression of the human glycoprotein IB alpha (*GPIBA*) gene from the murine locus. By disrupting the coding sequence of the murine *GPIba* gene, Ware and collaborators generated a murine model resembling the hallmark characteristics of the human Bernard-Soulier syndrome. In the next step, by replacing the missing murine gene with the human version of the *GPIba* gene, they were able to rescue the murine Bernard-Soulier phenotype and by this strategy demonstrated a direct link between the expression of a GPIb-IX-V complex with a normal megakaryopoiesis and platelet morphogenesis [205].

Zhang and colleagues developed a CD41-YFP mouse containing YFP-labeled megakaryocytes and platelets. In this case the yellow fluorescent protein (eYFP) cDNA was inserted just after the start site of the *GPIIb* gene by homologous recombination into embryonic stem cells [222]. In a different model, by inserting the cDNA of the green fluorescent protein (GFP) into the *growth factor-independent 1b*

(*Gfi1b*) gene locus, Vassen and collaborators were able to track the expression pattern of Gfi1b in erythroid cells, megakaryocytes, and their progenitor cells [200].

20.8 In Vivo Models Using Viral Vectors for Gene Transfer

For gene delivery into adult stem cells which cannot be cultured for prolonged times, transfer vectors have been developed that are derived from viruses and take advantage of the viral characteristics to infect target cells more easily than the non-viral transfer systems. For the stable integration of genetic material and therefore the long-term modification of cells, retroviral vectors are the transfer system of choice. They transduce their target cells by receptor-mediated binding and uptake and integrate their genome utilizing their viral integrase. By deletion of the viral coding genes from the wild-type viral genome, vectors were generated that cannot replicate and give space to transfer coding sequences of other genes of interest into cells (Fig. 20.3).

The most advanced retroviral vectors are derived from the gammaretroviral murine leukemia viruses [114, 123, 209] and the human immunodeficiency virus (HIV) [45, 139], known as the lentiviral vectors. Other retroviral vector systems are developed from the *equine infectious anemia viruses* (EIAV) [45, 127], the *simian immunodeficiency viruses* (SIV) [142], foamy viruses [192], and alpharetroviruses [184]. The simplest genome carries the gammaretrovirus, and all their genes are shared by the other members of the retrovirus family: (1) the *group-specific antigen* (gag), which encodes for the matrix, capsid, and nucleocapsid, (2) the *pol* gene (codes for the protease, reverse transcriptase, and integrase), and (3) the viral envelope (*env*). In the case of the complex lentivirus, additional genes are encoded in the viral RNA such as the *rev* gene and other accessory genes. The interaction of the Rev with the *rev responsive element* (RRE) contained in the viral RNA mediates the nuclear RNA export [115, 157]. To produce retroviral vectors, the viral genes are placed on the so-called helper plasmids which for the production of viral vector particles are cotransfected into the packaging cell lines [121, 216]. Packaging cells are cell lines that by expressing all necessary viral proteins produce the viral particles and release them into the cell culture supernatant.

Retroviral particles have a diameter of 100 nm and consist of an enveloped nucleocapsid and two copies of single-stranded RNA. The viral RNA contains a 5′ cap structure and a 3′ polyA tail. After infection of the target cells, the RNA is converted into double-stranded DNA by the reverse transcriptase and with the help of the integrase inserted into the host genome.

The integrated provirus contains two *long terminal repeats* (LTRs) on each site. The LTRs, contain the viral enhancer and promoter (in the *unique* U3 region), the cap and polyadenylation signals (in the R-region), and the *attachment site* (*att*). The retroviral 5′ untranslated region consists of the R- and U5-region, the attachment site (*att*), the primer binding site (PBS), and the packaging signal (ψ). R-region, att, PBS, and ψ are equally essential for the replicating virus and the vector as they

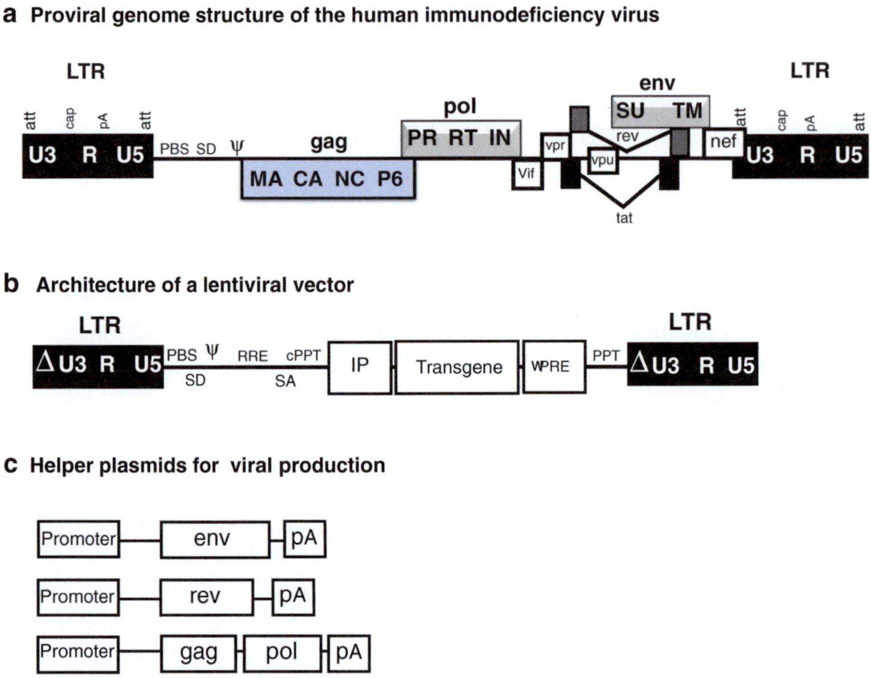

Fig. 20.3 Viral vector development. (**a**) Proviral genome structure of the human immunodeficiency virus. The proviral genome is flanked by long terminal repeats (LTR) (in *black boxes*) that provide the transcriptional control (by enhancer-promoter elements), start, and polyadenylation signals. The leader region contains the primer binding site (PBS) for the initiation of the reverse transcription, the splice donor site (SD), as well as the packaging signal Ψ. Downstream of the viral coding sequence is differentiated into gag (coding for structural protein matrix (MA), capsid (CA), nucleocaspid (NC) and P6 protein), the pol (coding for replication enzymes, *PR*, protease; *RT*, reverse transcriptase; *IN*, integrase), and env (*envelope protein*). The lentiviral genome contains accessory proteins (viral infectivity factor (vif), viral protein r (vpr), viral protein u (vpu), regulator of expression of virion proteins (rev), transactivator of transcription (tat), and negative factor (nef)) that are generated by alternative splicing and promote the maintenance of the viral infectivity. (**b**) Architecture of a lentiviral vector. For therapeutic purposes, the genome of the virus is modified, in order to generate replication-deficient viral particles. The gag, pol, and envelop proteins are removed and replaced by the coding sequence of the transgene. Elements like PBS, SD, Ψ, and splice acceptor (SA) are retained in the vector to guarantee packaging of the viral RNA (Ψ) and the initiation of reverse transcription (PBS). (**c**) Helper plasmids for viral production. For the generation of infectious viral particles, proteins like gag, pol, env, and rev are delivered in *trans* (from different expression plasmids). This allows the packaging and assembly of all structural proteins needed, but, because the RNA of these plasmids lack the packaging signal, just the RNA coding the transgene is packaged in the viral particles

mediate the packaging of the viral RNA (ψ), the initiation of the reverse transcription (PBS), and the integration (*att*). In contrast, enhancer and promoter sequences in the U3-regions can be deleted, and instead, a promoter can be inserted in an internal position just upstream of the cDNA. Retroviral vectors with this architecture are named self-inactivating (SIN) vectors because during the production the viral RNA is transcribed from the 5′LTR, but the deletion in the 3′LTR is copied to the 5′ position during the RT. For this reason, the U3-region is deleted in the integrated viral vector genome [99, 217, 224]. The SIN configuration minimizes the chance of replication-competent viral particles to occur during the treatment with gene vectors and allows the use of lineage-specific or inducible promoters in an internal position to express the transgene.

Retroviral vectors enter the cells by binding to specific cell surface molecules which are predefined by the viral envelope protein. Therefore, the tropisms of viral vectors can be modulated by the envelope protein [16]. For example, the murine ecotropic retrovirus binds the mCAT receptor, and the amphotropic virus binds the PIT-2 receptor. HIV-derived lentivirus recognizes CD4 and CXCR4/CXCR5 on the surface of T cells, but by pseudotyping with the *vesicular stomatitis virus glycoprotein* (VSVG), which binds the LDL receptor [53], the tropisms for transduction are very broad.

After entering the cell, the viral particle undergoes uncoating and the viral RNA is released but hidden in the pre-integration complex (PIC). The reverse transcriptase contained in the particle mediates the generation of the double-stranded DNA. The PIC travels along microtubules to the nuclear membrane. In the case of lentiviral vectors, the DNA can enter the nucleus via the nuclear pore, while gammaretroviral vectors rely on the breakdown of the nuclear membrane during mitosis [122]. As HSCs are only slowly cycling, lentiviral vectors are favored over gammaretroviral vectors for HSC transduction [66]. The integration of the double-stranded viral DNA is mediated by the integrase. The attachment to the genomic DNA is determined by specific tethering factors which bind to the integrases (LEDGF for lentiviruses [35], BET proteins for the gammaretroviruses [69, 41, 171]). These specific interactions explain the typical integration site preferences of the different members of the retoviruses; gammaretrovirus prefers to integrate into promoter regions and CpG islands, while lentivirus integrates within genes. Both of them tend to target actively transcribed gene regions [168]. In contrast alpha and foamy viruses integrate more randomly and therefore less often close to or in genes [126, 193].

To avoid integration-associated adverse reactions, lentiviral vector can also be used in a non-integrative form. However, the lentiviral genome will then not persist but will be diluted with increasing cell divisions. These lentiviral vectors contain an integrase with a mutation in the catalytic domain (D64, D116, E152) [156, 213]. Therefore, after entering in the nucleus, the lentiviral genomes stay in a non-integrated, episomal form, as one or two LTR circles [204]. These circles also form during the normal life cycle of lentiviruses but are considered to be nonproductive and will not integrate at later times either. Still, the transgene can be transcribed and produced, and therefore, integration-deficient lentiviral vectors (iDLVs) are very interesting delivery tools for transient expression, as they deliver the DNA to the cells via the lentiviral entry mechanism, e.g., for the delivery of reprogramming or transdifferentiating factors, transposase, zing finger nucleases, Cre, or Flp [111, 181].

20.9 Models for Megakaryopoiesis and Platelets Using Retroviral Vectors

As platelets do not contain a nucleus and a cellular genome, the application of gene transfer methods to platelets directly is limited. Therefore, genetic modifications must be performed at earlier stages in platelet differentiation, either in the progenitor cells or in the HSC. By transplantation of HSC into conditioned recipients, the hematopoiesis of the recipient mouse can be modified. Depending on the donor chimerism and transduction rates before transplantation, gammaretro-/lentiviral gene transfer to HSC allows the expression of ectopic proteins throughout the entire hematopoiesis.

For the analysis of gene functions in megakaryopoiesis and platelets, however, it may be important to restrict expression to the megakaryocytic lineage. By using tissue-specific promoters in retro-/lentiviral vectors, it is possible to direct expression to a certain blood lineage (Fig. 20.4). In the case of megakaryopoiesis, promoter fragments of MK-specific genes can be used. This concept has been demonstrated by using fragments of the platelet factor 4 (Pf4) [64, 65], GPIIb [220], or GPIba promoters [147, 148], as internal promoters in retro- or lentiviral vectors, summarized in Table 20.1.

The specificity from the MK promoters is mediated by typical transcription factor binding sites and regulatory regions. One study performed by Lemarchandel et al. [106] identified two important consensus sequences within the human glycoprotein IIB promoter (GPIIB), also identified later in the platelet factor 4 promoter (Pf4) [125, 160], in the glycoprotein IX promoter (GP9) [17], and in the glycoprotein VI promoter (GP6) [55]. One of these sequences, a –55 GATA binding site, and the second one, located at –40, contain an Ets consensus sequence. These two binding elements were also identified in the glycoprotein 1b alpha promoter (GP1bα) [74]. A further study performed by [150] identified Meis1/Pbx as important transcription factors for the megakaryocyte-specific expression of the *platelet factor 4* gene.

As all MK-specific promoters share similar binding sites, ideally, one consensus sequence for expression in MK could be determined and a synthetic promoter generated similarly to the approach published for hepatocyte-specific expression [138]. However, the promoter regulates only transcription, but RNA abundance is also controlled by posttranscriptional regulation via microRNAs. This concept has been successfully exploited by Brown, Gentner, and Naldini [25, 58] by introducing miR target sequences into the RNA transcribed from the vector after integration. They could demonstrate the successful detargeting of protein expression from antigen-presenting cells or HSC after lentiviral transduction [8, 60]. With high numbers of integrated vector copies, and as result of high levels of transcribed RNA containing the miR target sites, these could even delete miR from the system by excessing miR binding, inducing a miR knockout situation [59].

The importance of lineage-restricted expression was convincingly demonstrated in our study correcting the thrombocytopenia caused by Mpl deficiency in the Mpl–/– mouse. Human patients with MPL deficiency present with the hematological disorder named congenital amegakaryocytic thrombocytopenia (CAMT,

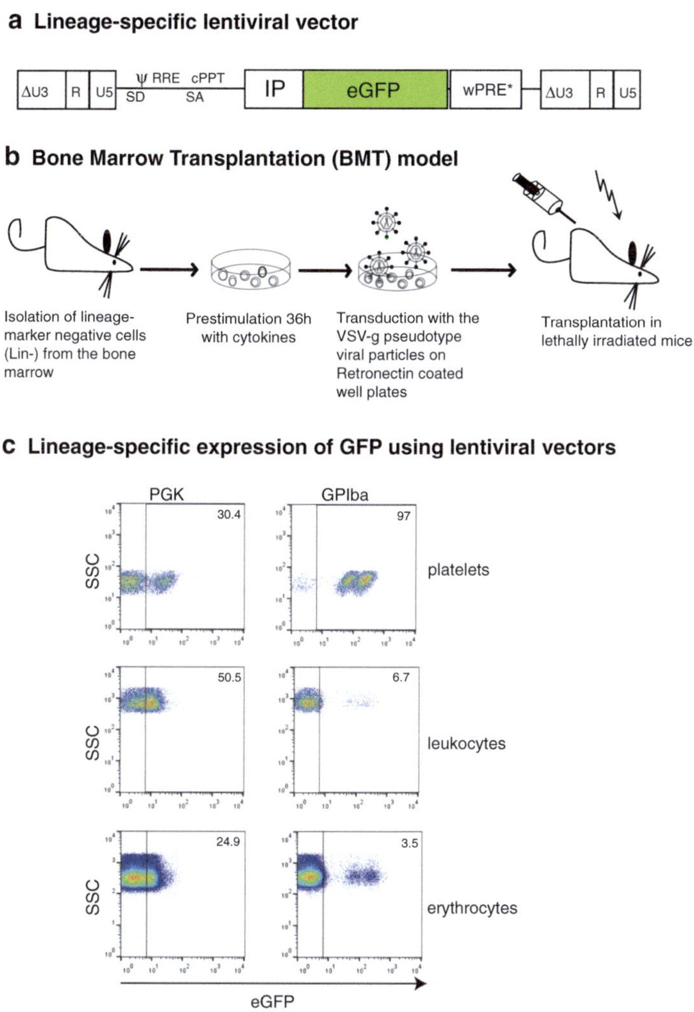

Fig. 20.4 Lentiviral vector for lineage-specific expression. (**a**) Lineage-specific lentiviral vector. The enhancer/promoter in the U3 region of the LTR is deleted (ΔU3). This modification allows the use of an internal cellular promoter to control the expression of the gene of interest, in this case eGFP (*wPre*, woodchuck hepatitis virus posttranscriptional regulatory element; *SD*, splice donor; *SA*, splice acceptor; *RRE*, rev responsive element; *cPPT*, central polypurine tract) (Modified from [77]). (**b**) Bone Marrow Transplantation model (BMT). For genetic modification in the hematopoietic stem cells (HSC) with lentiviral vectors, first the lineage marker-negative cells were isolated from the bone marrow of donor mice. Cells were cultured for 36 h for prestimulation with cytokines and then transduced by addition of VSVG pseudotyped lentiviral particles to the medium on retronectin-coated wells. Transduced cells were then transplanted into lethally irradiated recipient mice. (**c**) Lineage-specific GFP expression in the blood cell. Six weeks after transplantation, blood cells were analyzed by flow cytometry for eGFP expression. With the use of GP1b alpha promoter as internal promoter in the lentiviral vector, lineage-specific expression of eGFP was achieved: 97 % of the platelets expressed eGFP compare to the leukocytes (6.7 %) and erythrocytes (3.5 %). Using the ubiquitously expressing PGK (*phosphoglycerate kinase)* promoter as control, eGFP expression was observed in all blood cells without lineage specificity (30.4 % 50.5 %, and 24.9 % in platelets, leukocytes, and erythrocytes, respectively) (Data from D. Heckl, U. Modlich et al.)

Table 20.1 Examples of studies utilizing gammaretro- or lentiviral gene transfer

Gene	Vect.	Promoter	Model	Outcome	Reference
Hemophilia A models expressing FVIII					
Human B-domain deleted FVIII	SIV-LV	CMV	CB-CD34+, transplantation into Nod/Scid mice	~10 % human engraftment, efficient FVIII expression (1.2–3.5 ng/ml, 1–3 % of normal levels	[97]
eGFP, human B-domain deleted FVIII	SIV-LV	hGPIba (−258 to +330) hGPIIb (−554 to +33)	In vitro cell lines, – CD34+ in vitro MK differentiation, BMT in hemophilia A mouse model[a]	Platelet-specific expression, FVIII expression detected	[149]
Human B-domain deleted FVIII	LV	hGPIIb (−889 to +35)	BMT in hemophilia A mouse model	Sustained FVIII expression, correction of hemorrhagic symptoms, no inhibitors	[173]
Human inactivation resistant FVIII (IR8), human and canine B-domain deleted FVIII (hBF8, cBF8)	LV	mPf4 (−120 to +82)	BMT in hemophilia A mouse model	Improved clotting, IR8 more effective than hBF8, except for the cremaster model, probably because of delayed onset	[65]
Human B-domain deleted FVIII	LV	hGPIIb (−889 to +35)	BMT in pre-immunized $FVIII^{null}$ mice	Platelet-specific delivery improves the phenotype	[100]
Human/porcine B-domain deleted FVIII hybrid	SIV-LV, LV	CMV, EF1a, yeast PGK	BMT in hemophilia A mouse model	SIV-LV with CMV promoter selected as the best performing vector	[91]
Human B-domain deleted FVIII fused to the VWF D domain	LV	hGPIIb (−889 to +35) hGPIIb (−673 to +35)	Hemophilia A dog model[b]	Sustained long-term correction of the bleeding phenotype, granule targeting	[46]
Human and canine B-domain deleted FVIII	LV	mPf4 (−1,120 to +82)	BMT in hemophilia A mouse model	Canine more active than human, but also increases apoptosis in MK	[64]

Table 20.1 (continued)

Gene	Vect.	Promoter	Model	Outcome	Reference
Human B-domain deleted FVIII-MGMT(P140K)	LV	hGPIIb (−889 to +35)	BMT in hemophilia A mouse model non-myeloablative preconditioning	Enrichment to mean 5 % FVIII expression in platelets by BG/BCNU treatment	[169]
Human B-domain deleted FVIII	LV	hGPIIb (−889 to +35)	CD34+ into NSG or NSGF8KO mice	Improved hemostasis in NSGF8KO mice, lo contribution of human platelets	[172]
Human B-domain deleted FVIII	LV	MND or EF-1α-short, hGPIba	Intraosseal injection, hemophilia A mice	Long-term expression, platelet-specific (in the case of hGPIba), phenotype correction	[202]
Hemophilia A models expressing FVIIa					
Activated FVII, insertion of two RKR motives into the factor X activation-cleavage site	SIV-LV	hGPIba (−58 to +330)	BMT in hemophilia A mouse model	Correction of bleeding phenotype in mice with hemophilia A in the presence of neutralizing antibodies	[147]
Hemophilia B models expressing factor IX					
hFIX-PGK. MGMT	LV	Beta globin locus control region, reverse orientation	BMT in hemophilia B mouse model[c]	Erythrocyte-specific expression of FIX, (>500 ng/ml by vector copy number 1–2)	[33]
hFIX	LV	hGPIIb (−889 to +35)	BMT in hemophilia B mouse model	Release from the platelets, phenotype correction	[220]
hFIX	LV	hGPIIb (−889 to +35)	BMT in hemophilia B mouse model	Sustained phenotype correction in serial transplants, no inhibitors developed	[34]

(continued)

Table 20.1 (continued)

Gene	Vect.	Promoter	Model	Outcome	Reference
Models expressing cell surface receptors					
Human integrin beta3	GV	hGPIIb (−889 to +35)	CD34+ cells of Glanzmann's thrombasthenia patients with RSΔ9β3 and EAY115Cβ3 mutations	Ex vivo phenotype correction	[208]
Human GPIIb-MGMT(P140K)	LV	hGPIIb (−889 to +35)	BMT in the αIIbβ3 deficient dog model[d]	Improved hemostatic function, enrichment of transduced by drug selection	[50]
Human GPIBa, HA tagged	LV	hGPIIb (−889 to +35)	DAMI cells, CD34+ cells	Detection of the GPIBa on the cell surface	[174]
Human GPIBA	LV	hGPIIb (−889 to +35)	BMT in mouse model for Bernard-Soulier syndrome[e]	Phenotype correction, 70 % of the platelets pos.	[93]
Murine Mpl	GV	mMpl (−2,085 to +42)	BMT in wt mice, Mpl-/- cells into wt recipients	Lineage-specific Mpl expression prevents toxicity of ectopic expression	[206]
Murine Mpl	LV	mMpl (−2,085 to +42) hMpl (−770 to + 44) hGPIba (−258 to +330)	BMT in the Mpl-deficient mouse model[f]	Correction of aplastic anemia and thrombocytopenia	[77]

Vectors: *LV* HIV-based lentiviral vector, *SIV-LV* Simian immunodeficiency virus-based lentiviral vector, *GV* Gammaretroviral vector, *MGMT*(P140K) human O(6)-methylguanine-DNA-methyltransferase(P140K)

Promoters: *hGPIba* human glycoprotein Ib alpha, *hGPIIb* human glycoprotein IIb, also known as integrin alpha 2b and CD41, *mPf4* murine platelet factor 4, *EF1a* elongation factor 1 alpha, *CMV* cytomegalovirus, *PGK* phosphoglycerate kinase, *MND* modified MoMuLV promoter with myeloproliferative sarcoma virus enhancer, *Mpl* myeloproliferative leukemia oncogene, also known as the receptor for thrombopoietin

Mouse models

[a]Bi et al. [18]

[b]Lozier et al. [112]

[c]Wang et al. [203]

[d]Lipscomb et al. [108]

[e]Ware et al. [205]

[f]Alexander et al. [5]

[14, 198]. The gammaretro- or lentiviral overexpression of Mpl by ubiquitous promoters (viral promoter or the phosphoglycerate kinase (PGK) promoter) severely disturbed the balance between Mpl and its ligand thrombopoietin (Thpo). As a consequence, Thpo levels were reduced, causing severe thrombocytopenia in wild-type mice [206]. This occurred against all expectations, as Mpl is also a proto-oncogene, and therefore, high levels of expression will also cause activation of the downstream pathways and induce cell proliferation. The hyperproliferative phase, however, was only transient, and mice did not succumb due to myeloproliferative disorders but because of severe pancytopenia. Only the restricted expression of Mpl by the Mpl promoters and also the GPIba promoter allowed the correction of the thrombocytopenia without side effects. As Thpo/Mpl signaling is also essential for the maintenance of HSC, the phenotypic picture of CAMT is the aplastic anemia. Mpl expression from the GPIba or Mpl promoter also rescued the HSC defect [77]. In this respect, the GPIba promoter demonstrated surprisingly high activity in HSC similar to the MPL promoter.

A 2-kb fragment of the murine Mpl promoter was characterized [223] and has also been used to express Mpl in the Mpl−/− mouse by transgenic approaches [105, 189]. Due to the transfer technology generating the transgenic mice, in the study by Lannutti et al. 38 copies of the expression cassette were detected. Nevertheless, Mpl expression levels were lower than the endogenous levels in wild-type mice, but surprisingly, platelet counts were elevated (thrombocytosis). This can be explained by the high Thpo levels in Mpl−/− mice that strongly activate Mpl also when expressed at low levels. All these studies in the Mpl−/− mouse model highlight the importance to achieve physiological transgene expression levels to guarantee that phenotypes in murine models reflect the biological relevant situation.

20.10 Modification of Megakaryopoiesis by Adeno-Associated Vectors

The adeno-associated viral (AAV) vectors are developed from nonpathogenic human parvoviruses. They are small, non-enveloped viral vectors that carry a single-stranded DNA genome of about 5 kb (for review, see [27]). The AAV particles enter the cells by receptor-mediated uptake and their genomes persist as non-integrated episomes. For productive infectious cycles, the wild-type AAV depends on the coinfection with helper viruses such as adenoviruses. For the generation of vectors, similar to the retroviral system, all AAV-specific genes are removed and supplied to the producer cells by plasmid cotransfection. In the residual AAV genome, only the inverted terminal repeats (ITR) are left which do not contain enhancer/promoter sequences but have to flank the genomic information for productive packaging.

AAV vectors come in different serotypes that determine the tropism of infection, and some cells are rather resistant to AAV transduction including megakaryoblastic cell lines. However, by the use of a bispecific F(ab′)2 complex to megakaryocyte-specific receptors, transduction of megakaryoblastic cell lines and primary MK could be achieved [15]. AAV vectors can also display proteins on their capsid mediating vector binding to cell surface receptors and subsequent uptake [136]. In general, gene transfer by AAV vectors to dividing cells is transient and as the genome does not integrate, normally not applied to HSC. By in vivo vector delivery and targeting of post-mitotic hepatocytes, however, thrombopoietic growth factors or coagulation factors can be expressed with long-term impact on the in vivo phenotype. This concept is exploited for the treatment of hemophilia and discussed below.

20.11 Study of Gene Function in Human Platelets in Humanized Mouse Models

Species-specific differences make it desirable to analyze gene function directly in human rather than murine cells. To study gene functions, human HSC can be transduced and differentiated. Clinical trials aiming for the correction of monogenetic diseases of the blood system rely on the long-term modification by stable integration of the therapeutic gene and also the long-term engraftment of the modified HSC. Therefore, in vitro transduction protocols for human CD34+ cells are well developed and the transplantation and engraftment in humanized mouse models extensively studied [165, 212]. One hurdle is the efficient transduction of CD34+ cells without the loss of their repopulating potential. To induce proliferation by maintaining stemness in vitro, combinations of cytokines are applied (stem cell factor (SCF), thrombopoietin (THPO), Fms-like tyrosine kinase 3 ligand (FLT3L) [167]), or SCF, THPO, Angptl5 and insulin growth factor binding protein 2 [219]. In addition, small drugs can be added, e.g., the aryl hydrocarbon receptor antagonist Stemregenin (SR1) [20] and prostaglandin E2 [57]. All these methods are well suited to support in vitro genetic modifications of CD34+ cells. Retroviral transduction of CD34+ human hematopoietic cells can be further improved by the use of more suitable envelope proteins such as the baboon retroviral envelope [62], by displaying specific scFv on the envelope [23, 96] or by the inhibition of pathways responsible for blocking lentiviral transduction [155].

There are different immune-deficient mouse strains available as recipients that engraft human cell and establish human hematopoiesis in vivo including megakaryopoiesis and platelet release. The most widely used models are the NOD/SCID and NOD/SCID, IL2rg−/− (NOD.Cg-Prkdc^{scid}Il2rg^{tm1Wjl}, NSG) mouse strains [176]. There are other mice developed on different genetic backgrounds, the NOD.cg-Prkdc^{scid}Il2rg^{tm1Sug} (NOG), and Rag−/−, IL2rg−/− (C;129S4- Rag2$^{tm1.1Flv}$, BRG) mice [175]. In general, all models allow the analysis of human hematopoiesis, but with differences in lineage contributions of human cells. So far NSG mice show the highest human chimerism in the BM and supported human B, T, and NK cell development most efficiently [44].

Human platelets in the mouse are, however, difficult to obtain for reasons that are not well understood. In a recent study aiming for platelet-directed expression of human FVIII, the highest levels of human platelets within murine blood of NSG mice was detected 2–3 weeks after human HSPC transplantation (~50 %) and levels declined substantially after 5 weeks dropping to 1 % human contribution to the platelet pool after 7 months [172]. The major thrombopoietic cytokine THPO is not completely crossreactive between human and mouse with lower activity of the murine Thpo on the human MPL receptor. Replacing the murine *Thpo* gene by human *THPO* by a knock-in strategy in BRG mice supports human HSC engraftment and maintenance due to MPL signaling in HSC. However, *THPO* knock-in mice are thrombocytopenic due to lower cytokine levels (heterozygous) or complete lack of the murine Thpo (homozygous), and because the human THPO levels appear not to be sufficiently high to support human thrombopoiesis or replace the murine cytokine on murine platelets [163]. This finding demonstrates that the only murine environment may lack some of the important regulating factors for human megakaryopoiesis and platelet release.

20.12 Gene Therapy Approaches for Bleeding Disorders

Hematopoietic stem cell gene therapy has been employed in the clinic for the treatment of monogenetic hematopoietic disorders with severe disease courses with great therapeutic success [3, 31, 56, 72, 152]. This therapy is based on the isolation of autologous HSC, their in vitro genetic modification, and subsequent re-transplantation into the patient. For the long-term correction of hematopoiesis the gene correction also has to persist. Therefore, currently retro- or lentiviral vectors are the transfer systems of choice. Transduction with these vectors delivers the defective gene as an additional copy to the genome.

Although the concept of HSC-directed gene therapy is well established, the therapy-associated risks for the patients are still very high. These include the risk associated with the pre-transplant conditioning or insufficient engraftment of HSC that underwent in vitro cytokine stimulation and transductions, and the long-term prospect to develop leukemia due to insertional mutagenesis [30, 95]. Therefore, at present, transplantation of gene-modified HSC is only applied in patients with no other treatment options. However, in principle gene therapy could also cure bleeding disorders such as Glanzmann's thrombasthenia (GT) or Bernard-Soulier syndrome. Gene therapy for the treatment of Wiskott-Aldrich syndrome has been performed [21, 22] and is discussed in detail in the chapter by Braun and Klein in this book (Chapter 18).

Before gene therapy can be translated into the clinics, therapy efficacy has to be demonstrated in preclinical animal models. By using lentiviral vectors for expression of human GPIBA and transplantation of transduced HSC, the thrombocytopathia in the murine GPIba–/– mouse, a model for Bernard-Soulier syndrome, was corrected [93]. The treatment improved platelet counts, shortened the bleeding times, and normalized the platelet volume. In this study, expression from the self-inactivating lentiviral vector was driven by the human aIIb integrin promoter, providing specificity of expression to megakaryocytes and platelets. The correction showed an efficiency similar to that of the transgenic expression of the human *GPIBa* in *GPIba–/–* mice [205]. Gene therapy approaches were also developed to treat GT and tested in the Great Pyrenees dog model [50]. In these dogs, the inherited disorder is based on a 14-bp insertion in exon 13 of the *GPIIb* (integrin alpha

2b) gene [108]. The dogs develop symptoms similar to human GT patients and thus provide a very good model to study the human disease. Due to bleeding complications observed in previous transplantation trials in this canine model of GT, the transplantation of transduced peripheral blood CD34+ cells followed a non-myeloablative pretransplant conditioning with 100-cGy total body irradiation or 4-mg/kg busulfan [50]. As a consequence of the non-myeloablative precondition, the donor chimerism after transplantation was expected to be low, but could be increased by selection for the co-expressed drug resistance gene MGMTP140K by treatment with O^6-benzylguanine and carmustine. The contribution of gene-corrected platelets was enriched to 10% which remained stable over the observation period of up to 5 years in one of the dogs. Unfortunately, no information on the enrichment of donor cells in the BM or HSC level was provided.

The concept of enrichment of transduced cells by drug selection has been developed over the last 20 years [124, 132] and now successfully conduced in a clinical trial to treat glioblastoma patients who underwent autologous blood stem cell transplantation [2]. The enrichment of gene-modified cells post transplantation will allow reducing the toxicity of precondition regimes and may, therefore, open up the possibility to perform gene therapy treatment in patients with monogenetic disorders of the blood system presenting with milder phenotypes. However, it has to be kept in mind that cytotoxic drug selection is also accompanied by substantial side effects to patients. In addition, treatment with allogenic BM transplantation is one therapeutic option for GT (type I) with severe disease presentation [39, 201, 207] or also for Bernard-Soulier patients [161]; however, this is still the exception to the normal treatment.

The transplantation of gene-modified HSC targeting expression into platelets has also been explored for the treatment of hemophilias A and B. Platelet-specific expression of FVIII of FIX would supply the necessary clotting factors to the side where the coagulation process takes place. To ensure storage of these factors into the secretory alpha granules of platelets, fusion proteins with the vWF protein have been generated. Besides gene therapy, hemophilia is well treatable by substitution of the missing clotting factors with FVIII and FIX; however, in some cases, patients develop autoantibodies against the synthetic factors, especially FVIII. In these cases, the conventional treatment is ineffective, and platelet-directed delivery by gene therapy may solve these issues as the clotting factors are hidden within the platelet granules and therefore not recognized by the immune system. However, also the non-targeted expression in all blood cells from an ubiquitously active promoter in the lentiviral context can cure hemophilia by transduction and transplantation of HSC [91].

In the past, however, gene therapy approaches for the treatment of hemophilia have mostly been directed toward gene transfer to the liver using AAV or lentiviral vectors [78]. In a clinical trial run by the St. Jude Children's Hospital, Memphis, USA, and by the University Hospital London, UK, a single application of AAV type vector, serotype 8, coding for the human *FIX* gene resulted in long-term (up to 4 years reported) expression of ~6% of the normal FIX levels in the blood which was sufficiently high to reduce disease symptoms and in some cases even made the patients independent of any further replacement therapies [140, 141]. After AAV vector gene transfer, the genomic information is kept in a non-integrated, episomal form. This prevents any risk of interfering with the host genome, but it also does not persist on a long-term basis, as the episomes will not replicate with the cellular genome. As hepatocytes rarely divide, the episomal information may persist. In an approach making use of non-integrating lentiviral vectors, this has further been proven [117]. Immune recognition of the transduced liver cells or the viral capsid proteins is the major obstacle at the moment that has to be overcome by improved vector technology.

20.13 Advanced Technologies for Targeted Genetic Modifications and Gene Knockouts

In the recent years, methodologies have developed that allow side-specific modification in the genome by zinc finger nucleases (ZFN), transcription activator-like effector nucleases (TALEN) [19, 37, 133] or, most recently, the CRISPR/Cas9 approach [90, 137]. These technologies have in common that they target the genome by recognition sequences; in the case of ZFN and TALEN DNA-binding domains and in the CRISPR/Cas9 system, target site selection is determined by the guide RNA [43]. After binding to the target site, a DNA double-strand break is induced that can then be repaired by nonhomologous end joining, which is error-prone and results in nucleotide insertions or deletions, or by homologous recombination which allows the insertion of a new sequence. In this case, the homologous region has to be provided as a separate DNA template to the cells.

The CRISPR system was developed from the type II clustered, regularly interspaced, short palindromic repeats (CRISPR)/Cas9 system in bacteria where it serves as part of the adaptive immunity against viruses and plasmids. In this original system, the endonuclease Cas9 is guided to the DNA location by two RNAs CRISPR RNA (crRNA):transactivating crRNA (tracrRNA), which in the synthetic system are joined to a single guide RNA [90]. In the CRISPR system, therefore, the recognition is mediated by nucleotides, unlike the ZNF and TALEN systems which bind the DNA by protein-nucleotide interaction (DNA binding by zinc finger proteins or TAL proteins). This major difference also explains the easy use of the CRISPR/Cas9 system because the recognition sequence of the guide RNA can just be copied from the genomic sequence around the desired location following the Watson and Crick base pairing. However, also TAL proteins work on the *two amino acids/one nucleotide* base and can be rather flexibly designed. For the DNA cleavage, endonucleases are employed. In the case of ZNF and TALEN, the endonuclease FokI is used. Fok1 assembles from two inactive monomers when they meet at the desired site identified by two independent TAL or zinc finger arrays. In contrast, the Cas9 endonuclease is recruited by only one single-guide RNA, and therefore, the specificity is lower compared to the "double"-guided strategy. To overcome this limitation, a double Cas9 nickase system has also been developed [54, 67, 195]. By delivering more than one guide RNA together with the Cas9 endonuclease, multiple sites can be targeted in one cell (so-called multiplexing). Ideally, all players for this application can be encoded on one construct [92, 164]. By using lentiviral vectors for the delivery of the CRISPR/Cas9 into murine HSC, the multiplex approach has been successfully employed for a genetic screen of deletion mutants in murine leukemia [75].

A common feature of these three systems is the need to bring in all players into the same cells: The two ZFNs or TALENs in these two systems, or the guide RNA and Cas9 enzyme for the CRISPR/Cas9 system; when a homologous recombination in the targeted locus is desired, the homologous DNA as template for the recombination is also required. In some cells, such as pluripotent cells, transfection by electroporation can be performed. As pluripotent cells are long lived in their undifferentiated state, correct recombinants can be selected for these cells. However, in primary stem cells, such as HSC, this is a difficult task. In recent studies by the groups of L. Naldini and D. Kohn, proof-of-concept studies demonstrated the successful correction of the *IL2RG* gene or correction of the sickle cell mutation in the beta-globin gene, respectively, by ZFN-mediated approaches in human HSC of patients with the respective diseases and their subsequent transplantation into NSG mice [57, 80]. They developed protocols based on the delivery by integration-deficient lentiviral vectors (iDLV) and RNA electroporation. Similarly, this has been employed for the delivery into T cells [128]. Still, the use of targeted genome modification for mouse models based on the transplantation of in vitro-modified primary HSC is challenging; however, for the introduction of mutations in pluripotent stem cells and for the generation of mouse models with germline modifications or iPSC-based disease models, these new technologies open up new avenues in research.

20.14 Overcoming Unwanted Effects of Genetic Modification

Every genetic modification alters the genome and can induce unwanted (side) effects. In clinical gene therapy, this can lead to severe adverse events with life-threatening complications [210]. Furthermore, when working with animal models, the researcher wishes to obtain physiological models without skewed phenotypes. In both cases, gene transfer technologies have advanced very far and offer good strategies to overcome limitations.

The integration of expression cassettes into the genome containing enhancers/promoters can interfere with the regulation of neighboring genes by insertional mutagenesis. In this respect, this is independent of the transfer system used. However, with a higher likelihood to integrate into gene dense regions, into actively transcribed genes and close to promoters, which is characteristic for gammaretroviral vectors, the risk is increasing. As these vectors are derived from replicating viruses that were selected on the basis of good expression to ensure rapid spread, these regions in the genome are favored. The potentially most dangerous interaction from a vector is mediated by its enhancer/promoter through upregulating gene expression from cellular promoters ("enhancer interaction"). In the case of vector integrations close to proto-oncogenes, this has led to the development of leukemias in human gene therapy patients [22, 71, 82, 182] and animal models [129, 131, 170]. In fact, gammaretroviruses and gammaretroviral vectors are potent tools for gene fishing approaches by selecting insertional mutants [120, 191]. By deleting the viral enhancers/promoters and moving the promoter into the internal position, the insertional mutagenic (genotoxic) potential is reduced [130]. Lentiviral vectors have a slightly different integration preference and therefore less often interact with cellular promoters due to a further distance to them. On the other hand, lentiviral vectors tend to integrate into introns of genes and can, by splicing-mediated interaction, alter gene transcription. Downregulation of gene expression was observed after the integration of a lentiviral SIN vector with the megakaryocyte-specific GPIba promoter as internal promoter into the 8th intron of the early B-cell factor 1 (Ebf1) gene [76]. Haploinsufficiency of Ebf1 caused B-cell leukemia with a block in B-cell development. The GPIba promoter is widely used for megakaryocyte-specific expression. The promoter fragment (−258 to +330) spanning the entire first exon and first intron contains splice sites from the exon-intron boundaries. Splice interference generates transcripts with insufficient polyadenylation, which are, therefore, subjected to nonsense-mediated RNA decay. To overcome this problem, SIN vectors with cellular promoters devoid of unnecessary splice sites should be employed. Furthermore, some other vector platform may target integration into less active chromatin, for example, the alpharetroviral vectors or the nonviral sleeping beauty transposon system.

Another drawback in gene modification can be the silencing of transgene expression by epigenetic effects [10, 48]. This is especially problematic when introducing modifications into pluripotent cells. One option to avoid position effects is the targeting of so-called safe harbors, one of which is the AAVS1 locus, by use of the ZNF or TALEN approaches. The adeno-associated virus integration site 1 is located on chromosome 19, within the first intron of the PPP1R12C gene encoding the protein phosphatase 1 regulatory subunit 12. This region harbors open chromatin and is flanked by insulator elements that shield the inserted gene from activation from the outside. Gene expression from transgenic cassettes inserted into this locus was found to be active and not to have any effects on the host cell. Alternatively, other locations in the genome could be defined as “safe,” for example, the murine Rosa26 locus. Recently the group of M. Sadelain has defined criteria for a safe harbor as follows: (1) the distance to the next 5′ end of a gene to be at least 50 kb and (2) >300 kb away from cancer-related genes and (3) microRNAs, (4) to be located outside transcription units and (5) ultraconserved elements in the human genome [153]. When accepting these criteria, any cell with genetic modifications that fits this definition would be considered safely modified.

Alternatively, the expression cassette can be flanked by insulators. The best studied insulator is chicken hypersensitive site 4, derived from the chicken globin locus [38, 61]. Unfortunately, so far insulators cannot completely shut off interactions with the genome, but research in defining new and more accurate insulators is ongoing [109]. Besides, genetic elements are being investigated that allow for faithful expression from an inserted transgenic cassette by use of the ubiquitous chromatin opening (UCOE) element [221]. Until recently, this element was rather large, also containing additional splice sites, but more recent research has optimized the UCOE element [135].

20.15 Pluripotent Cells That Mimic Human Thrombocytopathias

The discovery of the technology to reprogram somatic cells to pluripotent cells, the so-called induced pluripotent stem cells (iPSC) by the seminal work of S. Yamanaka [186], has opened new options for the development of disease models. iPSC can be differentiated to megakaryocytes and platelets allowing us to study all intermediate steps in vitro. This makes it also possible to study gene functions directly in human cells, which is especially helpful if no suitable murine models exist, for example, when the murine phenotype does not or only partially reflects the human disease. One example is the murine Was knockout mouse which mimicks leukopenia and defects in T cell activation similar to human WAS patients, but mice also show altered B-cell responses and do not develop eczema or hematopoietic malignancies [178]. In addition, $Was^{-/-}$ mice have chronic colitis that is rarely found in human WAS patients. In respect to the defects in platelets, Was knockout mice only have moderate thrombocytopenia and no reduced platelet volumes. By utilizing iPS cells reprogrammed from human patients, the WAS deficient phenotype and underlying mechanisms can now be studied [85]. Similarly, iPS cells of GT patients have been generated, differentiated into platelets and their functionality compared to patient-derived platelets from peripheral blood [151]. By a gene correction approach delivering the defective *alpha 2b* gene via recombination into the AAVS1safe harbor locus, alphaIIb expression could be reestablished [185]. In this study, GPIIb was expressed under the control of the GPIba promoter, reaching the same expression levels as in equally differentiated wild-type iPS cells. Although these are very promising results, the regulation by the transgenic expression cassette can only partially mimic the physiological regulation. In a study to correct MPL deficiency in human MPL−/− iPS cells (CAMT iPS cells), MPL was expressed by a classical gammaretroviral vector with functional LTRs [79]. The expression levels in this approach did not accurately resemble the normal levels and in agreement with that the authors describe skewed differentiation in MPL overexpressing MPL−/− iPS cells. When aiming to correct defects in iPS cells or to overexpress genes, the controlled expression during all steps of differentiation, also avoiding potential silencing, has to be guaranteed and will be a challenge for the future [1].

References

1. Ackermann M, Lachmann N, Hartung S, Eggenschwiler R, Pfaff N, Happle C, Mucci A, Göhring G, Niemann H, Hansen G et al (2014) Promoter and lineage independent anti-silencing activity of the A2 ubiquitous chromatin opening element for optimized human pluripotent stem cell-based gene therapy. Biomaterials 35:1531–1542
2. Adair JE, Johnston SK, Mrugala MM, Beard BC, Guyman LA, Baldock AL, Bridge CA, Hawkins-Daarud A, Gori JL, Born DE et al (2014) Gene therapy enhances chemotherapy tolerance and efficacy in glioblastoma patients. J Clin Invest 124:4082–4092
3. Aiuti A, Biasco L, Scaramuzza S, Ferrua F, Cicalese MP, Baricordi C, Dionisio F, Calabria A, Giannelli S, Castiello MC et al (2013) Lentiviral hematopoietic stem cell gene therapy in patients with Wiskott-Aldrich syndrome. Science (New York, NY) 341:1233151

4. Akagi K, Sandig V, Vooijs M, Van der Valk M, Giovannini M, Strauss M, Berns A (1997) Cre-mediated somatic site-specific recombination in mice. Nucleic Acids Res 25:1766–1773
5. Alexander WS, Roberts AW, Nicola NA, Li R, Metcalf D (1996) Deficiencies in progenitor cells of multiple hematopoietic lineages and defective megakaryocytopoiesis in mice lacking the thrombopoietic receptor c-Mpl. Blood 87(6):2162–70
6. Anastassiadis K, Fu J, Patsch C, Hu S, Weidlich S, Duerschke K, Buchholz F, Edenhofer F, Stewart AF (2009) Dre recombinase, like Cre, is a highly efficient site-specific recombinase in E. coli, mammalian cells and mice. Dis Model Mech 2:508–515
7. Anastassiadis K, Schnütgen F, von Melchner H, Stewart AF (2013) Gene targeting and site-specific recombination in mouse ES cells. Methods Enzymol 533:133–155
8. Annoni A, Brown BD, Cantore A, Sergi LS, Naldini L, Roncarolo M-G (2009) In vivo delivery of a microRNA-regulated transgene induces antigen-specific regulatory T cells and promotes immunologic tolerance. Blood 114:5152–5161
9. Anton M, Graham FL (1995) Site-specific recombination mediated by an adenovirus vector expressing the Cre recombinase protein: a molecular switch for control of gene expression. J Virol 69:4600–4606
10. Antoniou MN, Skipper KA, Anakok O (2013) Optimizing retroviral gene expression for effective therapies. Hum Gene Ther 24:363–374
11. Araki K, Araki M, Yamamura K (1997) Targeted integration of DNA using mutant lox sites in embryonic stem cells. Nucl Acids Res 25:868–872
12. Aronovich EL, McIvor RS, Hackett PB (2011) The sleeping beauty transposon system: a non-viral vector for gene therapy. Hum Mol Genet 20:R14–R20
13. Babinet C, Cohen-Tannoudji M (2001) Genome engineering via homologous recombination in mouse embryonic stem (ES) cells: an amazingly versatile tool for the study of mammalian biology. An Acad Bras Cienc 73:365–383
14. Ballmaier M, Germeshausen M, Schulze H, Cherkaoui K, Lang S, Gaudig A, Krukemeier S, Eilers M, Strauss G, Welte K (2001) c-mpl mutations are the cause of congenital amegakaryocytic thrombocytopenia. Blood 97:139–146
15. Bartlett JS, Kleinschmidt J, Boucher RC, Samulski RJ (1999) Targeted adeno-associated virus vector transduction of nonpermissive cells mediated by a bispecific F(ab'gamma)2 antibody. Nat Biotechnol 17:181–186
16. Bartosch B, Cosset F-L (2004) Strategies for retargeted gene delivery using vectors derived from lentiviruses. Curr Gene Ther 4:427–443
17. Bastian LS, Yagi M, Chan C, Roth GJ (1996) Analysis of the megakaryocyte glycoprotein IX promoter identifies positive and negative regulatory domains and functional GATA and Ets sites. J Biol Chem 271:18554–18560
18. Bi L, Lawler AM, Antonarakis SE, High KA, Gearhart JD, Kazazian HH Jr (1995) Targeted disruption of the mouse factor VIII gene produces a model of haemophilia A. Nat Genet 10(1):119–121
19. Boch J, Scholze H, Schornack S, Landgraf A, Hahn S, Kay S, Lahaye T, Nickstadt A, Bonas U (2009) Breaking the code of DNA binding specificity of TAL-type III effectors. Science (New York, NY) 326:1509–1512
20. Boitano AE, Wang J, Romeo R, Bouchez LC, Parker AE, Sutton SE, Walker JR, Flaveny CA, Perdew GH, Denison MS et al (2010) Aryl hydrocarbon receptor antagonists promote the expansion of human hematopoietic stem cells. Science (New York, NY) 329:1345–1348
21. Boztug K, Schmidt M, Schwarzer A, Banerjee PP, Díez IA, Dewey RA, Böhm M, Nowrouzi A, Ball CR, Glimm H et al (2010) Stem-cell gene therapy for the Wiskott-Aldrich syndrome. N Engl J Med 363:1918–1927
22. Braun CJ, Boztug K, Paruzynski A, Witzel M, Schwarzer A, Rothe M, Modlich U, Beier R, Göhring G, Steinemann D et al (2014) Gene therapy for Wiskott-Aldrich syndrome--long-term efficacy and genotoxicity. Sci Transl Med 6:227ra33
23. Brendel C, Goebel B, Daniela A, Brugman M, Kneissl S, Schwäble J, Kaufmann KB, Müller-Kuller U, Kunkel H, Chen-Wichmann L et al (2015) CD133-targeted gene transfer into long-term repopulating hematopoietic stem cells. Mol Ther J Am Soc Gene Ther 23:63–70

24. Broach JR, Hicks JB (1980) Replication and recombination functions associated with the yeast plasmid, 2 mu circle. Cell 21:501–508
25. Brown BD, Cantore A, Annoni A, Sergi LS, Lombardo A, Della Valle P, D'Angelo A, Naldini L (2007) A microRNA-regulated lentiviral vector mediates stable correction of hemophilia B mice. Blood 110:4144–4152
26. Buchholz F, Angrand PO, Stewart AF (1998) Improved properties of FLP recombinase evolved by cycling mutagenesis. Nat Biotechnol 16:657–662
27. Büning H, Perabo L, Coutelle O, Quadt-Humme S, Hallek M (2008) Recent developments in adeno-associated virus vector technology. J Gene Med 10:717–733
28. Calaminus S, Guitart AV, Guitart A, Sinclair A, Schachtner H, Watson SP, Holyoake TL, Kranc KR, Machesky LM (2012) Lineage tracing of Pf4-Cre marks hematopoietic stem cells and their progeny. PLoS One 7:e51361
29. Capecchi M (1989) Altering the genome by homologous recombination. Science 244:1288–1292
30. Cavazzana-Calvo M, André-Schmutz I, Fischer A (2013) Haematopoietic stem cell transplantation for SCID patients: where do we stand? Br J Haematol 160:146–152
31. Cavazzana-Calvo M, Hacein-Bey S, de Saint Basile G, Gross F, Yvon E, Nusbaum P, Selz F, Hue C, Certain S, Casanova JL et al (2000) Gene therapy of human severe combined immunodeficiency (SCID)-X1 disease. Science (New York, NY) 288:669–672
32. Chandrashekran A, Sarkar R, Thrasher A, Fraser SE, Dibb N, Casimir C, Winston R, Readhead C (2014) Efficient generation of transgenic mice by lentivirus-mediated modification of spermatozoa. FASEB J Off Publ Federation Am Soc Exp Biol 28:569–576
33. Chang AH, Stephan MT, Lisowski L, Sadelain M (2008) Erythroid-specific human factor IX delivery from in vivo selected hematopoietic stem cells following nonmyeloablative conditioning in hemophilia B mice. Mol Ther J Am Soc Gene Ther 16:1745–1752
34. Chen Y, Schroeder JA, Kuether EL, Zhang G, Shi Q (2014) Platelet gene therapy by lentiviral gene delivery to hematopoietic stem cells restores hemostasis and induces humoral immune tolerance in FIX(null) mice. Mol Ther J Am Soc Gene Ther 22:169–177
35. Cherepanov P, Maertens G, Proost P, Devreese B, van Beeumen J, Engelborghs Y, de Clercq E, Debyser Z (2003) HIV-1 integrase forms stable tetramers and associates with LEDGF/p75 protein in human cells. J Biol Chem 278:372–381
36. Cho A, Haruyama N, Kulkarni AB (2009) Generation of transgenic mice. Curr Protocols Cell Biol/editorial board, Juan S. Bonifacino … et al. Chapter 19, Unit 19.11
37. Christian M, Cermak T, Doyle EL, Schmidt C, Zhang F, Hummel A, Bogdanove AJ, Voytas DF (2010) Targeting DNA double-strand breaks with TAL effector nucleases. Genetics 186:757–761
38. Chung JH, Whiteley M, Felsenfeld G (1993) A 5′ element of the chicken beta-globin domain serves as an insulator in human erythroid cells and protects against position effect in Drosophila. Cell 74:505–514
39. Connor P, Khair K, Liesner R, Amrolia P, Veys P, Ancliff P, Mathias M (2008) Stem cell transplantation for children with Glanzmann thrombasthenia. Br J Haematol 140:568–571
40. Cox MM (1983) The FLP protein of the yeast 2-microns plasmid: expression of a eukaryotic genetic recombination system in Escherichia coli. Proc Natl Acad Sci U S A 80:4223–4227
41. de Rijck J, de Kogel C, Demeulemeester J, Vets S, El Ashkar S, Malani N, Bushman FD, Landuyt B, Husson SJ, Busschots K et al (2013) The BET family of proteins targets moloney murine leukemia virus integration near transcription start sites. Cell Rep 5:886–894
42. Deng C-X (2012) The use of Cre–loxP technology and inducible systems to generate mouse models of cancer. In: Green JE, Ried T (eds) Genetically engineered mice for cancer research. Springer, New York, pp 17–36
43. Doudna JA, Charpentier E (2014) Genome editing. The new frontier of genome engineering with CRISPR-Cas9. Science (New York, NY) 346:1258096
44. Drake AC, Chen Q, Chen J (2012) Engineering humanized mice for improved hematopoietic reconstitution. Cell Mol Immunol 9:215–224
45. Dropulić B (2011) Lentiviral vectors: their molecular design, safety, and use in laboratory and preclinical research. Hum Gene Ther 22:649–657

46. Du LM, Nurden P, Nurden AT, Nichols TC, Bellinger DA, Jensen ES, Haberichter SL, Merricks E, Raymer RA, Fang J et al (2013) Platelet-targeted gene therapy with human factor VIII establishes haemostasis in dogs with haemophilia A. Nat Commun 4:2773
47. Dymecki SM (1996) Flp recombinase promotes site-specific DNA recombination in embryonic stem cells and transgenic mice. Proc Natl Acad Sci U S A 93:6191–6196
48. Ellis J (2005) Silencing and variegation of gammaretrovirus and lentivirus vectors // silencing and variegation of gammaretrovirus and lentivirus vectors. Hum Gene Ther 16:1241–1246
49. Emambokus NR, Frampton J (2003) The glycoprotein IIb molecule is expressed on early murine hematopoietic progenitors and regulates their numbers in sites of hematopoiesis. Immunity 19:33–45
50. Fang J, Jensen ES, Boudreaux MK, Du LM, Hawkins TB, Koukouritaki SB, Cornetta K, Wilcox DA (2011) Platelet gene therapy improves hemostatic function for integrin alphaIIb-beta3-deficient dogs. Proc Natl Acad Sci U S A 108:9583–9588
51. Feil R, Brocard J, Mascrez B, LeMeur M, Metzger D, Chambon P (1996) Ligand-activated site-specific recombination in mice. Proc Natl Acad Sci U S A 93:10887–10890
52. Feil R, Wagner J, Metzger D, Chambon P (1997) Regulation of Cre recombinase activity by mutated estrogen receptor ligand-binding domains. Biochem Biophys Res Commun 237:752–757
53. Finkelshtein D, Werman A, Novick D, Barak S, Rubinstein M (2013) LDL receptor and its family members serve as the cellular receptors for vesicular stomatitis virus. Proc Natl Acad Sci U S A 110:7306–7311
54. Fu Y, Sander JD, Reyon D, Cascio VM, Joung JK (2014) Improving CRISPR-Cas nuclease specificity using truncated guide RNAs. Nat Biotechnol 32:279–284
55. Furihata K, Kunicki TJ (2002) Characterization of human glycoprotein VI gene 5′ regulatory and promoter regions. Arterioscler Thromb Vasc Biol 22:1733–1739
56. Gaspar HB, Parsley KL, Howe S, King D, Gilmour KC, Sinclair J, Brouns G, Schmidt M, von Kalle C, Barington T et al (2004) Gene therapy of X-linked severe combined immunodeficiency by use of a pseudotyped gammaretroviral vector. Lancet 364:2181–2187
57. Genovese P, Schiroli G, Escobar G, Di Tomaso T, Firrito C, Calabria A, Moi D, Mazzieri R, Bonini C, Holmes MC et al (2014) Targeted genome editing in human repopulating haematopoietic stem cells. Nature 510:235–240
58. Gentner B, Naldini L (2012) Exploiting microRNA regulation for genetic engineering. Tissue Antigens 80:393–403
59. Gentner B, Schira G, Giustacchini A, Amendola M, Brown BD, Ponzoni M, Naldini L (2009) Stable knockdown of microRNA in vivo by lentiviral vectors. Nat Methods 6:63–66
60. Gentner B, Visigalli I, Hiramatsu H, Lechman E, Ungari S, Giustacchini A, Schira G, Amendola M, Quattrini A, Martino S et al (2010) Identification of hematopoietic stem cell-specific miRNAs enables gene therapy of globoid cell leukodystrophy. Sci Transl Med 2:58ra84
61. Ghirlando R, Giles K, Gowher H, Xiao T, Xu Z, Yao H, Felsenfeld G (2012) Chromatin domains, insulators, and the regulation of gene expression. Biochim Biophys Acta 1819:644–651
62. Girard-Gagnepain A, Amirache F, Costa C, Lévy C, Frecha C, Fusil F, Nègre D, Lavillette D, Cosset F-L, Verhoeyen E (2014) Baboon envelope pseudotyped LVs outperform VSV-G-LVs for gene transfer into early-cytokine-stimulated and resting HSCs. Blood 124:1221–1231
63. Graus-Porta D, Blaess S, Senften M, Littlewood-Evans A, Damsky C, Huang Z, Orban P, Klein R, Schittny JC, Müller U (2001) Beta1-class integrins regulate the development of laminae and folia in the cerebral and cerebellar cortex. Neuron 31:367–379
64. Greene TK, Lyde RB, Bailey SC, Lambert MP, Zhai L, Sabatino DE, Camire RM, Arruda VR, Poncz M (2014) Apoptotic effects of platelet factor VIII on megakaryopoiesis: implications for a modified human FVIII for platelet-based gene therapy. J Thrombosis Haemostasis JTH 12:2102–2112
65. Greene TK, Wang C, Hirsch JD, Zhai L, Gewirtz J, Thornton MA, Miao HZ, Pipe SW, Kaufman RJ, Camire RM et al (2010) In vivo efficacy of platelet-delivered, high specific activity factor VIII variants. Blood 116:6114–6122

66. Guenechea G, Gan OI, Inamitsu T, Dorrell C, Pereira DS, Kelly M, Naldini L, Dick JE (2000) Transduction of human CD34+ CD38- bone marrow and cord blood-derived SCID-repopulating cells with third-generation lentiviral vectors. Mol Ther J Am Soc Gene Ther 1: 566–573
67. Guilinger JP, Thompson DB, Liu DR (2014) Fusion of catalytically inactive Cas9 to FokI nuclease improves the specificity of genome modification. Nat Biotechnol 32:577–582
68. Guo F, Gopaul DN, van Duyne GD (1997) Structure of Cre recombinase complexed with DNA in a site-specific recombination synapse. Nature 389:40–46
69. Gupta SS, Maetzig T, Maertens GN, Sharif A, Rothe M, Weidner-Glunde M, Galla M, Schambach A, Cherepanov P, Schulz TF (2013) Bromo- and extraterminal domain chromatin regulators serve as cofactors for murine leukemia virus integration. J Virol 87:12721–12736
70. Gutiérrez L, Tsukamoto S, Suzuki M, Yamamoto-Mukai H, Yamamoto M, Philipsen S, Ohneda K (2008) Ablation of Gata1 in adult mice results in aplastic crisis, revealing its essential role in steady-state and stress erythropoiesis. Blood 111:4375–4385
71. Hacein-Bey-Abina S, Garrigue A, Wang GP, Soulier J, Lim A, Morillon E, Clappier E, Caccavelli L, Delabesse E, Beldjord K et al (2008) Insertional oncogenesis in 4 patients after retrovirus-mediated gene therapy of SCID-X1. J Clin Invest 118:3132–3142
72. Hacein-Bey-Abina S, Pai S-Y, Gaspar HB, Armant M, Berry CC, Blanche S, Bleesing J, Blondeau J, de Boer H, Buckland KF et al (2014) A modified γ-retrovirus vector for X-linked severe combined immunodeficiency. N Engl J Med 371:1407–1417
73. Hall B, Limaye A, Kulkarni AB (2009) Overview: generation of gene knockout mice. Curr Protocols Cell Biol/editorial board, Juan S. Bonifacino … et al. Chapter 19, Unit 19.12 19.12.1-17
74. Hashimoto Y, Ware J (1995) Identification of essential GATA and Ets binding motifs within the promoter of the platelet glycoprotein Ib alpha gene. J Biol Chem 270:24532–24539
75. Heckl D, Kowalczyk MS, Yudovich D, Belizaire R, Puram RV, McConkey ME, Thielke A, Aster JC, Regev A, Ebert BL (2014) Generation of mouse models of myeloid malignancy with combinatorial genetic lesions using CRISPR-Cas9 genome editing. Nat Biotechnol 32:941–946
76. Heckl D, Schwarzer A, Haemmerle R, Steinemann D, Rudolph C, Skawran B, Knoess S, Krause J, Li Z, Schlegelberger B et al (2012) Lentiviral vector induced insertional haploinsufficiency of Ebf1 causes murine leukemia. Mol Ther J Am Soc Gene Ther 20:1187–1195
77. Heckl D, Wicke DC, Brugman MH, Meyer J, Schambach A, Büsche G, Ballmaier M, Baum C, Modlich U (2011) Lentiviral gene transfer regenerates hematopoietic stem cells in a mouse model for Mpl-deficient aplastic anemia. Blood 117:3737–3747
78. High KH, Nathwani A, Spencer T, Lillicrap D (2014) Current status of haemophilia gene therapy. Haemophilia Off J World Fed Hemophilia 20(Suppl 4):43–49
79. Hirata S, Takayama N, Jono-Ohnishi R, Endo H, Nakamura S, Dohda T, Nishi M, Hamazaki Y, Ishii E, Kaneko S et al (2013) Congenital amegakaryocytic thrombocytopenia iPS cells exhibit defective MPL-mediated signaling. J Clin Invest 123:3802–3814
80. Hoban MD, Cost GJ, Mendel MC, Romero Z, Kaufman ML, Joglekar AV, Ho M, Lumaquin D, Gray D, Lill GR et al (2015) Correction of the sickle cell disease mutation in human hematopoietic stem/progenitor cells. Blood 125:2597–2604
81. Hoess RH, Abremski K (1984) Interaction of the bacteriophage P1 recombinase Cre with the recombining site loxP. Proc Natl Acad Sci U S A 81:1026–1029
82. Howe SJ, Mansour MR, Schwarzwaelder K, Bartholomae C, Hubank M, Kempski H, Brugman MH, Pike-Overzet K, Chatters SJ, de Ridder D et al (2008) Insertional mutagenesis combined with acquired somatic mutations causes leukemogenesis following gene therapy of SCID-X1 patients. J Clin Invest 118:3143–3150
83. Hug H, Costas M, Staeheli P, Aebi M, Weissmann C (1988) Organization of the murine Mx gene and characterization of its interferon- and virus-inducible promoter. Mol Cell Biol 8:3065–3079
84. Ikawa M, Tanaka N, Kao WW-Y, Verma IM (2003) Generation of transgenic mice using lentiviral vectors: a novel preclinical assessment of lentiviral vectors for gene therapy. Mol Ther J Am Soc Gene Ther 8:666–673

85. Ingrungruanglert P, Amarinthnukrowh P, Rungsiwiwut R, Maneesri-le Grand S, Sosothikul D, Suphapeetiporn K, Israsena N, Shotelersuk V (2015) Wiskott-Aldrich syndrome iPS cells produce megakaryocytes with defects in cytoskeletal rearrangement and proplatelet formation. Thrombosis Haemostasis 113(4):792–805
86. Ivics Z, Hackett PB, Plasterk RH, Izsvák Z (1997) Molecular reconstruction of sleeping beauty, a Tc1-like transposon from fish, and its transposition in human cells. Cell 91:501–510
87. Ivics Z, Mátés L, Yau TY, Landa V, Zidek V, Bashir S, Hoffmann OI, Hiripi L, Garrels W, Kues WA et al (2014) Germline transgenesis in rodents by pronuclear microinjection of sleeping beauty transposons. Nat Protoc 9:773–793
88. Izsvák Z, Ivics Z (2004) Sleeping beauty transposition: biology and applications for molecular therapy. Mol Ther J Am Soc Gene Ther 9:147–156
89. Jähner D, Stuhlmann H, Stewart CL, Harbers K, Löhler J, Simon I, Jaenisch R (1982) De novo methylation and expression of retroviral genomes during mouse embryogenesis. Nature 298:623–628
90. Jinek M, Chylinski K, Fonfara I, Hauer M, Doudna JA, Charpentier E (2012) A programmable dual-RNA-guided DNA endonuclease in adaptive bacterial immunity. Science (New York, NY) 337:816–821
91. Johnston JM, Denning G, Doering CB, Spencer HT (2013) Generation of an optimized lentiviral vector encoding a high-expression factor VIII transgene for gene therapy of hemophilia A. Gene Ther 20:607–615
92. Kabadi AM, Ousterout DG, Hilton IB, Gersbach CA (2014) Multiplex CRISPR/Cas9-based genome engineering from a single lentiviral vector. Nucleic Acids Res 42:e147
93. Kanaji S, Kuether EL, Fahs SA, Schroeder JA, Ware J, Montgomery RR, Shi Q (2012) Correction of murine Bernard-Soulier syndrome by lentivirus-mediated gene therapy. Mol Ther J Ame Soc Gene Ther 20:625–632
94. Kanatsu-Shinohara M, Toyokuni S, Shinohara T (2004) Transgenic mice produced by retroviral transduction of male germ line stem cells in vivo. Biol Reprod 71:1202–1207
95. Kaufmann KB, Chiriaco M, Siler U, Finocchi A, Reichenbach J, Stein S, Grez M (2014) Gene therapy for chronic granulomatous disease: current status and future perspectives. Curr Gene Ther 14:447–460
96. Kays S.-K, Kaufmann K.B, Abel T, Brendel C, Bonig H, Grez M, Buchholz C.J, Kneissl S (2015) CD105 Is a surface marker for receptor-targeted gene transfer into human long-term repopulating hematopoietic stem cells. Stem Cells Development 24(6):714–723
97. Kikuchi J, Mimuro J, Ogata K, Tabata T, Ueda Y, Ishiwata A, Kimura K, Kimura K, Takano K, Madoiwa S et al (2004) Sustained transgene expression by human cord blood derived CD34+ cells transduced with simian immunodeficiency virus agmTYO1-based vectors carrying the human coagulation factor VIII gene in NOD/SCID mice. J Gene Med 6:1049–1060
98. Kile BT (2014) The role of apoptosis in megakaryocytes and platelets. Br J Haematol 165: 217–226
99. Kraunus J, Schaumann DHS, Meyer J, Modlich U, Fehse B, Brandenburg G, von Laer D, Klump H, Schambach A, Bohne J et al (2004) Self-inactivating retroviral vectors with improved RNA processing. Gene Ther 11:1568–1578
100. Kuether EL, Schroeder JA, Fahs SA, Cooley BC, Chen Y, Montgomery RR, Wilcox DA, Shi Q (2012) Lentivirus-mediated platelet gene therapy of murine hemophilia A with pre-existing anti-factor VIII immunity. J Thromb Haemost JTH 10:1570–1580
101. Kufrin D, Eslin DE, Bdeir K, Murciano J-C, Kuo A, Kowalska MA, Degen JL, Sachais BS, Cines DB, Poncz M (2003) Antithrombotic thrombocytes: ectopic expression of urokinase-type plasminogen activator in platelets. Blood 102:926–933
102. Kuhn R, Schwenk F, Aguet M, Rajewsky K (1995) Inducible gene targeting in mice. Science 269:1427–1429
103. Kuno J, Poueymirou WT, Gong G, Siao C-J, Clarke G, Esau L, Kojak N, Posca J, Atanasio A, Strein J et al (2015) Generation of fertile and fecund F0 XY female mice from XY ES cells. Transgenic Res 24:19–29

104. Lacroix C, Giovannini D, Combe A, Bargieri DY, Späth S, Panchal D, Tawk L, Thiberge S, Carvalho TG, Barale J-C et al (2011) FLP/FRT-mediated conditional mutagenesis in pre-erythrocytic stages of Plasmodium berghei. Nat Protoc 6:1412–1428
105. Lannutti BJ, Epp A, Roy J, Chen J, Josephson NC (2009) Incomplete restoration of Mpl expression in the mpl-/- mouse produces partial correction of the stem cell-repopulating defect and paradoxical thrombocytosis. Blood 113:1778–1785
106. Lemarchandel V, Ghysdael J, Mignotte V, Rahuel C, Roméo PH (1993) GATA and Ets cis-acting sequences mediate megakaryocyte-specific expression. Mol Cell Biol 13:668–676
107. Limaye A, Hall B, Kulkarni AB (2009) Manipulation of mouse embryonic stem cells for knockout mouse production. Curr Protocols Cell Biol/editorial board, Juan S. Bonifacino … et al. Chapter 19, Unit 19.13 19.13.1-24
108. Lipscomb DL, Bourne C, Boudreaux MK (2000) Two genetic defects in alphaIIb are associated with type I Glanzmann's thrombasthenia in a great Pyrenees dog: a 14-base insertion in exon 13 and a splicing defect of intron 13. Vet Pathol 37:581–588
109. Liu M, Maurano MT, Wang H, Qi H, Song C-Z, Navas PA, Emery DW, Stamatoyannopoulos JA, Stamatoyannopoulos G (2015) Genomic discovery of potent chromatin insulators for human gene therapy. Nat Biotechnol 33:198–203
110. Lois C, Hong EJ, Pease S, Brown EJ, Baltimore D (2002) Germline transmission and tissue-specific expression of transgenes delivered by lentiviral vectors. Science (New York, NY) 295:868–872
111. Lombardo A, Genovese P, Beausejour CM, Colleoni S, Lee Y-L, Kim KA, Ando D, Urnov FD, Galli C, Gregory PD et al (2007) Gene editing in human stem cells using zinc finger nucleases and integrase-defective lentiviral vector delivery. Nat Biotechnol 25:1298–1306
112. Lozier JN, Dutra A, Pak E, Zhou N, Zheng Z, Nichols TC, Bellinger DA, Read M, Morgan RA (2002) The Chapel Hill hemophilia A dog colony exhibits a factor VIII gene inversion. Proc Natl Acad Sci USA 99:12991–12996
113. Lund AH, Duch M, Pedersen FS (1996) Transcriptional silencing of retroviral vectors. J Biomed Sci 3:365–378
114. Maetzig T, Galla M, Baum C, Schambach A (2011) Gammaretroviral vectors: biology, technology and application. Viruses 3:677–713
115. Malim MH, Hauber J, Le SY, Maizel JV, Cullen BR (1989) 6212-FIN.indd//The HIV-1 rev trans-activator acts through a structured target sequence to activate nuclear export of unspliced viral mRNA. Nature 338:254–257
116. Mansour SL (1990) Gene targeting in murine embryonic stem cells: introduction of specific alterations into the mammalian genome. Gene Anal Tech 7:219–227
117. Mátrai J, Cantore A, Bartholomae CC, Annoni A, Wang W, Acosta-Sanchez A, Samara-Kuko E, Waele L, de Ma L, Genovese P et al (2011) Hepatocyte-targeted expression by integrase-defective lentiviral vectors induces antigen-specific tolerance in mice with low genotoxic risk. Hepatology (Baltimore, Md) 53:1696–1707
118. McCrann DJ, Yezefski T, Nguyen HG, Papadantonakis N, Liu H, Wen Q, Crispino JD, Ravid K (2008) Survivin overexpression alone does not alter megakaryocyte ploidy nor interfere with erythroid/megakaryocytic lineage development in transgenic mice. Blood 111:4092–4095
119. Melton DW (2002) Gene-targeting strategies. Methods Mol Biol (Clifton, NJ) 180:151–173
120. Mikkers H, Berns A (2003) Retroviral insertional mutagenesis: tagging cancer pathways. Adv Cancer Res 88:53–99
121. Miller AD (1990) Retrovirus packaging cells/retrovirus packaging cells. Hum Gene Ther 1:5–14
122. Miller DG, Adam MA, Miller AD (1990) Gene transfer by retrovirus vectors occurs only in cells that are actively replicating at the time of infection. Mol Cell Biol 10:4239–4242
123. Miller AD, Rosman GJ (1989) Improved retroviral vectors for gene transfer and expression. Biotechniques 7:980-2–984-6, 989–90
124. Milsom MD, Williams DA (2007) Live and let die: in vivo selection of gene-modified hematopoietic stem cells via MGMT-mediated chemoprotection. DNA Repair 6:1210–1221

125. Minami T, Tachibana K, Imanishi T, Doi T (1998) Both Ets-1 and GATA-1 are essential for positive regulation of platelet factor 4 gene expression. Eur J Biochem/FEBS 258:879–889
126. Mitchell RS, Beitzel BF, Schroder ARW, Shinn P, Chen H, Berry CC, Ecker JR, Bushman FD (2004) Retroviral DNA integration: ASLV, HIV, and MLV show distinct target site preferences. PLoS Biol 2:E234
127. Mitrophanous K, Yoon S, Rohll J, Patil D, Wilkes F, Kim V, Kingsman S, Kingsman A, Mazarakis N (1999) Stable gene transfer to the nervous system using a non-primate lentiviral vector. Gene Ther 6:1808–1818
128. Mock U, Riecken K, Berdien B, Qasim W, Chan E, Cathomen T, Fehse B (2014) Novel lentiviral vectors with mutated reverse transcriptase for mRNA delivery of TALE nucleases. Sci Rep 4:6409
129. Modlich U, Kustikova OS, Schmidt M, Rudolph C, Meyer J, Li Z, Kamino K, von Neuhoff N, Schlegelberger B, Kuehlcke K et al (2005) Leukemias following retroviral transfer of multidrug resistance 1 (MDR1) are driven by combinatorial insertional mutagenesis. Blood 105:4235–4246
130. Modlich U, Navarro S, Zychlinski D, Maetzig T, Knoess S, Brugman MH, Schambach A, Charrier S, Galy A, Thrasher AJ et al (2009) Insertional transformation of hematopoietic cells by self-inactivating lentiviral and gammaretroviral vectors. Mol Ther J Am Soc Gene Ther 17:1919–1928
131. Modlich U, Schambach A, Brugman MH, Wicke DC, Knoess S, Li Z, Maetzig T, Rudolph C, Schlegelberger B, Baum C (2008) Leukemia induction after a single retroviral vector insertion in Evi1 or Prdm16. Leukemia 22:1519–1528
132. Moritz T, Mackay W, Glassner BJ, Williams DA, Samson L (1995) Retrovirus-mediated expression of a DNA repair protein in bone marrow protects hematopoietic cells from nitrosourea-induced toxicity in vitro and in vivo. Cancer Res 55:2608–2614
133. Moscou MJ, Bogdanove AJ (2009) A simple cipher governs DNA recognition by TAL effectors. Science (New York, NY) 326:1501
134. Müller U (1999) Ten years of gene targeting: targeted mouse mutants, from vector design to phenotype analysis. Mech Dev 82:3–21
135. Müller-Kuller U, Ackermann M, Kolodziej S, Brendel C, Fritsch J, Lachmann N, Kunkel H, Lausen J, Schambach A, Moritz T et al (2015) A minimal ubiquitous chromatin opening element (UCOE) effectively prevents silencing of juxtaposed heterologous promoters by epigenetic remodeling in multipotent and pluripotent stem cells. Nucleic Acids Res 43:1577–1592
136. Münch RC, Janicki H, Völker I, Rasbach A, Hallek M, Büning H, Buchholz CJ (2013) Displaying high-affinity ligands on adeno-associated viral vectors enables tumor cell-specific and safe gene transfer. Mol Ther J Am Soc Gene Ther 21:109–118
137. Mussolino C, Cathomen T (2012) TALE nucleases: tailored genome engineering made easy. Curr Opin Biotechnol 23:644–650
138. Nair N, Rincon MY, Evens H, Sarcar S, Dastidar S, Samara-Kuko E, Ghandeharian O, Man Viecelli H, Thöny B, de Bleser P et al (2014) Computationally designed liver-specific transcriptional modules and hyperactive factor IX improve hepatic gene therapy. Blood 123:3195–3199
139. Naldini L, Blömer U, Gallay P, Ory D, Mulligan R, Gage FH, Verma IM, Trono D (1996) In vivo gene delivery and stable transduction of nondividing cells by a lentiviral vector. Science (New York, NY) 272:263–267
140. Nathwani AC, Reiss UM, Tuddenham EGD, Rosales C, Chowdary P, McIntosh J, Della Peruta M, Lheriteau E, Patel N, Raj D et al (2014) Long-term safety and efficacy of factor IX gene therapy in hemophilia B. N Engl J Med 371:1994–2004
141. Nathwani AC, Tuddenham EGD, Rangarajan S, Rosales C, McIntosh J, Linch DC, Chowdary P, Riddell A, Pie AJ, Harrington C et al (2011) Adenovirus-associated virus vector-mediated gene transfer in hemophilia B. N Engl J Med 365:2357–2365
142. Nègre D, Duisit G, Mangeot PE, Moullier P, Darlix JL, Cosset FL (2002) Lentiviral vectors derived from simian immunodeficiency virus. Curr Top Microbiol Immunol 261:53–74

143. Ng AP, Kauppi M, Metcalf D, Hyland CD, Josefsson EC, Lebois M, Zhang J-G, Baldwin TM, Di Rago L, Hilton DJ et al (2014) Mpl expression on megakaryocytes and platelets is dispensable for thrombopoiesis but essential to prevent myeloproliferation. Proc Natl Acad Sci U S A 111:5884–5889
144. Nguyen HG, Yu G, Makitalo M, Yang D, Xie H-X, Jones MR, Ravid K (2005) Conditional overexpression of transgenes in megakaryocytes and platelets in vivo. Blood 106:1559–1564
145. Nowakowski A, Alonso-Martín S, Arias-Salgado EG, Fernández D, Vilar M, Ayuso MS, Parrilla R (2011) Megakaryocyte gene targeting mediated by restricted expression of recombinase Cre. Thromb Haemost 105:138–144
146. O'Gorman S, Fox DT, Wahl GM (1991) Recombinase-mediated gene activation and site-specific integration in mammalian cells. Science (New York, NY) 251:1351–1355
147. Ohmori T, Ishiwata A, Kashiwakura Y, Madoiwa S, Mitomo K, Suzuki H, Hasegawa M, Mimuro J, Sakata Y (2008) Phenotypic correction of hemophilia A by ectopic expression of activated factor VII in platelets. Mol Ther J Am Soc Gene Ther 16:1359–1365
148. Ohmori T, Kashiwakura Y, Ishiwata A, Madoiwa S, Mimuro J, Sakata Y (2007) Silencing of a targeted protein in in vivo platelets using a lentiviral vector delivering short hairpin RNA sequence. Arterioscler Thromb Vasc Biol 27:2266–2272
149. Ohmori T, Mimuro J, Takano K, Madoiwa S, Kashiwakura Y, Ishiwata A, Niimura M, Mitomo K, Tabata T, Hasegawa M et al (2006) Efficient expression of a transgene in platelets using simian immunodeficiency virus-based vector harboring glycoprotein Ibalpha promoter: in vivo model for platelet-targeting gene therapy. FASEB J Off Publ Fed Am Soc Exp Biol 20:1522–1524
150. Okada Y, Nagai R, Sato T, Matsuura E, Minami T, Morita I, Doi T (2003) Homeodomain proteins MEIS1 and PBXs regulate the lineage-specific transcription of the platelet factor 4 gene. Blood 101:4748–4756
151. Orban M, Goedel A, Haas J, Sandrock-Lang K, Gärtner F, Jung CB, Zieger B, Parrotta E, Kurnik K, Sinnecker D et al (2015) Functional comparison of induced pluripotent stem cell- and blood-derived GPIIbIIIa deficient platelets. PLoS One 10:e0115978
152. Ott MG, Schmidt M, Schwarzwaelder K, Stein S, Siler U, Koehl U, Glimm H, Kühlcke K, Schilz A, Kunkel H et al (2006) Correction of X-linked chronic granulomatous disease by gene therapy, augmented by insertional activation of MDS1-EVI1, PRDM16 or SETBP1. Nat Med 12:401–409
153. Papapetrou EP, Lee G, Malani N, Setty M, Riviere I, Tirunagari LMS, Kadota K, Roth SL, Giardina P, Viale A et al (2011) Genomic safe harbors permit high β-globin transgene expression in thalassemia induced pluripotent stem cells. Nat Biotechnol 29:73–78
154. Pertuy F, Aguilar A, Strassel C, Eckly A, Freund J-N, Duluc I, Gachet C, Lanza F, Léon C (2015) Broader expression of the mouse platelet factor 4-cre transgene beyond the megakaryocyte lineage. J Thromb Haemost JTH 13:115–125
155. Petrillo C, Cesana D, Piras F, Bartolaccini S, Naldini L, Montini E, Kajaste-Rudnitski A (2015) Cyclosporin a and rapamycin relieve distinct lentiviral restriction blocks in hematopoietic stem and progenitor cells. Mol Ther J Am Soc Gene Ther 23:352–362
156. Philippe S, Sarkis C, Barkats M, Mammeri H, Ladroue C, Petit C, Mallet J, Serguera C (2006) Lentiviral vectors with a defective integrase allow efficient and sustained transgene expression in vitro and in vivo. Proc Natl Acad Sci U S A 103:17684–17689
157. Pollard VW, Malim MH (1998) The HIV-1 rev protein/the HIV-1 rev protein. Annu Rev Microbiol 52:491–532
158. Rappaport A, Johnson L (2014) Genetically engineered knock-in and conditional knock-in mouse models of cancer. Cold Spring Harb Protoc 2014:897–911
159. Ravid K, Beeler DL, Rabin MS, Ruley HE, Rosenberg RD (1991) Selective targeting of gene products with the megakaryocyte platelet factor 4 promoter. Proc Natl Acad Sci U S A 88:1521–1525
160. Ravid K, Doi T, Beeler DL, Kuter DJ, Rosenberg RD (1991) Transcriptional regulation of the rat platelet factor 4 gene: interaction between an enhancer/silencer domain and the GATA site. Mol Cell Biol 11:6116–6127

161. Rieger C, Rank A, Fiegl M, Tischer J, Schiel X, Ostermann H, Kolb H-J (2006) Allogeneic stem cell transplantation as a new treatment option for patients with severe Bernard-Soulier Syndrome. Thromb Haemost 95:190–191
162. Robertson EJ (1991) Using embryonic stem cells to introduce mutations into the mouse germ line. Biol Reprod 44:238–245
163. Rongvaux A, Willinger T, Takizawa H, Rathinam C, Auerbach W, Murphy AJ, Valenzuela DM, Yancopoulos GD, Eynon EE, Stevens S et al (2011) Human thrombopoietin knockin mice efficiently support human hematopoiesis in vivo. Proc Natl Acad Sci U S A 108:2378–2383
164. Sakuma T, Nishikawa A, Kume S, Chayama K, Yamamoto T (2014) Multiplex genome engineering in human cells using all-in-one CRISPR/Cas9 vector system. Sci Rep 4:5400
165. Santoni de Sio, Naldini L (2009) Short-term culture of human CD34+ cells for lentiviral gene transfer. Methods Mol Biol (Clifton, NJ) 506:59–70
166. Sauer B, McDermott J (2004) DNA recombination with a heterospecific Cre homolog identified from comparison of the pac-c1 regions of P1-related phages. Nucleic Acids Res 32:6086–6095
167. Sauvageau G, Iscove NN, Humphries RK (2004) In vitro and in vivo expansion of hematopoietic stem cells. Oncogene 23:7223–7232
168. Schröder AR, Shinn P, Chen H, Berry C, Ecker JR, Bushman F (2002) HIV-1 integration in the human genome favors active genes and local hotspots. Cell 110:521–529
169. Schroeder JA, Chen Y, Fang J, Wilcox DA, Shi Q (2014) In vivo enrichment of genetically manipulated platelets corrects the murine hemophilic phenotype and induces immune tolerance even using a low multiplicity of infection. J Thromb Haemost JTH 12:1283–1293
170. Seggewiss R, Pittaluga S, Adler RL, Guenaga FJ, Ferguson C, Pilz IH, Ryu B, Sorrentino BP, Young WS, Donahue RE et al (2006) Acute myeloid leukemia is associated with retroviral gene transfer to hematopoietic progenitor cells in a rhesus macaque. Blood 107:3865–3867
171. Sharma A, Larue RC, Plumb MR, Malani N, Male F, Slaughter A, Kessl JJ, Shkriabai N, Coward E, Aiyer SS et al (2013) BET proteins promote efficient murine leukemia virus integration at transcription start sites. Proc Natl Acad Sci U S A 110:12036–12041
172. Shi Q, Kuether EL, Chen Y, Schroeder JA, Fahs SA, Montgomery RR (2014) Platelet gene therapy corrects the hemophilic phenotype in immunocompromised hemophilia A mice transplanted with genetically manipulated human cord blood stem cells. Blood 123:395–403
173. Shi Q, Wilcox DA, Fahs SA, Fang J, Johnson BD, DU LM, Desai D, Montgomery RR (2007) Lentivirus-mediated platelet-derived factor VIII gene therapy in murine haemophilia A. J Thromb Haemost JTH 5:352–361
174. Shi Q, Wilcox DA, Morateck PA, Fahs SA, Kenny D, Montgomery RR (2004) Targeting platelet GPIbalpha transgene expression to human megakaryocytes and forming a complete complex with endogenous GPIbbeta and GPIX. J Thromb Haemost JTH 2:1989–1997
175. Shultz LD, Ishikawa F, Greiner DL (2007) Humanized mice in translational biomedical research. Nat Rev Immunol 7:118–130
176. Shultz LD, Lyons BL, Burzenski LM, Gott B, Chen X, Chaleff S, Kotb M, Gillies SD, King M, Mangada J et al (2005) Human Lymphoid and Myeloid Cell Development in NOD/LtSz-scid IL2R null Mice Engrafted with Mobilized Human Hemopoietic Stem Cells. J Immunol 174:6477–6489
177. Smithies O, Gregg RG, Boggs SS, Koralewski MA, Kucherlapati RS (1985) Insertion of DNA sequences into the human chromosomal beta-globin locus by homologous recombination. Nature 317:230–234
178. Snapper SB, Rosen FS, Mizoguchi E, Cohen P, Khan W, Liu CH, Hagemann TL, Kwan SP, Ferrini R, Davidson L et al (1998) Wiskott-Aldrich syndrome protein-deficient mice reveal a role for WASP in T but not B cell activation. Immunity 9:81–91
179. Soriano P (1999) Generalized lacZ expression with the ROSA26 Cre reporter strain. Nat Genet 21:70–71
180. Southern PJ, Berg P (1982) Transformation of mammalian cells to antibiotic resistance with a bacterial gene under control of the SV40 early region promoter. J Mol Appl Genet 1:327–341

181. Staunstrup NH, Mikkelsen JG (2011) Integrase-defective lentiviral vectors – a stage for non-viral integration machineries. Curr Gene Ther 11:350–362
182. Stein S, Ott MG, Schultze-Strasser S, Jauch A, Burwinkel B, Kinner A, Schmidt M, Krämer A, Schwäble J, Glimm H et al (2010) Genomic instability and myelodysplasia with monosomy 7 consequent to EVI1 activation after gene therapy for chronic granulomatous disease. Nat Med 16:198–204
183. Sternberg N, Hamilton D (1981) Bacteriophage P1 site-specific recombination. I. Recombination between loxP sites. J Mol Biol 150:467–486
184. Suerth JD, Maetzig T, Galla M, Baum C, Schambach A (2010) Self-inactivating alpharetroviral vectors with a split-packaging design. J Virol 84:6626–6635
185. Sullivan SK, Mills JA, Koukouritaki SB, Vo KK, Lyde RB, Paluru P, Zhao G, Zhai L, Sullivan LM, Wang Y et al (2014) High-level transgene expression in induced pluripotent stem cell-derived megakaryocytes: correction of Glanzmann thrombasthenia. Blood 123:753–757
186. Takahashi K, Yamanaka S (2006) Induction of pluripotent stem cells from mouse embryonic and adult fibroblast cultures by defined factors. Cell 126:663–676
187. Takiguchi M, James C, Josefsson EC, Carmichael CL, Premsrirut PK, Lowe SW, Hamilton JR, Huang DCS, Kile BT, Dickins RA (2010) Transgenic, inducible RNAi in megakaryocytes and platelets in mice. J Thromb Haemost JTH 8:2751–2756
188. Thompson A, Zhao Z, Ladd D, Zimmet J, Ravid K (1996) A new transgenic mouse model for the study of cell cycle control in megakaryocytes. Stem Cells (Dayton, Ohio) 14 (Suppl 1):181–187
189. Tiedt R, Coers J, Ziegler S, Wiestner A, Hao-Shen H, Bornmann C, Schenkel J, Karakhanova S, de Sauvage FJ, Jackson CW et al (2009) Pronounced thrombocytosis in transgenic mice expressing reduced levels of Mpl in platelets and terminally differentiated megakaryocytes. Blood 113:1768–1777
190. Tiedt R, Schomber T, Hao-Shen H, Skoda RC (2007) Pf4-Cre transgenic mice allow the generation of lineage-restricted gene knockouts for studying megakaryocyte and platelet function in vivo. Blood 109:1503–1506
191. Touw IP, Erkeland SJ (2007) Retroviral insertion mutagenesis in mice as a comparative oncogenomics tool to identify disease genes in human leukemia. Mol Ther J Am Soc Gene Ther 15:13–19
192. Trobridge G, Josephson N, Vassilopoulos G, Mac J, Russell DW (2002) Improved foamy virus vectors with minimal viral sequences. Mol Ther J Am Soc Gene Ther 6:321–328
193. Trobridge GD, Miller DG, Jacobs MA, Allen JM, Kiem H-P, Kaul R, Russell DW (2006) Foamy virus vector integration sites in normal human cells. Proc Natl Acad Sci U S A 103:1498–1503
194. Tropel P, Roullot V, Vernet M, Poujol C, Pointu H, Nurden P, Marguerie G, Tronik-Le Roux D (1997) A 2.7-kb portion of the 5′ flanking region of the murine glycoprotein alphaIIb gene is transcriptionally active in primitive hematopoietic progenitor cells. Blood 90:2995–3004
195. Tsai SQ, Wyvekens N, Khayter C, Foden JA, Thapar V, Reyon D, Goodwin MJ, Aryee MJ, Joung JK (2014) Dimeric CRISPR RNA-guided FokI nucleases for highly specific genome editing. Nat Biotechnol 32:569–576
196. Turan S, Galla M, Ernst E, Qiao J, Voelkel C, Schiedlmeier B, Zehe C, Bode J (2011) Recombinase-mediated cassette exchange (RMCE): traditional concepts and current challenges. J Mol Biol 407(2):193–221
197. Utomo AR, Nikitin AY, Lee WH (1999) Temporal, spatial, and cell type-specific control of Cre-mediated DNA recombination in transgenic mice. Nat Biotechnol 17:1091–1096
198. van den Oudenrijn S, Bruin M, Folman CC, Peters M, Faulkner LB, de Haas M, von dem Borne AE (2000) Hm5074 441.448 // Mutations in the thrombopoietin receptor, Mpl, in children with congenital amegakaryocytic thrombocytopenia. Br J Haematol 110:441–448
199. van Keuren M, Gavrilina GB, Filipiak WE, Zeidler MG, Saunders TL (2009) Generating transgenic mice from bacterial artificial chromosomes: transgenesis efficiency, integration and expression outcomes. Transgenic Res 18:769–785
200. Vassen L, Okayama T, Möröy T (2007) Gfi1b:green fluorescent protein knock-in mice reveal a dynamic expression pattern of Gfi1b during hematopoiesis that is largely complementary to Gfi1. Blood 109:2356–2364

201. Walz A, Lenzen A, Curtis B, Canner J, Schneiderman J (2014) Use of allogeneic stem cell transplantation for moderate-severe Glanzmann thrombasthenia. Platelets 1–3
202. Wang X, Shin S.C, Chiang A.F, Khan I, Pan D, Rawlings D.J, Miao C.H (2015) Intraosseous delivery of lentiviral vectors targeting factor VIII expression in platelets corrects murine hemophilia A. Mol Ther J Am Soc Gene Ther
203. Wang L, Zoppè M, Hackeng TM, Griffin JH, Lee KF, Verma IM (1997) Factor IX-deficient mouse model for hemophilia B gene therapy. Proc Natl Acad Sci USA 94:11563–11566
204. Wanisch K, Yáñez-Muñoz RJ (2009) Integration-deficient lentiviral vectors: a slow coming of age. Mol Ther J Am Soc Gene Ther 17:1316–1332
205. Ware J, Russell S, Ruggeri ZM (2000) Generation and rescue of a murine model of platelet dysfunction: the Bernard-Soulier syndrome. Proc Natl Acad Sci U S A 97:2803–2808
206. Wicke DC, Meyer J, Buesche G, Heckl D, Kreipe H, Li Z, Welte KH, Ballmaier M, Baum C, Modlich U (2010) Gene therapy of MPL deficiency: challenging balance between leukemia and pancytopenia. Mol Ther J Am Soc Gene Ther 18:343–352
207. Wiegering V, Sauer K, Winkler B, Eyrich M, Schlegel PG (2013) Indication for allogeneic stem cell transplantation in Glanzmann's thrombasthenia. Hamostaseologie 33:305–312
208. Wilcox DA, Olsen JC, Ishizawa L, Bray PF, French DL, Steeber DA, Bell WR, Griffith M, White GC (2000) Megakaryocyte-targeted synthesis of the integrin beta(3)-subunit results in the phenotypic correction of Glanzmann thrombasthenia. Blood 95:3645–3651
209. Williams DA, Lemischka IR, Nathan DG, Mulligan RC (1984) Introduction of new genetic material into pluripotent haematopoietic stem cells of the mouse. Nature 310:476–480
210. Williams DA, Thrasher AJ (2014) Concise review: lessons learned from clinical trials of gene therapy in monogenic immunodeficiency diseases. Stem Cells Trans Med 3:636–642
211. Wong EA, Capecchi MR (1986) Analysis of homologous recombination in cultured mammalian cells in transient expression and stable transformation assays. Somat Cell Mol Genet 12:63–72
212. Woods NB, Fahlman C, Mikkola H, Hamaguchi I, Olsson K, Zufferey R, Jacobsen SE, Trono D, Karlsson S (2000) Lentiviral gene transfer into primary and secondary NOD/SCID repopulating cells. Blood 96:3725–3733
213. Yáñez-Muñoz RJ, Balaggan KS, MacNeil A, Howe SJ, Schmidt M, Smith AJ, Buch P, MacLaren RE, Anderson PN, Barker SE et al (2006) Effective gene therapy with nonintegrating lentiviral vectors. Nat Med 12:348–353
214. Yarovoi HV, Kufrin D, Eslin DE, Thornton MA, Haberichter SL, Shi Q, Zhu H, Camire R, Fakharzadeh SS, Kowalska MA et al (2003) Factor VIII ectopically expressed in platelets: efficacy in hemophilia A treatment. Blood 102:4006–4013
215. Ying Q-L, Wray J, Nichols J, Batlle-Morera L, Doble B, Woodgett J, Cohen P, Smith A (2008) The ground state of embryonic stem cell self-renewal. Nature 453:519–523
216. Yu SS, Han E, Hong Y, Lee J-T, Kim S, Kim S (2003) Construction of a retroviral vector production system with the minimum possibility of a homologous recombination. Gene Ther 10:706–711
217. Yu SF, von Rüden T, Kantoff PW, Garber C, Seiberg M, Rüther U, Anderson WF, Wagner EF, Gilboa E (1986) Self-inactivating retroviral vectors designed for transfer of whole genes into mammalian cells. Proc Natl Acad Sci U S A 83:3194–3198
218. Zha Y, Shah R, Locke F, Wong A, Gajewski TF (2008) Use of Cre-adenovirus and CAR transgenic mice for efficient deletion of genes in post-thymic T cells. J Immunol Methods 331:94–102
219. Zhang CC, Kaba M, Iizuka S, Huynh H, Lodish HF (2008) Angiopoietin-like 5 and IGFBP2 stimulate ex vivo expansion of human cord blood hematopoietic stem cells as assayed by NOD/SCID transplantation. Blood 111:3415–3423
220. Zhang G, Shi Q, Fahs SA, Kuether EL, Walsh CE, Montgomery RR (2010) Factor IX ectopically expressed in platelets can be stored in alpha-granules and corrects the phenotype of hemophilia B mice. Blood 116:1235–1243
221. Zhang F, Thornhill SI, Howe SJ, Ulaganathan M, Schambach A, Sinclair J, Kinnon C, Gaspar HB, Antoniou M, Thrasher AJ (2007) Lentiviral vectors containing an enhancer-less ubiquitously acting chromatin opening element (UCOE) provide highly reproducible and stable transgene expression in hematopoietic cells. Blood 110:1448–1457

222. Zhang J, Varas F, Stadtfeld M, Heck S, Faust N, Graf T (2007) CD41-YFP mice allow in vivo labeling of megakaryocytic cells and reveal a subset of platelets hyperreactive to thrombin stimulation. Exp Hematol 35:490–499
223. Ziegler S, Bürki K, Skoda RC (2002) A 2-kb c-mpl promoter fragment is sufficient to direct expression to the megakaryocytic lineage and sites of embryonic hematopoiesis in transgenic mice. Blood 100:1072–1074
224. Zufferey R, Dull T, Mandel RJ, Bukovsky A, Quiroz D, Naldini L, Trono D (1998) Self-inactivating lentivirus vector for safe and efficient in vivo gene delivery. J Virol 72:9873–9880

Erratum to: Transcriptional Regulation of Platelet Formation: Harnessing the Complexity for Efficient Platelet Production In Vitro

Marloes R. Tijssen, Thomas Moreau and Cedric Ghevaert

Erratum to:
Chapter 2 in: Harald Schulze and Joseph Italiano (eds), ***Molecular and Cellular Biology of Platelet Formation: Implications in Health and Disease,*** **https://doi.org/10.1007/978-3-319-39562-3_2**

The chapter 'Transcriptional Regulation of Platelet Formation: Harnessing the Complexity for Efficient Platelet Production In Vitro' has now been made available open access under a CC BY 4.0 license.

The copyright holder has been updated to The Author(s).

The updated online version of this chapter can be found at
https://doi.org/10.1007/978-3-319-39562-3_2

H. Schulze, J. Italiano (eds.), *Molecular and Cellular Biology of Platelet Formation*,
https://doi.org/10.1007/978-3-319-39562-3_21

MIX
Papier aus verantwortungsvollen Quellen
Paper from responsible sources
FSC® C105338

If you have any concerns about our products,
you can contact us on
ProductSafety@springernature.com

In case Publisher is established outside the EU,
the EU authorized representative is:
Springer Nature Customer Service Center GmbH
Europaplatz 3, 69115 Heidelberg, Germany

Printed by Libri Plureos GmbH
in Hamburg, Germany